D0824725

Studies in Surface Science and Catalysis 155

OXIDE BASED MATERIALS
New sources, novel phases, new applications

Studies in Surface Science and Catalysis

Advisory Editors: B. Delmon and J.T. Yates
Series Editor: G. Centi

Vol. 155

OXIDE BASED MATERIALS

New sources, novel phases, new applications

Edited by

Aldo Gamba

Dipartimento di Scienze Chimiche Fisiche e Mathematiche,
Università dell'Insubria at Como,
Como, Italy

Carmine Colella

Dipartimento d'Ingegneria dei Materiali e della Produzione,
Università Federico II
Napoli, Italy

Salvatore Coluccia

Dipartimento di Chimica I.F.M.,
Università di Torino,
Torino, Italy

2005

ELSEVIER

Amsterdam – Boston – Heidelberg – London – New York – Oxford – Paris – San Diego
San Francisco – Singapore – Sydney – Tokyo

ELSEVIER B.V.
Radarweg 29
P.O. Box 211, 1000 AE
Amsterdam, The Netherlands

ELSEVIER Inc.
525 B Street, Suite 1900
San Diego, CA 92101-4495
USA

ELSEVIER Ltd
The Boulevard, Langford Lane
Kidlington, Oxford OX5 1GB
UK

ELSEVIER Ltd
84 Theobalds Road
London WC1X 8RR
UK

First edition 2005

Library of Congress Cataloging in Publication Data
A catalog record is available from the Library of Congress.

British Library Cataloguing in Publication Data
A catalogue record is available from the British Library.

ISBN: 0-444-51975-0
ISSN: 0167-2991 (Series)

♾ The paper used in this publication meets the requirements of ANSI/NISO Z39.48-1992 (Permanence of Paper).
Printed in The Netherlands.

PREFACE

The International Workshop " New sources, novel phases, new applications", the third of the series on "Oxide based materials", was held September 13 to 16, 2004 at Società del Casino Sociale in Como, Italy. The attendance was high and motivated, in line with the previous two workshops organized at Villa Olmo in Como in 1996 and 2000.

As in the two previous events, the workshop brought together experimental and theoretical scientists of different origins and expertise to exchange information on common scientific research fields, especially on all those materials whose features and properties depend on the interaction between surface and ionic and/or molecular species. During the workshop the participants had the opportunity to compare their knowledge of familiar materials and share their experience on a varied range of different materials, often new materials, which included metal oxides, zeolites and other microporous compounds, mesoporous silicates and silica, hybrid inorganic-organic compounds, soil aggregates, layered materials, and bioactive glasses.

The organization of the workshop was made possible by generous financial support from the University of Insubria. Financial support from the Como Chamber of Commerce and Industry, the Department of Chemical and Environmental Sciences of Insubria University, and the Consortium INSTM and Micromerits, is also gratefully acknowledged.

The workshop was organized under the auspices of the Italian Zeolite Association (AIZ), which celebrated in Como its annual meeting (AIZ '04 Day). The full organization was supported by people of the University of Insubria (Varese/Como), University Federico II (Naples), and University of Turin, with the collaboration of the Centro di Cultura Scientifica of Como. Particular thanks are due to Lucia Gamba and Enrica Gianotti for their assistance in the editing of this book.

Carmine Colella
Salvatore Coluccia
Aldo Gamba

Contents

Studies in Surface Science and Catalysis 155
A. Gamba, C. Colella and S. Coluccia (Editors)

1

Preparation of zeolites via the dry-gel synthesis method

J. Weitkamp and M. Hunger

Institute of Chemical Technology, University of Stuttgart, D-70550 Stuttgart, Germany

The principles and advantages of the dry-gel conversion (DGC) method are described for the synthesis of zeolites Beta and EU-1 with different aluminum and gallium contents in the framework. By modern analytical techniques, the processes occurring during the conversion of a dry gel into zeolite particles were investigated. In the case of the dry-gel synthesis of zeolite [Ga]Beta, XRD indicated a fast formation of the long-range order as a function of the DGC time while, as revealed by NMR spectroscopy, the rearrangement of the local structure occurs during a period of up to 65 h. This rearrangement of the local structure during the dry-gel conversion process consists of different concerted mechanisms. The dominating step at the beginning of the dry-gel conversion is a breakage of chemical bonds leading to a strong increase in the concentration of defect SiOH groups. As the DGC time increases, the number of defect SiOH groups and of Q^1, Q^2, and Q^3 silicon species decreases significantly, indicating a condensation of bonds. In the synthesis of zeolite [Ga]EU-1 it was found that the most critical parameters are the amount of water present in the autoclave during the crystallization process and the contents of sodium cations and template molecules in the dry gel.

1. INTRODUCTION

The most common method for preparing zeolites is the hydrothermal crystallization. Generally, a viscous gel consisting of an aluminum and a silicon source crystallizes at autogenous pressure at temperatures between 373 and 523 K [1]. In 1990, Xu et al. [2] introduced the vapor-phase transport (VPT) method for the synthesis of zeolite ZSM-5. In this work, the gel was prepared from aluminum sulfate, sodium silicate and sodium hydroxide. After drying in air, the gel powder was placed on a porous plate in an autoclave. At its bottom, the autoclave contained a solution of ethylenediamine and triethylamine in water. There was no contact between the dry gel and the liquid phase (see Fig. 1a). After 5 to 7 days at 453 and 473 K, the dry gel crystallized under formation of ZSM-5 particles [2]. Meanwhile, the dry-gel conversion route has also been applied for the fabrication of membranes containing, e.g., zeolites FER [3] or MOR [4] on carriers such as alumina. Often, non-volatile quaternary ammonium compounds, e.g., tetraethylammonium hydroxide or tetrapropylammonium bromide, are used as templates. Such compounds must be incorporated into the dry gel, and in these cases only water is supplied from the gas phase. The corresponding method is called steam-assisted conversion, SAC (see Fig. 1b). For both techniques (VPT and SAC), the generic term dry-gel conversion (DGC) is commonly used [5]. A variety of zeolites, e.g., ZSM-5 (MFI) [2,6], ZSM-12 (MTW) [6], Beta (*BEA) [7], EMT (EMT) [8], faujasite (FAU) [5], ZSM-22 (TON) [9], or NU-1 (RUT) [10], have been

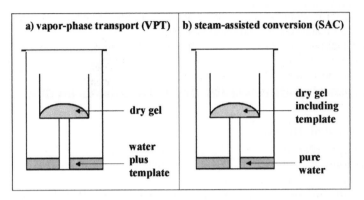

Fig. 1. Comparison of the vapor-phase transport (a) and the steam-assisted conversion method (b)

prepared via the DGC method. Bandyopadhyay et al. [11,12] applied this method for the preparation of aluminophosphate- and silicoaluminophosphate-type zeolites, such as AlPO-5 (AFI), AlPO-11 (AEL), SAPO-5 (AFI), SAPO-11 (AEL), and SAPO-34 (CHA). So far, however, only scant information is available about the chemistry of the conversion of a dry gel into a zeolite [13], while the formation of zeolitic membranes from dry gels has been investigated in more detail, mainly by using scanning electron microscopy (SEM), standard X-ray (XRD) and energy dispersive X-ray (EDX) diffraction, and pervaporation experiments [14].

The present review focuses on the conversion of dry gels into zeolites Beta (*BEA) and EU-1 (EUO) and presents some characteristic examples. The experimental techniques used for the characterization, such as SEM, XRD, and solid-state NMR spectroscopy, allowed to observe changes of the morphology, the long-range order, coordinations of framework atoms, bond connectivities, the state of the template molecules during the dry-gel conversion process, and to study the acidity of the surface sites in the obtained materials. The described synthesis routes lead to zeolites with a good crystallinity and a high concentration of Brønsted acid sites [15,16].

2. DRY-GEL SYNTHESIS OF GALLIUM-RICH ZEOLITE BETA

2.1. Preparation of zeolite [Ga]Beta by the DGC method

Zeolites [Ga]Beta were obtained, e.g., via a dry gel with an n_{Si}/n_{Ga} ratio of 8.5, which was prepared as follows: 3.0 g Cab-osil® M-5 (Fluka) were added to a suspension of the gallium source (0.58 g Ga_2O_3, Alfa Aesar) in 13.5 g of an aqueous solution of tetraethylammonium hydroxide (20 wt.-%, Acros). After stirring for 1 h, 0.95 g of a 4 N aqueous solution of NaOH followed by 5 g of water were added dropwise. The composition of the dry gel as determined by AES/ICP was: 17.01 SiO_2 : Ga_2O_3 : 0.50 Na_2O : 3.12 $(TEA)_2O$. After ageing of the gel for 2 h, it was heated in an oil bath to 353 K and dried for 4 h under stirring. Subsequently, the dry gel was ground into a fine powder, of which 1.5 g were placed in a Teflon beaker located in a stainless steel autoclave with a volume of 100 cm³ (see Ref. [15]), which contained a small amount (0.5 g) of water at the bottom. During the crystallization, the dry gel never came into direct contact with the liquid water. The dry-gel conversion was carried out at 453 K during periods of up to 65 h. The samples obtained after quenching the dry-gel conversion at 288 K were dried in air at 353 K. To remove the template molecules, the dried samples were calcined at 723 K in flowing nitrogen (58 l/h) with 5 vol.-% oxygen for 24 h and subsequently in air for 24 h.

The compositions of the dry gel and of the resulting zeolite [Ga]Beta were determined by AES/ICP. The morphology of the samples was investigated by SEM on a Cambridge CamScan CS44 instrument. X-ray powder diffraction patterns were recorded on a Siemens D5000 instrument using CuK$_\alpha$ radiation. In addition, the obtained materials were characterized by ^{1}H, ^{13}C, ^{29}Si, and ^{71}Ga NMR spectroscopy. Experimental parameters of the NMR studies are given in Ref. [15].

2.2. Long-range order of the dry-gel particles after increasing DGC times as studied by X-ray diffraction

A suitable way to study the progress of the crystallization process is to perform X-ray diffraction studies of samples taken after increasing DGC times. As an example, Fig. 2 shows the X-ray patterns of the as-synthesized materials obtained after dry-gel conversion times of 3.5, 16, 25, 50, and 65 h. The pattern obtained after 3.5 h does not show reflections at all indicating the presence of a totally amorphous material. After a conversion time of 16 h, characteristic X-ray reflections are clearly visible. A weak increase in the intensity of the main reflections at 2θ = 8.3 and 23.1° is observed only at elevated conversion times. However, significant changes can be observed for the weak reflections at ca. 2θ = 16.4 ° and 2θ = 25 to 31°, i.e., an intensity increase and a narrowing of the weak reflections with ongoing dry-gel conversion. This finding indicates that the long-range order is essentially established during an initial period of ca. 16 h, while thereafter only changes in the local structure occur.

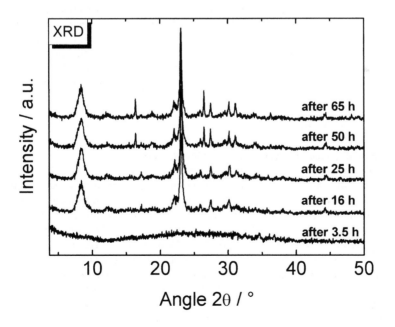

Fig. 2. X-ray patterns of as-synthesized materials obtained after DGC times of 3.5 h to 65 h [15]

The relative crystallinities of the samples obtained after different dry-gel conversion times were determined by a quantitative evaluation of the amplitude of the most intense X-ray reflection line at $2\theta = 23.1°$. As a reference, the amplitude of this line in the powder pattern of zeolite [Ga]Beta obtained after a conversion time of 65 h was used. This pattern (Fig. 2, top) agrees well with that of zeolite [Ga]Beta synthesized hydrothermally [17]. It has been reported that the DGC times for the synthesis of aluminum-containing zeolites Beta are significantly shorter than those for gallium-containing Beta samples, but longer for low-alumina zeolite Beta [18]. For example, Rao et al. [18] reported conversion times of 3 and 12 h for dry gels with n_{Si}/n_{Al} ratios of 15 and 365, respectively. In principle agreement with this observation, a decrease of the gallium content in the dry gel ($n_{Si}/n_{Ga} = 38$) led to a decrease of the crystallization rate with a relative crystallinity of only 60 % after a conversion time of 16 h, as opposed to a crystallinity of 82 % for the dry gel with $n_{Si}/n_{Ga} = 8.5$ after the same time [15].

2.3. Local structure of silicon atoms after increasing DGC times

By ^{29}Si MAS NMR spectroscopy of the as-synthesized materials, significant changes of the oxygen coordination of silicon atoms in the dry-gel particles as a function of the DGC time could be observed (Fig. 3). The spectrum of the fresh dry gel consists of signals at -78, -87, -97, and -107 ppm, due to Si(1OSi,3OH), Si(2OSi,2OH), Si(3OSi,1OH), and Si(4OSi) species corresponding to a Q^1, Q^2, Q^3, and Q^4 oxygen coordination, respectively [15]. With increasing DGC times, the Q^1, Q^2, and Q^3 signals strongly decrease (Fig. 3, left), and the spectra are dominated by Q^4 signals of silicon atoms with different numbers of silicon and gallium atoms in the first coordination sphere of T atoms (Fig. 3, right). The ^{29}Si MAS NMR spectrum of the sample obtained after a dry-gel conversion time of 16 h consists of signals at -114 to -111 ppm due to Si(4OSi) species (denoted Si(OGa)) on crystallographically non-equivalent framework positions. The low-field shoulder at -101 ppm is caused by a superposition of the signals of Si(3OSi,1OGa) species (denoted Si(1Ga)) and Si(3OSi,1OH) species (denoted SiOH). The weak signal at -95 ppm is originated by Si(2OSi,2OGa) species (denoted Si(2Ga)). The signal of Si(1Ga) species, occurring in the ^{29}Si MAS NMR spectrum of the sample obtained after a conversion time of 65 h (Fig. 3, right), is significantly narrower than in the spectra before. This finding indicates that the distribution of bond distances and bond angles in the local structure of the framework silicon atoms became smaller which is characteristic for the formation of a crystalline structure.

Considering the changes of the ^{29}Si MAS NMR spectra with increasing dry-gel conversion times, different processes can be observed. After dry-gel conversion times of up to 16 h, the intensities of the Q^1, Q^2 and Q^3 species decrease, while the signal intensity of Q^4 species increases. This indicates a healing of terminal bonds by a condensation of defect SiOH groups. After dry-gel conversion times of 16 h and more, the narrow Si(1Ga) and Si(2Ga) signals hint to an incorporation of gallium atoms at T positions into the framework of zeolite Beta.

From the relative intensities of the ^{29}Si MAS NMR signals of the final zeolite [Ga]Beta the number of gallium atoms incorporated at T positions into the zeolite framework can be calculated. In this case, the number of SiOH species contributing to the ^{29}Si MAS NMR signal at -101 ppm must be known, e.g., determined by ^{1}H MAS NMR spectroscopy of the calcined materials.

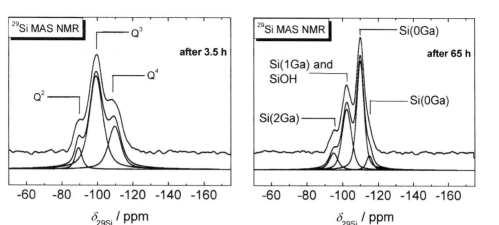

Fig. 3. ^{29}Si MAS NMR spectra of as-synthesized materials obtained after DGC times of 3.5 h (left) and 65 h (right) [15]

2.4. Incorporation of gallium atoms into the zeolite framework

^{71}Ga MAS NMR spectroscopy is a useful method for investigating the local environment of gallium species in zeolite crystals. Fig. 4a shows the ^{71}Ga MAS NMR spectrum of the gallium source Ga_2O_3 consisting of signals at ca. -6 to 24 ppm and ca. 174 ppm due to hexagonal and tetrahedrally coordinated gallium species, respectively [19]. The large line width of the low-field signal is due to second-order quadrupolar line broadening and the amorphous local structure of the gallium source. In the ^{71}Ga MAS NMR spectrum of the fresh dry gel (after 0 h), significant changes can be observed: The signal of octahedrally coordinated gallium loses intensity, while the signal of tetrahedrally coordinated gallium atoms is a superposition of a broad low-field signal at ca. 174 ppm and a narrow signal at ca. 150 ppm (Fig. 4b). These findings indicate a dissolution of the gallium source during the preparation of the dry gel and the formation of first gallium tetrahedra with a higher local symmetry. In the spectra of the as-synthesized materials obtained after increasing DGC times, a systematic decrease of the signal of octahedrally coordinated gallium species at ca. 2 ppm occurs, while the intensity of the signal of tetrahedrally coordinated framework gallium species at 150 ppm increases.

After a DGC time of 65 h (Fig. 4e), the ^{71}Ga MAS NMR spectrum consists of a single signal only at 150 ppm, which indicates a complete transformation of the gallium atoms to tetrahedrally coordinated species. The ^{71}Ga MAS NMR spectra clearly show that the incorporation of gallium atoms into the zeolite framework occurs during the whole dry-gel conversion process, which was not obvious from ^{29}Si and ^{1}H MAS NMR spectroscopy. However, the strongest changes of the gallium coordination occur between DGC times of 3.5 and 16 h, where most of the Q^1, Q^2 and Q^3 silicon species disappear in the ^{29}Si MAS NMR spectra and a strong decrease of the SiOH concentration from 2.5 to 1.4 mmol/g is observed by ^{1}H MAS NMR spectroscopy. Interestingly, the conversion of the dry-gel starts with a significant increase of the number of SiOH groups, in the present case from 0.8 mmol/g (fresh dry gel) to 2.5 mmol/g (after 3.5 h), which indicates a breakage of bonds as a prerequisite for the formation of crystalline domains [15].

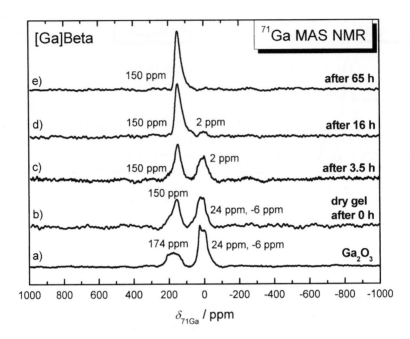

Fig. 4. ^{71}Ga MAS NMR spectra of as-synthesized materials obtained after different DGC times [15]

3. DRY-GEL SYNTHESIS OF ZEOLITE EU-1

3.1. Preparation of zeolite EU-1 by the DGC method

In a typical dry-gel synthesis of zeolite EU-1, such as a material with an n_{Si}/n_{Al} ratio of 23, a solution of 2.8 g of hexamethonium bromide (HMBr) in 30 g of water was added to 10 g of silica sol (VP AC 4038, Bayer). After stirring at room temperature for 2 h, a solution of 0.22 g sodium aluminate and 0.49 g NaOH in 3 g of water was added in the case of the preparation of zeolites [Al]EU-1. A modification of the n_{Si}/n_{Al} ratio of the dry gels was reached by changing the amount of sodium aluminate. For the adjustment of the n_{Na}/n_{Si} ratio, the amount of NaOH was varied. For the preparation of zeolites [Ga]EU-1, Ga_2O_3 was used as the gallium source. For a material with an n_{Si}/n_{Ga} ratio of 24, e.g., 0.22 g Ga_2O_3 were suspended in a solution of 0.61 g NaOH in 3 g of water. The n_{Si}/n_{Ga} ratio of the dry gels was varied by changing the amount of Ga_2O_3. For the adjustment of the n_{Na}/n_{Si} ratio, the amount of NaOH was modified (see Ref. [16]).

The resulting liquid mixtures were stirred for 2 h and dried to a powder at 353 K for 5 h under stirring. Subsequently, the dry gel was ground into a fine powder, of which 1.5 g were placed into a Teflon beaker located in a stainless steel autoclave with a volume of 100 cm^3 which contained 1.0 g water at the bottom. The dry-gel conversion was allowed to proceed at 453 K during periods of up to 7 days. The products obtained after quenching at 288 K were thoroughly washed with deionized water and dried in air at 353 K for 20 h.

To remove the template molecules, the dried samples were calcined at 813 K in flowing nitrogen with 15 vol.-% oxygen (58 l/h) for 24 h and, subsequently, in air for 24 h. The H$^+$-

forms of the zeolites were prepared by a fourfold ion exchange in an aqueous solution of NH$_4$NO$_3$ and subsequent thermal treatment at 723 K in vacuum ($p < 10^{-2}$ Pa) for 12 h.

3.2. Influence of the amount of template in the dry gel and the amount of water in the autoclave

The influence of the amount of template in the dry gel on the synthesis products was studied for the conversion of a dry gel with the molar composition 81.4 SiO$_2$: Al$_2$O$_3$: x Na$_2$O : y HMBr with x = 10.9 to 18.6 and y = 2.2 to 11.0. The results are shown in Fig. 5. According to these XRD patterns, highly crystalline zeolites [Al]EU-1 can be obtained with $n_{HMBr}/n_{Si} \geq 0.11$ and $n_{Na}/n_{Si} = 0.30$. A decrease of the template content in the dry gel (n_{HMBr}/n_{Si} = 0.09, 0.06, and 0.03) leads to the formation of α-quartz as impurity and to a loss of crystallinity. Even an increase of the sodium content in the dry gel, which favors the formation of zeolite EU-1 over zeolite EU-2 or an amorphous phase, cannot suppress the formation of α-quartz. For the synthesis of zeolite [Ga]EU-1, a similar influence of the template content on the quality of the synthesis products was observed.

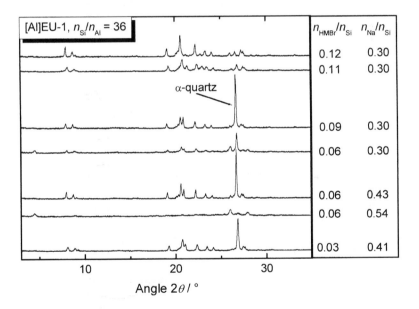

Fig. 5. X-ray patterns of as-synthesized samples obtained from a dry gel with an n_{Si}/n_{Al} ratio of 41 showing the influence of the template and sodium content on the dry-gel conversion process [16]

Fig. 6 shows the influence of the amount of liquid water in the autoclave on the conversion of a dry gel with the molar composition 61.0 SiO$_2$: Al$_2$O$_3$: 9.1 Na$_2$O : 7.4 HMBr. The XRD patterns clearly indicate that there is a minimum amount of water necessary for making highly crystalline zeolite EU-1. In comparison with the dry-gel synthesis of zeolite Beta, where the amount of water turned out to be a very critical factor for the crystallization process [5], zeolite EU-1 can be obtained over a large range of water contents, if $m_{water}/m_{gel} \geq 0.67$ (m_{water} = 1g, m_{gel} = 1.5 g). As calculated via the van der Waals equation, a mass of liquid water higher than 0.52 g in an autoclave with a volume of 100 cm^3 causes the presence of a liquid phase during the dry-gel conversion process at 453 K. This liquid water phase can be present either at the bottom of the autoclave or in the pores of the transformed solid material. Hence, the presence of liquid water is a prerequisite for the successful crystallization of zeolite EU-1 using the steam-assisted dry-gel conversion. This finding is at some variance to the dry-gel conversion process leading to zeolite Beta, where a water pressure in the range of the saturation pressure was sufficient to obtain materials with a high quality [5].

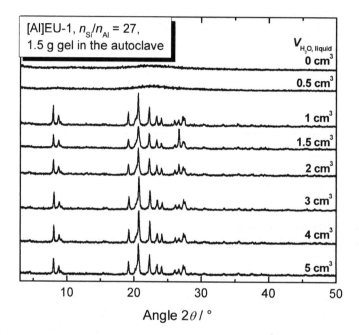

Fig. 6. X-ray patterns of as-synthesized samples obtained from a dry gel with an n_{Si}/n_{Al} ratio of 31 showing the influence of the water content on the dry-gel conversion process [16]

3.3. Synthesis and characterization of zeolites [Ga]EU-1

As shown in Table 1, the dry-gel synthesis of zeolites [Ga]EU-1 enables one to extend the range of compositions in comparison with [Ga]EU-1 materials synthesized hydrothermally towards both higher ($n_{Si}/n_{Ga} < 27$ [20]) and lower ($n_{Si}/n_{Ga} > 50$ [20]) gallium contents. As observed for the dry-gel synthesis of zeolites [Al]EU-1, the content of sodium cations in the dry gel plays an important role concerning the quality of the products. A high sodium content ($n_{Na}/n_{Si} > 0.3$) favors the formation of zeolites [Ga]EU-1 over an amorphous phase. In samples made from dry gels with high gallium contents ($n_{Si}/n_{Ga} \leq 34$, runs A to D), the deviations between the n_{Si}/n_{Ga} ratios of the dry gel, the as-synthesized and the H$^+$-forms of the zeolites are small. This finding indicates a complete conversion of the dry gel into a crystalline material without the formation of extra-framework species. In contrast, in samples made from dry gels with lower gallium contents ($n_{Si}/n_{Ga} \geq 63$, runs E and F), the n_{Si}/n_{Ga} ratios of the dry gel, the as-synthesized zeolite and its H$^+$-form differ significantly from each other.

This can be accounted for by extra-framework silicon species which are washed out, as also observed for aluminum-containing zeolites EU-1 [16].

The crystallinity of the as-synthesized zeolites [Ga]EU-1 was confirmed by their X-ray diffractograms. Characteristic reflections of zeolites [Ga]EU-1 [20] occurred for runs A to F, while no background reflections stemming from amorphous phases were present [16].

Table 1
n_{Si}/n_{Ga} and n_{Na}/n_{Si} ratios in the dry gels and n_{Si}/n_{Ga} ratios in the as-synthesized zeolites and in the H$^+$-forms of zeolites [Ga]EU-1 [16]

Run	n_{Si}/n_{Ga}[1]	n_{Na}/n_{Si}[1]	n_{Si}/n_{Ga}[2]	n_{Si}/n_{Ga}[3]	Product
A	11	0.30	12	12	[Ga]EU-1
B	19	0.30	15	14	[Ga]EU-1
C	24	0.30	18	24	[Ga]EU-1
D	34	0.30	29	34	[Ga]EU-1
	63	0.30	-	-	amorphous
E	63	0.47	45	51	[Ga]EU-1
	119	0.30	-	-	amorphous
F	119	0.50	86	72	[Ga]EU-1

[1] dry gel; [2] zeolite a.s.; [3] zeolite H$^+$-form

3.4. Characterization of the Brønsted acid sites in zeolites [Al]EU-1 and [Ga]EU-1 synthesized by the DGC method

A frequently applied technique for the quantitative determination of surface OH groups is ^{1}H MAS NMR spectroscopy of the calcined materials [21]. A typical ^{1}H MAS NMR spectrum of a dehydrated zeolite H-EU-1 consists of signals at 1.9, 2.4, and 4.0 ppm and a weak signal at 6.3 ppm, which are assigned to terminal silanol groups (SiOH), TOH groups (T: Al and Ga), bridging OH groups (SiOHT) and H-bonded SiOH groups, respectively. Generally, the bridging OH groups occurring in the ^{1}H MAS NMR spectra at 4.0 ppm are responsible for the Brønsted acidity of zeolites. The corresponding hydroxyl protons compensate the negative framework charges of tetrahedrally coordinated aluminum and gallium atoms in zeolites. In the present work, the concentration of SiOHAl and SiOHGa groups occurring at 4.0 ppm was determined by a comparison of the ^{1}H MAS NMR intensities with that of an external intensity standard.

Fig. 7 shows the concentration, c_{SiOHT}, of Brønsted acid sites plotted as a function of the concentration, c_T, of T atoms (T: Al and Ga) determined by AES/ICP. The graph shows a linear correlation between the concentration of the Brønsted acid sites and the aluminum or the gallium contents of the samples. This holds for the whole range of c_{Al} values investigated from 0.2 to 0.6 mmol/g (corresponding to n_{Si}/n_{Al} ratios of 23 and 83, see Ref. [16]) and c_{Ga} values from 0.3 to 1.1 mmol/g (corresponding to n_{Si}/n_{Ga} ratios of 12 to 72, runs A to F). The differences in the concentration of Brønsted acid sites in zeolites H-[Al]EU-1 and H-[Ga]EU-1 for the same c_T values may be due to different amounts of extra-framework aluminum and gallium species, respectively. The presence of extra-framework species in general was evidenced in Ref. [16] by ^{27}Al and ^{71}Ga MAS NMR spectroscopy. Generally, in zeolites H-[Ga]EU-1 a significant smaller number of Brønsted acidic SiOHT groups (T: Al and Ga) are formed at a given c_T value than in zeolites H-[Al]EU-1. This finding indicates, that in dehydrated zeolites [Ga]EU-1 a smaller number of gallium atoms exists in an intact tetrahedral oxygen coordination than in dehydrated zeolites H-[Al]EU-1.

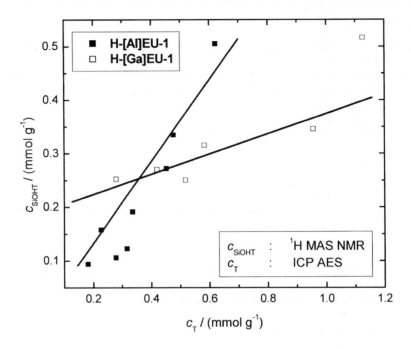

Fig. 7. Concentration of Brønsted acid sites (c_{SiOHT}), as determined by ^{1}H MAS NMR spectroscopy, plotted as a function of the concentration (c_T) of T atoms (T: Al and Ga), as obtained by AES/ICP [16]

A suitable way to study the strength of the Brønsted acidic SiOHT groups (T: Al and Ga) is the adsorption of ^{13}C-2-acetone as probe molecule and subsequent investigation of the adsorbate complexes by ^{13}C MAS NMR spectroscopy. The chemical shift of the carbonyl atoms in ^{13}C-2-acetone is sensitive with regard to an interaction of these molecules with surface sites on solid catalysts. A higher downfield shift of the ^{13}C MAS NMR signal of the carbonyl carbon atoms indicates an interaction with Brønsted sites characterized by a higher acid strength [22]. The ^{13}C MAS NMR spectra of zeolites [Al]EU-1 and [Ga]EU-1 with n_{Si}/n_{Al} and n_{Si}/n_{Ga} ratios of 23 and 24, respectively, loaded with one ^{13}C-2-acetone molecule per acid site, show signals at 214.9 and 213.3 ppm, respectively [16]. Comparing the aluminum- and the gallium-containing zeolites EU-1, therefore, the latter material has Brønsted sites with a lower acid strength. Generally, Brønsted acid sites in zeolites EU-1 have a significantly lower acid strength than those in zeolites [Ga]Beta, [Al]Beta, and [Al]ZSM-5, which causes chemical shifts of the ^{13}C MAS NMR signals of the carbonyl carbon atoms of adsorbed ^{13}C-2-acetone molecules of 219, 221, and 223 ppm, respectively.

4. CONCLUSIONS

The X-ray powder patterns of zeolites [Ga]Beta synthesized via the dry-gel conversion method and obtained after increasing conversion times show a fast formation of a long-range order during the first 16 h of the dry-gel conversion process. On the other hand, by ^{29}Si and ^{71}Ga MAS NMR spectroscopy it was evidenced that the rearrangement of the local framework structure is a more time-consuming process lasting up to 65 h.

The rearrangement of the local structure of the dry-gel particles starts with a dissolution of chemical bonds leading to a strong increase in the concentration of defect SiOH groups as indicated by ^{1}H MAS NMR spectroscopy. According to the results of ^{1}H and ^{29}Si MAS NMR spectroscopy, the further conversion of the dry gel leads to a significant diminution of defect SiOH groups and of Q^1, Q^2, and Q^3 species hinting to a condensation of Si-O-T bonds (T = Si or Ga). This silanol condensation is accompanied by the incorporation of gallium atoms into the zeolite Beta framework as evidenced by ^{71}Ga MAS NMR spectroscopy.

The dry-gel conversion method with hexamethonium bromide as template allows to prepare zeolites [Al]EU-1 and [Ga]EU-1 over a range of n_{Si}/n_{Al} ratios of 18 to 142 and n_{Si}/n_{Ga} ratios of 12 to 86, respectively. Especially for [Ga]EU-1 zeolites, the range of attainable n_{Si}/n_{Ga} ratios could be significantly extended over the one reached previously by hydrothermal syntheses. Compared to the dry-gel synthesis of zeolite Beta, the amount of water in the autoclave plays a less important role in the case of the dry-gel synthesis of zeolite EU-1, if $m_{water}/m_{gel} \geq 0.67$ is fulfilled. Critical parameters of a successful synthesis of zeolites EU-1 via the dry-gel route are the contents of sodium cations and template molecules in the dry gel. For aluminum-containing dry gels, a sodium content corresponding to an n_{Si}/n_{Al} ratio of larger than ca. 100 leads to the formation of zeolite EU-2 or to an amorphous product. The formation of α-quartz is caused by an amount of template cations in the dry gel corresponding to $n_{HMBr}/n_{Si} < 0.11$.

^{1}H MAS NMR spectroscopy of the aluminum- and gallium-containing zeolites EU-1 in the calcined H$^+$-form shows that a higher amount of gallium in comparison with aluminum is necessary to reach the same number of Brønsted acid sites in these materials. Their acid strength was characterized by adsorption of ^{13}C-2-acetone and subsequent investigation by ^{13}C MAS NMR spectroscopy. Due to the different electronegativities of aluminum and gallium atoms, the Brønsted acid sites in the gallium-containing zeolites H-EU-1 show a lower acid strength than those in aluminum-containing zeolites H-EU-1.

REFERENCES
[1] R.M. Barrer, Hydrothermal Chemistry of Zeolites, Academic Press, London, New York, 1982, pp. 105-182.
[2] W. Xu, J. Dong, J. Li, J. Li and F. Wu, J. Chem. Soc. Chem Commun., (1990) 755.
[3] N. Nishiyama, T. Matsufuji, K. Ueyama and M. Matsukata, Microporous Mater., 12 (1997) 293.
[4] N. Nishiyama, K. Ueyama and M. Matsukata, J. Chem. Soc. Chem. Commun., (1995) 1967.
[5] M. Matsukata, M. Ogura, T. Osaki, P.R.H.P. Rao, M. Nomura and E. Kikuchi, Top. Catal., 9 (1999) 77.
[6] R. Bandyopadhyay, Y. Kubota, N. Sugimoto, Y. Fukushima and Y. Sugi, Microporous Mesoporous Mater., 32 (1999) 81.
[7] P.R.H.P. Rao and M. Matsukata, Chem. Commun., (1996) 1441.
[8] M. Matsukata, K. Kizu, M. Ogura and E. Kikuchi, Cryst. Growth Des., 1 (2001) 509.
[9] S.G. Thoma, D.E. Trudell, F. Bonhomme and T.M. Nenoff, Microporous Mesoporous Mater., 50 (2001) 33.
[10] A. Bhaumik and T. Tatsumi, Microporous Mesoporous Mater., 34 (2000) 1.

12

[11] M. Bandyopadhyay, R. Bandyopadhyay, Y. Kubota and Y. Sugi, Chem. Lett., (2000) 1024.

[12] M. Bandyopadhyay, R. Bandyopadhyay, S. Tawada, Y. Kubota and Y. Sugi, Appl. Catal. A : General, 225 (2002) 51.

[13] M. Matsukata, T. Osaki, M. Ogura and E. Kikuchi, Microporous Mesoporous Mater., 56 (2002) 1.

[14] T. Matsufuji, N. Nishiyama, K. Ueyama and M. Matsukata, Microporous Mesoporous Mater., 32 (1999) 159.

[15] A. Arnold, S. Steuernagel, M. Hunger and J. Weitkamp, Microporous Mesoporous Mater., 62 (2003) 97.

[16] A. Arnold, M. Hunger and J. Weitkamp, Microporous Mesoporous Mater., 67 (2004) 205.

[17] M.L. Occelli, H. Eckert, A. Wölker and A. Auroux, Microporous Mesoporous Mater., 30 (1999) 219.

[18] P.R.H.P. Rao, K. Ueyama and M. Matsukata, Appl. Catal. A: General, 166 (1998) 97.

[19] H.K.C. Timken and E. Oldfield, J. Am. Chem. Soc., 109 (1987) 7669.

[20] G.N. Rao, V.P. Shiralkar, A.N. Kotasthane and P. Ratnasamy, in: M.L. Occelli and H.E. Robson (eds.), Synthesis of Microporous Materials, Vol. 1, Molecular Sieves, Van Nostrand Reinhold, New York, 1992, pp. 153-166.

[21] M. Hunger, Catal. Rev.-Sci. Eng., 39 (1997) 345.

[22] G.T. Kokotailo and D. White, J. Catal., 148 (1994) 779.

Studies in Surface Science and Catalysis 155
A. Gamba, C. Colella and S. Coluccia (Editors)

13

The Na-rich zeolites from Boron, California

William S. Wise

Department of Geological Sciences, University of California
Santa Barbara, California 93106 USA

Unusually Na-rich zeolites have been found in cavities in blocks of basalt on the waste dump at the US Borax Mine at Boron, 135 km northeast of Los Angeles, California. The basalt underlies Miocene basin fill deposits, consisting of sand, shale, tuff, and Na-borate deposits. The zeolites show two generations of growth. Early gmelinite-Na is overgrown by later chabazite-Na, and early clinoptilolite-Na is overgrown by heulandite-Na. Other zeolites in these cavities are phillipsite-Na, analcime, Na-rich mordenite, and mazzite-Na. The occurrence of searlesite and borax, attests to the saturation of the basalt with a sodium borate solution at least at some stage. The extreme Na-rich compositions of all the zeolites strongly indicate cation exchange in Na-borate brine, while the overgrowth crystals may have grown in the brine.

1. INTRODUCTION

When compiling zeolite compositions, one often wonders if the analyzed composition is same as when that particular zeolite crystallized. To be sure, there are instances where natural post-crystallization cation exchange probably occurred. For example, the occurrences of barrerite near the seashore suggest the Na probably came from the seawater, but evidence of crystallization in a Na-rich environment is lacking. There is evidence at the zeolite occurrence at Boron, California, that not only were the early formed crystals cation-exchanged, but growth continued in Na-rich solutions.

An unusual suite of Na-rich zeolites was found in cavities in blocks of basalt, sampled on the waste dump at the U.S. Borax Mine at Boron, 135 km northeast of Los Angeles, California. The blocks were exposed and removed by the ongoing quarrying operation. The basalt underlies basin fill deposits, consisting of silt, claystone, tuff, and borate deposits, all comprising the Kramer Beds.

1.1. Geologic Setting

The Mojave Desert region of southeastern California contains many basins that developed as early as the Miocene. These basins were formed by crustal extension, which was accompanied by scattered volcanic activity. Because most of the basins were hydrologically closed, alkaline and saline playa lakes were common. The western most was the Kramer basin, which was floored by two olivine basalt flows, each about 5 m thick. These flows were soon covered by a shallow, playa lake, from which several meters of silt and clay were deposited before the waters became enriched in sodium borate. The origin of the borate is uncertain, but is generally believed to have come from nearby thermal springs. Over the next several million years at least 100 m of playa deposits of claystone and sodium borate with a few interbeds of rhyolitic tuff were deposited in the basin. Fifty to 75 m of post-Miocene alluvial gravel covers the lake beds. Interestingly, only 55 km to the east is the contemporaneous Barstow basin that was only saline and alkaline. Rhyolite tuff that fell into that lake was altered to clinoptilolite and analcime [1]. Similar tuff beds in the Kramer borate-rich lake were converted to analcime and searlesite, $NaBSi_2O_5(OH)_2$. All of the zeolites, including the cavity minerals in the basalt, mostly likely are a result of diagenetic alteration [2], in which minerals formed by reactions of volcanic glass with playa lake water.

2. THE CAVITY MINERALS

Cavities in the basalt are all lined with saponite, which is overgrown by the iron sulfides, pyrrhotite and greigite; zeolites; borax and searlesite; and late Ca-bearing minerals that include calcite and Ca-borates. The zeolites show two generations of growth. An early set of crystals show the effects of alteration, e.g. replacement or coating by brown clay. The crystals of the later set are clear and transparent and tend to overgrow the early set. The early zeolites are phillipsite-Na, gmelinite-Na, clinoptilolite-Na, mazzite-Na, and Na-rich mordenite, while the later ones are analcime, chabazite-Na, heulandite-Na, and more phillipsite-Na.

2.1. Ferroan saponite

All fracture and cavity surfaces are coated with a layer of clay, which varies in appearance from pale green and waxy to green, ball-like masses. Upon exposure to the atmosphere the colors rapidly change to brown from oxidation of the ferrous iron. Electron microprobe analysis of the green form gives the following chemical composition: $(Ca_{0.10}Na_{0.32}K_{0.21})(Mg_{2.60}Fe^{2+}_{2.65}Al_{0.07}Ti_{0.41})(Si_{6.90}Al_{1.10})O_{20}(OH)_4 \cdot nH_2O$. The unusual aspect of this composition is the high amount of Na in the interlayer cation sites.

2.2. Phillipsite-Na

Phillipsite-Na, the first zeolite to form, commonly occurs as blocky groups of complexly twinned crystals up to 4 mm across. Many of these groups are partially to completely coated with saponite. Commonly the phillipsite has been replaced by saponite, leaving a solid pseudomorph or a shell of saponite.

A few of the saponite coated complex groups have overgrowths of colorless, transparent phillipsite. These overgrowths replicate the twinning of the host crystal, even though there is a substantial saponite layer between the host and growth crystals.

All of the phillipsite at Boron has Na as the dominant exchangeable cation with only minor amounts of K, Ca and Ba. The early formed crystals have slightly higher amount of cations other than Na, than do the later ones. Electron microprobe analyses give the following typical compositions: $Na_{3.07}K_{0.20}Ca_{0.08}Ba_{0.11}[Al_{3.64}Si_{12.35}O_{32}](H_2O)_n$ for the early crystals, and $Na_{4.27}K_{0.08}Ca_{0.01}[Al_{4.36}Si_{11.63}O_{32}](H_2O)_n$ for the late overgrowths. Note that the Si content of the framework is slightly lower in the later crystals.

2.3. Gmelinite-Na

Crystals of gmelinite-Na occur as single and twinned dipyramids 1 to 2 mm across, and in spherical clusters up to 6 mm in diameter. Commonly the gmelinite crystals are coated with a thin layer of clay, causing them to have a brownish, opaque appearance. Nearly half the crystals observed exhibit dissolution, ranging from very slight to nearly complete. The solution appears to start on the prism faces and proceeds to the crystal center. Many crystal remnants are merely shells consisting of pyramid faces.

Many of the gmelinite crystals with a clay coating are overgrown with uncoated, hexagonal plates of chabazite (herschelite habit). The two minerals are joined on their respective {0001} faces, commonly on both ends of the gmelinite crystal.

Electron microprobe analyses of these gmelinite crystals gives the compositions similar to $Na_{6.60}K_{0.10}Ba_{0.05}Mg_{0.16}[Al_{6.79}Si_{17.20}O_{48}](H_2O)_n$. Although the crystals were sectioned before analysis, the Mg may be from the clay coating rather than exchangeable cations.

2.4. Chabazite-Na

Most commonly chabazite-Na crystals occur as colorless, epitaxial overgrowths on gmelinite-Na. These overgrowths have the herschelite habit, that is twinned, pseudo-hexagonal plates 1-2 mm across. Some chabazite-Na occurs as complex clusters of crystals that are transparent and colorless. These clusters are typically 2-4 mm across. A typical composition of a platy overgrowth of chabazite-Na is $Na_{2.96}K_{0.24}[Al_{3.21}Si_{8.79}O_{24}](H_2O)_{9.3}$ [3]. An analysis of a clear, crystal cluster yielded $Na_{3.49}K_{0.16}[Al_{3.29}Si_{8.62}O_{24}](H_2O)_n$ with slightly less Si in the framework.

2.5. Clinoptilolite-Na

The most common zeolite in the basalt cavities is clinoptilolite-Na, which occurs in the typical heulandite-group habit of coffin-shaped, platy crystals. Lengths up to 3 mm are common, and most crystals occur with the forms {010}, {$\overline{1}11$}, {001}, {100}, {$20\overline{1}$}, and {110} [2]. All clinoptilolite-Na grew during the earlier stage, because all crystals are lightly to heavily coated with a layer of saponite. There are some samples showing dissolution, leaving a partial shell consisting of saponite.

The optical properties of the Boron clinoptilolite-Na crystals are similar to other clinoptilolite-Na, such as those from Agoura, California [4]. In these and several other examples of clinoptilolite the optic axial plane is parallel to (010). See Fig. 1.

Electron microprobe analyses show that crystals with a saponite coating are indeed clinoptilolite, i.e. the Si/Al is greater than 4.0, the lower limit allowed by the IMA zeolite nomenclature [5]. There is some range in composition from $Na_{6.48}K_{0.56}Ba_{0.06}Mg_{0.13}$ $[Al_{7.15}Si_{28.77}O_{72}](H_2O)_n$ [1], where Si/Al = 4.02, to $Na_{5.86}K_{1.06}[Al_{6.92}Si_{29.08}O_{72}](H_2O)_n$ and Si/Al = 4.20.

2.6. Heulandite-Na

In several blocks of basalt clear crystals of heulandite-Na occur as epitaxial overgrowths on the brownish clinoptilolite-Na. These overgrowths have the same crystal forms as the substrate crystal. Heulandite-Na also occurs as clear, isolated crystals with a distinctive morphology, elongated parallel to [102].

Crystals from both types of occurrences have distinctive optical properties, and are easily distinguished from clinoptilolite-Na (Fig. 1). Like all heulandite-Ca these crystals have the optic plane normal to (010), although here the optic axial angle is negative and large. These optical properties are close to but not identical with a Na-exchanged heulandite reported by Yang et al. [6].

Electron microprobe analyses of heulandite-Na show a slight difference between the two types of occurrence. An example of the overgrowths has the following composition: $Na_{6.68}K_{0.66}Ba_{0.27}Mg_{0.41}[Al_{8.06}Si_{27.69}O_{72}](H_2O)_n$, in which Si/Al = 3.43, and an isolated crystal has the composition, $Na_{6.57}K_{0.48}Mg_{0.30}[Al_{7.59}Si_{28.26}O_{72}](H_2O)_n$, where Si/Al = 3.72. Clearly both of these examples conform to the definition of heulandite with respect to clinoptilolite.

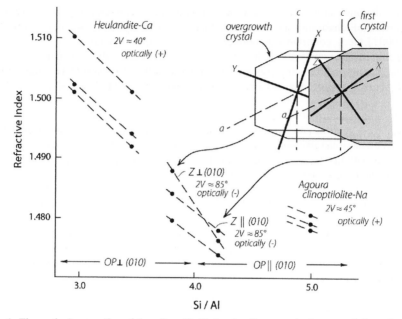

Fig. 1. The optical properties of the clinoptilolite/heulandite crystals. In general the refractive indices of the heulandite group minerals vary with the Si/Al ratio of the framework with a change in the orientation of the optic axial plane at about Si/Al = 4.0. The same change is observed in the Boron crystals, except both minerals are optically negative, possibly a result of the dominantly Na exchangeable cations (compare with [6]).

2.7. Analcime

Throughout the cavities analcime occurs as single crystals and rarely as complex balls of radial crystals. The single crystals have the common trapezohedron form, and are typically 2 mm across. The complex balls up to 3 mm in diameter consist of radial prisms terminated by a few trapezohedron faces. All analcime is clear and transparent, apparently having grown during the second stage.

All of the analcime that has been analyzed is unusually silica-rich. A typical analysis yields the composition, $Na_{14.02}K_{0.09}[Al_{13.33}Si_{34.44}O_{96}](H_2O)_n$, where Si/Al = 2.58.

2.8. Mazzite-Na

A very few cavities contain mats of very thin (10 μm), colorless, flexible fibers up to 2 mm. X-ray powder diffraction data show that this material has the structure of mazzite. Subsequent electron microprobe analysis has yielded a composition close to $Na_8[Al_8Si_{28}O_{72}](H_2O)_{30}$ [7]. The crystal structure determination of this mazzite reveals some extra-framework sites with low occupancy, which supports the possibility of post-crystallization cation exchange [7].

2.9. Na-rich mordenite

Mordenite occurs as sprays of white fibers up to 1 mm in length. These fibers are not as thin as those of mazzite-Na, and are brittle, rather than flexible. Some of the mordenite is brownish from a thin coating of saponite indicating at least some crystallization occurred during the early stage. Energy dispersive analysis shows only the presence of Na, Al, and Si.

2.10. Boron-bearing minerals

Scattered among the cavities are crystals of searlesite, $NaBSi_2O_5(OH)_2$, and borax, $Na_2B_4O_5(OH)_4(H_2O)_8$, each with a distinctive morphology. Searlesite crystals are colorless, glassy, tapered blades up to 4 mm in length with a recognizable monoclinic aspect. Borax crystals are glassy, blocky crystals less than 1 mm across. Upon exposure of more than a few days, borax dehydrates to opaque white tincalconite. Where these minerals occur with zeolites, both are late in the growth sequence.

Ulexite, $NaCaB_5O_6(OH)_6(H_2O)_5$, and colemanite, $Ca_2B_6O_{11}(H_2O)_5$, also occur in some cavities, but appear to have formed later during a third stage from Ca-bearing fluids.

3. ORIGIN OF THE ZEOLITE PHASES AND THE UNUSUAL COMPOSITIONS

Zeolites in the basalt underlying the Na-borate lake beds at Boron are probably a result of diagenetic reactions of basaltic glass with groundwater rather than hydrothermal alteration. The rock shows no effect of pervasive hydrothermal alteration, and the flows appear to have been saturated with lake water as soon as the basin formed.

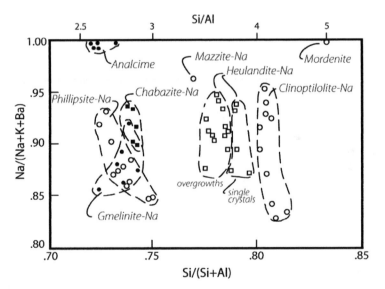

Fig. 2. Compositional plot of analyses of zeolites from the basalt at Boron, California. Phillipsite-Na, gmelinite-Na, clinoptilolite-Na, Na-rich mazzite, and Na-rich mordenite crystallized during the first stage. Some phillipsite-Na, chabazite-Na, heulandite-Na and analcime formed during the second stage.

The diagenetic clay and zeolite minerals all have unusually Na-rich compositions. The compositional ranges of each zeolite are plotted in Fig. 2. Each of the zeolite species with the exception of analcime has Na-compositions at the limit of known ranges in world-wide occurrences [2]. Those zeolites formed in the first generation, phillipsite, gmelinite, and clinoptilolite, have the wider ranges of composition. Overgrowth phillipsite-Na tends to have the higher Na and lower Si contents. Because overgrowth chabazite-Na overlaps much of the

18

range of earlier gmelinite-Na, the change from one crystal structure to other is difficult to explain. Overgrowth heulandite-Na has significantly less Si than the early clinoptilolite. The slight, but persistent Si difference between the overgrowth crystals and the isolated ones, is also difficult to explain. The high Si content of the analcime crystals is similar to that in the analcime–searlesite replacement of rhyolitic tuff within the Na-borate sediment and to that replacing tuff in the nearby Barstow basin [1].

I conclude that the initial zeolites were phillipsite, gmelinite, clinoptilolite, and possibly mazzite and mordenite, which had compositions typical of basalt occurrences, i.e. there was substantially more Ca and some K. The influx of Na-borate waters caused exchange of the cations to the Na-rich compositions and in many instances dissolution of the pre-existing crystals. Further crystallization did occur in the Na-rich water and resulted in the transparent crystals of phillipsite-Na, chabazite-Na, heulandite-Na, and analcime.

REFERENCES
[1] R.A. Sheppard and A.J. Gude, 3rd, U. S. Geol. Surv., Prof. Pap. 634 (1969).
[2] W.A. Deer, R.A Howie, W.S. Wise and J. Zussman, Rock Forming Minerals, Vol. 4B, Second ed.. Framework Silicates: Silica Minerals, Feldspathoids and the Zeolites, The Geological Society, London, 2004.
[3] W.S. Wise and W.D. Kleck, Clays Clay Miner., 36 (1988) 131.
[4] W.S. Wise, W.J. Nokleberg, and M. Kokinos, Am. Mineral., 54 (1969) 887.
[5] D.S. Coombs, A. Alberti, T. Armbruster, G. Artioli, C. Colella, E. Galli, J.D. Grice, F. Liebau, J.A. Mandarino, H. Minato, E.H. Nickel, E. Passaglia, D.R. Peacor, S. Quartieri, R. Rinaldi, M. Ross, R.A. Sheppard, E. Tillmanns and G. Vezzalini, Mineral. Mag., 62 (1998) 533.
[6] P. Yang, J. Stolz, T. Armbruster and M. Gunter, Am. Mineral., 82 (1997) 517.
[7] R. Arletti, E. Galli, G. Vezzalini, and W.S. Wise, Am. Mineral., (in press).

Studies in Surface Science and Catalysis 155
A. Gamba, C. Colella and S. Coluccia (Editors)

Phase transformations and structural modifications induced by heating in microporous materials

Alberto Alberti and Annalisa Martucci

Dipartimento di Scienze della Terra, Università di Ferrara
Corso Ercole I° d'Este, 32, 44100 Ferrara, Italy

The thermal behaviour of microporous materials is relevant both for their characterization and for their application, and varies dramatically from one material to another. This paper intends to give a picture of their behaviour and to highlight analogies and differences in these materials as a function of their chemistry and topology. The information on their response to heating and dehydration/calcinations, even if still scanty and sometimes controversial, shows that some materials undergo dehydration-induced decrease in volume, which can be dramatic or very scarce as a function of the extraframework content. These changes in volume can be partially or completely irreversible if they involve structural modifications of the tetrahedral framework. It is to highlight that topological changes are always due to the so called "face sharing tetrahedral" process.

1. INTRODUCTION

One of the most remarkable properties of microporous materials is their thermal behaviour (i.e., stability, phase transformations, rate and temperature of dehydration, and rehydration). It is of particular importance owing to the wide industrial applications of these materials. However, for many zeolite-type materials, detailed and accurate information on their response to heating is still scanty and, even then, controversial or unreliable.

It is well known that zeolite-like materials lose water and/or organic molecules by heating. It is also known that the behaviour of these compounds, and consequently the efficiency in their applications, often varies remarkably as a function of many parameters: topology, synthesis conditions, structure direct agents, exchangeable cations, exchange modalities, chemistry and order-disorder in the framework, extraframework ion distribution and occupancy, location of acid sites, and many others. In some cases this process causes a change in the topology that may not be reversible, at least for quite a long time. In other cases no remarkable modification of the framework occurs.

The aim of this work is to give a picture of their thermal behaviour. Particular attention will be paid to the typical microporous materials, i.e. framework alumino-silicates and alumino-phosphates.

It is to note that the peculiarity of their response to the heating is present in their own name "zeolite" (i.e. boiling stone) created by Cronstedt in 1756 [1] for minerals which expel water when heated and hence seem to boil. Apart from this curious coincidence, since 1890, Rinne [2] observed that when natrolile is heated, it changes into a new phase, which he called metanatrolite. On exposure to moist air this metanatrolite absorbs water and reconverts to natrolite. Hey [3] showed that metanatrolite is not an independent species 'but merely the final dehydration product of natrolite'. The polymorphic changes of the natural zeolite heulandite by heating were first studied by Rinne already in 1923 [4]. He showed that if heulandite is heated to about 350°C and then allowed to cool to room temperature, it transforms into a new stable phase that Slawson [5] called heulandite B.

Afterwards many authors studied the thermal behaviour of zeolites, but it is only since 1972-73 [6,7] that the structural modifications induced by heating in dehydrated natural zeolites were studied by means of X-ray single crystal structure analysis. In the subsequent years the same experimental procedure was applied to many zeolites, thus obtaining interesting and innovative results. This topic was tackled for many years 'ex situ', i.e. by collecting the intensity data after the sample was dehydrated – usually in vacuum – at selected temperatures in a furnace and subsequently sealed in a glass capillary, therefore in a status not far from the equilibrium conditions. In the recent years the diffraction data have frequently been collected 'in situ' in order to follow the transformation process 'in real time', i.e. in conditions far from their equilibrium conditions. This new approach is now possible owing to the development of new X-ray diffraction techniques on powders or single crystals.

2. DISCUSSION

2.1. Alumino-silicates

A large number of open framework alumino-silicates are present in nature. They are characterized by an Si/Al ratio, which varies between 1.0 (e.g. gismondine [8]) and 7.6 (mutinaite [9]). The negative charge of the framework is compensated by alkaline or alkaline-earth cations (mainly Na, K, Ca, Mg) in extraframework sites.

The greatest part of zeolitic topologies found in nature has also been synthesized in hydrothermal conditions usually using alkali as countercations. In addition, also zeolitic topologies not found in nature have been synthesized. In these materials the Si/Al ratio is usually quite low, and varies between 1 (e.g. zeolite A, zeolite X) and 9-10 (e.g. synthetic mordenite [10]). More recently, after the discovery of ZSM-5 [11], a large number of new microporous materials have been synthesized where the structure direct agent (SDA) or the template material is an organic molecule, sometimes associated to an alkali cation. These compounds are usually characterized by a high Si/Al ratio ($10 < Si/Al \leq \infty$). Almost all compounds with a high Si/Al ratio are characterized by a high thermal stability ($\approx 1000°C$), a property exhibited only by few natural zeolites (e.g. mordenite [12], boggsite [13]) that have a high (for a natural zeolite) Si/Al ratio. In all these alumino-silicates both Al and Si are fourfold coordinated.

Alberti and Vezzalini [14] in a study on the behaviour of zeolites on heating concluded that the structural effects of dehydration on these materials can be divided into three groups characterized by:

I   reversible dehydration accompanied by rearrangement of the extraframework cations and residual water molecules without remarkable changes in the framework and in the cell volume

II   reversible dehydration accompanied by a strong distortion of the framework and a large decrease of the cell volume

III   dehydration accompanied by topological changes in the framework, as a consequence of the breaking of T-O-T bridges.

Some zeolites can give more than one heat-induced phase under different thermal conditions (heating temperature, heating time, time elapsed after heating), which can be classified within different categories of the above scheme. The exchangeable cations and/or the SDA or template molecules present inside the cages and channels of the structure are extremely important on the response to heating. Table 1 lists some zeolites of the three groups.

Table1
Examples of the three groups of zeolites according to the Alberti and Vezzalini scheme[14].

Material	Framework Type Code	References
GROUP I		
Chabazite	(CHA)	[15]
Mordenite	(MOR)	[12]
Ferrierite	(FER)	[16]
Boggsite	(BOG)	[13]
Bikitaite	(BIK)	[17]
Zeolite A	(LTA)	[18]
Faujasite	(FAU)	[19]
Mazzite	(MAZ)	[20]
GROUP II		
Rho	(RHO)	[21]
Gismondine	(GIS)	[22]
Amicite	(GIS)	[23]
Garronite	(GIS)	[24]
Natrolite	(NAT)	[25]
Mesolite	(NAT)	[26] [27]
Scolecite	(NAT)	[27]
Epistilbite	(EPI)	[28]
Yugawaralite	(YUG)	[29] [30]
Laumontite	(LAU)	[31]
Ba-phillipsite	(PHI)	[32]
Clinoptilolite	(HEU)	[33] [34]
GROUP III		
Heulandite	(HEU)	[35]
Stilbite	(STI)	[36]
Stellerite	(STI)	[37] [38]
Barrerite	(STI)	[39]
Brewsterite	(BRE)	[40] [41]
Li-Losod	(LOS)	[42]
Na-TMA-EAB	(EAB)	[43]

In 1992 Baur [44] observed that the framework of zeolites should be classified inflexible or flexible. An inflexible framework is so rigid that no appreciable structural modifications or remarkable changes in the cell parameters occur, whichever its chemical composition. A framework classified as flexible can be described as collapsible or noncollapsible. Noncollapsible framework is such if parts of it are stretched while other parts are compressed, and viceversa; therefore, an equilibrium is reached in this self-limiting framework. An example of a flexible noncollapsible framework is given by zeolite A [44], where the changes in angle around framework oxygens O1 and O2 compensate each other; as they distort in opposite directions, i.e. antirotate. Collapsible framework is where the unit cell volume can be strongly compressed. An example of a flexible collapsible framework is given by natrolite [44], where all T-O-T angles vary in the same direction when cell parameters and volume change, that is corotate. Frameworks of group I according to the classification by Alberti and Vezzalini [14] can be inflexible, like mazzite [20], or flexible noncollapsible, i.e. with antirotating T-O-T angles , such as that of bikitaite [19]. Viceversa, flexible collapsible frameworks correspond to group II of Alberti and Vezzalini scheme. In this case the framework can collapse until the smallest chemical possible values of the corotating T-O-T angles (around 125°) are reached. This is the case, for example, of gismondine [22]. Zeolites of group III. can have an inflexible framework, as is the case of zeolites EAB [43] and Li-Losod [42], or a flexible collapsible framework, as in the case of zeolites of the heulandite group [14].

As concerns the behaviour of zeolites of group II it is to be noted that the decrease in volume can be quite low, as in the case of natural laumontite (~ 4%) [31], or very large, as for Ba-exchanged phillipsite (27%) [32]. The exchangeable cations play a fundamental role in the behaviour of this group of zeolites; for example, the cell volume decrease is more than 10%, about 6% and only 3% in Na- [33]], K- [34] and Cs-exchanged [33] clinoptilolite, respectively. Moreover, often materials with the same topology, like gismondine, amicite, and garronite, or natrolite, mesolite, and scolecite display relevant differences among them. In fact, the cell volume can be reduced by 21% in natrolite [25], but only by 3% in mesolite and scolecite [27]. It must be pointed out that natrolite can lose all its water content without collapse of the framework, whereas the amorphisation of mesolite and scolecite occurs when the water content comes below 50%. Analogously, in the phases with GIS framework amicite can be completely dehydrated (under vacuum at ambient temperature !) [23], whereas gismondine [22] and garronite [24] become amorphous when more than half of their water content is lost. It is evident that this different behaviour is related to the predominant presence of monovalent cations in natrolite (Na) and in amicite (Na and K) and of divalent cations (Ca) in the other phases. It is frequent that zeolites of group II modify two or more times their space group during the dehydration process, or even that their unit cell volume is doubled at a given water loss or temperature.

In our opinion, the zeolites of group III are particularly interesting. As mentioned before, the structures of heat-induced phases of these zeolites, differ from those of hydrated phases for the breaking of T-O-T oxygen bridges and by the migration of the T cations to new 4-coordinated sites. The new polyhedra have three vertices as before (i.e. common to the previously occupied tetrahedra), the fourth vertex being an oxygen, when it is shared by two tetrahedra, or an hydroxyl, when it is unshared. This mechanism was described by Taylor [45] as "the migration of silicon from an initially filled to an initially empty tetrahedron", a process which occurs in many silicates. In these structures the oxygen packing provides the five vertices of the two tetrahedra with a face in common, whereas in group III zeolites, usually water molecules or hydroxyls migrating during dehydration, are attracted by the tetrahedral cations to form the new vertex. Following Alberti and Vezzalini [14], for convenience, the couple of tetrahedra, occupied before and after the migration of the T cation, will be called "face sharing tetrahedra" from now on; it is obvious that only one tetrahedron can be filled at a time. At present this phase transformation, in alumino-silicates, has been found only in some zeolites, the topology of which can be described as a stacking sequence of 6-rings (EAB and Losod) and in zeolites of the heulandite group.

The (Na-TMA)-form of synthetic zeolite EAB, with an ABBACC stacking sequence of 6-rings, changes around 360°C to a sodalite-type product, where the sequence is ABCABC. This change of sequence is explained by Meier an Groner [43] with the breaking of three oxygen bridges in one 6-ring of the D6R (see Fig.1).

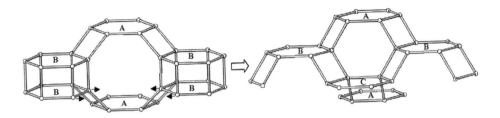

Fig. 1. The figure highlights the ABBA stacking sequence of zeolite EAB (at the left) and the ABCA sequence of sodalite-type structure (at the right). The arrows in EAB show the migration of the tetrahedral cations of the D6R due to the breaking of the oxygen bridges in one 6-ring.

The migration of the tetrahedral cations of the D6R's generates new T-O-T bridges and new 6-rings according to the following scheme

$$A \; B \; B \; A \; C \; C$$
$$\downarrow \quad \downarrow$$
$$A \; B \; C \; A \; B \; C$$

Similarly the framework of Li-Losod, which consists of 6-rings in a ABAC sequence, when treated with a dilute LiOH solution, changes into Li-cancrinite, based on an ABAB stacking sequence, due to the breaking of T-O-T bridges in the C 6-ring [42].

Stilbite, barrerite, and stellerite, compounds characterized by the same topology (STI) but different chemical composition and symmetry, and also heulandite and isostructural clinoptilolite (HEU), and brewsterite (BRE) are zeolites with a framework characterized by the same $4^2 5^4$ cage (Fig. 2), or by the interconnection of the same 4-4-1-1 structural units [46]. They all constitute a very interesting example of group III zeolites. In all these compounds the loss of H_2O leads to a distortion of the tetrahedral framework, which causes a compression and elongation of their channel system, so that they act as zeolites of group II. In addition, in all these materials dehydration causes the rotation of the structural units around the centre of gravity and a pronounced zig-zag of the chains of these units. Consequently there is severe strain on some T-O-T bridges that can possibly cause their breaking (see Fig. 2).

24

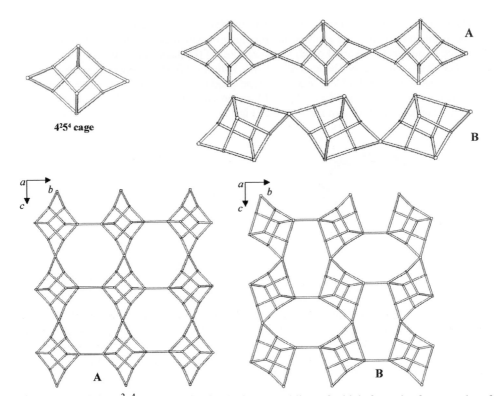

Fig. 2. (Top, left). $4^2 5^4$ polyhedral subunit the assembling of which form the frameworks of zeolites of heulandite group. (Top, right) Chains of $4^2 5^4$ cages present in HEU and STI topologies; A, hydrated phases, B, dehydrated phases. (Bottom). The sheet of $4^2 5^4$ cages present in STI topology. (Left; A) hydrated phases. (Right; B) dehydrated phases.

Natural heulandite, heated in the range of 250-500°C, transforms into a contracted phase, (called phase B by Slawson [5]), which remains stable over a long period of time (in the time span of years, depending on the heating temperature and the chemical composition of sample). Phase B is generated by the breaking of a T-O-T bridge in the four rings of the $4^2 5^4$ cage, which occurs with a frequency depending on the chemical composition, heating time and temperature of heating of the sample [35]. Only one of the two T atoms migrates to a new site to form a "face sharing tetrahedron"; the new fourth vertex can be shared, on a statistical basis, by an analogous tetrahedron of an adjacent structural unit layer (see Fig. 3). The migration of the other T atom of the broken T-O-T bridge, to form another "face sharing tetrahedron" is prevented by geometrical constraints; therefore, the framework is necessarily interrupted when the T-O-T is broken [35]. Phase B rehydrates into a phase, which we call phase I, so stable that no change is observed even after years [47]. Near Infrared Spectroscopy (NIR) proved that the unshared vertices in B and I phases are hydroxyls [47]. Also in Cd-exchanged heulandite the loss of water molecules causes a change in its topology by the breaking of the same T-O-T oxygen bridge of the 4-ring of the $4^2 5^4$ cage [48]. However, in disagreement with what happens in natural heulandite, both T cations migrate to form new "face sharing tetrahedra".

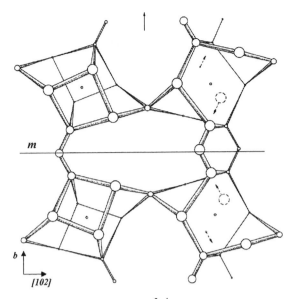

Fig. 3. Framework of heulandite B. The two $4^2 5^4$ polyhedral subunits on the left show the uninterrupted framework and the two $4^2 5^4$ polyhedral subunits on the right the interrupted framework. Dashed circles and arrows show the shift of tetrahedral cations as a consequence of the breaking of the oxygen bridge. Only the oxygens bridging the $4^2 5^4$ chains are represented.

'Ex situ' X-ray single crystal structure analysis showed that barrerite [39] and stellerite [38], on heating in the range 200-400°C, transform into a new contracted phase (phase B). In this phase B an oxygen bridge of the four ring is broken and both tetrahedral cations of the broken T-O-T bridge migrate to form "face sharing tetrahedra". After a few days the B phase of barrerite and stellerite transform into a new phase called C, by rehydration at ambient conditions. These are stable even after years. However, there is an important difference between barrerite C and stellerite C. Whereas the unit cell volume of barrerite C is strongly reduced with respect to the unit cell volume of non-heated barrerite, the unit cell volume of stellerite C is quite similar to that of the untreated stellerite. The explanation of this different behaviour resides in the modalities of the rupture of the T-O-T bridge. In both minerals the broken T-O-T bridge is the same but in barrerite the fourth vertices of the new tetrahedra are shared by analogous tetrahedra of adjacent units of the same chain, whereas in stellerite these new fourth vertices cannot be shared by analogous tetrahedra; thus the framework of stellerite B is more flexible with respect to that of barrerite B, which is blocked by new T-O-T bridges (see Fig. 4). Consequently the rehydrated phase C of stellerite can be more similar, in unit cell volume, to its non-heated phase than the phase C of barrerite to its non-heated phase. NIR spectra confirm the presence of hydroxyl groups in heat collapsed phases of barrerite [49]. The 'ex situ' dehydration behaviour of stilbite strongly resembles that of stellerite, suggesting [38] that the two materials, which have a very similar chemical composition, display similar polymorphic transformations.

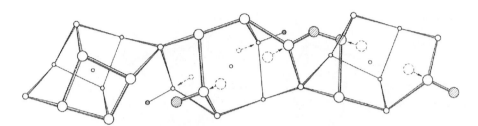

Fig. 4. Chains of 4^25^4 cages in barrerite B. The three units show their possible behaviour in the structure. Dashed circles and arrows show the shift of the tetrahedral cations as a consequence of the breaking of oxygen bridges in the four rings of the cages. Shaded circles represent the new oxygen sites generated by the migration of the T cations.

Arletti et al. [37] and Cruciani et al. [36] performed an 'in situ' study on the thermal behaviour of stellerite and stilbite, using X-ray synchrotron radiation. Their results confirmed that heating causes the breaking of a T-O-T bridge in the 4-ring of the 4^25^4 cage but demonstrated that the breaking occurs on an edge of the 4-rings different from that found in phase B of stellerite B by the 'ex situ' study. Moreover, only one of the two T atoms involved in the rupture of the bridge migrates to form a new "face sharing tetrahedron", so that the framework of these new phases is necessarily interrupted. This means that the behaviour of stellerite and stilbite upon dehydration is different if heated at conditions near to or far from the equilibrium conditions. This behaviour could be present in other microporous materials, with very important differences also for their industrial applications

Investigations on the thermal behaviour of brewsterite were performed both on 'ex situ' single crystals X-ray diffraction - previously dehydrated in vacuum at selected temperature [40,41] - and on 'in situ' synchrotron X-.ray powder diffraction [50]. Single crystal studies show that as the dehydration is almost complete a T-O-T bridge of the 4-ring of the 4^25^4 cage breaks; one of these T atoms migrates to a new tetrahedral site to form a "face sharing tetrahedron" with the previous one, the second migrates to a new site, which is characterized by a fivefold coordination. This coordination polyhedron shares three vertices with the previous occupied one, the fourth vertex joins two adjacent layers of the framework, the fifth one is unshared. In disagreement with these results, when the dehydration process of brewsterite is performed by 'in situ' X-ray powder diffraction no indication of any T-O-T bridge breaking exists [50]. Once again, the behaviour of these zeolites is different if the heating process is carried out at conditions near to or far from the equilibrium. A detailed study on the stability of dehydrated brewsterite [51] showed that after four years from the heating of the mineral held at ambient conditions, it tends to regain its non-heated structure. The peculiarity of this transformation resides in the breaking at ambient temperature of the T-O-T bridges generated by dehydration. This behaviour is obviously related to the rehydration of the compound and we do not exclude that also other apparently stable phases, such as partially rehydrated barrerite [39] and heulandite [47] , could regain the original topology in a longer period of time.

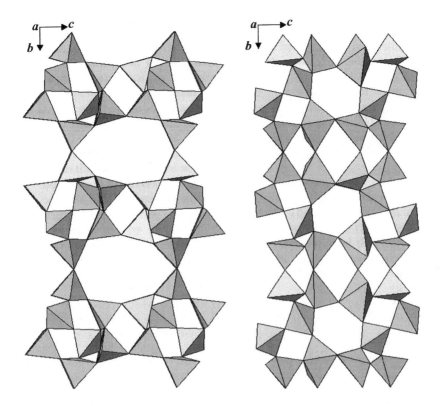

Fig. 5. Framework of collapsed brewsterite without tetrahedral cations migrations (left) and of collapsed brewsterite resulting from the migration of tetrahedral cations (right).

The question is: what is the engine of these topological transformations?.
Wyart and Sabatier [52] emphasized the influence of water on the Si,Al mobility in the solid state in feldspars. Donnay et al. [53] proposed a transformation mechanism with water as a necessary catalyst for silicates. At a high temperature the oxygen bridges under strain, due to cationic movements, are attacked by H^+ ions. Therefore a T-O-T bridge is broken, thus forming a TO_4 complete tetrahedron and a TO_3 incomplete tetrahedron. This last T cation migrates to form a complete "face sharing tetrahedron" attracting an OH group. This last vertex can be shared by another tetrahedron, and under these circumstances a water molecule is regenerated. Moreover, it is probable that the breaking is favoured, as observed by Meier and Groner [43], in the T-O-T bridges where one T is aluminium. The extraframework cations play a fundamental role in these processes. The synthetic zeolite EAB in its (Na-TMA)-form transforms into sodalite by heating, whereas its (K-TMA)-form does not [43]. Barrerite and stellerite, zeolites with the same framework but different exchangeable cations (Na and Ca respectively) strongly differ (as discussed above) in their thermal behaviour. In natural heulandite, T-O-T bridges are broken if Ca is the predominant exchangeable cation

[35], but this topological transformation does not occur if the mineral is very rich in monovalent cations [33,34].

Particular attention must be given to the high silica zeolites synthesized in the presence of fluoride ions. Pure siliceous zeolite IFR [54] was prepared from fluorine-containing media and organic cations acting as templating agent. Single crystal X-ray diffraction studies indicated that F^- is part of a 5-coordinated $[SiO_4F]^-$ unit in a quite trigonal-bipyramidal geometry. If the structure of IFR compound is calcined (thermally removing template and fluoride) the topology of IFR does not change and all Si atoms become fourfold coordinated. A similar situation also occurs in the ITQ-4 structure [55].

2.2.   Alumino-phosphates

The peculiarity of the framework of alumino-phosphate compounds is the variable coordination number of the Al cations, which can be 4, 5, or 6. Another feature of these materials is the influence of the composition of the synthesis moisture on the coordination number of framework Al atoms. For example, the presence of fluoride ions in the reaction moisture produces fluoro-alumino-phosphates where some Al framework ions increase their coordination number by bonding F^- anions.

The history of AlPOs as microporous materials begins in 1982 when Wilson and coworkers [56] synthesized some 3D 4-connected $AlPO_4$ compounds. As a matter of fact, it has long been known that some natural AlPO materials can build open frameworks. Moreover, already in 1961 d'Yvoire synthesized some AlPOs phases the open framework structure of which was determined many years after [57]. However this knowledge was underestimated by the researchers until the Wilson et al's. [56] work. After this paper a very large number of 3D open framework alumino-phosphates have been synthesized. They are often structurally related to zeolite alumino-silicates but differ markedly, since in these compounds Al and P atoms alternate regularly and aluminium exhibits the ability to also have five- and sixfold coordination via additional bonding of water molecules, hydroxyl groups or fluoride ions. In many compounds, usually synthetic but also natural, phosphorous is partially substituted by silicon or by germanium, whereas Al is partially, or even fully, substituted by gallium or transition metals. As a general rule, but not as a law, P can be substituted by tetravalent cations, Al by divalent or trivalent cations. In these cases the negative charge of the framework can be compensated by alkaline cations or ammonium ions or charged organic molecules.

On the basis of the present knowledge, these materials can be classified as follows:

I.   frameworks consisting only of corner linked PO_4 and AlO_4 tetrahedra alternating Al and P sites, with an Al/P ratio equal to one

II.   hydrated compounds that contain aluminium octahedrally coordinated to four framework oxygens – shared with P atoms - and two terminal water molecules. Al/P =1.

III.   structures that include one or more OH groups or fluoride anions bridging two Al atoms and making them 5-coordinated. Al/P=1.

IV.   structures containing Al polyhedra, coordinating F or OH groups, which share edges. These extraframework species are often bridged to three Al polyhedra. Al/P=1.

V.   3D open framework alumino-phosphates with an Al/P ratio lower than unity. When this ratio is quite low (less than 0.9) these structures show terminal P=O or P-OH bonds. Al 5- or 6-coordinated polyhedra can be present

VI.   3D open AlPO frameworks with an Al/P ratio greater than unity. In this case P atoms are always 4-coordinated to Al polyhedra, which are usually sixfold coordinated.

Note that different templates or structure directing agents not only can synthesize materials with different topologies but also materials with the same 3D 4-connected topology that must be classified in different groups.

This classification generalizes that of Parise [58] and does not differ remarkably from the scheme proposed by Kongshang et al. [59], from which it mainly differs because the fluorinated compounds are not considered as a different group, and open frameworks with Al/P greater than one are.

Our interest will now be focused on the structural modification and phase transformations of the AlPO$_4$ materials belonging to these groups in order to verify whether different behaviours can be associated to structural peculiarities of the different groups. As a rule, we will discuss about pure AlPOs or substances where the substitution of Al or P with other cations is quite low.

GROUP I

Many AlPO$_4$ materials belong to this group, where corner linked PO$_4$ and AlO$_4$ tetrahedra regularly alternate in the framework. An exhaustive sample of such compounds is listed in Table 2.

Table 2

AlPO$_4$ materials the framework of which consists only of alternating corners linked PO$_4$ and AlO$_4$ tetrahedra.   Al/P=1

Material	Framework Type Code	References	Notes
AlPO$_4$-5	(AFI)	[60] [62]	[60] Al^V with F$^-$
AlPO$_4$-11	(AEL)	[63]	
AlPO$_4$-12-TAMU	(ATT)	[64]	Calcined at 600°C
AlPO$_4$-16	(AST)	[65][66]	
AlPO$_4$-31	(ATO)	[67] [68] [69]	[68] Al^V with OH
AlPO$_4$-34	(CHA)	[61]	
AlPO$_4$-36	(ATS)	[70] [71]	[70] Calcined at 450°C [71] Al^V, Al^{VI} Calcined-rehydrated
AlPO$_4$-41	(AFO)	[72]	Calcined at 600°C
AlPO$_4$-50	(AFY)	[73]	
AlPO$_4$-56	(AFX)	[74]	Calcined at 600°C
STA-2	(SAT)	[75]	
STA-6	(SAS)	[76]	
STA-7	(SAV)	[77]	

The most relevant peculiarity of these compounds is their thermal stability, in some cases up to 1000°C. In the majority of the cases AlPOs of group I transform above 1000°C in AlPO-tridymite or AlPO-cristobalite. Their thermal stability decreases if Al or P atoms are partially substituted by other cations. It is quite common that the as-synthesized materials display a real symmetry lower than their topochemical symmetry. The topochemical symmetry normally occurs when the material is calcined and the extraframework species are expelled. Note, however, that some of the crystal structures listed in Table 1 have been determined after the calcination of the material, so that it is not sure that the as-synthesized phase is only characterized by 4-coordinated Al (Al^{IV}) and P cations. Moreover, some of these materials if synthesized in a medium containing fluoride atoms can have Al sites with a coordination number greater than four (i.e. five, Al^{V}, or six, Al^{VI}) with F⁻ as additional bonding species. This is the case, for example, of $AlPO_4$-5 [60] or $AlPO_4$-34 [61]. In any case calcination restores the 3D 4-connected linkages.

GROUP II

This group is formed by the open $AlPO_4$ materials where some of the Al atoms have a coordination number greater than four, normally six, but the additional coordinating species are not shared with other polyhedra of the framework (Table 3). Both natural and synthetic AlPOs can display these features. The natural alumino-phosphates variscite and metavariscite, both with chemical formula $AlPO_4 \cdot 2H_2O$, are formed by alternating Al octahedra and P tetrahedra to form an open framework [78,79]. ^{27}Al MAS NMR data indicate that $AlPO_4$-52 in its calcined form is built up by alternating Al and P tetrahedral, whereas some 6-coordinated Al atoms – with two water molecules – are present both in calcined-rehydrated and as-synthesized forms with Al^{VI}/Al^{IV} ratios of about 1/3 and 2/3 in the as-synthesized and calcined-rehydrated forms respectively [80,81]. Unfortunately, it was not possible to establish whether these water molecules are bonded to a particular Al site or are distributed, with partial occupancy, over all the Al sites of the framework.

Table 3
$AlPO_4$ materials containing aluminium octahedrally coordinated to four framework oxygens and two water molecules.   Al/P=1

Material	Framework Type Code	References
$AlPO_4$-Variscite		[78]
$AlPO_4$-Metavariscite		[79]
$AlPO_4$-52	(AFT)	[80] [81]
$AlPO_4$-H3 ↔ $AlPO_4$-C	(APC)	[57] [82] [83] [84]
$AlPO_4$-D ↔ $AlPO_4$-H6	(APD)	[83] [84] [85]
VPI-5 ↔ $AlPO_4$-H1	(VFI)	[86] [87] [88]
$AlPO_4$-8	(AFT)	[89] [90] [91]
$AlPO_4$-H2	(AHT)	[92] [93]

An interesting case of topological transformation in this group of AlPOs is given by the APC-APD topologies. D'Yvoire [57] had already synthesized four hydrated materials identified as $AlPO_4$-Hi (i=1,4) in 1961. Pluth and Smith [82] found that a synthetic phase - with composition $AlPO_4 \cdot 1.5H_2O$ – shows an XRPD pattern, which strongly resembles that of $AlPO_4$-H3 of d'Yvoire. The topology of $AlPO_4 \cdot 1.5H_2O$ can be described as a 3D framework formed by 4-, 6- and 8-rings of tetrahedra. In this structure PO_4 tetrahedra alternate between AlO_4 tetrahedra and $AlO_4(H_2O)_2$ octahedra. The 4-coordinated Al atoms are in the node of the 4-rings whereas 6-coordinated Al atoms are in the nodes of the 6-rings.

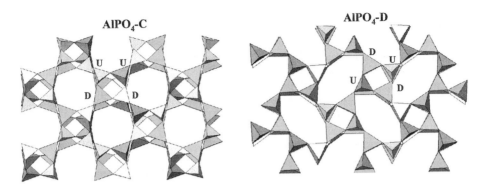

Fig. 6. The framework topology of $AlPO_4$-C and $AlPO_4$-D

$AlPO_4$-H3 is converted into AlPO-C [83] at about 100°C by the removal of the water molecules without changes in the framework topology. This process is completely reversible. AlPO-C transforms irreversibly into a new phase, AlPO-D, at about 250°C through a reconstructive polymorphic process. In the transformation AlPO-C→AlPO-D, the UUDD (double-crankshaft) chains of AlPO-C change to the UDUD (narsarsukite-type) chains of AlPO-D through a 'face sharing tetrahedra' (or topotactic) transformation (see Fig. 6). As AlPO-C, also AlPO-D is a 3D 4-connected net of tetrahedra. If AlPO-D is cooled at about 50°C, it rehydrates to the $AlPO_4$-H6 phase of d'Yvoire [57]. The transformation of AlPO-D to $AlPO_4$-H6 is as fast and as reversible as the transformation of AlPO-C to $AlPO_4$-H3. It is unknown if the rehydration of AlPO-D to AlPO-H6 gives rise to Al-cations with a coordination number greater than four, as occurs in the AlPO-C → AlPO-H3 transformation. According to Keller et al [83], if AlPO-D is heated to 700°C, it transforms into a new similar, but uncharacterised, thermally stable phase. Finally, if AlPO-H3 is heated to about 900°C for some hours, it changes to AlPO-tridymite. It would be extremely interesting to follow these phase transformations 'in situ', in order to ascertain whether they are due to successive topotactic transformations or to amorphousization and recrystallization of AlPO-H3 materials. A new alumino-phosphate (AlPO-CJ3), the framework of which is analogous to AlPO-D, was synthesized in an aqueous system using ethanolamine (ETA) as a template [85]. The most interesting feature of AlPO-CJ3 is - besides the sixfold coordination of an Al site - that ETA not only acts as a template but also as a ligand, donating its O atoms to increase the coordination of Al. Upon removal of the template above 450°C, AlPO-CJ3 transforms to the 3D 4-connected structure of AlPO-D.

Davis and coworkers in 1988 [86] reported the first synthesized molecular sieve, which contains rings consisting of more than twelve T atoms. This material, VPI-5, is characterized by an 18-ring 1D channel system. This 18-ring is circumscribed, with hexagonal symmetry, by six 6-rings alternating two fuses 4-rings that form a triple cranckshaft chain. Thereafter, McCusker et al [87] demonstrated that two water molecules complete the octahedral coordination of one of the three symmetrically independent Al

Atoms. This Al is located in between the fused 4-rings. In 1990, Dessau et al described the crystal structure of AlPO-8, a synthetic molecular sieve characterized by a 1D 14-ring channel system [89]. In the subsequent years, many authors put in evidence that VPI-5 is the analogue of AlPO-H1 of d'Yvoire and may transform into the AlPO-8 phase. This transformation is influenced by many parameters, heating rate, degassing under vacuum of VPI-5, and so on. In any case, the transformation of VPI-5 into AlPO-8 seems irreversible, whereas the dehydration process of VPI-5 seems reversible. It is not the purpose of this work to determine the conditions that lead to the phase transformation VPI-5 → AlPO-8 but to highlight the transformation process. As in the case of the topological transformation AlPO-C → AlPO-D, the transformation VPI-5 → AlPO-8 can be described as the migration of the 6-coordinated Al and one 4-coordinated P atom to new tetrahedral sites in a process described before as "face sharing tetrahedra" transformation.

Another interesting compound of this group is AlPO-H2 [92,93]. Its structure is constructed exclusively by two fused four rings forming triple cranckshaft chains which circumscribe a 1D 10-ring channel system, in the same way as the same cranckshaft chains circumscribe the 18-membered rings of VPI-5. X-ray diffraction indicated that the Al atom at the central cranckshaft of the chain is octahedrally coordinated with two water molecules and four framework oxygens., as occurs in VPI-5. AlPO-H2 is not thermally stable; upon heating to 100°C, it transforms into AlPO-tridymite. It is however possible to dehydrate AlPO-H2 under vacuum at room temperature without topological changes even if strong distortion of the framework occurs. This compound easily transforms into AlPO-tridymite also in its dehydrated phase. Also the topological transformation AlPO-H2→tridymite occurs through the process described before, i.e. the "face sharing tetrahedra" transformation. In both phase transformations AlPO-H2→tridymite and VPI-5→AlPO-8, the dimensions of the rings are reduced by four tetrahedral units, from 10 to 6 for AlPO-H2, from 18 to 14 for VPI-5. It is interesting to note that both for VPI-5 and AlPO-H2 the synthesis was carried out in presence of an amine molecule and in both cases the structures do not contain organic molecules in the channels but only water molecules.

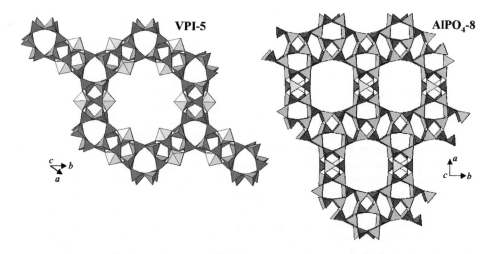

Fig. 7. The framework topology of VPI-5 and AlPO$_4$-8. The octahedra in VPI-5 represent the 6-coordinated Al atoms.

GROUP III

In this group an OH, or H_2O, or F^- bridges two framework Al atoms to form an Al-(OH, H_2O, F)-Al bond. This situation can occur only if the two Al are in the same 4- or 6-ring of alternating Al-P cations. In the first case a couple of 3-rings composed by two 5-coordinated Al atoms and one 4-coordinated P atom are formed, in the second case an analogous 3-ring and a 5-ring composed by two 5-coordinated, one 4-coordinated Al, and two P atoms are generated (Table 4). The microporous compounds of group III display an odd feature: when the two Al atoms involved in the bridge are symmetrically independent they are 5-coordinated, when they are symmetry related one Al is 5-coordinated, the other is 6-coordinated, to a terminal water molecule or a F^- ion. Particularly interesting is the evidence that every time an Al site is fivefold coordinated, the three oxygen atoms in the triangular of the trigonal bipyramid are closer to Al than the two apical oxygens; this can be explained by an sp^2 hybridization.

Table 4
$AlPO_4$ materials that include one or more OH groups or fluoride anions bridging two Al atoms and making them 5-coordinated.   Al/P=1

Material	Framework Type Code	References
$AlPO_4$-EN3 ↔ $AlPO_4$-53(B)	(AEN)	[58] [94] [95] [96]
$AlPO_4$-53(C)		[94] [95]
$AlPO_4$-21	(AWO)	[97] [98]
$AlPO_4$-25	(ATV)	[99]
$AlPO_4$-18	(AEI)	[100]
$AlPO_4$-17	(ERI)	[101]
$AlPO_4$-40	(AFR)	[102]
$AlPO_4$-ZON	(ZON)	[103] [104] [105]
IST-1	(PON)	[106]
$AlPO_4$-CJ2		[107] [108]
Mu-13	(MSO)	[109]
ULM-3		[110] [111]
ULM-4		[111]
SSZ-51		[112]
$AlPO_4$-NH2		[113]

The crystal structure of the open framework AlPO-EN3 (AEN) was solved by Parise [58] in 1985. Thereafter a number of AlPOs with the same topology of AlPO-EN3 have been synthesized and characterized. The 3D 4-connected framework can be described as an assembly of layers, consisting of 4-, 6-, and 8-rings, in an ABAB sequence. Two of the three Al sites of the same 6-ring are in fivefold coordination by sharing an (OH) group. One of the AlPOs materials isostructural with AlPO-EN3 is AlPO-53(A) described by Kirchner et al. [94]. If AlPO-53(A) is calcined and dehydrated, a new phase, AlPO-53(B), is formed where the (OH) groups are lost and a 3D 4-connected topology exists. If AlPO-53(A) is calcined at high temperature (around 700°C) and this temperature is maintained for some days, the as-synthesized material transforms into a new phase with a different topology, called AlPO-53(C) by Kirchner et al.[94].

34

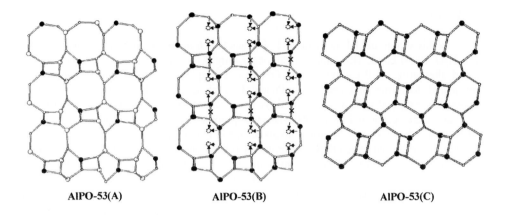

| AlPO-53(A) | AlPO-53(B) | AlPO-53(C) |

Fig. 8. The orthogonal view of the 4-, 6-, 8-rings layer of AlPO$_4$-53 phases. Black and small empty circles represent P and Al atoms. The shaded lines between two Al in AlPO$_4$-53(A) represent the Al-OH-Al bridge bonding 5-coordinated Al atoms.

This new topology differs from the first one as one bond of the four rings is broken and the two cations (Al and P) move to generate a couple of "face sharing tetrahedra". As a result, chains of alternating 4-, and 6-rings substitute the 8-ring chains and new distorted 8-ring chains substitute the 4-, and 6-ring chains present in the as-synthesized compound (see Fig. 8).

In 1985, Parise and Day [97], and Bennett et al. [98] solved the crystal structure of a new alumino-phosphate called AlPO-21 (AWO). Its building unit is a double crankshaft arrangement and the 3D 4-connected topological symmetry contains 4-, 6-, and 8-rings. As for AlPO-53(A), a (OH) group bridges two of the three Al sites of the 6-rings, which are now 5-coordinated in a trigonal-bipyramidal geometry. The loss of water and organic molecules by calcination transforms AlPO-21 in a new topological distinct phase, AlPO-25 [99] (ATV). The double crankshaft chain disappears and the UUDD connectivity of the four rings is substituted by an UDUD connectivity, giving rise to a new 4-ring chain of narsarsukite type. Also in this case, the topological transformation can be explained by a "face sharing tetrahedra" process.

As we have seen, AlPO-21 and AlPO-EN3 are characterized by a change in the framework topology after calcination. It is possible that such transformations occur in other AlPO$_4$ materials of the group III, but the lack of information on many compounds prevents a more detailed description of their thermal behaviour. It seems that after calcination, the more common structural modifications of this group is the loss of the additional species bonded to Al atoms, with a formation of a 3D 4-connected framework, without any topological change.

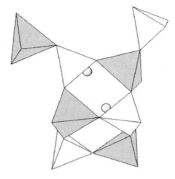

Fig. 9. The "open cube" present in AlPO$_4$-40, AlPO$_4$-ZON and SSZ-51 microporous materials. White and shaded polyhedra represent Al and P tetrahedra, spheres represent F atoms or (OH) groups which form a bridge between two Al atoms.

The fundamental building unit of many compounds of this group is an "open cube" or also a D4R with one disconnected (or ring-opened) edge. An (OH) group or a fluoride ion is encapsulated within this building unit to form a bridge between two Al atoms of one 4-ring of the "open cube", thus generating a couple of 5-coordinated Al atoms (Fig. 9). The other two symmetrically independent Al, which concur to form the other 4-ring of the "open cube", are 4-coordinated. This situation is present in AlPO-40 [102], AlPO-ZON [103,104] and SSZ-51 [112]. If AlPO-ZON is calcined at 600°C, the fluoride ions that form the bridge between two Al atoms are lost and all framework cations become 4-coordinated without topological changes in the structure. Surprisingly, if the calcined AlPO-ZON is rehydrated, the two Al atoms, which were 5-coordinated in the as-synthesized material, are now 4-coordinated, whereas the two other Al atoms, 4-coordinated in the as synthesized compound, are now 5- and 6-coordinated, with one or two water molecules respectively [103,105] (see Fig. 10). A similar behaviour probably also occurs in AlPO-40 and SSZ-51 [112].

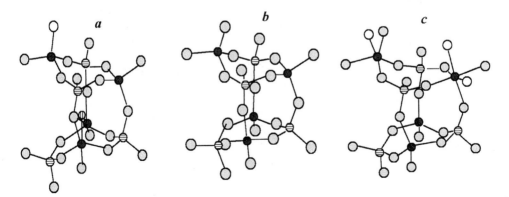

Fig. 10. The fundamental building unit ("open cube") in as synthesized (a), calcined (b), and rehydrated (c) AlPO$_4$-ZON. Black, shaded, horizontal dashed, vertical dashed, and empty circles represent Al, O, P, F atoms, and H$_2$O molecules respectively.

GROUP IV

    This group is formed by $AlPO_4$ materials containing Al coordination polyhedra that share edges. Almost all these compounds are characterized by a tetramer of Al polyhedra – with Al atos 4-, 5- or 6-coordinated - where a couple of polyhedra shares an edge, the vertices of which can be O, OH or F. These vertices always bond three different Al polyhedra. Scarce information exists about the thermal behaviour of these materials. We know that UiO-26 can loose its unshared water molecules reversibly, but if calcined and dehydrated at 400°C and 600°C, it transforms irreversibly into two new phases, the structures of which are unknown [59]. AlPO-14 [117] and AlPO-JDF [126], microporous materials of this group, can loose the additional bonding, by calcinations and dehydration, without topological changes (Table 5).

Table 5
$AlPO_4$ materials that contain Al polyhedra that share edges. Al atoms can be 4-, 5- or 6-coordinated.   Al/P=1

Material	Framework Type Code	References
$AlPO_4$-34	(CHA)	[114]
$AlPO_4$-12		[115] [116]
$AlPO_4$-14	(AFN)	[117] [118] [119]
$AlPO_4$-15		[120] [121]
Tinsleyite		[122] [123]
UiO-26		[59]
ULM-6		[124]
MIL-27		[125]
$AlPO_4$-JDF		[126]

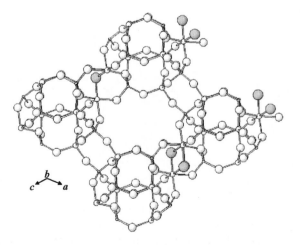

Fig. 11. The framework of CoAPO-34. The shaded circles represent the extra-ligand F⁻ ions of the 6-coordinated Al atoms

The 'in situ' time resolved X-ray structure refinement, by Rietveld analysis, of a CoAPO-34 material, synthesized in a mixture containing morpholine molecules and HF [114], highlighted that two Al atoms are 6-coordinated with two extra-ligand F$^-$ ions, and that these two polyhedra share an edge (Figure 11). During the heating process, upon deflurinaxation, these Al atoms assume a 5-fold coordination, and when the residual fluorine atoms and organic molecules are expelled, a 3D 4-connected chabasite-type topology is restored. However the real symmetry of the material remains triclinic, as in the as-synthesized phase, even if cell parameters take on strictly metrically rhombohedral values, indicating that the structure maintains a memory of the distortion imposed by the unusual coordination of Al [114].

GROUP V

In the microporous AlPO compounds considered up to now, all oxygen atoms of T-O$_4$ tetrahedra (T = Al or P) are shared by an Al and a P cations; this dictates the 1:1 ratio of Al/P. However, the existence of terminal P=O and/or P-OH groups leads to an Al/P ratio less than unity. So far 3D an open framework has been found with an Al/P ratio varying in the range 1/2 - 12/13 (Table 6). Unlike 3D AlPOs with Al/P ratio equal to unity and with neutral framework, the framework of AlPOs with Al/P < 1 bears negative charges, which will be compensated by protonated groups. This implies that these last compounds can have Bronsted acidity after removing the template.

Table 6
3-Dimensional AlPO materials with an Al/P less than unity.

Material	Framework Type Code	References	Al/P
AlPO-CJB1		[127]	12/13
AlPO-CJB2		[128]	11/12
AlPO-CJ11		[129]	11/12
JDF-20		[130]	5/6
AlPO-HDA		[131]	4/5
AlPO-x3		[132]	3/4
AlPO-DETA		[133]	2/3
AlPO-CJ4		[134]	1/2
Product K		[135] [136]	1/2

In this group of open frameworks we can distinguish two different behaviours, depending on the Al/P ratio.
a) $11/12 \leq$ Al/P < 1. In this subgroup both Al and P atoms share their vertices with framework oxygen atoms. The minor amount of Al with respect to P is compensated by a coordination number of some Al atoms larger than four (5 or 6). The thermal stability of these materials is always greater than 600°C.
b) $1/2 \leq$ Al/P < 5/6. In this subgroup Al atoms can be 4-, 5-, or 6-coodinated but share their vertices only with framework oxygens. The only exception is given by product K [135] where one of the two symmetrically independent Al sites coordinates three water molecules. Contrary to Al atoms, some of the P atoms, always 4-coordinated, are bonded to terminal OH groups or oxygen atoms. The thermal stability of this subgroup is always less than 600°C. It

is easy to suppose that the different thermal stability of the two subgroups is due to the presence of terminal O or OH vertices in P tetrahedra, which could produce a reduced thermal stability.

No topological transformation of these materials is known; upon dehydration and calcinations they become amorphous or give an AlPO-tridymite phase.

GROUP VI

Open framework alumino-phosphates with an Al/P ratio greater than one belong to this group (see Table 7).

Table 7
AlPO materials with an Al/P greater than unity.

Material	Framework Type Code	References	Notes
Augelite		[137]	Al/P=2/1
Fluellite		[138]	Al/P=2/1
Cacoxenite		[139]	(Al, Fe)/P=25/17

To date only natural materials are known with this feature. Their peculiarity is that P atoms are always 4-coordinated to framework oxygens whereas Al atoms are always 6-coordinated with framework oxygens, bridging P atoms, fluorine anions and hydroxyls. Examples of this group are fluellite, where the symmetrically independent Al site coordinates four oxygens and two fluorine atoms [137], and angelite, where both the symmetrically independent Al sites coordinate water molecules and framework oxygens [138].

Particularly interesting, and well known, is the tetrahedral-octahedral framework of the hydrated iron-alumino-phosphate cacoxenite [139]. Its crystal structure is characterized by two different building units. Pairs of $(Fe^{3+},Al)(O,OH)_6$ octahedra share an edge to form dimers, and three dimers share octahedron corners to form a ring that has a (PO_4) group at its core. The second unit consists of a tetramer analogous to that found in $AlPO_4$-15 [120] and in tinsleyite [122,123]. The most intriguing feature of the cacoxenite structure is the enormous 1D channel system (with a free diameter larger than 14Å) formed by 30-membered rings of P tetrahedra and (Fe^{3+},Al) octahedra. The stability and modifications of this framework are still unknown even if it appears related to the loss of the water molecules located in the large channel. No information is known on the thermal behaviour of fluellite and angelite.

CONCLUSIONS

1.   The dehydration process in alumino-silicates and alumino-phosphates develops along many different routes, and no possibility exists to foresee which of these routes will be followed by a given compound.

2.   In natural or as-synthesized alumino-silicates, where both Al and Si cations are usually 4-coordinated, the dehydration process can result in: a) no remarkable modification in framework atoms position and unit cell volume, b) relevant distortion of the framework and remarkable decrease of the unit cell volume, c) topological change in the framework topology due to a 'face sharing tetrahedra' process. A high Si/Al ratio seems to favour an a) behaviour. Re-hydration usually restores the unheated structure.

3. In as synthesized or natural alumino-phosphates Al atoms are frequently 5- or 6-coordinated. A first step of heating usually causes the loss of the additional species (H_2O, OH, F^-) without topological changes. In some cases a second step of heating results in topological changes due to a 'face sharing tetrahedra' process. Re-hydration can restore the initial structure but also modify the polyhedral configuration of some Al atoms of the framework; the dehydration-rehydration process is sometimes irreversible.

To conclude, thermal behaviour in microporous materials usually shows a witty response with respect to our forecast.

References

[1] A.F. Cronstedt, Kongl. Vetenskaps Acad. Handl, Stockholm, 17 (1756) 120.
[2] F. Rinne, Sitzungsber. Preuss. Akad. Wiss., 14 (1890) 1163. After Hey (1932).
[3] M.H. Hey, Miner. Mag. 23 (1932) 243.
[4] F. Rinne, Z. Kristallogr., 59 (1923) 230.
[5] C.B. Slawson, Am. Miner., 10 (1925) 305.
[6] J.J. Pluth, J.V. Smith, Mater. Res. Bull., 7 (1972) 1311.
[7] A. Alberti, TMPM Tschermaks Min. Petr. Mitt., 19 (1973) 173.
[8] G. Gottardi, E. Galli, Natural Zeolites, Springer-Verlag, Berlin, 1985, pp 122-133.
[9] E. Galli, G. Vezzalini, S. Quartieri, A. Alberti, M. Franzini, Zeolites, 19 (1997) 318.
[10] D.W. Breck, Zeolites Molecular Sieves, Wiley, New York, 1974.
[11] R.J. Argauer, G.E. Landolt. U.S. Pat. 3 702 886 (1972).
[12] A Martucci, M. Sacerdoti, G. Cruciani, C. Dalconi, Eur. J. Mineral., 15 (2003) 485.
[13] S. Zanardi, G. Cruciani, A. Albert, E. Galli, Am. Miner., 89 (2004) 1033.
[14] A. Alberti, G. Vezzalini, in: Proc. 6th Int. Zeolite Conf., Reno 1983 (1984) (Eds. D. Olson, A. Bisio) Butterworths, Guildford, UK, 834.
[15] W.J. Mortier, J.J. Pluth, J.V. Smith, Mat. Res. Bull., 12 (1977) 97.
[16] M.C. Dalconi, A. Alberti, G. Cruciani, J. Phys. Chem. B, 107 (2003) 12973.
[17] O. Ferro, S. Quartieri, G. Vezzalini, C. Ceriani, E. Fois, A. Gamba, G. Cruciani, Am. Mineral., 89 (2004) 94.
[18] J.J. Pluth, J.V. Smith, J. Am. Chem. Soc., 102 (1980) 4704.
[19] F. Porcher, M. Sonhasson, Y. Dusansoy, C. Lecomte, Eur. J. Miner., 11 (1999) 333.
[20] R. Rinaldi, J.J. Pluth, J.V. Smith, Acta Crystallogr., B31 (1975) 1603.
[21] A. Bieniok, W.H. Baur, Mat. Sci. Forum, 79-82 (1991) 721.
[22] G. Vezzalini, S. Quartieri, A. Alberti, Zeolites 13 (1993) 34.
[23] G. Vezzalini, A. Alberti, A. Sani, M. Triscari, Micr. Mes. Mat., 31 (1999) 253.
[24] L. Schröpfer, W. Joswig, Eur. J. Mineral., 9 (1997) 53.
[25] W.H. Baur, W. Joswig, N. Jb. Miner. Mh., 1996 (1996) 171.
[26] K. Ståhl, R. Thomasson, Zeolites, 14 (1994) 12.
[27] K. Ståhl, J. Hanson, J. Appl. Cryst., 27 (1994) 543.
[28] G. Cruciani, A. Martucci, C. Meneghini, Eur. J. Mineral., 15 (2003) 257
[29] A. Alberti, S. Quartieri, G. Vezzalini, Eur. J. Mineral., 8 (1996) 1273.
[30] G. Artioli, K. Ståhl, G. Cruciani, A. Gualtieri, J.C. Hanson, Am. Mineral., 86 (2001) 185.
[31] K. Ståhl, G. Artioli, J.C. Hanson, Phys. Chem. Minerals, 23 (1996) 328.
[32] A. Sani, G. Cruciani, A. Gualtieri, Phys. Chem. Minerals, 29 (2002) 351.

40

[33] M. Johnoson, D. O'Connor, P. Barnes, C.R.A. Catlow, SL. Owens, G. Sankar, R. Bell, S.J. Teat, R. Stephenson, J. Phys. Chem. B, 107 (2003) 942.

[34] E. Galli, G. Gottardi, H. Mayer, A. Preisinger, E. Passaglia, Acta Cryst. B39 (1983) 189.

[35] A. Alberti, G. Vezzalini, TMPM Tschermak Min. Petr. Mitt., 31 (1983) 259.

[36] G. Cruciani, G. Artioli, A. Gualtieri, K. Ståhl, J.C. Hanson, Am. Miner., 82 (1997) 729.

[37] R. Arletti, E. Mazzuccato, G. Vezzalini, Proc. 32th Int. Geological Congress, Florence, 2004.

[38] A. Alberti, R. Rinaldi, G. Vezzalini, Phys. Chem. Minerals, 2 (1978) 365.

[39] A. Alberti, G. Vezzalini, in: Natural Zeolites. Occurrence, Properties, Use. (Eds. L.B. Sand, F. Mumpton) (1978) Pergamon Press, Oxford. pp. 85-98.

[40] A. Alberti, M. Sacerdoti, S. Quartieri, G. Vezzalini, Phys. Chem. Minerals, 26 (1999) 181.

[41] M. Sacerdoti, G. Vezzalini, S. Quartieri, Micr. Mes. Mat., 41 (2000) 107.

[42] W. Sieber, W.M. Meier, Helv. Chim. Acta, 57 (1974) 1533.

[43] W.H. Meier, M. Groner, J. Solid State Chemistry, 37 (1981) 204.

[44] W.H. Baur, J. Solid State Chemistry, 97 (1992) 243.

[45] H.F.W. Taylor, J. Appl. Chem., 10 (1960) 317

[46] A. Alberti, Am. Mineral., 64 (1979) 1188.

[47] A. Alberti, G. Vezzalini, F. Cariati, L. Erre, P. Piu, Zeolites, 5 (1985) 289-291.

[48] N. Doebelin, T. Armbruster, Micr. Mes. Mat., 61 (2003) 85-103.

[49] A. Alberti, F. Cariati, L. Erre, P. Piu, G. Vezzalini, Phys. Chem. Minerals, 9 (1983) 189-191.

[50] K. Ståhl, J.C. Hanson, Micr. Mes. Mat., 32 (1999) 147.

[51] A. Alberti, G. Vezzalini, S. Quartieri, G. Cruciani, S. Bordiga, Micr. Mes. Mat., 42 (2001) 277.

[52] J. Wyart, G. Sabatier, Bull. Soc. Fr. Miner. Crist., 81 (1958) 223.

[53] G. Donnay, J. Wyart, G. Sabatier, Zeit. Krist., 112 (1959) 161.

[54] I. Bull, L.A. Villaescusa, S.J. Teat, M.A. Camblor, P.A. Wright, P. Lighfoot, R.E. Morris, J. Am. Chem. Soc., 122 (2000) 7128.

[55] G. van de Goor, C.C. Freyhardt, P.Z. Behrens, Z. Anorg. Allg. Chem., 621 (1995) 311.

[56] S.T. Wilson, B.M. Lok, C.A. Messina, T.R. Cannan, E.M. Flanigen, J. Am. Chem. Soc., 102 (1982) 1146.

[57] F. d'Yvoire, Bull. Soc. Chim. Fr., (1961) 1762.

[58] J.B. Parise, Zeolites: Synthesis, structure, technology and application: Studies in Surface Science and Catalysis (Eds. B. Držaj, S. Hočevar, S. Pejovnik), Elsevier, Amsterdam, vol. 24 (1985) p. 271.

[59] K.O. Kongshaug, H. Fjellvåg, K.P. Lillerud, Micr. Mes. Mat., 40 (2000) 313.

[60] R.D. Gougeon, E.B. Brouwer, P.R. Bodart, L. Delmotte, C. Marichal, J-M. Chézeau, R.K. Harris, J. Phys. Chem. B, 105 (2001) 12249.

[61] A. Martucci, A. Alberti, G. Cruciani, A. Frache, S. Coluccia, L. Marchese, J. Phys. Chem. B, 107 (2003) 9655.

[62] J.M. Bennett, J.P. Cohen, E.M. Flanigen, J,J, Pluth, J.V. Smith, Intrazeolite Chemistry. (Eds., G.D. Stucky, G.D. Dwyer) Am. Chem. Soc. Symp. Ser., 218 (1983) 109.

[63] J.W. Richardson Jr., J.J. Pluth, J.V. Smith, Acta Cryst. B44 (1988) 367.

[64] P.R. Rudolf, C. Saldarriaga-Molina, A. Clearfield, J. Phys. Chem., 90 (1986) 6122.

[65] J.M. Bennett, R.M. Kirchner, Zeolites, 11 (1991) 502.

[66] C. Schott-Darie, J. Patarin, P.Y. Le Goff, H. Kessler, E. Benazzi, Micr. Mat., 3 (1994) 123.

[67] J.M. Bennett, R.M. Kirchner, Zeolites, 12 (1992) 338.

[68] W.H. Baur, W. Joswig, D. Kassner, J. Kornatowski, G. Finger, Acta Cryst., B50 (1994) 290.

[69] G. Mali, A. Meden, A. Ristić, N. Novak Tušar, V. Kaučič, J. Phys. Chem. B, 106 (2002) 63.

[70] J.V. Smith, J.J. Pluth, K.J. Andries, Zeolites, 13 (1993) 166.

[71] M.H. Zahedi-Niaki, G. Xu, H. Meyer, C.A. Fyfe, S. Kaliaguine, Micr. Mes. Mat., 32 (1999) 241.

[72] R.M. Kirchner, J.M. Bennett, Zeolites, 14 (1994) 523.

[73] N. Novak Tušar, A. Ristić, A. Meden, V. Kaučič, Micr. Mes. Mat., 37 (2000) 303.

[74] S.T. Wilson, R.W. Broach, C. Scott Blackwell, C.A. Bateman, N.K. McGuire, R.M. Kirchner, Micr. Mes. Mat., 28 (1999) 125.

[75] G.W. Noble, P.A. Wright, Å, Kvick, J. Chem. Soc., Dalton Trans., (1997) 4485.

[76] V. Patinec, P.A. Wright, P. Lighfoot, R.A. Aitken, P.A. Cox, J. Chem. Soc., Dalton Trans., (1999) 3909.

[77] P.A. Wright, M.J. Maple, A.M.Z. Slawin, V. Patinec, R.A. Aitken, S. Welsh, P.A. Cox, J. Chem. Soc., Dalton Trans., (2000) 1243.

[78] R. Kniep, D. Mootz, A. Vegas, Acta Cryst., B33 (1977) 263.

[79] R. Kniep, D. Mootz, Acta Cryst., B29 (1973) 2292.

[80] J.M. Bennett, R.M. Kirchner, S.T. Wilson, Zeolites: Facts, Figures, Future: Studies in Surface Science and Catalysis (Eds. P.A. Jacobs, R.A. van Santen), Elsevier, Amsterdam, vol. 49B (1989) p. 731.

[81] N.K. McGuire, C.A. Bateman, C. Scott Blackwell, S.T. Wilson, R.M. Kirchner, Zeolites, 15 (1995) 460.

[82] J.J. Pluth, J.V. Smith, Acta Cryst., C42 (1986) 1118.

[83] E.B. Keller, W.M. Meier, R.M. Kirchner, Solid State Ionics, 43 (1990) 93.

[84] L. Canesson, I. Arcon, S. Caldarelli, A. Tuel, Micr. Mes. Mat., 26 (1998) 117.

[85] K. Wang, J. Yu, G. Zhu, Y. Zou, R. Xu, Micr. Mes. Mat., 39 (2000) 281.

[86] M.E. Davis, C. Saldarriaga, C. Montes, J. Garces, C. Crowder, Nature, 331 (1988) 698.

[87] L.B. McCusker, Ch. Baerlocher, E. Jahn, M. Bülow, Zeolites, 11 (1991) 308.

[88] D.M. Poojary, J.O. Perez, A. Clearfield, J. Phys. Chem., 96 (1992) 7709.

[89] R.M. Dessau, J.L. Schlenker, J.B. Higgins, Zeolites, 10 (1990) 522.

[90] J.W. Richardson Jr., E.T.C. Vogt, Zeolites, 12 (1992) 13.

[91] J.A. Martens, E. Feijen, J.L. Lievens, P.J. Grobet, P.A. Jacobs, J. Phys. Chem., 95 (1991) 10025.

[92] H. Li, M.E. Davis, J. Chem. Soc., Faraday Trans., 89 (1993) 951.

[93] G.J. Kennedy, J.B. Higgins, C.F. Ridenour, H. Li, M.E. Davis, Solid State Nuclear Magnetic Res., 4 (1995) 173.

[94] R.M. Kirchner, R.W. Grosse-Kunstleve, J.J. Pluth, S.T. Wilson, R.W. Broach, J.V. Smith, Micr. Mes. Mat., 39 (2000) 319.

[95] K.O. Kongshaug, H. Fjellvåg, B. Klewe, K.P. Lillerud, Micr. Mes. Mat., 39 (2000) 333.

[96] M. Soulard, J. Patarin, B. Marler, Solid State Sciences, 1 (1999) 37.

[97] J.B. Parise, C.S. Day, Acta Cryst., C41 (1985) 515.

[98] J.M. Bennett, J.M. Cohen, G. Artioli, J.J. Pluth, J.V. Smith, Inorg. Chem., 24 (1985) 188.

[99] J.W. Richardson Jr., J.V. Smith, J.J. Pluth, J. Phys. Chem., 94 (1990) 3365.

[100] A. Simmen, L.B. McCusker, Ch. Baerlocher, W.M. Meier, Zeolites, 11 (1991) 654.

[101] J.J. Pluth, J.V. Smith, J.M. Bennett, Acta Cryst., C42 (1986) 283.

42

[102] V. Ramaswamy, L.B. McCusker, Ch. Baerlocher, Micr. Mes. Mat., 31 (1999) 1.
[103] M. Roux, C. Marichal, J. Le Meins, Ch. Baerlocher, J.M. Chézeau, Micr. Mes. Mat., 63 (2003) 163.
[104] D.E. Akporiaye, H. Fjellvåg, E.N. Halvorsen, J. Hustveit, A. Karlsson, K.P. Lillerud, J. Phys. Chem., 100 (1996) 16641.
[105] H. Fjellvåg, D.E. Akporiaye, E.N. Halvorsen, A. Karlsson, K.O. Kongshaug, K.P. Lillerud, Solid State Sciences, 3 (2001) 603.
[106] J.L. Jordá, , L.B. McCusker, Ch. Baerlocher, C.M. Morais, J. Rocha, C. Fernandez, C. Borges, J.P. Lourenco, M.F. Ribeiro, Z. Gabelica, Micr. Mes. Mat., 65 (2003) 43.
[107] G. Férey, T. Loiseau, P. Lacorre, F. Taulelle, J. Solid State Chemistry, 105 (1993) 179.
[108] N.Z. Logar, L. Golič, V. Kaučič, Micr. Mat., 9 (1997) 63.
[109] J.-L. Paillaud, P. Caullet, L. Schreyeck, B. Marler, Micr. Mes. Mat., 42 (2001) 177.
[110] F. Taulelle, V. Munch, C. Huguenard, A. Samoson, T. Loiseau, N. Simon, J. Renaudin, G. Ferey, Proc. 12th Int. Zeolite Conf. (Eds. M.M.J. Treacy, B.K. Marcus, M.E. Bisher, J.B. Higgins) Materials Research Soc., Warrendale, PA, U.S.A. (1999) p.2409.
[111] T. Loiseau, F. Taulelle, G. Férey, Micr. Mes. Mat., 5 (1996) 365.
[112] R.E. Morris, A. Burton, L.M. Bull, S.I. Zones, Chem. Mater., 16 (2004) 2844.
[113] R.P. Bontchev, S.C. Sevov, J. Mater. Chem., 9 (1999) 2679.
[114] A. Martucci, A. Alberti, G. Cruciani, A. Frache, L. Marchese, H.O. Pastore (submitted)
[115] J.B. Parise, J. Chem. Soc., Chem. Commun., (1984) 1449.
[116] J.B. Parise, Inorg. Chem., 24 (1985) 4312.
[117] R.W. Broach, S.T. Wilson, R.M. Kirchner, Proc. 12th Int. Zeolite Conf. (Eds. M.M.J. Treacy, B.K. Marcus, M.E. Bisher, J.B. Higgins) Materials Research Soc., Warrendale, PA, U.S.A. (1999) p. 1715.
[118] R.W. Broach, S.T. Wilson, R.M. Kirchner, Micr. Mes. Mat., 57 (2003) 211.
[119] M. Helliwell, V. Kaučič, G.M.T. Cheetham, M.M. Harding, B.M. Kariuki, P.J. Rizkallah, Acta Cryst., B49 (1993) 413.
[120] J.B. Parise, Acta Cryst., C40 (1984) 1641.
[121] J.J. Pluth, J.V. Smith, J.M. Bennett, J.P. Cohen, Acta Cryst., C40 (1984) 2008.
[122] P.J. Dunn, R.C. Rouse, T.J. Campbell, W.L. Roberts, Am. Mineral., 69 (1984) 374.
[123] S. Dick, T. Zeiske, J. Solid State Chemistry, 133 (1997) 508.
[124] N. Simon, T. Loiseau, G. Férey, J. Chem. Soc., Dalton Trans., (1999) 1147.
[125] N. Simon, T. Loiseau, G. Férey, Solid State Sciences 1 (1999) 339.
[126] Q. Gao, S. Li, R. Xu, Y. Yue, J. Mater. Chem., 6 (1996) 1207.
[127] W. Yan, J. Yu, R. Xu, G. Zhu, F. Xiao, Y. Han, K. Sugiyama, O. Terasaki, Chem. Mater., 12 (2000) 2517.
[128] W. Yan, J. Yu, Z. Shi, P. Miao, K. Wang, Y. Wang, R. Xu, Micr. Mes. Mat., 50 (2001) 151.
[129] K. Wang, J. Yu, Z. Shi, P. Miao, W. Yan, R. Xu, J. Chem. Soc., Dalton Trans., (2001) 1809.
[130] Q. Huo, R. Xu, S. Li, Z. Ma, J.M. Thomas, R.H. Jones, A.M. Chippindale, J. Chem. Soc., Chem. Comm., (1992) 875.
[131] J. Yu, K. Sugiyama, S. Zheng, S. Qiu, J. Chen, R. Xu, Y. Sakamoto, O. Terasaki, K. Hiraga, M. Light, M.B. Hursthouse, J.M. Thomas, Chem. Mater., 10 (1998) 1208.
[132] Y-H. Xu, B-G. Zhang, X-F. Chen, S-H. Liu, C-Y. Duan, X-Z. You, J. Solid State Chemistry, 145 (1999) 220.
[133] B. Wei, G. Zhu, J. Yu, S. Qiu, F-S. Xiao, O. Terasaki, Chem. Mater., 11 (1999) 3417.
[134] W. Yan, J. Yu, Z. Shi, R. Xu, Chem. Commun., (2000) 1431.

[135] P.B. Moore, T. Araki, Am. Mineral., 64 (1979) 587.
[136] O.V. Yakubovich, M.S. Dadashov, Sov. Phys. Crystallogr., 33 (1988) 500.
[137] T. Araki, J.J. Finney, T. Zoltai, Am. Mineral., 53 (1968) 1096.
[138] B.B. Guy, G.A. Jeffrey, Am. Mineral., 51 (1966) 1579.
[139] P.B. Moore, J. Shen, Nature, 306 (1983) 356.

Studies in Surface Science and Catalysis 155
A. Gamba, C. Colella and S. Coluccia (Editors)

V-MCM-22: Synthesis and Characterization of a Novel Molecular Sieve

A. Albuquerque[a,b], H. O. Pastore[a] and L. Marchese[b]

[a]Grupo de Peneiras Moleculares Micro e Mesoporosas, Instituto de Química, UNICAMP, CP 6154, CEP 13084-862, Campinas-SP, Brasil, gpmmm@iqm.unicamp.br

[b]Dipartimento di Scienze e Tecnologie Avanzate, Università del Piemonte Orientale "A. Avogadro", V. Bellini 25G, I-15100, Alessandria, Italia, leonardo.marchese@mfn.unipmn.it

Abstract: This work presents the synthesis and spectroscopic characterization of V-MCM-22, a novel molecular sieve. The synthesis was performed under static hydrothermal crystallization, using $VOSO_4$ as source of vanadium at $SiO_2/V_2O_3 = 66$ and hexamethyleneimine as structure-directing agent. Aluminum was also incorporated in the structure at SiO_2/Al_2O_3 ratios (SAR) of 50 and 80, named V-MCM-22 (50) and V-MCM-22 (80) respectively. Only V-MCM-22 (50) showed higher crystallinity than the parent MCM-22. The insertion of vanadium ions in the framework sites was confirmed by diffuse reflectance (DR) UV-Vis and FTIR spectroscopy, the latter using CO adsorption at 100K. The V-MCM-22 (50) sample presented V in the 5+ state after calcination of the template; two families of vanadium sites were found: a family of distorted tetrahedral oxovanadium $(SiO)_3V=O$ species absorbing at 290, 250 and 330 nm which underwent a reduction to V^{IV} (d-d transition at 550 nm) after treatment in H_2 at 500°C, and a family of tetrahedral oxovanadium with a lower distortion degree, which showed bands at 225, 245 and 265 nm. The latter species were stable after reduction. Hydroxyls bound either to the Lewis vanadium centers or to partially extra-framework Al ions were detected by FTIR and their acidity monitored by CO adsorption. The stretching frequency of these hydroxyls showed a red-shift of ca. 200 cm^{-1} by CO adsorption which suggests an acidity intermediate between silanols (90 cm^{-1}) and bridged SiO(H)Al groups (320 cm^{-1}). This material is a good candidate for selective oxidation reactions of organic molecules.

1. INTRODUCTION

Vanadium-containing microporous molecular sieves are of special interest because of their structural and red/ox properties, which afford a structure displaying channels and/or cavities with restricted space where selective oxidation reactions may occur [1]. Two families of vanadium-containing molecular sieves are of special relevance, one based on aluminophosphate frameworks (VAPOs and VAPSOs) [2] and the other based on silicate networks (V-containing zeolites) [1].

In this work, the synthesis and the characterization of a vanadoaluminosilicate, V-MCM-22, with MWW (IZA code) structure, is described for the first time. The MCM-22 zeolite was first synthesized in the laboratories of Mobil Oil Corporation, using hexamethyleneimine as structure-directing agent [3]. Its structure (Fig. 1) is built of two independent channel systems, both accessible through windows composed of 10-membered (10-MR) TO_4 rings (T = Si or Al). One channel system is made of sinusoidal 10-MR channels (Fig. 1c) and the other by the stacking of supercavities, whose 7.1 Å internal diameter is defined by 12- MR, with a height of 18.2 Å (Figs. 1a and 1b). The unit cell has hexagonal symmetry P6/mmm and contains 72 T atoms, with $Na_x[Al_xSi_{72} - xO_{144}].nH_2O$ chemical composition.

It has been often reported that MCM-22 samples with good crystallinity and phase purity are obtained only by hydrothermal synthesis under stirring, in periods varying from 3 to 12 days [4,5,6]. However, it was recently showed that static synthesis with lower crystallization times also led to materials with comparable cristallinity [7]. V-MCM-22 was synthesized in this work following the procedure by Marques et al. [7], introducing VOSO$_4$ as the vanadium source.

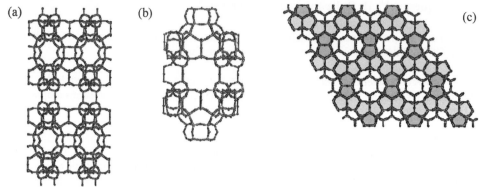

Fig. 1. (a) Schematic representation of MCM-22 zeolite; (b) the MWW supercavity in the center of (a), and (c) projection in the plane (001) showing the bidimensional sinusoidal channel system.

The insertion of vanadium ions within zeolite frameworks has been attempted following several preparation routes, including hydrothermal direct (one-pot) synthesis and post synthesis treatments (ion-exchange, chemical vapour deposition, impregnation, etc...). Since the earlier reports of the V-silicalites in the 1990's [8], vanadium ions have been introduced in several zeolite structures, including MFI, BEA, MEL, ZSM-48, etc.... [9-13]. Multi-technique approaches have been often used for the characterization of structure, oxidation state and dispersion of the vanadium species and, depending on the vanadium loading, a variety of species have been detected. Isolated pseudo-tetrahedral, square pyramidal or octahedral V^V and V^{IV} ions, VO^{2+} ions and V-oxide aggregates or clusters are the most commonly detected intra-zeolite vanadium species [9-13].

Several geometries have been proposed for framework vanadium sites, however they are invariably connected to defect sites of the zeolite where they interact with (or are stabilized by) one or two hydroxyl groups and may bind extra ligands such as H$_2$O, NH$_3$, etc. Among these, the structures reported in Scheme 1 have been often proposed for framework V^V, the difference between them being the presence of an OH group directly bound to the vanadium.

Despite the lack of unequivocal structural data, non-hydroxylated (SiO)$_3$V=O species (Structure A, Scheme 1) appear more reliable for framework V^V sites and account for the enhancement of Brønsted acidity observed in V-silicalite samples [9]. Hydroxylated (SiO)$_2$(OH)V=O species (Structure B, Scheme 1), have been proposed by Dzwigaj et al. [14] to explain an absorption at 3650 cm^{-1} in the IR spectra of V incorporated in Si-β zeolite. These authors justify their assignment on the basis of the presence of similar absorptions in V-supported on bulk oxides. However, it should be taken into account that if V^V ions occupy a zeolite framework position, the presence of vicinal silanols interacting by H-bond with the V-OH groups is very likely; these species should give broad absorptions at frequencies lower that 3650 cm^{-1} and probably be masked by other H-bonded silanols present in defective Siβ zeolite.

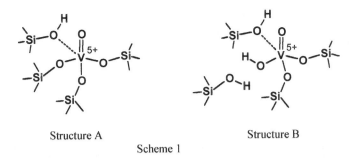

Structure A   Structure B

Scheme 1

Non-hydroxylated $(SiO)_3V=O$ species appear to be the most probable active sites for the oxidation reactions as they may transform into V^{IV} via hydrolysis of a V-O bond in a red/ox process as that represented in Scheme 2. However, clear-cut evidences of this transformation are still lacking, in that the acidic properties of the hydroxyls surrounding the vanadium centers, and their possible modification on passing from structure A to structure B, have not yet been monitored directly. It would be very relevant, therefore, to investigate the presence of hydroxyls bound to the vanadium centers and to evaluate their acidity after oxidation and reduction treatments. FTIR of molecular probes, especially of weak bases which do no perturb/modify significantly the catalytic site, may give a significant contribution in this direction [15].

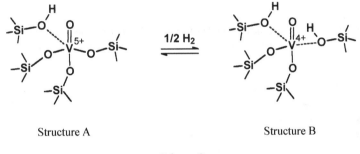

Structure A   Structure B

Scheme 2

In this work, diffuse reflectance UV-Vis and FTIR of CO adsorbed at liquid nitrogen temperature were used to characterize V-MCM-22 with the aim to shed some light on the structure of the vanadium species formed after both oxidation and reduction treatments, e.g. to elucidate the structure of the species which undergo redox processes.

2. EXPERIMENTAL

1.1. Synthesis
Synthesis gels were prepared with two SiO_2/Al_2O_3 ratios (SAR):

$$1.0\ SiO_2: 0.02\ Al_2O_3: 0.03\ VOSO_4: 0.30\ NaOH: 0.60\ HMI: 30\ H_2O$$
$$1.0\ SiO_2: 0.0125\ Al_2O_3: 0.03\ VOSO_4: 0.20\ NaOH: 0.60\ HMI: 30\ H_2O$$

where HMI stands for hexamethyleneimine, the structure-directing agent (SDA). The synthesis procedure used was adapted from Marques *et al.* [7]. Hydrothermal treatments were carried out under static conditions at 150°C for 192h for sample with SAR=50, named V-MCM-22 (50), and 264h for sample with SAR=80, named V-MCM-22 (80).

The HMI was removed by heating under vacuum from room temperature to 500°C at 1°C/min and left at this temperature for 10-12 h; subsequently, the sample was heated up to 580°C, dry oxygen (60-80 Torr O_2) was added and left at this temperature for 6 h. These treatments were performed within the cells for spectroscopic measurements.

1.2. Materials Characterization

1.2.1 X-ray diffraction (XRD)
Data were collected for the as-synthesized and calcined samples as hand-pressed wafers on a Shimadzu XRD 6000 diffractometer at room temperature with CuKα radiation, generated at 40 kV and 30 mA from 2θ = 1.4 to 50° at a rate of 2° min⁻¹ and slits of 0.5°, 0.5° e 0.3mm for exit, reception e divergence, respectively.

1.2.2. Scanning electron microscopy (SEM)
SEM was performed on a JEOL 6360LV microscope. The samples were graphite sputtered on a Bal-Tec MED020 before analysis.

1.2.3. Diffuse reflectance (DR) UV-Vis spectroscopy
DR UV-Vis spectroscopy was carried out on a Perkin Elmer Lambda 900 UV/Vis/NIR spectrophotometer, using a quartz cell that permitted analysis both under vacuum conditions (residual pressure ≤ 10⁻⁵ mbar) and under controlled gas atmospheres. Spectra were taken from 190 to 800 nm. *Oxidized samples* were obtained under O_2 at 580°C (2 hours) and *reduced samples* were obtained at 500°C under H_2 (2 hours); in both cases the last treatment before the analysis was under vacuum at 500°C.

1.2.4. Fourier Transformed Infrared Spectroscopy (FTIR)
FTIR analysis was carried out using a Bruker Equinox 55 spectrophotometer equipped with a pyroelectric detector (DTGS type); a resolution of 4 cm⁻¹ was adopted. Pelletized samples were treated by the same treatments described above in the case of DR-UV-Vis experiments, and using a quartz cell suitable for adsorption-desorption experiments at room and liquid nitrogen temperature; temperatures of around 100K could be reached with this cell [16].

3. RESULTS AND DISCUSSION

V-MCM-22 was synthesized with two different SiO_2/Al_2O_3 ratios (SAR), 50 and 80. Phase purity and crystallinity of the samples were monitored by XRD. Fig. 2 shows the diffractograms of the as-synthesized and calcined sample with SAR = 50, in comparison with pure MCM-22 (Fig. 2a). The XRD of the V-MCM-22 (50) displays a pure crystalline phase (Fig. 2b). Conversely, the XRD of the sample with SAR=80 indicated that the as-synthesised material was highly disordered, and that after calcination an amorphous phase was formed (data not shown). For these reasons, all the spectroscopic analyses were performed on the sample with SAR=50.

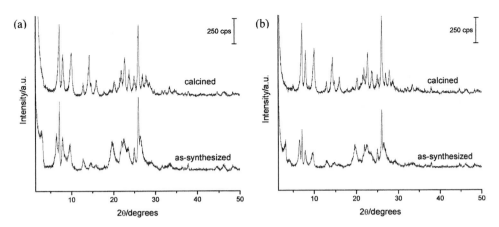

Fig. 2. XRD of MCM-22 (a) and V-MCM-22 (b) with SAR=50.

The diffractograms of the V-MCM-22(50) sample displayed sharper peaks in comparison to the parent MCM-22, which is an indication that V insertion within the MWW framework does not cause an overall distortion of the structure. It is known that larger amounts of heteroatoms within the MWW framework, Al for instance, lead to higher crystalline samples when HMI is used as SDA. MCM-22 with SAR = 30 is, in fact, more crystalline than that with SAR = 50 [7]. In the case of V-MCM-22, the ratio $SiO_2/(Al_2O_3+V_2O_5)$ is 29, which means that the total amount of heteroatoms (Al+V) is similar to the amount of Al in MCM-22 (30). The higher crystallinity of the V-MCM-22 (50) in comparison to MCM-22 (50) can be therefore taken as an indication that vanadium ions are inserted within the MWW framework and that, similarly to Al, V insertion has beneficial effects on the crystallinity.

Both V-MCM-22 and the parent MCM-22 showed a similar evolution on passing from the as-synthesized to the calcined form. The removal of HMI led, in fact, to the transformation from a layered structure to a tridimensional zeolite network. This evolution is clearly monitored by the disappearance of the (001) basal reflection at $2\theta° = 3.20$ of the layered MCM-22 precursors along with the appearance of the reflections at $2\theta° = 14.30$ and 23.77 of the MWW framework [17].

Fig. 3 shows a scanning electron micrograph of the as-synthesized sample with SAR=50. The presence of aggregates with a *doughnut* shape reveals a pattern similar to that already observed for the aluminosilicate MCM-22 synthesized under static conditions [7].

DR-UV-Vis spectra of the calcined and reduced V-MCM-22 (50) sample are displayed in Fig. 4a. The spectrum of the calcined sample is dominated by a complex band which is composed by several absorptions. Beside a maximum at around 265 nm shoulders are found at around 225, 245, 330 and 405 cm^{-1}. All these absorptions are related to V^{5+} species as argued by the lack of any EPR signal (data not shown); this result is in agreement with several studies in the literature which report that, after calcination, essentially V^{5+} ions are found in V-containing zeolites [1,4, 8-14].

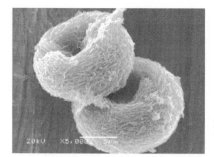

Fig. 3. Scanning electron micrograph of as-synthesized V-MCM-22 (SAR=50).

Apart from the absorption at 405 cm^{-1}, which can be assigned either to $(VO_{2.5})_x$ oxide clusters or isolated octahedral species [10,14], all other absorptions are consistent with oxygen-to-vanadium charge transfer (CT) transitions in distorted (pseudo)tetrahedral oxovanadium $(SiO)_3V=O$ specie; the absorption frequency being related to the degree of distortion of the tetrahedral group [10]. Hydroxylated $(SiO)_2(OH)V=O$ species, already proposed in the literature, cannot be ruled out at this stage. For a detailed description of the electronic transition states of these species, the elegant work of Tran *et al.* can be taken as a reference [18].

The spectrum of the reduced sample showed an absorption in the 350-200 nm range with reduced intensity and complexity in comparison with that of the oxidised material: maxima were found at 225, 245 and 265 nm, accounting for the presence of oxovanadium species similar to those found with the oxidised sample. The absorptions at 330 and 405 nm disappeared after reduction, whereas a new, very broad band, extending from 350 to 800 nm (maximum at around 550 nm), was formed. A subtraction of the spectrum obtained after reduction from that obtained after oxidation was computed (dotted line) to detect more clearly the absorption spectrum of the species which underwent modifications after the reduction treatment. Although this procedure may introduce artefacts, the computed spectrum represents, with a reasonable approximation the oxidisable species in the V-MCM-22 sample. There is a broad absorption with maximum at 290 nm and shoulders at 250 and 330 nm, representing the spectroscopic features of the most distorted tetrahedral oxovanadium $(SiO)_3V=O$ species [10].

Upon re-oxidation of the reduced sample, the 290, 330 and 405 nm absorptions reappeared while simultaneously the large band at 550 nm disappeared, indicating that a reversible red/ox cycle occurred under thermal oxidation/reduction conditions. The 550 nm band is a d-d transition of a d^1 metal complex [19], and its presence clearly indicates that V^{4+} ions were formed upon reduction. The presence of V^{4+} has been also detected in several V-containing zeolites after thermal reduction [8,13]. The structure of the reduced species, which are present at the framework sites (Structure B, Scheme 2), was recently proposed by an EPR study on VS-1 zeolites [12]. It was suggested that the reduction process occurred via hydrolysis of one V-O bond, and it is very likely that this involves highly distorted oxovanadium species with strained Si-O-V bonds.

An indirect prove of this process is that tetrahedral oxovanadium $(SiO)_3V=O$ species with a lower degree of distortion, which absorb at lower wavelength (225, 245 and 265 nm), are present in both calcined and reduced V-MCM-22 materials thus indicating that they are stable upon reduction. The presence of more than one V-site is not unexpected since three different Al sites are described for MCM-22 [20].Structural studies, however, are needed for a description of the local geometry of these sites and of their distortion degree from a perfect tetrahedral environment.

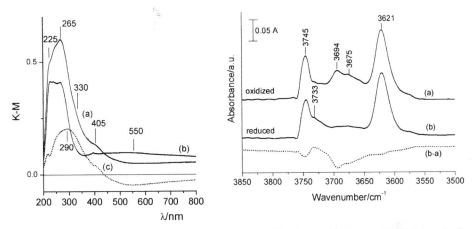

Fig. 4. (left panel) DR UV-Vis spectra of VMCM-22 (50) after oxidation at 500°C (a) and after reduction under hydrogen at 500°C (b); difference spectrum (a-b) is also reported in dotted line (c).
Fig. 5 (right panel) FTIR spectra in the hydroxyl stretching region of oxidized (a) and reduced (b) V-MCM-22 (50) at 500°C; the difference spectrum (dotted line) obtained after subtraction of spectrum (a) from (b), is also reported.

In the absence of structural data, the DR-UV-Vis study presented here indicate that at least two families of tetrahedral oxovanadium $(SiO)_3V=O$ species exist in the MWW framework sites and that they display different reducibility upon treatment with hydrogen. It is proposed that the reducible sites have a V-O bond which can be hydrolysed upon reduction, this picture is consistent with the IR results of CO adsorbed at liquid nitrogen temperature (*vide infra*). It is of note that three kinds of tetrahedral vanadium species were proposed in different framework sites of VSiβ zeolite, and some of them were almost insensitive to dehydration-rehydration processes depending on their accessibility to water molecules [14].

In order to better understand the changes that occur during reduction of V-MCM-22, Figure 5 shows the FTIR spectra in the hydroxyl stretching region of oxidized (curve a) and reduced (curve b) V-MCM-22 which resemble those of mildly dealuminated MCM-22 [21]. The spectrum of the calcined sample (Fig. 5a) displays bands at 3745 and 3621 cm^{-1} which are unambiguously assigned to the ν(OH) stretching modes of silanols and bridged SiO(H)Al groups, respectively. A weak absorption at 3675 cm^{-1} was also found and assigned to partially extra-framework Al-OH species, which have been probably formed by dealumination of the sample [21].

A band at 3694 cm^{-1} was also found with the oxidized sample and tentatively assigned to hydroxyl groups H-bonded to V^{5+} ions in highly distorted tetrahedral oxovanadium $(SiO)_3V=O$ species (Structure A, Scheme 2). This band is almost completely absent in the reduced V-MCM-22 (Fig. 5b), which showed only bands due to silanols, bridged hydroxyls, Al-OH and an additional broad band at 3733 cm^{-1} heavily overlapped to the band of silanols.

The bands that disappeared upon reduction are more clearly evidenced as negative bands by subtracting the spectrum of the oxidized sample from that of the reduced one (Fig. 5, dotted line). The main absorption is found at 3694 cm^{-1} with a shoulder at 3675 cm^{-1}. The species that were produced upon reduction, are observed as positive bands (main absorption at 3733 cm^{-1} with a tail extending to lower wavenumbers). It may be noticed that the population of bridged hydroxyls change only slightly after reduction at 500°C (broad features at 3650-3550 cm^{-1}), meaning that framework Al sites are stable under these conditions, as expected.

The band at 3733 cm^{-1} has been assigned either to terminal SiOH species located in framework defects or to octahedral Al-OH species in Al$_2$O$_3$ particles [21]. In the present case, however, this absorption, or part of it, seems to be related to the presence of V^{4+} ions formed after reduction, and might account for the silanols interacting with the vanadium centres represented in Scheme 2 (Structure B).

The experiment of CO adsorption at around 100K was especially designed to monitor the acidity of the hydroxyls surrounding the vanadium centers of both oxidized and reduced V-MCM-22 zeolite. Fig. 6 shows the difference spectra in the stretching hydroxyl region obtained by subtracting the spectrum of the bare sample obtained after oxidation or reduction (the same spectra of Fig. 5) and cooled to ~ 100 K. Positive bands correspond to species that are formed upon adsorption, whereas negative bands are related to surface hydroxyls species that are bound to CO after adsorption.

CO adsorption at ~ 100K produced a downward shift of some 320 cm^{-1} of the bridged hydroxyls absorbing at 3621 cm^{-1}, both on the oxidized and reduced sample which showed, in fact, a strong band centered at ~ 3300 cm^{-1}. A shift of 320 cm^{-1} was also found by previous studies on pure MCM-22 [21] and this suggests that the vanadium sites have no significant effect on the bridged SiO(H)Al groups, which are the strongest Brønsted acid sites in the MCM-22 zeolite. A band at 2172 cm^{-1}, which clearly correlated with that at 3300 cm^{-1} and corresponds to the stretching of the CO bound to bridged hydroxyls, was also found (Fig.7a).

Besides the strong (positive) band at ~ 3300 cm^{-1}, due to CO H-bonded to bridged hydroxyls, a very broad and composite absorption, extending from 3600 to 3350 cm^{-1}, is formed upon CO adsorption at ~ 100K. At high CO doses, the overall absorption is different in the case of oxidized and reduced samples, being more intense in the former one (compare curves (1) of Figs. 6a and 6b), especially for the presence of a broad band centered around 3500 cm^{-1}. Position and shape of this broad absorption were determined more clearly by subtracting the difference spectrum of CO adsorbed on the reduced sample (Fig. 6b, curve 1) from that of the oxidized sample (Fig. 6a, curve 1). The result of this subtraction (inset of Fig. 6a) shows that a broad band centered at ~ 3490 cm^{-1}, along with a narrower one at ~ 3290 cm^{-1}, is formed upon CO adsorption on the oxidized V-MCM-22.

It is proposed that the hydroxyls responsible for the broad band at 3490 cm^{-1} are those bound to oxovanadium species, which absorb at 3694 cm^{-1} before CO interaction, whereas the band at 3290 cm^{-1} is due to CO bound to bridged hydroxyls, whose concentration is slightly larger in the case of the oxidized sample. However, a contribution of partially extra-framework Al-OH cannot be ruled out in that these species show broad absorption at 3700-3600 cm^{-1} and downward shift of ~ 200 cm^{-1} upon CO adsorption [21].

The bands at 3490 and 3694 cm^{-1} are clearly correlated as determined by decreasing progressively the CO doses: the former decreased whereas the latter became less negative (see the difference spectra of Fig. 6a). This strong correlation hints about the acidity of the hydroxyls (silanols) bound to the V^{5+} centers (structure A, Schemes 1 and 2). The downward shift of these hydroxyls ($\Delta v_{OH\cdots CO} \approx 200$ cm^{-1}) upon CO adsorption suggests, in fact, that these species have an acidity intermediate between that of isolated silanols in amorphous silicas ($\Delta v_{OH\cdots CO} = 90$ cm^{-1} [15]) and the strong Brønsted acid sites in the bridged hydroxyls of MCM-22 ($\Delta v_{OH\cdots CO} = 320$ cm^{-1}). Scheme 3 shows the structure of the different OH$\cdots$CO complexes and the relevant OH downward shifts upon CO adsorption.

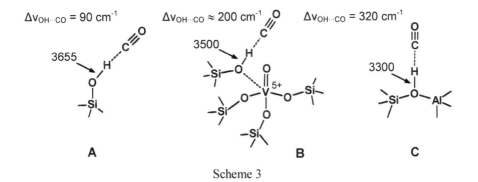

Scheme 3

When the absorption at 3490 cm^{-1} is completely removed by decreasing the CO coverage (Fig. 6a, dashed line), almost all the bridged OH are still bound to CO (band at 3300 cm^{-1}), along with a very broad absorption at 3400-3500 cm^{-1}, which heavily overlaps to the bands at 3490 and 3300 cm^{-1}. This absorption, which is also present in the spectra of CO adsorbed on the reduced sample (Fig. 6b) should be, therefore, related to hydroxyls with acidity stronger than that of silanols bound to oxovanadium species.

A close inspection at the behavior of the band at 3675 cm^{-1}, assigned to partially extra-framework Al-OH species, led to the conclusion that these hydroxyls decrease by CO adsorption in parallel to the increase of the absorption at 3400-3500 cm^{-1}. A downward shift of 275-175 cm^{-1} was therefore estimated for these acid groups; a more precise value could not be determined for both the broadness of the absorption and the overlapping with spectroscopic features which belong to other hydroxyls with strong acid protons. The presence of different bridged hydroxyls, which can be clearly inferred by the broadness and asymmetry of the band at 3621 cm^{-1}, was proposed by Onida et al. [21].

In the CO stretching frequency region (Fig. 7), besides the CO bound to bridged hydroxyls (band at 2172 cm^{-1}) and the absorption of physisorbed, liquid-like CO (band at 2138 cm^{-1}), a broad band at 2165 cm^{-1} was also found at high CO doses on the oxidized sample (Fig. 7a). A weaker band at 2167 cm^{-1} was also detected on the reduced material (Fig. 7b). Absorptions in this position were assigned to CO bound to Al-OH in partially dealuminated MCM-22 [21]. Still, both the intensity and the width of the band found in the oxidized sample suggest that this absorption has a composite nature. It is proposed that the stretching of CO adsorbed on the silanols bonded to the oxovanadium species also falls in this region.

The nature of the hydroxyls absorbing at 3733 cm^{-1} and formed after reduction of V-MCM-22 is more difficult to be inferred, as they can be composed by species of more than one type. It is, in fact, clear that this band (or part of it) shifted to 3653 cm^{-1} upon CO adsorption ($\Delta\nu_{OH\cdots CO} = 80$ cm^{-1}), which is a strong indication that these hydroxyls are related to weakly interacting silanols (the isolated silanols in amorphous silicas shifted by 90 cm^{-1} [15, 22].

Moreover, in addition to the broad bands in the 3600-3300 cm^{-1} region, which are also present in the oxidized sample, a weak and broad absorption was also found at around 3560 cm^{-1} upon CO adsorption on the reduced material. It has to be considered that the only hydroxyls which remain to be assigned are the ones absorbing within the tail of the band at 3733 cm^{-1}, and these may interact with CO shifting to 3560 cm^{-1} ($\Delta\nu_{OH\cdots CO} \approx 160$ cm^{-1}). The observed shift indicates that these hydroxyls have a medium acidity, lower than that of the bridged hydroxyls and of that of the hydroxyls bonded to Lewis V^{5+} centers in oxovanadium species. V^{4+} of the type represented in structure B of Scheme 2 might account for this acidity in that they should have a lower Lewis acidity than V^{5+}, thus inducing a lower charge-drawing with respect to the hydroxyl groups.

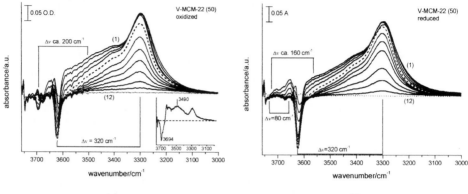

(a)   (b)

Fig. 6. FTIR spectra of V-MCM-22 in the hydroxyl stretching region after CO adsorption at ~ 100K on oxidized (a) and reduced (b) V-MCM-22 (50). Difference spectra are presented after subtraction of the spectrum of the bare sample obtained after oxidation or reduction, respectively. The curves correspond to decreasing CO doses from 30 Torr (curves 1) to 1.10^{-2} Torr (curves 12). The inset reports the spectrum obtained by subtracting curve (1) of the reduced sample from curve (1) of the oxidized one.

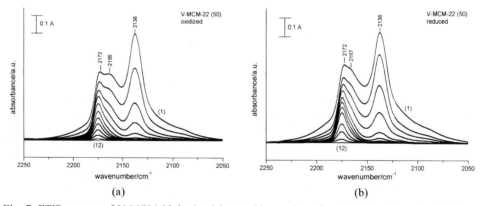

(a)   (b)

Fig. 7. FTIR spectra of V-MCM-22 in the CO stretching region after CO adsorption at ~ 100K on oxidized (a) and reduced (b) V-MCM-22 (50). Difference spectra are presented after subtraction of the spectrum of the bare sample, and the curves correspond to decreasing CO doses as in Fig. 6.

Further experimental data are needed to monitor the acidity of the hydroxyls bound to the Lewis V^{4+} centers formed by reduction of V-MCM22, especially on materials where all the bridged hydroxyls have been exchanged and became O^- Na^+. This will allow to explore more accurately the region 3650-3550 cm^{-1} where the V^{4+}-bound hydroxyls might absorb further to CO adsorption. In addition, theoretical computations are needed to define the most realistic structure/geometry of both V^{5+} and V^{4+} ions at the framework zeolite site, i.e. the ones with minimum energy. We are presently moving in both these directions, in order to get more insights on the structural model of vanadium redox sites proposed in the present spectroscopic study for V-MCM-22 zeolite.

REFERENCES

[1] G. Bellussi and M.S. Rigutto, Stud. Surf. Sci. Catal., 137 (2001) 911

[2] M. Hartmann and L. Kevan, Res. Chem. Intermed., 28 (2002) 625.

[3] M. K. Rubin and P. Chu, U. S. Patent 4 959 325 (1990)

[4] R. Ravishankar, T. Sem, R. Ramaswamy, H. S. Soni, S. Ganapathy and S. Sivasanker, Stud. Surf. Sci. Catal., 84 (1994) 331.

[5] M. Hunger, S. Ernst, and J. Weitkamp, Zeolites, 15 (1995) 188.

[6] A. Corma, C. Corell, and J. Pérez-Pariente, Zeolites, 15 (1995) 2.

[7] A.L.S.Marques, J.L.F. Monteiro, and H.O. Pastore, Micropor. Mesopor. Mater., 32 (1999) 131.

[8] M.S. Rigutto and H. van Bekkum, Appl. Catal., 68 (1991) L1.

[9] G. Centi, S. Perathoner, F. Trifirò, A. Abukais, C.F: Aissi and M. Guelton, J. Phys. Chem., 96 (1992) 2617.

[10] M. Wark, M. Koch, A. Brückner and Grünert, J. Chem. Soc., Faraday Trans., 94 (1998) 2033.

[11] S. Dzwigaj, M. Matsuoka, M. Anpo and M. Che, J. Phys. Chem. B, 104 (2000) 6012.

[12] A.M. Prakash and L. Kevan, J. Phys. Chem. B, 104 (2000) 6860.

[13] P. R. Hari Prasad Rao, A. V. Ramaswamy and P. Ratnasamy, J. Catal., 137 (1992) 225

[14] S. Dzwigaj, P. Massiani, A. Davidson and M. Che, J. Mol. Catal. A, 155 (2000) 169.

[15] S. Coluccia, L. Marchese and G. Martra, Micropor. Mesopor. Mater., 30 (1999) 43.

[16] L. Marchese, S. Bordiga, S. Coluccia, G. Martra and A. Zecchina, J. Chem. Soc., Faraday Trans., 89 (1993) 3483.

[17] Ch. Berlocher, W.M. Meier and D.H. Olson, Atlas of Zeolite Framework Types, Elsevier Science B.V. Amsterdam, The Netherlands, 2001; http://www.iza-online.org/

[18] K. Tran, M.A. Hanning-Lee, A. Biswas, A.E. Stiegman and G.W. Scott, J. Am. Chem. Soc., 117 (1995) 2618.

[19] A.B.P. Lever, Inorganic Electronic Spectroscopy, Elsevier Science Publishers B.V., Amsterdam, The Netherlands, 1984.

[20] S.L. Lawton, A.S. Fung, G.J. Kennedy, L.B. Alemany, C.D. Chang, G.H. Hatzikos, D.N. Lissy, M.K. Rubin, H.K.C. Timken, S. Steurenagel, and D.E. Woessner, J. Phys. Chem., 100 (1996) 3788.

[21] B.Onida, F. Geobaldo , F. Testa , F. Crea , and E. Garrone, Micropor. Mesopor. Mater., 30 (1999) 119.

[22] G. Ghiotti, E. Garrone, C. Morterra, and F. Boccuzzi, J. Phys. Chem., 83 (1979) 2863.

Studies in Surface Science and Catalysis 155
A. Gamba, C. Colella and S. Coluccia (Editors)

57

Molybdenum oxide-based systems, prepared starting from Anderson-type polyoxometalates precursors, as catalysts for the oxidation of i-C$_4$ hydrocarbons

D. André[a], **A. Belletti**[b], **F. Cavani**[b*], **H. Degrand**[b], **J.-L. Dubois**[a], **F. Trifirò**[b]

[a]ATOFINA, CRRA BP 63, Rue Henri Moissan, 69493 Pierre Bènite, France
[b]Dipartimento di Chimica Industriale e dei Materiali, Viale Risorgimento 4, 40136 Bologna,
Italy. *cavani@ms.fci.unibo.it

Te/Mo and V/Te/Mo mixed oxides, prepared starting from Anderson-type polyoxometalates precursors, were characterized and used as catalysts for the oxidation of isobutene and of isobutane to methacrolein. Catalysts were active and moderately selective in olefin oxidation; the addition of V slightly increased the activity, but led to a worsening of the selectivity to methacrolein. The dispersion over high-surface-area silica improved the activity, due to the higher number of active sites available, and to the generation of more reducible species, but worsened the selectivity to methacrolein. Diffusional effects were likely responsible for the enhanced degree of combustion. Catalysts were instead non-active in the oxidation of isobutane.

1. INTRODUCTION

Molybdenum oxide-based systems represent an important class of heterogeneous catalysts for the selective oxidation of hydrocarbons [1-6]. Of considerable relevance are Te/Mo mixed oxides, which are able to perform the allylic oxidation of unsaturated hydrocarbons, to yield the corresponding aldehydes. Specifically, Te/Mo/O catalysts were claimed to be active and selective in the oxidation of propylene to acrolein. An efficient cooperation between the catalyst components is necessary in order to obtain the best catalytic performance [7,8].

Te/Mo mixed oxides can be prepared by solid state reaction, but in this case the segregation of Te oxide may occur. This represents a problem, since the volatility of Te, especially in its lowest oxidation states, is an important drawback of these catalytic systems. One more effective procedure involves the thermal structural decomposition of molecular-type precursors, and specifically of Te/Mo Anderson-type polyoxometalates [5,6]. (NH$_4$)$_6$TeMo$_6$O$_{24}$, a Te^{6+}-containing salt of an heteropolyacid, can be prepared starting from aqueous solutions of ammonium heptamolybdate and telluric acid [9]; the structure can be described from the (Mo$_7$O$_{24}$)$^{6-}$ unit, in which the central Mo atom has been replaced for Te^{6+}, in a planar arrangement of the primary unit. The thermal structural decomposition of these compounds yields oxides in which the dispersion of Te inside MoO$_3$ provides the best conditions to avoid segregation of TeO$_2$, and to achieve the best element-cooperation. Also, it is possible to replace or to add other elements in the polyoxometalate precursor, to adapt the catalyst composition to the reaction requirements. In such a way, elements are forced to stay in the same framework, and synergic interactions can develop. We have adopted this strategy for the preparation of V/Te/Mo mixed oxides, starting from the corresponding Anderson-type polyoxometalates. The active components have been dispersed onto high-surface-area silica, in order to improve the activity in hydrocarbons oxidation, and over low-surface-area steatite.

Materials prepared have been checked as catalysts for the selective oxidation of isobutene and isobutane. The industrial catalyst for the oxidation of isobutene to methacrolein is based on Bi/Mo oxides, and gives 85% yield to the desired product. We have used V/Te/Mo/O based-systems as alternative catalysts for this reaction, with the aim of studying them as multifunctional systems for the single step oxidation of isobutane to methacrolein and/or methacrylic acid, the reaction mechanism of which includes the oxidehydrogenation of the alkane to an adsorbed unsaturated intermediate [10]. The need for a multifunctional catalyst which includes different elements, and specifically (i) V for the activation of the alkane and the formation of the olefinic intermediate, and (ii) Te and Mo for the formation of the allylic intermediate and for O^{2-} insertion, is well exemplified by P/Mo/V Keggin-type polyoxometalates, which to date are the best performing catalysts for the direct oxidation of isobutane to methacrylic acid [10-13].

2. EXPERIMENTAL

The Te/Mo polyoxometalate was prepared by dissolution of $(NH_4)_6Mo_7O_{24}.4H_2O$ and H_6TeO_6 in water (Te/Mo atomic ratio 1/6), followed by addition of ammonium hydrate until a pH of 5.5 developed [9]. Under these conditions the insoluble $(NH_4)_6TeMo_6O_{24}7H_2O$, Anderson-type polyoxometalate, precipitated. To prepare the V/Te/Mo/O compound, the Te/Mo polyoxometalate was first re-dissolved in water, at pH 4.5, and then NH_4VO_3 was added in the desired amount. The suspension was heated at 40°C to favour the complete dissolution of the vanadate. Samples having Te/V atomic ratio equal to 1/0.5, 1/1 and 1/2 were prepared. Thereafter, the compound was precipitated by water evaporation. In the case of supported samples, the aqueous solution containing dissolved ammonium vanadate and the Te/Mo polyoxometalate (Te/V 1/1), has been used for the impregnation of high-surface-area silica (360 m^2/g) and of low-surface-area steatite. The amount of active phase loaded corresponds either to the 10 or to the 20 wt.%, with respect to the overall sample weight.

All these materials are referred to as "catalysts precursors". Precursors were then dried and calcined in flowing air from room temperature to 300°C (3°/min, isothermal step 1 h), and then to 500°C (3°/min, isothermal step 2 h), to obtain the catalytic material.

Characterization of precursors and of calcined samples was carried out by means of X-ray diffraction, FT-IR spectroscopy, Raman spectroscopy. Thermal-Programmed-Reduction and Oxidation (TPR, TPO) tests were done with a ThermoQuest TPDRO 1100 Instrument. Catalytic tests were carried out in a stainless steel, continuous flow reactor, loading 3.5 g of catalyst; residence time was equal to 4.5 s. CSi was added for catalyst dilution (catalyst/CSi volume ratio 1/1). Feed composition was: (i) for the oxidation of isobutene, 3 mol.% isobutene, 18 % oxygen, 8% water, remainder helium; (ii) for the oxidation of isobutane, 26% isobutane, 13% oxygen, 12% water, remainder helium. Reactants and products were analyzed by means of GC. An OV351 column (FID) was used for the separation of oxygenates (methacrolein, methacrylic acid, acetone, acetic acid, acetaldehyde, etc.); a Molecular Sieve 5A column (TCD) was used for the separation of O_2, CO and Kr (internal standard); a SilicaPlot column (TCD) was used for the separation of CO_2, isobutane and isobutene.

3. RESULTS AND DISCUSSION

3.1 Characterization of bulk and supported V/Te/Mo/O samples

Table 1 summarizes the main characteristics of catalyst precursors and of the corresponding oxides after thermal treatment.

Table 1

Main characteristics of catalysts prepared

Samples	Precursors structure	Calc. catalysts, cryst. phases	Calc. catalysts, s. area, m^2/g
Bulk catalysts			
Te/Mo/O (Te/V 1/0)	$(NH_4)_6TeMo_6O_{24}.7H_2O$	MoO_3	3
V/Te/Mo/O (Te/V 1/0.5)	$(NH_4)_6TeMo_6O_{24}.7H_2O + (NH_4)VO_3$	nd	6
V/Te/Mo/O (Te/V 1/1)	$(NH_4)_6TeMo_6O_{24}.7H_2O + (NH_4)VO_3$	MoO_3, V_2O_5	7
V/Te/Mo/O (Te/V 1/2)	$(NH_4)_6TeMo_6O_{24}.7H_2O + (NH_4)VO_3$	MoO_3, V_2O_5	6
Supported catalysts			
Te/Mo/O (Te/V 1/0) (10%)-SiO_2	nd	nd	225
V/Te/Mo/O (Te/V 1/1) (10%)-SiO_2	nd	nd	257
V/Te/Mo/O (Te/V 1/1) (20%)-SiO_2	$(NH_4)_6TeMo_6O_{24}$, SiO_2	MoO_3, SiO_2	125
V/Te/Mo/O (Te/V 1/1) (10%)-steatite	$(NH_4)_6TeMo_6O_{24}$, steatite	MoO_3, steatite	< 3
V/Te/Mo/O (Te/V 1/1) (20%)-steatite	$(NH_4)_6TeMo_6O_{24}$, steatite	MoO_3, steatite	< 3

The Anderson-type structure was present in all precursors; the XRD patterns (Figure 1 shows some selected patterns for precursors and catalysts) and FT-IR spectra correspond to those reported in literature for the ammonium salt of telluromolybdic acid: $(NH_4)_6TeMo_6O_{24}.H_2O$ (JCPDS 26-0080). Reflections relative to NH_4VO_3 (JCPDS 25-0047) were weak, and their intensity increased when the Te/V atomic ratio was decreased. The pattern of the calcined Te/Mo/O sample evidenced the presence of MoO_3 (molybdite, JCPDS 35-0609); in V/Te/Mo/O samples the decomposition generated MoO_3 and V_2O_5 (JCPDS 85-0601). Evidences for the formation of V/Mo mixed oxides were not obtained.

In supported V/Te/Mo/O precursors, only the reflections of the Anderson compound were evident, while in corresponding calcined samples, reflections attributable to MoO_3 were present. The absence of reflections relative to vanadium oxide can be attributed to the low amount of this compound; however, the dispersion of vanadium oxide over the support, especially in the case of high-surface-area silica, can not be excluded.

Figure 2 reports the TPR profiles of calcined, unsupported catalysts. The Te/Mo/O sample was characterized by a broad reduction peak beginning at around 400°C, and covering the entire temperature range up to the isothermal step (650°C); the reduction did not reach completion at the latter temperature, within the time of the isothermal step. The apparent peak at high temperature is indeed an artefact due to the decreased rate of reduction, with the beginning of the isothermal step. The presence of increasing amounts of V generated an additional reduction peak, with maximum at $\approx$ 600°C, which overlapped to the broader one attributable to Mo oxide, and the intensity of which was approximately proportional to the V content in sample. In the case of the sample having Te/V atomic ratio equal to 1/2, thus

60

containing the greater amount of V, additional shoulders were evident at lower temperatures. Even though the complexity of the profile does not allow reliable assignments to be made, the main additional peak at $\approx 600°C$ can be attributed to the reduction of V_2O_5, and peaks at lower temperatures to the reduction of vanadium species other than that present in bulk vanadia, which originate from the interaction with molybdenum trioxide, or, to the formation of new Mo species, more easily reducible than those in V-free bulk samples.

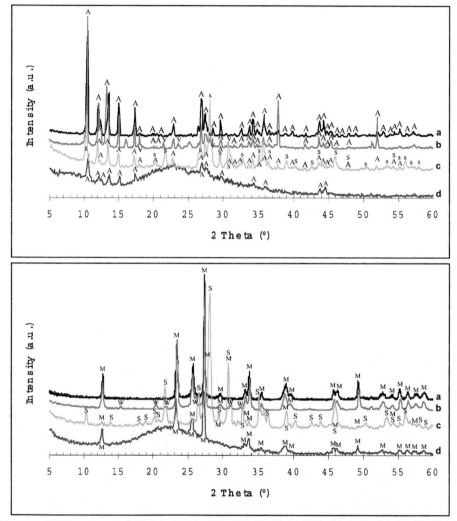

Figure 1. XRD patterns of: Te/Mo/O (a); V/Te/Mo/O (Te/V 1/1) (b); V/Te/Mo/O (Te/V 1/1) (20%)-Steatite (c); V/Te/Mo/O (Te/V 1/1) (20%)-SiO$_2$ (d), after drying (top figure) and after thermal treatment at 500°C in air (bottom figure). Symbols: A: (NH$_4$)$_6$TeMo$_6$O$_{24}$ (JCPDS 26-0080); s: Steatite; V: NH$_4$VO$_3$ (JCPDS 25-0047); M: MoO$_3$ (JCPDS 35-0609); W: V$_2$O$_5$ (JCPDS 85-0601).

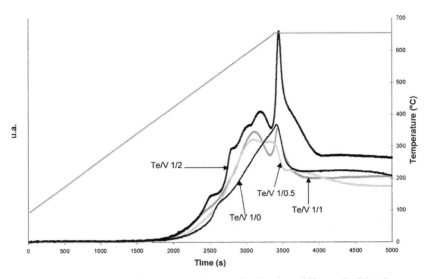

Figure 2. TPR profiles of calcined V/Te/Mo/O samples, having different Te/V ratios.

Catalysts were downloaded after reaction, and characterized by means of TPR and TPO measurements, in order to gain information on the catalyst working state during catalytic tests. In all cases, samples were downloaded after reaction under conditions at which total isobutene conversion was reached, and by cooling from reaction temperature to room temperature in an helium flow. The following indications were obtained:

a) The catalyst average oxidation state under working conditions was lower than that one in calcined catalysts. In fact, in TPO tests the consumption of oxygen was non-negligible, and in TPR tests the consumption of hydrogen was lower than for the corresponding calcined catalysts.

b) Differences were observed between Te/Mo/O and V/Te/Mo/O samples. With the latter samples, TPR tests evidenced that in downloaded catalysts all most easily reducible species (identified in TPR of calcined catalysts, Figure 2) were present in a reduced state (in fact were no longer reducible). In correspondence, TPO tests indicated a considerable O_2 consumption for the same samples, with a net oxidation peak in the profile in the 500-to-600°C temperature range, of intensity approximately proportional to the V content. In downloaded Te/Mo/O catalyst, instead, the TPR profile evidenced that only a small fraction of Mo species having intermediate reducibility was in a reduced state. In correspondence, the O_2 consumption in TPO tests was minimal.

These tests demonstrate that in the case of Te/Mo/O catalyst, the re-oxidation step in the redox process is quick, and therefore it is not the rate-limiting step; in fact, the catalyst downloaded after reaction was as oxidized as the calcined one. In spent V/Te/Mo/O samples, instead, V was at least in part present in a reduced state, and its oxidation was probably slow.

Figure 3 shows the TPR profile for silica-supported V/Te/Mo/O samples having increasing amounts of active phase (5, 10 and 20 wt.%) (Te/V 1/1), in comparison with the corresponding unsupported one.

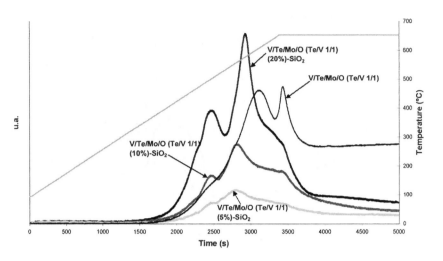

Figure 3. TPR profiles of calcined silica-supported V/Te/Mo/O (Te/V 1/1) samples, with increasing amount of active phase loading.

The dispersion on the supports greatly modified the reduction profile of samples. The amount of H_2 consumed was higher than for the unsupported catalyst, despite the lower amount of reducible oxide; also, the reduction was completed within the time of the isothermal step. This indicates that the dispersion over the high-surface-area support considerably improved the amount of active phase available for the reaction. Moreover, main reduction peaks fell at considerably lower temperature with respect to the unsupported catalyst; this clearly indicates an enhanced reducibility, likely due to the interaction between the active components and the support. The profiles of all silica-supported samples were similar, despite the different amounts of active phase loaded.

3.2 Catalytic performance of bulk V/Te/Mo/O catalysts in isobutene oxidation

Figure 4 plots the conversion of isobutene, and the selectivities to the products as function of temperature, for sample V/Te/Mo/O (Te/V 1/0.5).

Under the conditions employed, isobutene was the limiting reactant; total conversion of the hydrocarbon was reached at around 380°C. At low temperature (340-350°C), the main by-products were carbon oxides, and methacrolein was obtained with a selectivity around 35-40%. Small amounts of methacrylic acid (selectivity lower than 1%) also formed. Other by-products were acrolein, acetic acid and acetone (with overall selectivity lower than 10%). C balance, as calculated from the comparison of converted isobutene and formed products, was close to 80% in the temperature range 330 to 380°C, and then worsened down to 65% at high temperature. The lack in C balance was due to the formation of heavy products which were not eluted in our GC column, but which were then identified as arising (i) from condensation reactions occurring on isobutene, followed by dehydrociclization and oxidation reactions to yield alkylaromatics and oxygenated alkylaromatics, and (ii) from oligomerization and polymerization reactions. When the reaction temperature was raised above 350-360°C, the selectivity to methacrolein declined, while that to carbon oxides correspondingly increased.

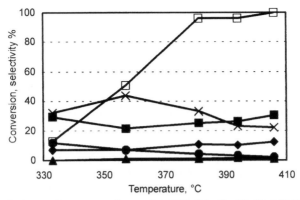

Figure 4. Catalytic performance of sample V/Te/Mo/O (Te/V 1/0.5) as a function of temperature. Isobutene conversion (□), and selectivity to methacrolein (✖), methacrylic acid (▲),CO (◆), CO$_2$ (■), and others (acetic acid, acetaldehyde, acetone, acrolein) (●).

Figure 5 compares the conversion of isobutene and the selectivity to methacrolein + methacrylic acid for bulk, unsupported samples having increasing amounts of vanadium oxide.

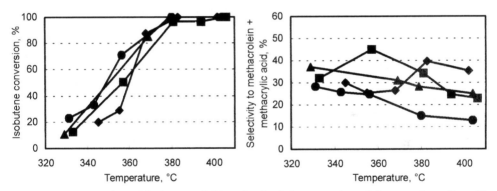

Figure 5. Conversion of isobutene (left) and selectivity to methacrolein + methacrylic acid (right) as functions of temperature for bulk V/Te/Mo/O catalysts, having increasing amounts of V: Te/V 1/0 (◆), Te/V 1/0.5 (■), Te/V 1/1 (▲), Te/V 1/2 (●).

The addition of V had the following effects on catalytic performance:
a) The activity of V/Te/Mo/O samples was higher than that of the Te/Mo/O catalyst. The increase of activity may be attributed either to a direct role of V in the activation of the hydrocarbon, or to the higher surface area (see Table 1).
b) The selectivity to methacrolein slightly increased for the sample having Te/V ratio 1/0.5 (maximum selectivity 45%, at 360°C), if compared to the Te/Mo/O sample; the maximum in selectivity shifted towards lower temperature. This indicates that V promoted the formation of methacrolein at low temperature, but also catalyzed its decomposition at

T>360°C. This is also demonstrated by the performance of samples having higher V content (Te/V 1/1 and 1/2): the selectivity to methacrolein decreased all over the entire range of temperature examined, and was considerably lower than that obtained with the Te/Mo/O sample. In correspondence, the formation of carbon oxides was strongly enhanced. Moreover, the selectivity to methacrylic acid slightly increased, but remained nevertheless very low. The formation of heavy by-products (indirectly determined as lack in C balance) was not affected by the V content; the C balance was for all samples comprised between 65 ad 85-90%, being the highest at both the lower (below 340°C) and the higher (above 380°C) temperatures.

3.3 Catalytic performance of supported V/Te/Mo/O catalysts in isobutene oxidation

Figure 6 compares the performance of the following catalysts: bulk V/Te/Mo/O (Te/V 1/1), silica-supported Te/Mo/O (10% of active phase), silica-supported V/Te/Mo/O (Te/V 1/1, 10% and 20% of active phase), steatite-supported V/Te/Mo/O (Te/V 1/1, 10% of active phase).

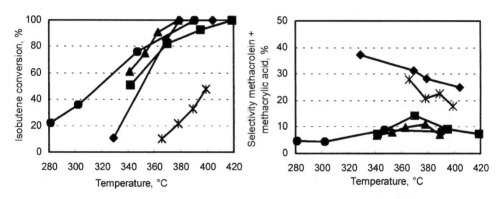

Figure 6. Conversion of isobutene (left) and selectivity to methacrolein + methacrylic acid (right) as functions of temperature for bulk and supported (V)/Te/Mo/O catalysts (Te/V 1/1). Catalysts: Bulk V/Te/Mo/O (◆), silica-supported Te/Mo/O (10 wt.% of active phase) (■), silica-supported V/Te/Mo/O (10 wt.% of active phase) (▲), silica-supported V/Te/Mo/O (20 wt.% of active phase) (●), steatite-supported V/Te/Mo/O (10 wt.% of active phase) (✕).

The comparison of the catalytic performances evidences the following:
a) The dispersion of V/Te/Mo/O over silica increased the activity, since the same levels of conversion were obtained at considerably lower temperature than with the unsupported catalyst. The higher activity may be due to either the increased surface area, and hence to the higher availability of active sites, or to the easier reducibility of the latter. However, the trend of conversion for increasing temperatures clearly evidenced the presence of a lower activation energy (15 kcal/mole against 28 kcal/mole for the unsupported sample); in fact, total isobutene conversion was obtained at comparable temperature for both unsupported and supported samples. This different activation energy may be associated either to the different nature of the active species (as inferred from TPR measurements), or to the presence of diffusional limitations which decrease the particle efficiency. Supported V/Te/Mo/O (10% of active phase) catalyst was slightly more active than the supported Te/Mo/O one, as it was for the corresponding unsupported samples (Figure 5).

b) All silica-supported catalysts gave a selectivity to methacrolein not higher than 10-15%, with negligible formation of methacrylic acid. This was only in part due to the higher degree of combustion, and mostly to the higher formation of acetic acid and acetone (with also formation of acrolein and acetaldeyde, each one with selectivity lower than 5%). It is likely that isobutene or methacrolein itself underwent consecutive oxidative degradation reactions in the pores of the catalyst, before counter-diffusion towards the gas phase might occur. These degradation phenomena may be favoured by the presence of diffusional limitations, and by higher intra-particle effective residence times.

c) In the case of the steatite-supported V/Te/Mo/O sample (Te/V 1/1), the activity was much lower than that of the corresponding bulk catalyst. The selectivity to methacrolein was intermediate between that of the bulk catalyst and that of the silica-supported catalysts.

3.4 Catalytic performance in isobutane oxidation

Table 2 summarizes the catalytic performance in isobutane oxidation of unsupported and supported V/Te/Mo/O samples. Catalysts had a low activity in the oxidation of isobutane; moreover, main products were carbon oxides. Since no isobutene was found amongst the reaction products, despite the very low conversions achieved, this means that the catalyst is unable to convert selectively the alkane to the olefinic intermediate, while it directly transforms the reactant to carbon oxides.

Table 2
Catalytic performance of V/Te/Mo/O catalysts in isobutane oxidation.

Temperature, °C	Isobutane conv., %	Select. to methacrolein, %	Select. to CO_x, %	Select. to others[a], %
V/Te/Mo/O (Te/V 1/1)				
394	1	9	85	5
419	2	8	89	2
432	3	6	92	1
Silica-supported V/Te/Mo/O (10% active phase)				
376	1	4	90	6
400	3	3	93	3
426	5	4	93	3
439	6	1	97	2

[a]others: acetic acid, acetone, acrolein, acetaldehyde.

4. CONCLUSIONS

Te/Mo and V/Te/Mo mixed oxides, prepared starting from Anderson-type polyoxometalates precursors, were active and moderately selective in the oxidation of isobutene to methacrolein; the addition of V slightly increased the activity, but led to a worsening of the selectivity to methacrolein. The dispersion over high-surface-area silica improved the activity, due to the higher number of active sites available, and to the generation of more reducible species, but worsened the selectivity to methacrolein. Diffusional effects were likely responsible for the enhanced degree of combustion. Catalysts were non-active in the oxidation of isobutane.

66

ACKNOWLEDGEMENTS

Atofina is acknowledged for financial support. Prof. Edmon Payen and Prof. Michel Fournier, from Univ. of Lille (F), are acknowledged for discussion and suggestions.

REFERENCES
1. R.K. Grasselli, G. Centi and F. Trifirò, Appl. Catal., 57 (1990) 149.
2. W. Ueda, Y. Moro-oka and T. Ikawa, J. Catal., 88 (1984) 214.
3. H. Hayashi, Catal. Surveys Japan, 3 (1999) 43.
4. L. Moens, P. Ruiz, B. Delmon and M. Devillers, Appl. Catal. A, 180 (1999) 299.
5. P. Botella, J.M. Lopez Nieto and B. Solsona, J. Mol. Catal. A., 184 (2002) 335.
6. N. Fujikawa, K. Wakui, K. Tomita, N. Ooue and W. Ueda, Catal. Today, 71 (2001) 83.
7. R.K. Grasselli, Topics Catal., 21 (2002) 79.
8. R.K. Grasselli, J.D. Burrington, D.J. Buttrey, P. DeSanto, C.G. Lugmair, A.F. Volpe and T. Weingand, Topics Catal., 23 (2003) 5.
9. H.T. Evans, Acta Cryst. B, 30 (1974) 2095.
10. G. Busca, F. Cavani, E. Etienne, E. Finocchio, A. Galli, G. Selleri and F. Trifirò, J. Mol. Catal. A, 114 (1996) 343.
11. F. Cavani, R. Mezzogori, A. Pigamo, F. Trifirò and E. Etienne, Catal. Today, 71 (2001) 97.
12. F. Cavani and F. Trifirò, in "Basic Principles in Applied Catalysis", M. Baerns (Ed.), Springer, Berlin, Series in Chemical Physics 75 (2003) 21.
13. M. Langpape, J.M.M. Millet, U.S. Ozkan and P. Delichere, J. Catal., 182 (1999) 148.

Studies in Surface Science and Catalysis 155
A. Gamba, C. Colella and S. Coluccia (Editors)
© 2005 Elsevier B.V. All rights reserved

Nanostructured Ba-Al-oxides prepared by cogel procedure for NOx-Storage-Reduction catalysts having enhance performances and hydrothermal stability

G.E Arena, L. Capitò, and G. Centi*

University of Messina (Dept. of Industrial Chemistry and Eng. of Materials, and ELCASS, European Lab. for Surface Science and Catalysis), Salita Sperone 31, 98166 Messina. INSTM (Consortium of Materials Science and Technology), Research Unit of Messina, Italy.

The reactivity of $Pt/Ba(O)/Alumina$ catalysts prepared by sol-gel (1% wt. Pt, 15% wt. BaO) in the reaction of NO_x storage-reduction (NO_xSR) before and after severe hydrothermal treatment (800°C) has been studied as a function of the preparation method (type of catalysis during the sol-gel, nature of the precursors, ageing, calcination temperature). The results are compared with those of a reference sample prepared by wet impregnation method. The large spreading of the catalytic performances, in terms of both NO_x storage and resistance to deactivation by hydrothermal treatment, evidences that a complex surface chemistry is present and that the performances are considerably depending on the catalysts nanostructure. It is suggested that Ba doping and/or decoration of alumina nanoparticles increases their thermal stability, but crystallization of Ba-aluminate occurs in the presence of water vapour with a drastic lowering of the NO_x storage properties. The analysis of the results also suggests that an amorphous nonstoichiometric $BaAl_2O_4$ like surface phase is present together with Ba-carbonate surface species. The two species are active in different range of temperatures and their relative amount depends on the preparation. Their amount also influences the hydrothermal stability of the samples in terms of both activity after this treatment and severity of the deactivation after this treatment. The sol-gel preparation method leads in general to improved resistance to deactivation by hydrothermal treatment and better performances at reaction temperatures above about 400°C with respect to samples prepared by wet impregnation method.

1. INTRODUCTION

NO_x storage-reduction (NO_xSR) type catalysts are one of the most viable solution for the elimination of NO_x from vehicle emissions containing O_2 (lean burn gasoline and diesel engines), in order to meet future requirements on emissions levels (Tier 2 in US and Euro V in Europe) [1,2]. Although nearly-commercial catalysts already exist, an improvement of the low temperature activity in NO_x storage and of the hydrothermal stability is necessary [3].

Duprez et al. [4] studying the deactivation of a $Pt/Ba/Al_2O_3$ NO_x-trap model catalyst submitted to thermal ageing at 800°C evidenced that the thermal ageing leads to the formation of

$BaAl_2O_4$ having a lower NO_x storage capacity compared to barium carbonate. In addition, they noted also that the platinum activity for the NO to NO_2 oxidation was lowered by thermal treatments. Uy et al. [5] characterizing by in situ UV and visible Raman spectroscopy fresh and thermally aged NO_x storage-reduction $Pt/Ba/Al_2O_3$ catalysts also found that Ba nitrate particles in aged catalyst are more crystalline and is responsible of the segregation of Pt particles aside from the Barium particles. Close contact between these two main components (Pt and Ba) is necessary especially during the reduction step [6], because decreases the poisoning effect of CO on the noble metal [7] making so more effective the reduction of the stored NO_x.

Increasing the hydrothermal stability of the catalysts requires to improve catalyst design and preparation methodologies [8]. Few studies have been focused on this issues, despite the large number of publications on NO_xSR catalysts, but dealing especially on the reaction mechanism and using the same methodology reported in the earlier Toyota patents (impregnation of alumina with barium acetate). Recent developments indicated that materials derived from Mg/Al hydrotalcite precursors show interesting performances in terms of high hydrothermal stability, low temperature (below 200°C) performances and SO_2 poisoning resistance [8-13]. However, a limit are the low performances at high temperatures (above 450°) and the difficulty in preparing multilayer-type and mechanically resistant monoliths. The use of multilayer monolithic NO_x catalysts is a topic of growing interest for their interesting performances [8,14,15], but suitable preparation methodologies should be used such as sol-gel type techniques which interest in the field of the manufacture of materials for automotive applications is fast growing [16].

The lowering of the performances of NO_xSR catalysts after hydrothermal treatment is due the sintering of both oxides (support and Barium phases) and noble metal particles. A possibility to limit this effect is to prepare a mixed $BaO-Al_2O_3$ oxide using sol-gel procedures [17,18]. The Ba may results stabilized from the higher dispersion and stronger interaction with alumina which is possible using this procedure. Furthermore, it is known that alkaline-earths stabilize the alumina surface area from the high temperature sintering. Balint et al. [19] reported a high thermal stability of alumina nanoparticles containing small amounts of barium. The insertion of large cations, such as La or Ba in the lattice of Al_2O_3 has 2 effects. The $\gamma \rightarrow \alpha$ phase transition of alumina is shifted to higher temperature because the ionic diffusion is hindered. An other effect is the prevention of the sintering of nanoparticles due to the anisotropic layered structures formed by the Ba located on the surface of nanoparticles. Also Rossignol and Kappenstein [20] evidenced that the doping of alumina with Ba through the sol-gel process enhances the surface area and increase the $\theta \rightarrow \alpha$-$Al_2O_3$ phase transition temperature up to 1315°C. This effect is strongly related to the methodology of preparation of the catalyst. Worsen results have been obtained by impregnating with Ba salts the alumina particles. The stabilization of the surface area of the support is a very important parameter to improve the hydrothermal stability.

On the other hand, the sol-gel procedure may enhances the formation of barium-aluminate. The formation of a $BaAl_2O_4$ spinel structure is one of the key mechanisms responsible of the lowering of the performances during thermal and hydrothermal ageing [4,5,21], but the formation of this phase depends on the type of contact between the Ba phases (carbonates, nitrates, oxide) and the alumina support. The formation of Ba-doped alumina nanoparticles or surface layers may prevent the large surface reconstruction and recrystallization to $BaAl_2O_4$ spinel structure[19]. Busca et al. [22] also find that the surface properties of a Ba-alumina sample (Ba:Al ratio 1:12) are due to the preferential exposition of Ba-O-containing "mirror planes", parallel to 0001 planes, and that this behaviour may be associated to the resistance to sintering of this material, due to a blocking of the crystal growth along the crystallographic c axis. This blocking mechanism would strongly depend on the preparation methodology.

In conclusion, there is no obvious relationship expected between methodology of preparation of Pt-Ba(O)/Alumina NO_xSR catalysts and their performances and hydrothermal stability. The nature and surface dispersion of the Barium phases as well as their interaction with Pt particles would affect the NO_x storage properties and regenerability of the catalyst, but it also effect the sintering and recrystallization mechanisms during the thermal and hydrothermal ageing and obviously the catalytic performances after these stability tests. It may be expected that the doping of alumina particles (possible using sol-gel methods) limits the mechanism of surface reconstruction to a Barium-alumina spinel phase responsible for the loss of NO_x storage activity after thermal ageing. On the other hand, Ba can be less accessible for NO_x storage in samples prepared by Ba-Al cogel with respect to samples in which Ba is deposited on the surface only, and in general it is known that the characteristics of the materials prepared by sol-gel considerably depend on the details of synthesis procedure [23].

The aim of the work here reported is the study of the properties and reactivity in NO_xSR of Pt/BaO-Al$_2$O$_3$ samples where the BaO-Al$_2$O$_3$ mixed oxide support is prepared by cogel, and analyze their performances before and after hydrothermal treatment (800°C, 6h) in comparison with a reference system prepared by impregnation method (as in patented Toyota systems [1,3,24]). The following specific aspects in the preparation were analyzed: (i) the use of acid or basic catalytic in preparing the Ba-Al cogel, (ii) the nature of the Ba and Al precursors for the sol-gel preparation of the BaO-Al$_2$O$_3$ mixed oxide, (iii) the temperature of calcinations in the 500-900°C temperature range, and (iv) the type of precursor to add Pt by impregnation on the BaO-Al$_2$O$_3$ mixed oxide, because Pt dispersion and interaction with the support is another crucial factor to limit deactivation after hydrothermal treatment.

In all studied samples, the amount of Pt and Ba (% wt.) was maintained the same, i.e. 1% and 15%, respectively, in order to have more comparable data. The study was focused mainly in the identification of the catalytic properties related to the preparation parameters, while the advanced characterization of these catalysts to identify the reasons of the relationship will be the topic of further studies, because due to the complexity of the problem it is preferable to make the advanced characterization studies on samples having well defined catalytic performances.

2. EXPERIMENTAL

2.1. Preparation of the catalysts

The Ba-Al-oxide samples were prepared by sol-gel method using either basic (NH$_4$OH 1N) or acid (HCl 37%) catalysis for the gelification. In the basic catalysis procedure (samples 1B-4B; Table 1), the Aluminum precursor (Al tri-*sec*-butoxide; [C$_2$H$_5$CH(CH$_3$)O]$_3$Al) is maintained in a stirred reactor under dry nitrogen atmosphere (all the apparatus is in a dry-box) and heated at 80°C. Then Ba-acetate in 2-butanol solution (the Ba-acetate is solubilized using ultrasounds) is added in the right amount to obtain the aimed Ba:Al molar ratio. After 40 min, a NH$_4$OH 1N solution (6 ml) is very slowly added drop by drop up to reach basic pH and complete gel formation. The gel is then aged for five days, filtered and washed, and the dried in an air-circulating oven at 60°C (24 h). After fine grounding the sample is then calcined at 700°C or 900°C (10h), using an heating ramp of 5°C/min. In the acid catalyzed procedure (samples 1A-4A; Table 1), the procedure is similar to that described before, but (i) Ba-nitrate is used instead of Ba-acetate/2-butanol solution, and (ii) HCl 37% (0,6 ml) is added instead of the NH$_4$OH solution. Table 1 summarizes the characteristics of the prepared by this sol-gel procedure.

Some samples have been also prepared using Al-*iso*-propoxide [Al(OCH(CH₃)₂)₃] instead of Al tri-*sec*-butoxide (samples 5-8; Table 2). In this case, the Al-*iso*-propoxide is pre-hydrolized adding defined amounts of H_2O (see Table 2), then Ba-acetate is added. Other preparation procedure aspects were analogous to those previously described, apart those different indicated in the Table 2.

On the calcined Ba-Al-oxide samples, Pt is added by incipient wet impregnation using either H_2PtCl_6 or $Pt(NH_3)_2(NO_2)_2$ as starting salts. Intermediate drying steps at 100°C are made up to complete addition of the solution. After the last drying step, the catalysts are calcined at 500°C overnight. The Pt loading was 1% wt. in all samples.

Reference catalyst (Ref.) is made using as support γ-Al_2O_3 (110 m²/g) impregnated with a solution of $Ba(CH_3COO)_2$ and adding Pt as for the other samples. Calcinations temperature was 550°C.

Table 1 Summary of the characteristics of the NOxSR catalysts prepared by sol-gel procedure using Al tri-*sec*-butoxide as the aluminum source.

Code	Ba precursor	Pt precursor	Type catalysis	T calc., °C
1A	Ba(NO₃)₂		acid	700
1B	(CH₃COO)₂Ba	H₂PtCl₆	basic	
2A	Ba(NO₃)₂		acid	900
2B	(CH₃COO)₂Ba		basic	
3A	Ba(NO₃)₂		acid	700
3B	(CH₃COO)₂Ba	Pt(NH₃)₂(NO₂)₂	basic	
4A	Ba(NO₃)₂		acid	900
4B	(CH₃COO)₂Ba		basic	

Table 2 Summary of the characteristics of the NOₓSR catalysts prepared by sol-gel procedure using Al-*iso*-propoxide as the aluminum source.

Code	H₂O/Al-iso-propoxide molar ratio	Ba precursor	Pt precursor	Type catalysis	Ageing, days	T calc., °C
5	125	(CH₃COO)₂Ba		HNO₃ (8N, 10 ml)	3	900
6			H₂PtCl₆			
7	55,55	Ba(NO₃)₂		HNO₃ (1N, 10 ml)	8	500
8						900

2.2. Hydrothermal ageing treatment

The hydrothermal ageing treatment was made in a fixed bed reactor apparatus at 800°C (6 h) feeding a mixture of air saturated with water at 60°C (about 10% water in air). The catalyst was in the same form of that used for the NOₓ storage-reduction tests, e.g. small particles in the 30-60 mesh (0.25-0.5 mm) range size.

2.3. Characterization of the samples

The surface area of the sample was determined using the B.E.T. method (N₂ adsorption) and an ASAP 2010 (Micromeritics) instrument. Table 3 reports the surface area of the samples before and after the hydrothermal treatment. The percentage of surface area reduction after hydrothermal treatment ranges from 10 to 30%, apart the case of sample 7 which, however, was calcined only at 500°C instead of 700° or 900°C as the other samples. No significant change in the surface area was detected after the catalytic tests.

Table 3 Surface area of the samples before and after hydrothermal treatment.

Code	Surface area, m^2/g		% reduction after hydrothermal treatment
	Before	After	
ref.	110	82	25,6
1A	128	118	7,8
2A	82	72	12,2
3A	128	96	25
4A	82	76	7,3
1B	277	217	27,6
2B	211	206	2,4
3B	277	214	22,7
4B	211	152	27,9
5	98	66	32,6
6	117	72	38,4
7	172	79	54,0
8	66	58	12,1

The X-ray diffraction (XRD) powder pattern of the samples were recorded in the $2\theta=10-80$ range with an ItalStructure APD2000 diffractometer (Cu K_α radiation). All the samples before hydrothermal treatment show the presence of intense line of Ba-carbonate and very weak lines of BaO, together with the broad lines typical of γ-Al_2O_3. No evidence were found of the presence of $BaAl_2O_3$, although they may be masked from those of alumina.

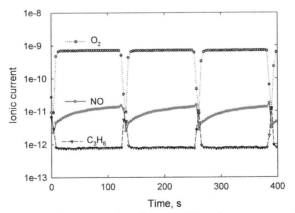

Figure 1 Change in the reactor outlet concentration of oxygen, NO and propene as monitored during the sequence of cyclic changes in the feed composition during NO_xSR testing experiments.

2.4. NOx storage-reduction tests

The NO_xSR activity was studied using a series of cyclic sequences of feed changes from lean conditions (120s: 5% O_2, 10.8% CO_2, 954 ppm NO, remaining He) to rich conditions (6s: 3.3% CO, 1.1% H_2, 6000 ppm C_3H_6, 5% O_2, 10.8% CO_2, 954 ppm NO, remaining He). Space-velocity was set to 60,000 h^{-1} using the catalyst in the form of powder (30-60 mesh). Figure 1 reports an example of the change in the concentration of oxygen, NO and propene during this sequence of cyclic tests. After evaluating the behaviour during a sequence of cycles (at least 10 cycles) at a given temperature, the reactor temperature is raised to another temperature where the behaviour is further monitored in a series of cycles.

NO_x storage activity is given as the mean conversion of NO_x during a cycle (e.g. the mean NOx conversion value estimated by integration of the NO conversion curve during the 120s of the lean part of a cycle), further mediated over the series of cycles (at least 10 cycles).

RESULTS AND DISCUSSION

Figure 2 reports the comparison of the behavior of the reference sample (e.g. 1% Pt on 15% barium/alumina, the latter prepared by incipient wet impregnation) with that of an analogous sample (in terms of composition and Pt and Ba precursors for the preparation; see table 2) prepared by the sol-gel method. The behavior of the samples after the hydrothermal treatment is also reported. The two samples have comparable surface area before and after the hydrothermal treatment.

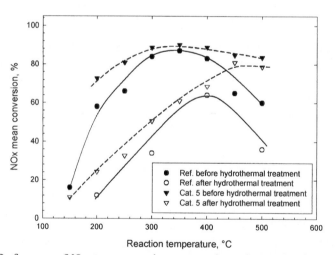

Figure 2 Performances (NOx mean conversion; see experimental part) of Ref. and Cat. 5 samples before and after hydrothermal treatment.

The reference sample shows the typical performances of this type of catalysts. The activity is low at temperatures < 200°C, there is a maximum activity in the 300-400°C, while the activity decreases at temperature above 400°C, due to the start of the competition with the desorption of NO_x adsorbed species (note that the presence of CO_2 during the tests significantly lower the thermal stability of adsorbed NO_x species with respect to thermal stability of Ba-nitrate, for example). The hydrothermal treatment significantly depress the activity in all the range of temperatures with a shift of the maximum in NO_x conversion at slightly higher temperatures.

The sample prepared by sol-gel method (Cat. 5) before the hydrothermal treatment shows similar maximum NO_x conversion of the Ref. sample, but better activities at low and higher temperatures, reasonably in relation to a better dispersion of the NO_x storage component. Also in this case there is a strong depression of the NOx conversion activity after hydrothermal treatment, but the activity maintains slightly better than that of the Ref. sample, especially at high temperatures ($\geq 450°C$).

An overview of the performances of all the samples is reported in Figure 3 which shows the mean conversion at 250°C (top graph) and 400°C (bottom graph) before and after the hydrothermal treatment as a function of the surface area of the sample (before and after the hydrothermal treatment, respectively).

There are several information which may be evidenced from these graphs. The first indication is that there is no correlation between surface area of the samples are reactivity in NO_xSR experiments, neither before and after hydrothermal treatment. After this treatment, there is a decrease of the surface area (as shown in Table 3) and a relevant loss of catalytic reactivity in NO_xSR tests, but there is no direct correlation between the two aspects.

It could be also noted that notwithstanding the same composition of all these samples (1% Pt, 15% Ba), the activity may vary in a quite large range, indicating the high dependence of the performances from the preparation even if XRD characterization, for example, does not reveals significant differences between the samples.

The samples prepared by sol-gel with basic catalysis (samples 1B-4B) lead generally to a higher surface area both after calcination at 700°C and 900°C, but this would not correspond to better reactivity performances. Better results are obtained using acid catalysis (samples 1A-4A) for the sol-gel transition, notwithstanding the lower surface area. The use of Al-*iso*-propoxide as the aluminum source (samples 5-8) leads to better results with respect to the use of Al-*iso*-propoxide (samples 1A-4A), although a direct comparison is not possible, because other parameters were changed in the preparation. It may be noted also that in general there is a high dependence of the catalytic performances from the preparation procedure (starting compounds, type of catalysis for the sol-gel transition, aging of the gel, calcination).

The nature of the Pt source (Pt is added by incipient wet impregnation in all samples) also influence the performances. The use of the Pt amino-nitro compound [$Pt(NH_3)_2(NO_2)_2$] leads to better results than the use of the Pt hexa-chloro compound [H_2PtCl_6], although the differences are not dramatic (compare the activity of the samples 1A, 2A, 1B, 2B with that of the samples 3A, 4A, 3B, 4B, respectively). The use of the Pt amino-nitro complex allows also to obtain slightly more stable samples after hydrothermal treatment. Figure 4 reports the comparison of the mean loss of activity at 250°C and 400°C, e.g. the mean percentage of loss of NO_x conversion at 250°C and 400°C with respect to the NO_x conversion of the sample before the hydrothermal treatment, for the samples prepared by sol-gel using the acid catalysis procedure (samples 1A - 4A).

The partial retention of chloride by Pt/γ-alumina catalysts prepared from chlorine-containing precursors quickly reduces activity in the reduction of NO by propene [25]. When Ba is present, stable $BaCl_2$ species may also form. Chlorine inhibits Pt activity toward methane combustion due to a strongly held fraction of residual chlorine species around the platinum particles [26]. EXAFS studies indicate that the presence of chlorine atoms inhibit the formation of Pt-O bonds [27]. Chlorine may also inhibit the mobility of spillover species [28].

74

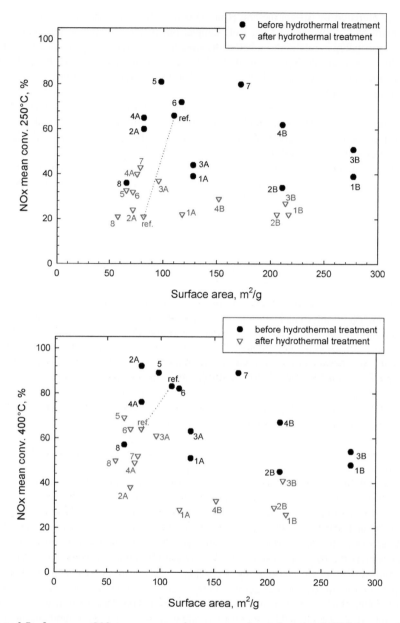

Figure 3 Performances (NOx mean conversion; see experimental part) at 250°C (top graph) and 400°C (bottom graph) of all samples before and after hydrothermal treatment.

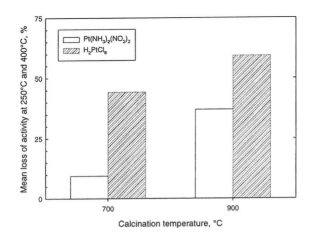

Figure 4 Mean percentage of loss of NOx conversion at 250°C and 400°C with respect to the NO_x conversion of the sample before the hydrothermal treatment, for the samples prepared by sol-gel using the acid catalysis procedure (samples 1A - 4A).

The difference in the activity of the 1A-4A samples before hydrothermal treatment is relatively low. Using the amino complex the activity is about 5-10% higher (see Figure 3). After the hydrothermal treatment this difference increases, as shown in Figure 4. This indicates that in NO_xSR reaction the inhibition effect of Cl on the Pt activity has a minor influence, whereas residual Cl ions have instead a more relevant effect on the sintering of Pt and/or Ba species.

A positive effect on the activity, before hydrothermal treatment (HT), is noted when calcination temperature moves from 700°C to 900°C. The same temperature produces minor effect in activity after HT. Figure 5a reports the example for the 1A and 2A samples (calcination temperatures of 700° and 900°C, respectively) before and after HT. A similar effect is also present in the samples prepared by basic catalysis for the gelation step (samples 3B and 4B; Figure 5b), although in this case after HT is slightly more active the sample calcined at 700°C (above 400°C).

The increase of the calcination temperature from 700°C to 900°C decreases the surface area (Figure 3), but data in Fig. 3 suggest that there is no direct relationship between activity and surface area. It may be noted, however, that a higher surface area should eventually promote Ba dispersion, while in this case the decrease of the surface area (due to calcination temperature) leads to a large increase in the activity (Figure 5). It could be also observed that the increasing of activity related to the calcination temperature is more pronounced in the sample prepared by acid catalysis (Fig. 5a) than in that prepared by basic catalysis (Fig. 5b) which has a surface area around two times higher.

It was noted earlier for Pt supported on $(CeO_2\text{-}ZrO_2)/(Ba\text{-doped alumina})$ [17] that during calcination, Ba ions migrated over the catalyst surface. The effect of the calcination is thus the promotion of barium surface migration and therefore the increase in NO_x storage properties. Sintering of Ba and formation of Ba-aluminate are two competitive reactions, but their occurrence is secondary during calcination in air. Instead, when steam is present in the feed (hydrothermal treatment; note that the temperature of this treatment and time are lower than those for the calcination at 900°C) determines a very severe loss of activity. This indicates that the surface reconstruction responsible for the deactivation (reasonably, the formation of a spinel-like Ba-aluminate surface layer) is induced by the presence of water in the feed.

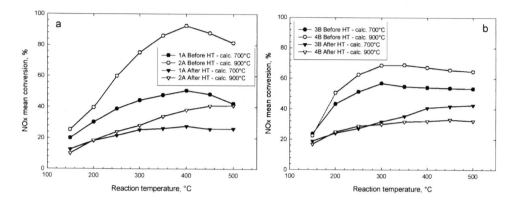

Figure 5 Mean NOx conversion before and after the hydrothermal (HT) treatment as a function of the reaction temperature and calcination temperature of the samples:
(a) 1A and 2A samples; (b) 3B and 4B samples.

In alumina nanoparticles the insertion of Ba in the lattice of Al_2O_3 hinders the ionic diffusion due to the anisotropic layered structures formed by the Ba located on the surface of nanoparticles [19]. It is thus reasonable that this mechanism, which is also responsible for the higher thermal stability of Ba-doped alumina with respect to alumina itself, prevents also the reconstruction by the formation of Ba-aluminate. The formation of $BaAl_2O_4$ by reaction of Ba-carbonate with alumina was previously reported [29]. Ba-carbonate is the main phase present on the surface of these catalysts at room temperature, as XRD analysis show. However, decomposition of surface Ba-carbonate is nearly complete at temperatures above 700°C [8] and the kinetic of solid state reaction to form $BaAl_2O_4$ is too slow at temperatures below 700°C.

When water vapour is present in the feed, BaO convert to $Ba(OH)_2$ as surface species [30] and this has been reported to promote $BaAl_2O_4$ formation in the 600-800°C temperature range [31]. $BaAl_2O_4$ nanocrystals with pure spinel structure were prepared. at 650°C by decomposition of Ba-Al hydrate precursor [32]. Hydrothermal treatment has been also reported to promote the formation of Ba-aluminate [33].

In conclusion, Ba doping on alumina particles prevent their phase transformation and sintering, but in the presence of water vapour the formation of surface Ba hydroxide and/or hydroxylation of alumina surface leads probably to a ionic mobility of Ba surface species which inhibit the Ba effect and therefore allowing the surface reconstruction during hydrothermal treatment with formation of Ba-aluminate phase. This transformation reduce the amount of Ba which can react with NO_x species and thus considerably depress NO_x storage activity.

It is interesting to note that examining the properties of $BaAl_2O_4$ [34] it has been reported the formation at low temperatures of amorphous nonstoichiometric phases with excess of Ba due to the replacement of two Ba^{2+} for an $[Al-O-Al]^{4+}$ link. The formation of an intermediate product $Ba_3Al_2O_6$ during the synthesis of $BaAl_2O_4$ by solid-solid reaction at 900° between barium carbonate and alumina has been also demonstrated [29]. Diffusion experiments using barium carbonate-alumina couples have further evidenced the formation of various barium aluminates phases with excess of barium, as a function of the distances of diffusion [35].

The structure of $BaAl_2O_4$ is characterized from Al_2O_4 units in distorted tetrahedra with covalent bonding, while Ba-O and O-O bonds are ionic [36]. Pacchioni et al. [37] evidenced by DFT cluster model calculations that in the adsorption of NO and NO_2 at the BaO surface the low-coordinated sites exhibit a much larger reactivity than the flat terraces. The formation of O^- ions in the case of NO_2 adsorption can be very important for the further reactivity of the surface. Miletic et al. [39] using local orbital DFT calculations also pointed out the role of locally charged states in the mechanism of NO_x storage. On the other hand, due to the presence of large amounts of CO_2 in the emissions of car engines, the BaO is quickly transformed to Ba-carbonate like species which are supposed to be the effective NO_x storage species in Pt/BaO/Alumina catalysts [8,38], notwithstanding its lower NO_x storage properties [40], reasonably due to the inhibition of the presence of the ionic oxygen surface sites mentioned before. However, other surface Ba species less able to forming Ba-carbonate may have a lower reactivity under "clean" NO_xSR tests (e.g. without the other typical feed components such as CO_2 and H_2O), but more relevant role in "real" conditions, e.g. in the presence of CO_2 and H_2O. For this reason, some author believe that are not Ba-carbonate surface species the main storage center for NO_x, but instead surface $BaAl_2O_4$ [41]. Deactivation of the sample results from the conversion of the $BaAl_2O_4$ phase to the $BaCO_3$ phase [41b].

On the other hand, formation of crystalline $BaAl_2O_4$ has been indicated as the responsible for deactivation during thermal and hydrothermal treatment [4,5], as discussed in the introduction. The discrepancy may be probably related to the differences between surface $BaAl_2O_4$ like phase and crystalline $BaAl_2O_4$. As discussed before, a non-stoichiometric $BaAl_2O_4$ like surface phase, with an excess of Ba, can be present on the surface. In this phase, the substitution of $[Al-O-Al]^{4+}$ with two Ba^{2+} ions [34] would create surface charged oxygen species and therefore it is likely that this phase would be active in NO_x storage. This phase could form more thermally stable adsorbed NO_x species (nitrate-like) than Ba-nitrate which in the presence of CO_2 in the feed start to be decomposed at temperatures $\geq 450°C$, explaining the decrease of NO_x storage performances on increasing the reaction temperature in Pt/Ba(O)/Alumina catalysts prepared by wet impregnation (Figure 2).

By favouring the formation of the nonstoichiometric $BaAl_2O_4$ like surface phase with respect to a surface Ba-carbonate phase the NO_x storage properties at temperatures above about 400-450°C may thus by enhanced, a clearly relevant aspect from the application point of view. It should be noted that nearly "commercial" NO_xSR catalyst contain also alkaline metals such as K to promote the performances in the 450-550°C temperature range, but these alkaline ions (i) may diffuse into alumina and (ii) make more difficult the regeneration of the catalyst deactivated by SO_2. An improvement of the high temperature NO_x storage properties without adding alkaline metals is thus interesting.

Figure 2 shows that the samples prepared by sol-gel may have effectively improved properties in terms of high-temperature NO_x storage. Figure 3 also shows that depending on the preparation a large range of variation in the NO_x storage properties is possible. Some interesting information is given from the comparison of the results of samples 2A and 4B before hydrothermal treatment (Figure 5A and 5B, respectively). These two samples have very close performances in the 150-300°C temperature range. However, quite different properties in the 350-500°C temperature range. A similar effect is also present after the hydrothermal treatment. The interpretation is related to two different type of NO_x storage sites which are present; the first one responsible for the lower temperature NO_x storage performances (assigned to Ba-carbonate surface species) and the second one active during the higher temperature NO_x storage performances (tentatively assigned to the nonstoichiometric $BaAl_2O_4$ like surface phase). The relative ratio between these species is depending on the preparation, explaining the range variation in the catalytic properties shown in Figure 3, notwithstanding the sample formal composition. The proof of this hypothesis, however, is difficult, because (i) these species and in particular the nonstoichiometric $BaAl_2O_4$ like surface phase are amorphous and therefore XRD data do not allow to obtain good information, and (ii) due to the presence of a complex overlap of bands, IR data do not allow to obtain a clear differentiation of the nature of surface NO_x species and their stability. Further studies are in progress to demonstrate the nature and presence of multiple type of surface Ba species and their role in different temperature range regimes.

The presence of two type of sites would also explain the great dependence from the preparation of the deactivation after hydrothermal treatment. In order to summarize all the results, two indexes of activity were chosen: (1) mean NO_x conversion in the 150°-500°C temperature range after hydrothermal treatment, e.g. the mean NO_x conversion value in all this range of temperature, and (2) mean reduction of the NO_x conversion in the same temperature range after the hydrothermal (HT) treatment, e.g. the percentage of reduction of the NOx conversion after HT treatment with respect to the same sample before HT treatment, mediating this percentage value over the 150°-500°C temperature range. The value of these two parameters for all the catalysts studied is reported in Figure 6. Better catalysts are those showing the higher mean NO_x conversion and /or lower mean reduction of the NO_x conversion after the catalytic tests, e.g. those located in the right /top area of the graph.

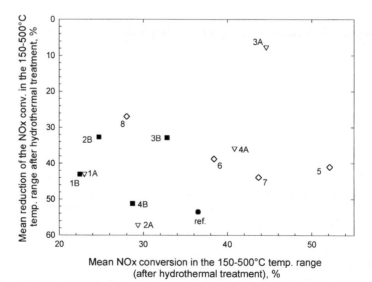

Figure 6 Mean NO_x conversion in the 150°-500°C temperature range after hydrothermal treatment and mean reduction of the NOx conversion in the same temperature range after the hydrothermal (HT) treatment (see text) for all tested NO_xSR catalysts.

With respect to the Ref. catalyst, e.g. that prepared following earlier Toyota patents, various samples prepared by sol-gel method show better performances after the hydrothermal treatment and/or lower reduction of the mean activity after the hydrothermal treatment. Almost all the samples, apart from the samples 2A and 4B, allow to obtain catalysts more stable to hydrothermal deactivation. Therefore, Figure 6 clearly evidences that the sol-gel preparation method allow to improve the performances of Pt/Ba(O)/Alumina catalysts with respect to wet impregnation method, even if there is a great dependence of the properties from the preparation method itself.

The large spreading of the catalytic performances, in terms of both NO_x storage and resistance to deactivation by hydrothermal treatment, further evidences that a more complex surface chemistry is present in these catalysts with respect to what suggested in the literature, although further characterization studies are necessary to support the interpretation. Ba doping and/or decoration of alumina nanoparticles increase their thermal stability, but in the presence of water vapour crystallization of Ba-aluminate occurs with a drastic lowering of the NO_x storage properties. Analysis of the results also suggest that together with Ba-carbonate surface species also an amorphous nonstoichiometric $BaAl_2O_4$ like surface phase is present. The two species are active in different range of temperatures and their relative amount depends on the preparation. Their amount influences also the hydrothermal stability of the samples in terms of both activity and deactivation amount after HT. The sol-gel preparation method, however, leads in general to improved resistance to deactivation by hydrothermal treatment, probably due to an enhanced formation of the amorphous nonstoichiometric $BaAl_2O_4$ like surface phase. This formation explains also the general better performances of the catalysts prepared by this method at reaction temperatures above about 400°C with respect to samples prepared by wet impregnation method.

In conclusion, the investigation of the reactivity of Pt/Ba(O)/Alumina catalysts prepared by sol-gel in the reaction of NO_x storage-reduction (NO_xSR) shows that the nanostructure of these catalysts, which depends considerably on the preparation method (type of catalysis during the sol-gel, nature of the precursors, ageing, calcination temperature), has a marked influence both on the reactivity and the hydrothermal stability. More detailed studies are necessary to understand these aspects and to characterize in more detail the surface chemistry and the nature of the species involved, but the results evidences that this preparation methodology is promising to prepare new improved NO_xSR catalysts

REFERENCES

[1] M. Takeuchi, S. Matsumoto, Topics in Catal., 28 (2004), 151.
[2] R. Burch, J.P. Breen, F.C. Meunier, Appl. Catal., B: Env., 39 (2002) 283.
[3] S. Matsumoto, Catal. Today. 90 (2004) 183.
[4] S. Elbouazzaoui, X. Courtois, P. Marecot, D. Duprez, Topics in Catal., 30/31 (2004) 493.
[5] D. Uy, A.E. O'Neill, J. Li, W.L.H. Watkins, Catal. Letters, 95 (2004) 191.
[6] Z. Liu, J.A. Anderson, J. Catal., 224 (2004) 18.
[7] K. Yamazaki, N. Takahashi, H. Shinjoh, M. Sugiura, Appl. Catal. B: Env., 53 (2004) 1.
[8] G. Centi, G.E. Arena and S. Perathoner, J. Catal., 216(1/2) (2003) 443-454.
[9] G. Centi, G. Fornasari, C. Gobbi, M. Livi, F. Trifiro, A. Vaccari, Catal. Today, 73 (2002) 287.
[10] G. Fornasari, F. Trifiro, A. Vaccari, F. Prinetto, G. Ghiotti, G. Centi, Catal. Today, 75 (2002) 421.
[11] F. Basile, G. Fornasari, M. Livi, F. Tinti, F. Trifiro, A. Vaccari, Topics in Catal., 30/31 (2004) 223.
[12] M. Guyon, J.C. Beziat, F. Figueras, H. Kochkar, J.M. Clacens, Fr. Demande FR 2831468 (2003), assigned to Renault Fr.
[13] H. Cheng, G. Chen, S. Wang, D. Wu, Y. Zhang, H. Li, Korean J. Chem. Eng., 21 (2004) 595.

[14] S. Takeshima, K. Yoshida, Y. Nakano, US Patent 2003216254 (2003), assigned to Toyota.

[15] Y. Li, US Patent 2004151645 (2004), assigned to Engelhard Co.

[16] C.K. Narula, ACS Symposium Series, 727(Inorganic Materials Synthesis (1999) 144.

[17] L.F. Liotta, A. Macaluso, G.E. Arena, M. Livi, G. Centi, G. Deganello, *Catal. Today*, 75 (2002) 439.

[18] L.F. Liotta, G. Deganello, J. Molec. Catal. A: Chem. 204-205 (2003) 763.

[19] I. Balint, Z. You, K.-i. Aika, Phys. Chem. Chem. Phys., 4 (2002) 2501.

[20] S. Rossignol, C. Kappenstein, Int. J. Inorg. Materials, 3 (2001) 51.

[21] B.-H. Jang, T.-H. Yeon, H.-S. Han, Y.-K. Park, J.-E. Yie, Catal. Letters, 77 (2001) 21.

[22] G. Busca, C. Cristiani, P. Forzatti, G. Groppi, Catal. Letters, 31 (1995) 65.

[23] G. Centi, M. Marella, L. Meregalli, S. Perathoner, M. Tomaselli, T. La Torretta, in Advanced Catalysts and Nanostructured Materials, W.R. Moser Ed., Academic Press Pub. 1996; Ch. 4, p. 63-89.

[24] N. Takahashi, H. Shinjoh, T. Iijima, T. Suzuki, K. Yamazaki, K. Yokota, H. Suzuki, N. Miyoshi, S. Matsumoto, Catal. Today, 27 (1996) 63.

[25] I.V. Yentekakis, R.M. Lambert, M. Konsolakis, N. Kallithrakas-Kontos, Catal. Letters, 81 (2002) 181.

[26] J.F. Lambert, E. Marceau, B. Shelimov, J. Lehman, V. Le Bel De Penguilly, X. Carrier, S. Boujday, H. Pernot, M. Che, M., Studies in Surface Science and Catal., 130B (2000) 1043.

[27] F.J. Gracia, E.E. Wolf, J.T. Miller, A.J. Kropf, Studies in Surface Science and Catal., 139 (2001) 471.

[28] S. Salasc, V. Perrichon, M. Primet, M. Chevrier, F. Mathis, N. Moral, Catal. Today, 50 (1999) 227.

[29] L. Perier-Camby, G. Thomas, Solid State Ionics, 63-65 (1993) 128.

[30] H.S. Zeng, K. Inazu, K.-i. Aika, J. Catal., 211 (2002) 33.

[31] J. Wang, Y. Ren, R. Yang, Zhongguo Fenti Jishu, 8 (2002) 23.

[32] C. Zhang, L. Wang, L. Cui, Y. Zhu, J. Crystal Growth, 255 (2003) 317.

[33] A.H.M. Ahmed, L.S.D. Glasser, Lesley J. Appl. Chem. & Biotechn., 21 (1971) 103.

[34] S.-Y. Huang, R. Von Der Muehll, J. Ravez, P. Hagenmuller, J. Phys. and Chem. of Solids, 55 (1994) 119.

[35] L. Perier-Camby, G. Thomas, Solid State Ionics, 93 (1997) 315.

[36] H. Uchikawa, K. Tsukiyama, Yogyo Kyokaishi, 74 (1966) 13.

[37] M.M. Branda, C. Di Valentin, G. Pacchioni, J. Phys. Chem. B, 108 (2004) 4752.

[38] F. Rodrigues, L. Juste, C. Potvin, J.F. Tempere, G. Blanchard, G. Djega-Mariadassou, Catal. Letters, 72 (2001) 59.

[39] M. Miletic, W.F. Schneider, K.C. Hass, J.L. Gland, Abstracts of Papers, 222nd ACS National Meeting, Chicago, IL, United States, August 26-30, 2001 (2001) COLL-286. Publisher: American Chemical Society, Washington, D. C

[40] I. Nova, L. Castoldi, L. Lietti, E. Tronconi, P. Forzatti, Catal. Today, 75 (2002) 431.

[41] (a) X. Li, M. Meng, P. Lin, Y. Fu, T. Hu, Y. Xie, J. Zhang, Topics in Catal., 22 (2003) 111. (b) M. Meng, X. Li, P. Lin, Y. Fu, G. Yue, Xinshiji De Cuihau Kexue Yu Jishu, Quanguo Cuihuaxue Jihuiyi Lunwenji, 10th, Zhangjiajie, China, Oct. 15-19, 2000 (2000) 947. Editor: B. Zhong. Publisher: Shanxi Kexue Jishu Chubanshe, Taiyuan, Peop. Rep. China.

Studies in Surface Science and Catalysis 155
A. Gamba, C. Colella and S. Coluccia (Editors)

The synthesis, characterization and use of metal niobates as catalysts for propane oxidehydrogenation

N. Ballarini[1]**, G. Calestani**[2]**, R. Catani**[3]**, F. Cavani**[1]**, U. Cornaro**[4]**, C. Cortelli**[1]**, M. Ferrari**[1]

[1]Dipartimento di Chimica Industriale e dei Materiali, Viale Risorgimento 4, 40136 Bologna
[2]Dipartimento di Chim. Gener. Inorg., Chim. Anal., Chim. Fis., Parco Area delle Scienze 17A, 43100 Parma, Italy
[3]Snamprogetti SpA, Viale De Gasperi 16, 20097 S. Donato MI, Italy
[4]Snamprogetti SpA, Via Maritano 26, S. Donato MI, Italy

Metal (Cr, Fe and V) / Nb mixed oxides were prepared with the co-precipitation technique, and then characterized and used as catalysts for the oxidehydrogenation (ODH) of propane to propylene, under both co-feed and cyclic conditions. $VNbO_5$ formed after thermal treatment in air at 500-550°C, but decomposed if treated at higher temperature, while $CrNbO_4$ (a rutile-type compound) and $FeNbO_4$ were stable up to 700°C. In the case of the Ga/Nb sample, instead, the formation of the mixed oxide $GaNbO_4$ occurred only at a minor extent. The most active and selective catalyst in propane ODH under cyclic conditions was $VNbO_5$, which however decomposed under reaction conditions yielding Nb-rich V/Nb mixed oxides and VO_x; the latter was the active phase in the reaction. However, no advantage was found in terms of selectivity to propylene with respect to the co-feed conditions. Fe/Nb and Cr/Nb samples acted as oxidehydrogenation catalysts, but gave very poor performance, while the Ga/Nb sample rather dehydrogenated propane, giving high selectivity to propylene under anaerobic conditions in the cyclic operation, but was quite unselective in the co-feed mode.

1. INTRODUCTION

Multimetal or metal/non-metal mixed oxides find application as heterogeneous catalysts for several chemical transformations, including redox and acid/base reactions [1,2]. In the former case, the formation of mixed oxides makes possible a better control of the oxidative properties of the active species, through the formation of compounds in which the chemical-physical properties, and finally the catalytic performance, do not simply represent an average of the properties of each single component. In some cases, the need for a mixed oxide does not only come from the necessity of having suitable redox properties for a defined reaction, but also from the need of combining quite different properties inside the same structure. This is necessary when complex transformations have to be performed, which require (i) different kind of oxidizing attacks on the organic substrate, or (ii) the combination of acid and redox characteristics. This is the case of the oxidative transformation of light alkanes to olefins or to oxygenated compounds, in which a proper Lewis-type acidity or basicity is required in order to activate the molecule, and proper (oxi)dehydrogenating or O-insertion properties are necessary to obtain the desired product. Catalytic systems which are known to operate efficiently in the oxidative transformation of light alkanes are (i) rutile-type V/Sb/O, catalyst for the ammoxidation of propane to acrylonitrile, (ii) V/Mg/O, active in the oxidehydrogenation (ODH) of propane to propylene and of isobutane to isobutene, (iii) Mo/V/Te(Sb)/(Nb)/O, claimed to be very active and selective in propane ammoxidation and oxidation, (iv) Keggin-type polyoxometalates P/Mo/V/O, for the oxidation of isobutane to methacrylic acid, and others as well [3].

One class of compounds which has received little attention is that of metal niobates [4-6]. V/Nb mixed oxides may potentially represent suitable systems for the ODH of light alkanes, since Nb^{5+} is known to possess Lewis-type acidity, and V is effective in the ODH of saturated hydrocarbons; indeed, Nb oxide-based systems have been investigated in the past as catalysts for the oxidation of light alkanes [7-11]. Other elements which are known to be active in the (oxidative) DH of alkanes are Fe, Ga and Cr [12]. The formation of a mixed compound in which the different elements are present, can lead to a better co-operation of the species involved in the reaction mechanism.

Aim of the present work was the synthesis and characterization of V/Nb, Fe/Nb, Cr/Nb and Ga/Nb mixed oxides. The materials prepared were used as heterogeneous catalysts for the ODH of propane; the reaction was carried out under either co-feed or cyclic conditions (redox-decoupling operation, with alternate feeding of the alkane and of oxygen). The latter procedure has been claimed to give better selectivity to the olefin than traditional, co-feed operation, due to the lower contribution of combustion reactions which occur in the presence of molecular oxygen [13,14].

2. EXPERIMENTAL

All catalysts were prepared with the co-precipitation technique, starting from alcoholic solutions of the different components, $Cr(NO_3)_3 6H_2O$, $Ga(NO_3)_3$, $VO(acac)_2$, $NbCl_5$, which were dropped into an aqueous solution maintained at pH 7 by NH_4OH addition; after precipitation, the solvent was evaporated. The solids obtained were calcined either in air or in N_2 flow at 700°C, unless otherwise specified.

The catalysts were characterized before and after reaction by Raman spectroscopy (Renishaw 1000 instrument, Ar laser at 514 nm, power 25 mW), and by X-ray diffraction (Phillips PW 1050/81, CuKα radiation). Ex-situ characterization of downloaded catalysts was done by cooling the samples in an inert flow, from reaction to room temperature.

Catalytic tests of propane ODH were carried out under co-feed conditions using the following reaction conditions: temperature 550°C, gas residence time 2 or 4 s (the amount of catalyst loaded was 1.8 cm^3), feed composition: 20 mol.% propane, 20 mol.% oxygen, remainder helium. Tests under cyclic conditions were carried out at gas residence times of 2 and 4 s, and at the temperature of 550°C. Half-cycle reducing periods of variable time-length (typically 0.5, 1.5, 2.5, 4.5 and 14.5 minutes), with a feed containing 20% propane in helium, were alternated with the half-cycle oxidizing step, lasting 30-40 min (feed air, residence time 2 s, temperature 550°C). The oxidation step was the first one to be applied when cyclic operation was started.

Reactants and products were analyzed by means of gas-chromatography. Propane, propylene, acetic acid and acrolein were separated with a Porapak Q column (FID), while hydrogen, oxygen, carbon monoxide and carbon dioxide were separated with a Carbosieve S column (TCD). The carbon balance was determined by comparing the effective propane conversion with the sum of the yields as inferred from GC analysis. Missing C thus corresponds to the C unbalance. Blank tests were done by carrying out the reaction with an inert component (acid-washed corindone) filling the reactor. Tests made by feeding only diluted propane led to a very low conversion of propane (between 0.2 and 0.4% at 550°C), while tests made by feeding 20% propane and 20% oxygen led to a propane conversion of 2.5% at 550°C, with formation of propylene and carbon dioxide.

3. RESULTS AND DISCUSSION

3.1 Characterization of metal niobates

Table 1 reports the samples prepared, and summarizes their main characteristics. Several V/Nb/O samples were prepared, having different atomic ratio between components, and treated under different conditions.

Table 1
Samples prepared, and their main characteristics.

Catalyst	Me/Nb, at. ratio	Thermal treatment	Crystalline phases after calcination at 700°C	Surface area, m^2/g
V/Nb/O	1/1	3h, air	$VNbO_5$, VNb_9O_{25}, V_2O_5	6
V/Nb/O	1/1	12 h, air	VNb_9O_{25}, V_2O_5	2
V/Nb/O	3/17	12 h, air	$V_3Nb_{17}O_{50}$	nd
V/Nb/O	1/2	12 h, air	$V_4Nb_{18}O_{55}$, V_2O_5	nd
V/Nb/O	1/9	12 h, air	VNb_9O_{25}	nd
V/Nb/O	1/20	12 h, air	Nb_2O_5, $V_2Nb_{23}O_{62}$	nd
V/Nb/O	1/1	3 h, nitrogen	$VNbO_4$ (tr Nb_2O_5)	48
Fe/Nb/O	1/1	3 h, air	$FeNbO_4$ (Fe_2O_3)	18
Cr/Nb/O	1/1	3 h, air	$CrNbO_4$ (rutile-type), Cr_2O_3, Nb_2O_5	8
Ga/Nb/O	1/1	3 h, air	Nb_2O_5, $GaNbO_4$	22

The V/Nb/O system develops a high number of mixed compounds: (a) rutile $VNbO_4$ [15], which indeed is a non-stoichiometric compound, and accepts valence states different from V^{3+} and Nb^{5+}, and in which the ratio between components can be different from 1; (b) orthorhombic $VNbO_5$, which can be prepared by the sol-gel route starting from alkoxides [16-18], and calcination at 500-600°C. The compound is an orthovanadate, in which V^{5+} is tetrahedrally coordinated and the V tetrahedra do not have O atoms in common, while Nb^{5+} is octahedrally coordinated; (c) other compounds with V/Nb ratios different from 1, and in which V can have mixed valence states, such as VNb_9O_{25} (tetragonal), $V_2Nb_{23}O_{62}$ (monoclinic), $V_4Nb_{18}O_{55}$ (orthorhombic) [19], and others as well: $V_3Nb_9O_{29}$, $V_3Nb_{17}O_{50}$, $V_2Nb_6O_{19}$.

Figure 1 reports the X-ray diffraction patterns of V/Nb 1/1 after thermal treatment at different temperatures, in air. At 500°C the pattern fits the one of $VNbO_5$ (JCPDS 46-0046); in ref [13], it was erroneously reported that this compound was synthesized by calcination at 700°C, while the correct temperature was 500°C. It is worth mentioning that this was the first example of preparation of this compound with a procedure other than the sol-gel method [13]. This means that a true co-precipitation makes possible to develop a mixed oxo-hydrate, which is the precursor of the mixed oxide. Patterns of Figure 1 show that $VNbO_5$ was not stable, and decomposed at temperatures higher than 500°C into V_2O_5 (41-1426) and two Nb-rich mixed oxides, VNb_9O_{25} (18-1447) and $V_4Nb_{18}O_{55}$ (46-0087). The reflections relative to the latter compound disappeared at 750°C; at this temperature the decomposition of $VNbO_5$ was complete, and only reflections attributable to V_2O_5 and VNb_9O_{25} were present.

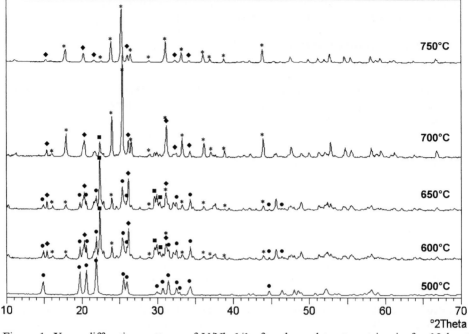

Figure 1. X-ray diffraction patterns of V/Nb 1/1 after thermal treatment in air, for 12 h, at increasing temperatures. ◆ V_2O_5 (41-1426); * VNb_9O_{25} (18-1447); ● $VNbO_5$ (46-0046); ■ $V_4Nb_{18}O_{55}$ (46-0087).

Figure 2 shows the Raman spectrum recorded at room temperature for V/Nb 1/1 after thermal treatment for 3 hours at increasing temperatures. The spectrum of $VNbO_5$ (after calcination at 500°C) showed two intense bands at 970-980 and ≈ 730-750 cm^{-1}. In literature, a weak IR band at 1020 cm^{-1} is attributed to the V=O stretching vibration due to V_2O_5 impurities [16]; in fact, no vanadyl bond is associated to V in $VNbO_5$, due to the tetrahedral coordination of V in the orthovanadate. However, the very intense band in the Raman spectrum at 970-980 cm^{-1} can be likely attributed to the niobyl moiety [20]. Calcination at increasing temperatures led to development of an additional band at 990-1000 cm^{-1}, relative to V=O, which became the prevailing one after calcination at 700°C; also other bands developed, attributable to V_2O_5. A spectrum similar to that one obtained at 650°C, relative to a partially decomposed $VNbO_5$, was previously attributed erroneously to pure $VNbO_5$ [13].

The niobyl moiety is also observed in Nb oxide, but only after calcination at high temperatures. Indeed, Nb_2O_5 is characterized by different crystalline structures [20-25]. Raman spectrum of the hydrated niobium oxide, if calcined at temperatures comprised between 500 and 700°C (the temperature range at which either the orthorhombic $T-Nb_2O_5$ or the monoclinic TT form are stable), shows a broad band at 690 cm^{-1}, and less intense bands in the 200 to 400 cm^{-1} region. The thermal treatment at temperatures higher than 700°C yields a completely different spectrum, with a strong band at 1000 cm^{-1} assigned to terminal niobyl species, two intense bands between 700 and 800 cm^{-1}, and one below 300 cm^{-1} [25]. Above 900°C, the monoclinic form of niobium oxide ($H-Nb_2O_5$), the usual commercial compound, is stable.

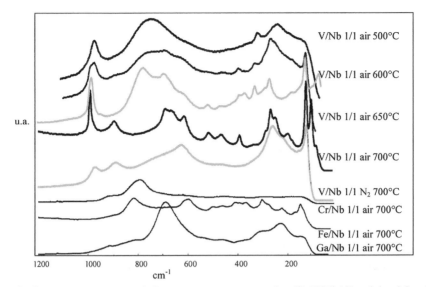

Figure 2. Raman spectra recorded at room temperature for (i) V/Nb1/1 calcined in air at 500°C for 12 h, and at 700°C for 3h; (ii) V/Nb 1/1 calcined in N₂ at 700°C for 3 h; (iii) Cr/Nb 1/1 calcined at 700°C for 3h; (iv) Fe/Nb 1/1 calcined at 700°C for 3h; and (v) Ga/Nb 1/1 calcined at 700°C for 3h.

After calcination at 700°C, the Raman spectrum of V/Nb 1/1 evidenced all the bands attributable to V_2O_5; the band at 690-700 cm^{-1} can be attributed either to the Nb_2O_5 form which is stable at this temperature, or to some Nb-rich V/Nb mixed oxide. The bands at 900 and 610 cm^{-1}, instead, are not relative to any single oxide; the same bands were also observed in sample V/Nb 1/1 after treatment in N₂ at 700°C.

The method adopted for the preparation of catalysts, the co-precipitation from homogeneous, alcoholic solutions of the different metal salts, made possible to obtain the mixed oxides already after mild calcination treatment. In fact, the formation of $VNbO_5$ was not possible when the preparation was carried out by first grinding together V_2O_5 and Nb_2O_5 (the latter having been prepared by thermal treatment of Nb oxohydrate at 700°C), and then by heating the mixture in air. The evolution of the in-situ Raman spectrum of this mixture in a heated cell is compared in Figure 3 with in-situ spectra recorded by heating the V/Nb 1/1 co-precipitate at temperatures higher than 500°C (short pre-heating of the co-precipitate for 1 h at 400°C in muffle was necessary in order to avoid the release of corrosive fumes inside the cell itself).

Results show that: (i) the co-precipitate (Figure 3, left), after a short treatment at 500°C, already showed the bands attributable to $VNbO_5$, but also those relative to V_2O_5 and Nb_2O_5. The latter however disappeared already at 600°C, and at 700°C the spectrum corresponded to that of $VNbO_5$. The discrepancy of temperature between spectra recorded in the heated cell and those recorded ex-situ, after prolonged treatment in muffle, was likely due to the different conditions employed for the two thermal treatments. At 750°C the compound decomposed, and bands developed which indicate the formation of the single oxides and of the Nb-rich V/Nb oxide. Under the same conditions, the spectra recorded for the mixture of vanadia and niobia (Figure 3, right) evidence the disgregation of the crystalline habit of the single oxides

86

only for temperature higher than 650°C. Under these conditions, however, the formation of $VNbO_5$ did not occur, and the spectrum corresponded to that obtained at above 700°C for the co-precipitated sample.

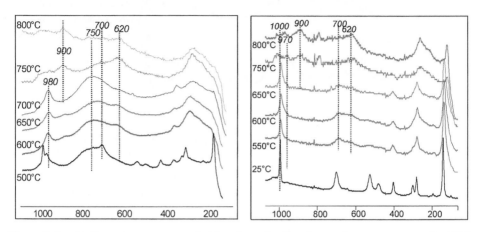

Figure 3. In-situ Raman spectra recorded in a heated cell at increasing temperatures for V/Nb 1/1 prepared by co-precipitation (left), and for a mixture of V_2O_5 and Nb_2O_5 (right).

The evolution of XRD patterns with temperature of V/Nb 1/2 was analogous to that of V/Nb 1/1. After calcination at 600°C, a mixture of $VNbO_5$ and $V_4Nb_{18}O_{55}$ formed (the latter compound was present in larger amount). For increasing calcination temperatures (650 and 700°C), the decomposition of $VNbO_5$ yielded V_2O_5 and VNb_9O_{25}, while reflections relative to $V_4Nb_{18}O_{55}$ were left. Finally, at 750°C the latter compound decomposed, and reflections relative to VNb_9O_{25} predominated, while V_2O_5 was the minor phase. In the case of V/Nb 1/20, the different crystalline structures of Nb_2O_5 were obtained, depending on the calcination temperature; the form stable at lower temperature (71-0336, orthorombic) was the only one after treatment at 600°C, but converted into the more stable forms (32-0711 and hexagonal 28-0317), for increasing temperatures.

When the thermal treatment for V/Nb 1/1 was done in nitrogen, the rutile-type compound $VNbO_4$ developed. $VNbO_4$ is stable in a wide range of temperature under inert atmosphere, but when thermally treated in air it converts into $VNbO_5$ (46-0046) [13].

Figure 4 shows the XRD diffraction patterns of samples Fe/Nb 1/1, Cr/Nb 1/1, Ga/Nb 1/1 and V/Nb 1/1 after calcination in air at 700°C for 3 h, and of V/Nb 1/1 after calcination in N_2, at 700°C; the corresponding Raman spectra are reported in Figure 2. The development of metal niobate, $MeNbO_4$, by thermal treatment in air was possible with those elements which possess a high-temperature stable Me^{3+} valence state: Cr^{3+} and Fe^{3+}. In the case of Ga/Nb 1/1, however, the formation of the mixed oxide occurred only at a minor extent.

Cr and Nb form niobates having different ratio between the two elements. $CrNbO_4$, as well as $CrVNbO_6$ and $FeVNbO_6$, are all characterized by the rutile-type structure [26-28]. In the ideal stoichiometric composition, they correspond to a solid solution between VO_2 and $Cr(Fe)NbO_4$; however, these systems are often characterized by non-stoichiometry. The preparation of well crystallized materials usually involves the mixing of the required amounts of the single oxides, followed by thermal treatment for several days at 1000-1100°C. With our procedure of preparation, the formation of the tetragonal rutile $CrNbO_4$ (34-0366) was not complete after calcination at 550°C, and major phases were Cr_2O_3 (38-1479) and Nb_2O_5 (28-0317). At above 650°C, instead, rutile was the prevailing compound, with minor amounts of Cr_2O_3 (38-1479), as evidenced by both XRD and Raman spectroscopy. Raman spectrum of the sample after calcination at 700°C showed an intense broad band at 790 cm^{-1}, which does not correspond to any band neither in chromium oxide nor in niobium oxide, and a less intense one at 900 cm^{-1}; other bands were at wavelength lower than 700 cm^{-1}.

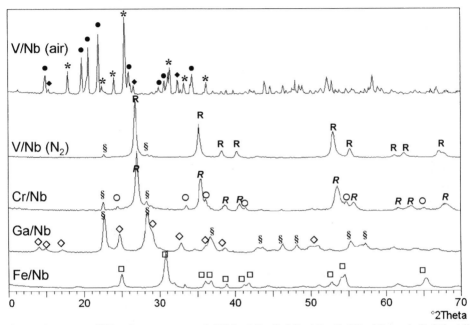

Figure 4. X-ray diffraction patterns of V/Nb 1/1, Cr/Nb 1/1, Fe/Nb 1/1 and Ga/Nb 1/1 calcined at 700°C in air for 3h, and of V/Nb 1/1 calcined in N_2 at 700°C. * VNb_9O_{25} (18-1447); ◆ V_2O_5 (41-1426); ● $VNbO_5$ (46-0046); O Cr_2O_3 (38-1479); □ $FeNbO_4$ (71-1349); **R** rutile $VNbO_4$; ◇ $GaNbO_4$ (16-0739); § $Nb_{16.8}O_{42}$ (71-0336); **R** rutile $CrNbO_4$ (34-0366).

When Cr_2O_3 and commercial Nb_2O_5 (the high temperature stable form) were ground together, the Raman band relative to the niobyl species shifted towards lower wavelength; the thermal treatment of the mixture at increasing temperatures (in the 550 to 700°C range) led to some further minor changes in the Raman spectrum of Nb oxide, but only after treatment at 800°C did the bands attributable to Cr_2O_3 disappear. This indicates a profound modification of the latter compound, as a consequence of the spreading over niobium oxide or of the formation of a mixed oxide. However, the spectrum was different from that of $CrNbO_4$, suggesting that the formation of the rutile compound is possible at relatively mild temperature only starting from the co-precipitate.

Fe and Nb develop different types of mixed oxides, including $FeNbO_4$ (either orthorhombic or rutile), $FeNb_2O_6$, $Fe_4Nb_2O_9$, Nb-rich phases (e.g., $FeNb_{49}O_{124}$), and Nb-Fe spinel phase, depending on O_2 partial pressure and temperature. All these phases have non-stoichiometric composition [29]. Our Fe/Nb 1/1 sample corresponded to the monoclinic form $FeNbO_4$ (16-0374), with traces of Fe_2O_3 (33-0664) and Nb_2O_5 (28-0317). The formation of the mixed oxide was possible already after treatment at 500°C, and the compound was stable even after treatment at 800°C. The Raman spectrum of the sample was more complex than that of the Cr/Nb 1/1; two intense bands at 810 and 600 cm^{-1} are attributable to vibration modes of the mixed oxide, but other less intense are due to Fe and Nb oxides.

Few information is available in literature concerning the characteristics of Ga/Nb mixed oxides. $GaNbO_4$ having the monoclinic (72-1666, 16-0739) structure may form, but also systems having large excess of Nb in structure can develop, e.g., $GaNb_{11}O_{29}$. In our case, the XRD pattern and the Raman spectrum indicate that still Nb_2O_5 (71-0336) was the main crystalline compound, and that the formation of $GaNbO_4$ (16-0739) had occurred only at a minor extent.

Table 1 summarizes the crystalline phases identified in each sample after calcination at 700°C.

3.2 Reactivity in propane ODH: co-feed conditions

Samples Fe/Nb 1/1, Cr/Nb 1/1, Ga/Nb 1/1 (all calcined at 700°C) and V/Nb 1/1 (calcined at 500°C) were tested as catalysts for the ODH of propane to propylene, under both co-feed and cyclic (redox-decoupling) conditions. Before running the tests under cyclic conditions, samples were pretreated in air at 550°C (which is also the maximum temperature used in catalytic tests), for 30 min. Under such preliminary treatment, the characteristics of compounds were left unaltered with respect to the calcined catalysts.

Table 2 summarizes the catalytic performance of samples under co-feed conditions, at the temperature of 550°C and residence time 2 s.

Table 2
Summary of catalytic performance in propane ODH under co-feed conditions (all samples calcined in air at 700°C for 3h, unless otherwise specified).

Catalyst	Propane conv., %	Select. to propylene, %	H_2 conc., mol. %
V/Nb 1/1, 500°C, 3h	32	20	0.25
Cr/Nb 1/1	23	16	0
Fe/Nb 1/1	21	13	nd
Ga/Nb 1/1	23	5	0.30
V/Nb 1/1, 12h	33	22	0.39
V/Nb 3/17, 12h	30	31	nd

Under these conditions, all catalysts reached total conversion of the limiting reactant (O_2), with a corresponding propane conversion which was a function of the distribution of products. V/Nb 1/1, for instance, gave 32% propane conversion due to the high CO/CO_2 ratio (equal to 1.9), much higher than that obtained with Cr/Nb (CO/CO_2 0.4), or with Fe/Nb (0.15) and Ga/Nb (0.2). The selectivity to propylene was nevertheless very low for all samples; it was the lowest in those samples which had the highest combustion activity (i.e., the lowest CO/CO_2 ratio). The formation of H_2 also was non-negligible for Ga/Nb and V/Nb 1/1; with the former catalyst, the amount of hydrogen produced was comparable to that of propylene, while with V/Nb 1/1 it was approximately 1/3rd of the propylene produced. This suggests that under these conditions the dehydrogenation of propane to propylene, rather than its ODH, is the true mechanism for the formation of propylene, at least with Ga/Nb and V/Nb 1/1. Table 2 also reports the results obtained with V/Nb 1/1 calcined at 700°C for 12 h; in this case, therefore, the catalyst consisted of Nb-rich V/Nb mixed oxide and of vanadia.

3.3 Reactivity in propane ODH: cyclic conditions

Figure 5 shows the catalytic behavior of V/Nb 1/1 under cyclic conditions. The instantaneous conversion of propane, and the selectivity to propylene, carbon monoxide and carbon dioxide are plotted as functions of the reaction time during the half-cycle, reduction step; during this step, the catalyst run in a stream of diluted propane, under anaerobic conditions. Also, the molar concentration of H_2 in the exit stream is reported. Other by-products obtained in minor amount were light hydrocarbons (ethane, ethylene, methane), with an overall selectivity which was for all catalysts lower than 3%. It is worth mentioning that the selectivity was calculated considering only the mentioned C-containing products (propylene, CO_x, light hydrocarbons), thus without taking into account the accumulation of coke on catalysts. The amount of coke was evaluated indirectly through the instantaneous C balance, that is the ratio between the sum of yields and the conversion of propane. In the case of V/Nb 1/1, the C balance was close to 100% for tests measured during the initial 3 min, while was 85% for the test after 4.5 min, and was 45% for the final test, after 14.5 min reaction time.

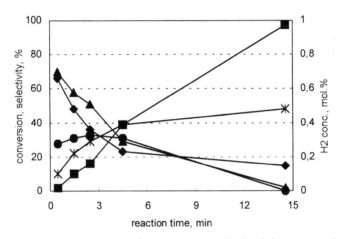

Figure 5. Instantaneous conversion of propane (♦) and selectivity to propylene (■), to CO (▲), to CO_2 (●) as functions of the half-cycle reaction time, under cyclic conditions (reductive step). Temperature 550°C, gas residence time 2 s. Catalyst V/Nb 1/1, calcined in air at 500°C 3h.

The initial propane conversion was very high; therefore, the fresh, fully oxidized catalyst was active towards the alkane. Under these conditions, however, the main products were CO and CO_2, while the selectivity to propylene was quite low; a small amount of H_2 was also detected. When the reaction time increased, the conversion rapidly declined, and finally stabilized at the value of 15%, which is close to the thermodynamic value for propane dehydrogenation at 550°C. Correspondingly, the selectivity to propylene increased, while the selectivity to CO and CO_2 became very low. The formation of H_2 increased, and the C balance worsened, indicating an unsteady accumulation of coke. Therefore, while the catalyst became progressively more and more reduced, due to the absence of gas phase oxygen, its behavior was that of a dehydrogenation catalyst.

The performance of Me/Nb 1/1 catalysts is compared in Figure 6: the conversion of propane and the selectivity to propylene are plotted as functions of the half-cycle, reduction time, at 550°C and gas residence time 4 s.

The following differences were found:

1) V/Nb 1/1 was the most active catalyst; all samples, but Ga/Nb, showed a decrease of the conversion for increasing reaction times, indicating that the catalysts oxidized the substrate, and that the progressive reduction decreased their activity. The final activity of Cr/Nb and Fe/Nb was practically nil, while V/Nb 1/1 maintained a dehydrogenation activity even after prolonged exposure to the reducing stream. The activity of Ga/Nb, instead, was independent on time-on-stream; the conversion was close to 20% for whichever value of reaction time. In this case, therefore, no oxidant activity was present, and the material acted as a dehydrogenating catalyst. It is worth mentioning that the same catalyst, if exposed to co-feed conditions, gave very low selectivity to propylene, with carbon oxides as the prevailing products (Table 2).

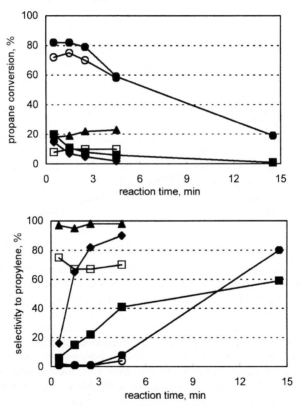

Figure 6. Instantaneous propane conversion (top) and selectivity to propylene (bottom) as functions of the half-cycle reaction time (reductive step). T 550°C, gas residence time 4 s. Symbols: V/Nb 1/1 calc. 500°C (●); Cr/Nb 1/1 calc. 700°C (♦); Fe/Nb 1/1 calc. 700°C (■); Ga/Nb 1/1 calc. 700°C (▲); V/Nb 1/1 calc. 700°C 12 h (○); V/Nb 3/17 calc. 700°C 12h (□).

2) The selectivity to propylene was very high (close to 95%) for Ga/Nb, with negligible formation of carbon oxides. This confirms the dehydrogenating properties of this material. On the other hand, the C balance was very low (65% at the beginning of reaction time, and then 55-60% for longer reaction times), indicating the accumulation of coke during the test. The same catalyst also produced the highest amount of H_2, with a concentration equal to 1.1 mol% at the beginning, rapidly increasing to 1.7-1.8% for longer reaction times.

3) With V/Nb, Fe/Nb and Cr/Nb, the selectivity to propylene was initially very low, and increased when the reaction time increased (and the conversion of propane lowered). This indicates that the progressive reduction of the metal ions makes the catalysts more selective to the olefin, due to (i) the lower oxidizing ability, (ii) the increased contribution of dehydrogenation as compared to oxidehydrogenation, and (iii) the decreased level of conversion, which makes consecutive reactions less important. Correspondingly, for all samples the C balance worsened. The concentration of H_2 could not be estimated in the case of Cr/Nb and Fe/Nb, due to the very low conversions attained. For V/Nb 1/1, the concentration of H_2 increased for progressively longer reaction times (see also Figure 5).

4) The behavior of V/Nb 1/1 was quite similar (also for what concerns H_2 production and C balance) when the sample was calcined either at 500°C or at 700°C. In the case of V/Nb 3/17 calcined at 700°C, instead, the catalyst was poorly active, and behaved like a dehydrogenating catalyst; in fact, the conversion of propane and the selectivity to propylene were stable, and did not depend on propane conversion. This difference can be attributed to the characteristics of catalysts downloaded after reaction (see below).

5) Table 3 summarizes the main structural features of catalysts downloaded after reaction under cyclic conditions. The following samples had XRD pattern and Raman spectrum which were the same as those of the corresponding samples before reaction: Cr/Nb 1/1, Fe/Nb 1/1, Ga/Nb 1/1 and V/Nb 3/17 calcined at 700°C. This indicates that the mixed niobates of Cr, Fe and Ga are stable in the reaction environment. $V_3Nb_{17}O_{50}$ also was stable, while the same was not true for $VNbO_5$, which had decomposed to vanadium oxide and to VNb_9O_{25}. Therefore, V_2O_5 played the redox cycle, and the high catalytic activity of V/Nb 1/1 (either calcined at 500°C or at 700°C) was due to the presence of vanadium oxide. The latter is generated under reaction conditions for the sample which had been calcined at 500°C ($VNbO_5$), while was already present in the fresh catalyst for the sample which had been calcined at 700°C. Since $VNbO_5$ is stable in air up to 550°C (the same temperature at which the catalytic tests were carried out), its decomposition in the reaction environment is either due to the development of local temperatures well above 550°C, or to the presence of the strongly reducing conditions. Moreover, catalytic tests demonstrate that reduced vanadium oxide has dehydrogenating properties, as already pointed out by other authors [30]. On the contrary, the stable $V_3Nb_{17}O_{50}$ has a very poor oxidizing power, and mainly acts a dehydrogenating catalyst.

Figure 7 compares the selectivity to propylene as a function of propane conversion, for tests run under co-feed conditions and under cyclic conditions. For all samples, but Ga/Nb, the selectivity declined on increasing the propane conversion, due to the higher contribution of consecutive combustion reactions, and to the fact that under cyclic conditions highest conversions were achieved with more oxidized (and less selective) catalysts.

Table 3
Main structural characteristics of samples downloaded after reaction under cyclic conditions.

Catalyst	Me/Nb, at. ratio	Thermal treatment	Crystalline phases after reaction (last step, oxidation, unless otherwise specified)
V/Nb/O	1/1	500°C, 3h, air	$V_2O_5 + VNb_9O_{25}$; $V_2O_3 + V_2O_4 + VNb_9O_{25}$ (last step, reduction)
V/Nb/O	1/1	700°C, 12 h, air	$V_2O_5 + VNb_9O_{25}$
V/Nb/O	3/17	700°C, 12 h, air	$V_3Nb_{17}O_{50}$
Fe/Nb/O	1/1	3 h, air	$FeNbO_4$ (tr Fe_2O_3)
Cr/Nb/O	1/1	3 h, air	$CrNbO_4 + Cr_2O_3 + Nb_2O_5$
Ga/Nb/O	1/1	3 h, air	$Nb_2O_5 + GaNbO_4$

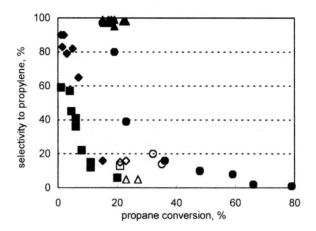

Figure 7. Selectivity to propylene as a function of propane conversion under cyclic conditions (full symbols) and under co-feed conditions (open symbols). Symbols: V/Nb 1/1 calc. 500°C (●); Cr/Nb 1/1 calc. 700°C (♦); Fe/Nb 1/1 calc. 700°C (■); Ga/Nb 1/1 calc. 700°C (▲). Reaction temperature 550°C; gas residence time 2 and 4 s.

Different was the case for Ga/Nb, which acted as a dehydrogenating catalyst. Moreover, in the case of Cr/Nb and Fe/Nb, which had comparable performance, high selectivity to propylene could be obtained only for very low conversion. On the contrary, V/Nb 1/1 had high selectivity even for 15% propane conversion (selectivity higher than 95%). This was due to the fact that the reduced catalyst (which had been exposed to the reducing stream for prolonged reaction times, during cyclic operation) had a non-negligible dehydrogenating activity, with a high selectivity to the olefin. Under these conditions, the catalyst thus behaved like Ga/Nb. The same was not true for Cr/Nb and Fe/Nb; the catalysts did not have an appreciable dehydrogenating activity under these conditions. It is worth mentioning that the selectivity plotted refers to gas-phase products, and does not include the amount of coke accumulated on catalysts; the latter was relevant in the case of Ga/Nb and Cr/Nb, while with V/Nb 1/1 it was non-negligible only for prolonged reaction times during the reduction step. Taking this into account, it is even more evident that V/Nb 1/1 was the most selective catalyst under cyclic conditions.

The comparison between cyclic and co-feed conditions evidences that, with the catalysts employed, there was no advantage in the separation of the two steps of the redox mechanism. The exception is Ga/Nb, which was quite unselective to propylene in the presence of O_2. However, in this case the high selectivity in the absence of oxygen was not due to a more selective ODH pathway, but rather to the occurrence of a dehydrogenative transformation of the alkane.

4. CONCLUSIONS

Metal (Cr, Fe, Ga and V) / Nb mixed oxides were prepared with the co-precipitation technique, and characterized. The formation of $VNbO_5$ was possible only under mild conditions of thermal treatment (not higher than 550°C), while higher temperatures led to the decomposition into Nb-rich V/Nb mixed oxides and vanadium oxide. Rutile-type $CrNbO_4$ and $FeNbO_4$ formed after thermal treatment at 700°C, with minor amounts of Me and Nb oxides, while in the case of the Ga/Nb sample the formation of $GaNbO_4$ only occurred at a minor extent.

Samples were tested as catalysts for the ODH of propane under both co-feed and cyclic (redox-decoupling) conditions. The most active and selective catalyst was $VNbO_5$, which however decomposed under reaction conditions yielding Nb-rich V/Nb mixed oxides and vanadium oxide. The latter was the active compound in the reaction, performing initially the ODH of propane for shorter reaction times during the reductive step (as long as it was able to furnish ionic oxygen), and then acting as a dehydrogenating catalyst for longer reaction times. However, no advantage was found in terms of selectivity to propylene with respect to the co-feed conditions. Fe/Nb and Cr/Nb mixed oxides acted as ODH catalysts, but gave very poor performance, while the Ga/Nb sample acted as a dehydrogenating catalyst, giving high selectivity to propylene under anaerobic conditions, while being quite unselective in co-feed operation.

ACKNOWLEDGEMENTS

Snamprogetti and CNR (Progetto Coordinato CNRC00307A-001) are gratefully acknowledged for financial support.

REFERENCES
1. A. Cimino and F.S. Stone, Adv. Catal., 47 (2002) 141.
2. G. Centi and S. Perathoner, Curr. Opin. Solid State Mater. Sci., 4 (1999) 74.
3. G. Centi, F. Cavani and F. Trifirò, "Selective Oxidation by Heterogeneous Catalysis", Kluwer Academic/Plenum Publ., New York, 2001.
4. I.E. Wachs, J.-M. Jehng, G. Deo, H. Hu and N. Arora, Catal. Today, 28 (1996) 199.
5. K. Tanabe and S. Okazaki, Appl. Catal. A, 133 (1995) 191.
6. J.-M. Jehng and I.E. Wachs, Catal. Today, 8 (1990) 37.
7. R. Burch and R. Swarnakar, Appl. Catal., 70 (1991) 129.
8. K. Ruth, R. Kieffer and R. Burch, J. Catal., 175 (1998) 16.
9. O. Desponds, R.L. Keiski and G.A. Somorjai, Catal. Lett., 19 (1993) 17.
10. K. Ruth, R. Burch and R. Kieffer, J. Catal., 175 (1998) 27.
11. H. Watanabe and Y. Koyasu, Appl. Catal. A, 194 (2000) 479.
12. F. Cavani and F. Trifirò, in "Basic Principles in Applied Catalysis", M. Baerns (Ed.), Springer, Berlin, Series in Chemical Physics 75, 2003, p. 21.

13. N. Ballarini, F. Cavani, C. Cortelli, C. Giunchi, P. Nobili, F. Trifirò, R. Catani and U. Cornaro, Catal. Today, 78 (2003) 353.
14. N. Ballarini, F. Cavani, M. Ferrari, R. Catani and U. Cornaro, J. Catal., 213 (2003) 95.
15. V.W. Rüdorff and J. Märklin, Z. Anorg. Allg. Chemie, 334 (1964) 142.
16. J.M. Amarilla, B. Casal and E. Ruiz-Hitzky, Mater. Lett., 8 (1989) 132.
17. J.M. Amarilla, B. Casal, J.C. Galvan and E. Ruiz-Hitzky, Chem. Mater., 4 (1992) 62.
18. J.M. Amarilla, B. Casal and E. Ruiz-Hitzky, J. Mater. Chem., 6 (1996) 1005.
19. O. Yamaguchi, Y. Mukaida and H. Shigeta, Adv. Powder Techn., 1 (1990) 3.
20. L.J. Burcham, J. Datka and I.E. Wachs, J. Phys. Chem. B, 103 (1999) 6015.
21. K. Naito and T. Matsui, Solid State Ionics, 12 (1984) 125.
22. I. Nowak and M. Zoliek, Chem. Rev., 99 (1999) 3603.
23. A.A. McConnell, J.S. Anderson and C.N.R. Rao, Spectrochim. Acta, 32A (1976) 1067.
24. T. Ushikubo, Y. Koike, K. Wada, L. Xie, D. Wang and X. Guo, Catal. Today, 28 (1996) 59.
25. J.-M. Jehng and I.E. Wachs, Chem. Mater., 3 (1991) 100.
26. A. Petersen and H. Muller-Bushbaum, Z. Anorg. Allg. Chem., 609 (1992) 51.
27. K. Ravindran Nair and M. Greenblatt, Mat. Res. Bull., 18 (1983) 1257.
28. C.J. Chen, M. Greenblatt, K. Ravindran Nair and J.V. Waszczak, J. Solid State Chem., 64 (1986) 81.
29. K. Kitayama, J. Solid State Chem., 69 (1987) 101.
30. M. Volpe, G. Tonetto and H. de Lasa, Appl. Catal. A, 272 (2004) 69.

Studies in Surface Science and Catalysis 155
A. Gamba, C. Colella and S. Coluccia (Editors)
© 2005 Elsevier B.V. All rights reserved

Room temperature interaction of CO with alkali-metal cations in M-ZSM-5 zeolites as studied by joint FT-IR spectroscopy and microcalorimetry

B. Bonelli,[a] B. Fubini,[b] B. Onida,[a] G. Turnes Palomino,[c] M. Rodríguez Delgado,[c] C. Otero Areán[c] and <u>E. Garrone</u>.[a,*]

[a] Dipartimento di Scienza dei Materiali ed Ingegneria Chimica, Politecnico di Torino, I-10129 Torino, Italy

[b] Dipartimento di Chimica Inorganica, Fisica e dei Materiali, Università di Torino, I-10125 Torino, Italy

[c] Departamento de Química, Universidad de las Islas Baleares, 07071 Palma de Mallorca, Spain

*corresponding author: e-mail edoardo.garrone@polito.it.

ABSTRACT

The interaction of CO with K-ZSM-5 zeolite has been studied at 303 K by means of FT-IR spectroscopy and adsorption microcalorimetry, and compared with previous results concerning CO adsorption on Li- and Na-ZSM-5 zeolites. In all three cases, IR spectroscopy indicates the presence of two M^+ sites, less distinguishable on Na-ZSM-5, and clearly observable on Li-ZSM-5 and K-ZSM-5 zeolites. The case of Li-ZSM-5 offers peculiar features, probably due to the smaller size of Li^+ ions. For the main species, the heat of adducts formation and the extinction coefficient of the corresponding IR absorption band have been determined.

1. INTRODUCTION

Zeolites and related microporous materials are currently used in a wide range of technological processes, including gas separation, molecular sieving and heterogeneous catalysis.[1] ZSM-5 zeolites, the object of the present study, are characterized by an MFI-type[2] structure and a low Al content, which implies a low density of extraframework cations. The cation sites in MFI-type zeolites may be considered as sufficiently far away from each other to be non-interacting. Hence, such systems may be considered as being ideal in the thermodynamic sense, and allow therefore relatively easy spectroscopic and calorimetric characterisation.

Carbon monoxide is, by far, the most extensively used probe molecule for IR spectroscopic studies of zeolites.[3] The interaction of CO with alkali-metal cations in zeolites is weak and is usually studied at liquid nitrogen temperature:[4] besides M^+---CO adducts, also M^+---OC species are formed, in a temperature-dependent equilibrium.[5] IR measurements conducted by means of variable temperature IR spectroscopy allowed determination of the enthalpy difference between the two isomers.[5]

We have found, however, that room-temperature adsorption on such systems is possible, and is therefore possible to run joint microcalorimetric and spectroscopic studies. These allow the thorough determination of the thermodynamic features of cation-CO interaction. Aim of this work is the detailed study of the adsorption of CO on K-ZSM-5, as compared with similar studies already carried out concerning the adsorption on Li-ZSM-5[6] and Na-ZSM-5[7] at room temperature, and with previous results concerning the system CO/K-ZSM-5 followed by means of a variable-temperature IR cell.[8]

2. EXPERIMENTAL SECTION

The K-ZSM-5 zeolite (Si/Al= 14), prepared following standard procedures, and checked by powder X-ray diffraction, showed good cristallinity and the absence of any diffraction lines not assignable to the MFI-type structure.[2] The resulting K/Al ratio was close to 1.

For IR measurements, thin self-supporting wafers were prepared and outgassed at 723 K under a dynamic vacuum inside an IR cell, which allowed *in situ* gas dosage. The same activation conditions were adopted prior to calorimetric measurements on the zeolite powder. IR spectra were collected by means of a Bruker IF66 FTIR spectrometer, at 2 cm^{-1} resolution, on the zeolite blanks and after dosing CO at increasing equilibrium pressures, in the 0.5 – 80 Torr range. Since at pressures higher than 2 Torr a substantial contribution from gas-phase CO was present, parallel measurements were conducted at the same equilibrium pressures on a silicalite wafer. The IR spectra taken at room temperature on this purely siliceous MFI-type zeolite showed only a broad band at *ca.* 2140 cm^{-1} with rotational wings (P- and R-branch). These spectra were subtracted from those corresponding to CO adsorbed on K-ZSM-5.

The heats of adsorption were determined by means of a Tian-Calvet microcalorimeter (Setaram) operated at 303 K, connected to a volumetric apparatus allowing simultaneous measurement of adsorbed amount (uptake, n_a), heat released (Q), and equilibrium pressure (p) for small increments of CO dosed to the sample. Before CO adsorption, the zeolite powder has been outgassed at 723 K.

A first adsorption run (yielding the primary isotherm) was carried out, up to equilibrium pressures of about 60 Torr, for technical reasons. This was followed by prolonged evacuation at room temperature, to remove the reversible fraction of adsorbate, and a second adsorption run was performed up to the same equilibrium pressure, so obtaining the secondary isotherm. Note that the term isotherm comprises both the adsorption isotherm (adsorbed amount n_a vs. quasi-equilibrium pressure p) and calorimetric isotherms (total evolved heat Q vs. p). The secondary isotherm allows to evaluate the reversible adsorption, as, at a given p value, the difference between primary and secondary isotherms yields the irreversible quantity (either adsorbed amount or evolved heat). A compact representation of data is the plot of differential heats $q_{diff} = (dQ/dn_a)$ as a function of n_a.[9]

3. RESULTS AND DISCUSSION

Figure 1 shows IR spectra recorded for K-ZSM-5. The band at 2227 cm^{-1} is due to CO adsorbed on partially extra-framework Al^{3+} species, and will not be discussed further. The bands at 2163 and 2117 cm^{-1} have been already observed and interpreted in the course of variable-temperature studies as the K^{+}---CO and K^{+}---OC adducts, respectively, on an exposed K^{+} site: the O-bonded isomer resulted to be less stable than that C-bonded by 3.2 kJ mol^{-1}.[8] The band at 2149 cm^{-1} was not discussed in low-temperature studies, because it had a small intensity. We assigned it to CO on K^{+} ions more shielded by zeolite framework oxygen atoms than those giving rise to the band at 2163 cm^{-1}.[8] Both EXAFS[10] and theoretical studies[11] show in Cu^{+}-ZSM-5 zeolites the presence of two different extra-framework Cu^{+} sites, which can be coordinated to either two or three-four framework oxygen atoms. We assume here a similar situation: those ions with a larger coordination number would show a smaller net electric field, reflected by the smaller shift in the frequency of adsorbed CO. Spectra recorded at lower temperature show that both the band at 2149 cm^{-1} and the band due to K^{+}---OC adducts are minor features: spectra in Figure 1 show that they become prominent at room temperature. The reason lies in thermodynamic, according to which at higher temperatures also higher energy levels become populated, whereas at low temperatures only the ground state is so.

Computer simulation have been performed of the band envelope in the 2189-2100 cm^{-1} range by using three bands, with Gaussian profiles and maxima at 2163, 2149 and 2117 cm^{-1}, an example of the excellent fit obtained is reported in the inset to Figure 1. Comparison with the results at variable temperature allows to establish that the intensity ratio between the 2163 and 2117 cm^{-1} bands is that expected at room temperature, thus confirming that the C-bonded counterpart of band at 2117 cm^{-1} is that at 2163 cm^{-1}, in agreement with a rule-of-thumb according to which the location of bands due to M^{+}---CO and M^{+}---OC species are approximately symmetrical with respect to the gas-phase CO stretch, with Δv = + 26 and -20 cm^{-1} in the present case. Being the second K^{+}---CO band at 2149 cm^{-1}, the corresponding K^{+}---OC band is expected at around 2137 cm^{-1}, where no absorption is seen. The reasons for the absence of this OC-bonded counterpart are currently under investigation.

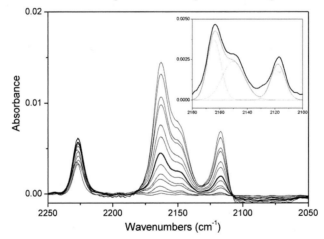

Fig. 1. FT-IR spectra of CO (p = 0.5-50 mbar) adsorbed on K-ZSM-5 at 303 K.

To illustrate the occurrence of two types of extra-framework sites, Figure 2 compares difference spectra recorded on Li-ZSM5, Na-ZSM-5 and K-ZSM-5 at comparable CO coverage. With Li-ZSM-5,[6] the two sites, hereafter referred to as Li$^{+}_A$ and Li$^{+}_B$, form monocarbonyl species with bands at 2193 (Li$_A$(CO)$^{+}$) and 2187 (Li$_B$(CO)$^{+}$) cm^{-1}; on Li$_A$$^{+}$ sites the formation of dicarbonyls species was also found, with a band located around 2185 cm^{-1} and thus superimposed to that of Li$_B$(CO)$^{+}$ species at 2187 cm^{-1}. With Na-ZSM-5,[7] a weak band is seen at 2157 cm^{-1}, besides that at 2176 cm^{-1} due to CO adsorbed on exposed Na^{+} ions.

The occurrence of the more shielded sites seems to be, in this case, less extensive. Also available are data concerning the sample Rb-ZSM-5:[12] variable-temperature IR spectroscopy showed, beside the band at 2161 cm^{-1} assigned to Rb$^+$---CO adducts, a shoulder at about 2148 cm^{-1}.

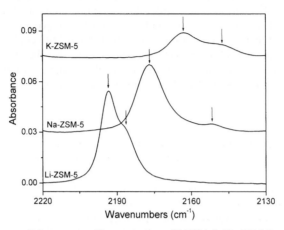

Fig. 2. Spectra recorded at comparable coverage on Li-ZSM-5, Na-ZSM-5 and K-ZSM-5 outgassed at 723 K: arrows point out the occurrence of two sites for CO coordination through the C atom.

The occurrence of two sites seems thus to be a general phenomenon: in the following, the site giving rise to bands at higher frequency will be referred to as M$_A^+$ and the other one as M$_B^+$. Plotting the frequencies of bands due to M$_A$(CO)$^+$ as a function of $1/(R_M + R_{CO})^2$, where R$_M$ is the cation radius and R$_{CO}$ = 0.21 nm, a linear correlation is obtained (Figure 3).[13] This latter quantity being proportional to the electric field at the CO centre of mass, this fact is usually taken as evidence that electrostatics plays a basic role in the interaction.[13]

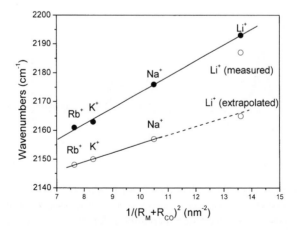

Fig. 3. C-O stretching frequency versus $1/(R_M + R_{CO})^2$ for alkali-metal exchanged ZSM-5 zeolites (M = Li, Na, K, Rb). Black symbols refer to M^+_A---CO adducts, open symbols to M^+_B---CO ones.

A similar behaviour is observed also for the B-type cations, as the values for Rb^+, K^+ and Na^+ line up. The $Li_B(CO)^+$ adduct makes exception, as the measured frequency 2187 cm^{-1} is well above that expected according to the linear correlation found with other cations (2165 cm^{-1}) sitting at M^+_B sites, whereas no band at about 2165 cm^{-1} was observed on Li-ZSM-5.[6] This indicates that Li_B^+ has a peculiar behaviour with respect to other M^+_B sites in M-ZSM-5 zeolites, which will be confirmed by other data below.

Computer fitting of the spectra in Figure 1 showed that the intensity of all bands increase in a proportional way, notwithstanding the difference in the energy of interaction. This is strong indication that we are dealing with the Henry region of all three "local" isotherms concerning species $K_A(CO)^+$, $K_A(OC)^+$ and $K_B(CO)^+$. Accordingly, the overall adsorption isotherm in Figure 4 is given by a straight line passing through the origin

$$n_a = N_{max} * K_{app} * p \tag{1}$$

where N_{max} and K_{app} are the maximum adsorbed amount (μmol g^{-1}) and the overall equilibrium constant (Torr^{-1}), respectively, the latter comprising all three modes of adsorption. In a previous work on exactly the same sample,[14] a value of N_{max} = 775 μmol g^{-1} has been found for the adsorption of CO_2: such value is not too far from that of 1000 μmol g^{-1} expected on the basis of the chemical composition of the zeolite (Si/Al = 14; K/Al = 1), and rightly below that value because of some dealumination, as shown in Figure 1. The overall equilibrium constant K_{app} results to be $1.34*10^{-3}$ Torr^{-1} (Table 1).

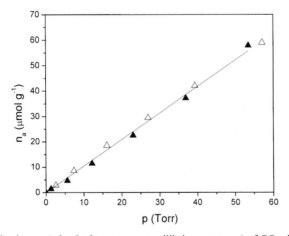

Fig. 4. Volumetric isotherms (adsorbed amount vs equilibrium pressure) of CO adsorbed on K-ZSM-5 zeolite outgassed at 723 K. Black symbols: primary isotherm; empty symbols: secondary isotherm.

Figure 5a reports the differential heat of adsorption, q_{diff} (kJ mol^{-1}), as a function of the amount of adsorbed CO on K-ZSM-5: the initial value at about 44 kJ mol^{-1} is due to adsorption of CO on extra-framework Al^{3+} species; then, q_{diff} is rather constant at about 25 kJ mol^{-1} and decreases with increasing coverage, because of the heterogeneity of K^+ centres. The value of 25 kJ mol^{-1} is close to that of 27 ± 1 kJ mol^{-1} determined by Savitz et al. on K-ZSM-5 at 195 K.[15] Note that at low temperatures, adsorption on stronger sites prevails, so that a slightly lesser value is expected.

Figure 5b compares differential heats of adsorption measured on K-ZSM-5 with those already available for Li-ZSM-5 and Na-ZSM-5 zeolites, in the same equilibrium pressure range.[6,7] On K-ZSM-5 and Li-ZSM-5 zeolites, differential heat of adsorption decreases markedly with coverage, in contrast with the Na-ZSM-5 case: for this latter system, the constancy of differential heat is ascribed to a less pronounced heterogeneity of Na^+ sites, as already discussed.

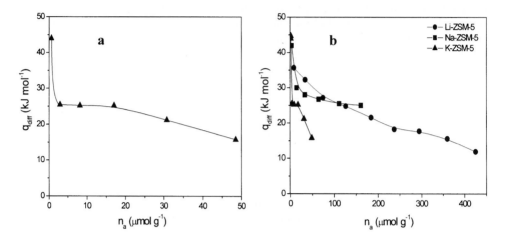

Fig. 5 Section a): differential heat of adsorption as a function of the amount of adsorbed CO on K-ZSM-5 outgassed at 723 K. Section b): comparison among differential heats of adsorption measured on Li-ZSM-5, Na-ZSM-5 and K-ZSM-5. Data for Li-ZSM-5 and Na-ZSM-5 are from ref. 6 and 7, respectively.

For about the same CO final pressure, the coverage attained is in the order Li- ZSM-5 > Na-ZSM-5 > K-ZSM-5: this is due to the strength of the interaction, which is usually assumed to be higher for cations with higher charge to radius ratio, *i.e.* Li^+ ions. Indeed, values of differential heats concerning K-ZSM-5 are always smaller than those concerning Na-ZSM-5. The case of Li-ZSM-5 is puzzling, because in the first stages of adsorption differential heats are highest (36 ± 1 kJ mol^{-1} in excellent agreement with the value of 37 ± 1 kJ mol^{-1} reported by Savitz et al.[15]), then fall to surprisingly low values. In a previous work,[6] this has been ascribed to a particularly low heat of formation of $Li_B(CO)^+$, involving an endothermic step consisting in the "extraction" of the partially sunk cation. The enthalpy of formation of the $Li_B(CO)^+$ species was estimated to be 18 kJ mol^{-1}, as contrasted with that of $Li_A(CO)^+$ adducts, 44 kJ mol^{-1}.

The average values of extinction coefficients $\langle\varepsilon\rangle$ have been calculated from the plot of normalised IR intensities as a function of adsorbed amounts, coming from volumetric measurements (Figure 6): proportionality is seen, at least at low coverage, whereas at higher adsorbed amounts deviations are seen, to be discussed elsewhere.

Linear regressions around the origin lead to values of 12, 14, and 10.5 km mol^{-1} on Li-ZSM-5, Na-ZSM-5 and K-ZSM-5, respectively. The value found on Li-ZSM-5 is smaller than that found on Na-ZSM-5, in apparent contrast with the common knowledge that absorption coefficient of adsorbed CO should increase almost linearly with frequency. Indeed, the value for K-ZSM-5 is smaller than that of Na-ZSM-5. We ascribe also this feature to a special status of the $Li_B(CO)^+$ species, which in our opinion has an abnormally low extinction coefficient. This point is of interest per se, in that it could be evidence that partially sunk cations may form carbonylic species with specific intensity which depends on the degree of protrusion of the cation: this would explain why often an erratic behaviour of extinction coefficients is observed on oxides.

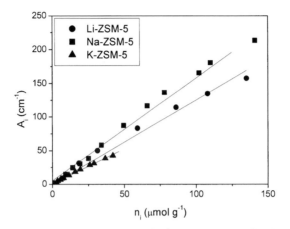

Fig. 6. Integrated intensity of the IR absorption band of M^+--CO adducts as a function of the amount of adsorbed CO.

Table 1
Spectroscopic and thermodynamic data for CO adducts on different ZSM-5 zeolites (Data for Li-ZSM-5 and Na-ZSM-5 are from refs. 6 and 7).

M-ZSM-5	Li-ZSM-5	NaZSM-5	K-ZSM-5	Rb-ZSM-5
ν of $M_A(CO)^+$ adduct (cm^{-1})	2193	2176	2163	2161
ν of $M_B(CO)^+$ adduct (cm^{-1})	2187	2157	2149	2148
ν of $M_A(OC)^+$ adduct (cm^{-1})	2112	2112	2117	2119
ε $(km\ mol^{-1})$	17.0 $(Li_A(CO)^+)$ 6.5 $(Li_B(CO)^+)$	14.0	10.5	-
K $(Torr^{-1})$	$Li_A(CO)^+$ 0.46 ± 0.001 $Li_B(CO)^+$ 1.90 ± 0.04*10^{-3}	6.04 ± 0.71*10^{-3}	1.34 ± 0.03*10^{-3}	-
ΔH^0 $(kJ\ mol^{-1})$	$Li_A(CO)^+$ 52 $Li_B(CO)^+$ 18	27	25	-

In the present case, this point has another bearing: as the enthalpies of formation of $Li_A(CO)^+$ and $Li_B(CO)^+$ have been calculated in a previous paper by assuming the same specific intensity, these values, reported above, are in error. The detailed re-calculation will be reported elsewhere: as a reasonable conclusion, it results that $\varepsilon_A = 17$ km mol^{-1}; $\varepsilon_B = 6.5$ km mol^{-1}, and accordingly $\Delta H°_A = 52$ kJ mol^{-1} and $\Delta H°_B = 18$ kJmol^{-1} (Table 1).

4. CONCLUSIONS

Joint volumetric, microcalorimetric and FT-IR measurements at room temperature allowed the determination of average values of equilibrium constant, extinction coefficient and heat of adsorption of CO on K-ZSM-5. Comparison has also been made with Li-ZSM-5 and Na-ZSM5 systems. In all cases, the occurrence of two M^+ absorbing sites has been determined: with Na-ZSM-5, such an heterogeneity of sites is less pronounced as compared to K-ZSM-5 and Li-ZSM-5. This has been ascribed to the existence of two sites in the structure, also evidenced in the case of Cu^+. The case of Li-ZSM-5 is peculiar, since two different Li^+ sites have been observed, one of them, referred to as $Li^+{}_B$, has an exceptionally low extinction coefficient and heat of adsorption: this behaviour has been explained by assuming that B-type Li^+ cations remain sunk among framework oxygen anions and that adsorption of CO involves an endothermic step, with a small movement of the cation from its equilibrium position.

REFERENCES

[1] R.P. Towsend, Properties and Applications of Zeolites, The Chemical Society, London, 1980.
[2] W.M. Meier, D.H. Olson, Atlas of Zeolite Structure Types, Butterworth-Heinemann, London, 1992.
[3] A. Zecchina, C. Otero Areán, Chem. Soc. Rev. 25 (1996) 187.
[4] A. Zecchina, S. Bordiga, C. Lamberti, G. Spoto, L. Carnelli, C. Otero Areán, J. Phys. Chem. 98 (1994) 9577.
[5] C. Otero Areán, A.A. Tsyganenko, E. Escalona Platero, E. Garrone, A. Zecchina, Angew. Chem. Int. Ed. 37 (1998) 3161.
[6] B. Bonelli, E. Garrone, B. Fubini, B. Onida, M. Rodríguez Delgado, C. Otero Areán, Phys. Chem. Chem. Phys. 5 (2003) 2900.
[7] E. Garrone, B. Fubini, B. Bonelli, B. Onida, C. Otero Areán, Phys. Chem. Chem. Phys. 1 (1999) 513.
[8] O.V. Manoilova, M. Peñarroya Mentruit, G. Turnes Palomino, A.A. Tsyganenko, C. Otero Areán, Vib. Spectroscop. 26 (2001) 107.
[9] B. Fubini, Thermochimica Acta 135 (1988) 19.
[10] C. Lamberti, S. Bordiga, M. Salvalaggio, G. Spoto, A. Zecchina, F. Geobaldo, G. Vlaic, M. Bellatreccia, J. Phys. Chem. B 101 (1997) 344.
[11] D. Nachtigallová, P. Nachtigall, M. Sierka, J. Saure, Phys. Chem. Chem. Phys. 1 (1999) 2019.
[12] C. Otero Areán, M. Peñarroya Mentruit, M. Rodríguez Delgado, G. Turnes Palomino, O.V. Manoilova, A.A. Tsyganenko, E. Garrone, Stud. Surf. Sci. and Catal. 142 (2002) 207.
[13] C. Lamberti, S. Bordiga, F. Geobaldo, A. Zecchina, C. Otero Areán, J. Chem. Phys. 103 (1995) 3158.
[14] B. Bonelli, B. Civalleri, B. Fubini, P. Ugliengo, C. Otero Areán, E. Garrone, J. Phys. Chem. B 104 (2000) 10978.
[15] S. Savitz, A.L. Myers, R.J. Gorte, Microporous Mesoporous Mater. 37 (2000) 33.

Studies in Surface Science and Catalysis 155
A. Gamba, C. Colella and S. Coluccia (Editors)

Properties of zeolitized tuff/organic matter aggregates relevant for their use in pedotechnique
I. Chemical and physical-chemical properties

M. Bucci[a], **A. Buondonno**[a], **C. Colella**[b], **E. Coppola**[a], **A.P. Leone**[c] and **M. Mammucari**[a]

[a]Dipartimento di Scienze Ambientali, Seconda Università degli Studi di Napoli,
Via Vivaldi 43, 81100 Caserta, Italy

[b]Dipartimento di Ingegneria dei Materiali e della Produzione, Università Federico II,
Piazzale V. Tecchio, 80125 Napoli, Italy

[c]CNR – ISAFoM, Via Patacca 85, 80056 Ercolano (NA), Italy

A study aiming at evaluating the technical and economic suitability of zeolitized tuffs as mineral substrate for soil re-building in pedotechnique strategies was carried out. The chemical and physical-chemical properties of aggregates ped models, obtained by interaction between tannic acid, i.e., an organic matrix precursor of humic substances, and a zeolitized material, namely Neapolitan Yellow Tuff, using Ca ions as "bridge", were investigated. Ped models were characterized according to chemical and physical standard methods for soil analysis, color analysis by visual interpretation and by spectro-radiometry, as well as by FT-IR spectrometry. The results suggest that the interaction between zeolite, Ca ions and tannic acid leads to the formation of stable aggregates with peculiar chemical and chemical-physical properties.

1. INTRODUCTION

It is well known that natural zeolites are a widely-spread class of crystalline minerals belonging to tectosilicates and structurally defined as microporous materials. Thanks to their unique properties [1, 2], i.e., (i) high ion exchange capacity, greater than that of soil clay minerals, albeit smaller than that of humic substances; (ii) large porosity, implying high water retention capacity and (iii) great surface reactivity and catalytic activity, natural zeolites are advantageously utilized for various industrial, environmental and agricultural purposes [3].

As regards the agronomic field, the traditional research topics have dealt with the application of zeolites to improve the crop production under both qualitative and quantitative aspects. Quite a lot of studies carried out over the last thirty years, published in proceeding of the most important conferences on natural zeolites [4-9], showed that both soil and plants can benefit from zeolite amendments: about thirty different crops, including vegetables and cereals, appear to take advantage from soil treatment with zeolitized materials. Such improvement of plant performance depends on the complex reactions and interactions of zeolites in the soil/plant system, which increases the global soil fertility, particularly in the rhizosphere, by optimizing the environmental soil conditions. Some of the accounting reasons for these positive effects are the following: (i) when charged with NH_4^+ or K^+, zeolites behave like slow-release N- or K- fertilizers; moreover zeolites preserve NH_4^+, added as fertilizer or produced by biochemical reactions in soil, which, once selectively retained by zeolite, is not exposed to nitrification pathway; (ii) as rigid porous media, zeolites can absorb and retain large amounts of water, thus acting as a reservoir of available water for crops; (iii) in moist soil regimes and through leaching of bases and desaturation of exchange complex, soil tends to become acid; this condition enhances the mobility of potential toxic elements, PTE, such as Al, Mn, Pb, Cd and others; natural zeolites effectively reduce the mobility of PTE by both uptake by exchange reaction and by buffering soil pH through their alkaline hydrolysis. Recent studies showed that polluted soils can be successfully restored by treatment with organo-zeolite mixtures [10-12].

Such considerations stimulated to open up new inviting aspects dealing with the use of zeolites in the applied pedology and pedotechnique fields, with special reference to reclaim and re-build damaged and degraded soils. The term "degradation" encompasses various factors and processes, either natural or anthropic, including environmental pollution, that lead to "the decline of soil quality" [13] and to "the reduction in the capacity of land to produce benefits" [14]. The ultimate consequence of soil degradation is desertification, which leads soil to loose its own capability to produce "goods and services" [15]. In the last years, both the global extension of the surfaces and the quality of the soils decreased: recent data suggest that the surface of available fertile soil per person will be reduced from 0.27 ha to less than 0.14 ha within the next 40 years [16]. Is has been calculated in the European Union that about 52 million ha of soil (more than 16% of the total surface) are stricken by forms of deterioration [17], whereas desertification is spreading in all Euro-Mediterranen countries, including southern and insular Italy [18]. Such phenomena are caused by both natural environmental factors, as well as by anthropic activities. As main natural causes of desertification with severe effects, loss of structural stability, soil erosion, and recurrent droughts have reduced in the last twenty years the quantity and quality of available agricultural lands. Among the anthropic causes of degradation, soil misuse, environmental pollution, land abandon, uncontrolled quarry mining as well as spreading of illegal activities, remain prevalent.

The purposes of the reclamation of polluted sites, the rehabilitation of soil functioning in the degraded areas, the rebuilding of soil lost by erosion or anthropic removal can be reached applying suitable pedotechnical procedures. Pedotechnique is an innovative branch of soil science, defined as "the design and realization of very specific soil (and substrate) volumes and profiles through very intensive use of advanced technology and very intensive soil manipulation" [19]. From this standpoint, we must consider that degradation induces a severe reduction of soil organic matter and soil structural stability, with subsequent decrease of surface reactivity and soil exchange activity. The soil exchange activity is due to the exchange complex formed by stable organo-mineral aggregates [20], which govern the fundamental physical-chemical reactions in soil, such as mass-energy exchanges between soil and environment, sorption/desorption of water and nutrient or toxic substances, and ion exchange [21, 22].

With the purpose to recover degraded soils and to control the mobility and toxicity of the contaminants, a convenient and applicable pedotechnique strategy is that of recovering the functioning of the exchange complex using materials with high surface activity. It is therefore appropriate to look at the "co-utilization" [23] of mineral and organic by-products for the amendment and the rehabilitation of polluted soils, as well as for the preparation of "proto-horizons" to rebuild degraded soils. Natural or recycled organic matrices are characterized by high exchange capacity, but in the absence of mineral phases with aggregating and stabilizing power, they tend to degrade in time, thus reducing their activity. It is fundamental then to couple organic matrices to mineral phases with suitable mineralogical and chemical-physical features, able to undergo natural pedogenic processes. Zeolites, being characterized by a large surface area and a statistically uniform surface charge distribution may be worth candidates as support for organo-mineral aggregates formation.

On these bases, a research program was started aiming at evaluating the possible use of zeolitic materials and organic waste matrices in strategies for soil restoration. In fact, tuffaceous materials meet two fundamental requirements for their possible use in pedotechnique: (i) they are likely susceptible to be pedogenized, and (ii) they are admissible as soil amendment according to the Italian and European legislation. Indeed, zeolites frequently occur in pyroclastic rocks, such as zeolitized tuffs. It is well known that zeolitized rocks are widely spread in large areas of earth surface and represent the pedogenic substrate for various soils; furthermore, the zeolitic quarry spoil is registered in the European Waste Catalogue and it is recognized by the Italian law as "materials utilizable for the environmental restoration". It is also interesting to note that the World Reference Base for Soil Resources (WRBSR) from FAO [24] identifies as "diagnostic materials" of the "anthropogenic soils", i.e., the soils "made" by the human action, the so-called "anthropogeomorphic soil materials", which refer to "unconsolidated mineral or organic material ... produced by human activities". In particular, the term "spolic anthropogeomorphic materials" refers to "earthy material resulting from industrial activities, ... mine spoil, river dredging, highway constructions, ...". The WRBSR then points out and recognizes that such materials can be the constituents of "anthropogenic" soils, i.e., those soils, the nature and evolution of which are primarily determined by human action.

Preliminary laboratory investigation [25] showed that organo-zeolite mixtures can evolve towards re-organized systems in which the organic matter is more stable against spontaneous air-oxidation, thus suggesting that an intimate interaction between zeolitized materials and organic matter occurred as a symptom of an early-stage pedogenesis. It is widely recognized that pedogenesis act via an almost countless number of processes. On the whole they can be grouped as chemical, physical or biological, but without a clear-cut division. The aggregation is the primary physical process whereby a number of particles are held or brought together to form soil structure units known as *peds* [26]. The ped formation is a crucial process of the pedogenesis [27], that take place by the formation of soil colloid associations, also referred to as conglomerates or organo-mineral complexes. These are formed by mutual interaction among organic and mineral phases and cations [28-30], and can be represented as $[(C-P-SOM)_x]_y$ [30], where C indicates clay fraction, P a polyvalent metal ion (Ca^{2+}, Mg^{2+}, Fe^{3+}, Al^{3+}, etc.) acting as bridge-cation, and SOM is the humidified soil organic matter; x and y are finite numbers dictated by the size of the primary particle. From the considerations above, it appears that studies on the mechanisms for ped aggregation and expression represent a fundamental tool not only to elucidate the pedogenic processes, but also to better understand the phenomena of soil degradation related to the loss of structure stability.

Indeed, the long time needed for the pedogenesis makes practically impossible to observe the ped formation *in situ*. A good approach attempting to understand and clarify the complex phenomena of ped evolution is to model conglomerates from mineral phase, bridge-cations, humus components [28, 31, 32]. This approach would be relevant since detailed knowledge is available on the nature and amount of each component, thus allowing to discriminate their peculiar effects on the final properties of the produced models.

The present paper deals with the chemical and physical-chemical properties of aggregates formed by interaction of a zeolitized tuff and a polyphenol representative of low-molecular weight humic substances. To prevent zeolites alteration, zeolitized tuff and polyphenol were allowed to react in a neutral-subalkaline suspension by using Ca^{2+} as bridge-cation. Aggregates were characterized through the following analyses: oxidable organic carbon, water-extractable polyphenols, color, Fourier-transformed infrared spectrometry (FTIR).

2. MATERIALS AND METHODS

2.1. Starting materials

Starting materials for the preparation of aggregate models were:
(i) Neapolitan Yellow Tuff, NYT, from Marano, (Napoli, Italy); this material represents the more recent (12,000 a b.p.) tuffaceous formations of the Phlegraean Fields (Napoli, Italy), covering an area of about 13 km^2 [33]. The NYT sample used for the present investigation, with the following chemical composition (wt %): SiO_2 = 52.15, Al_2O_3 = 18.56, Fe_2O_3 = 0.20, MgO = 0.20, CaO = 2.35, K_2O = 7.54, Na_2O = 3.30, P_2O_5 = 0.11, H_2O = 15.73, had a cation exchange capacity (CEC), measured by the ammonium acetate method [34], equal to 2.12 meq g^{-1}, and a zeolite content, estimated by quantitative X-ray diffractometry, using the Reference Intensity Ratio procedure [35], equal to 58%, with phillipsite = 44%, chabazite = 4% and analcime = 10%. Smectite was also present (10%).
(ii) tannic acid, AT, a soluble polyphenol ($C_{76}H_{52}O_{46}$, FW 1.702 kD, Fluka);
(iii) calcium ion, as $Ca(OH)_2$ (FW 74.09 D, Sigma) in a 0.5 M suspension.

2.1.1. Conditioning NYT with Ca and preparation of NYT suspension

NYT was ground to pass a 0.200 mm sieve. To avoid any relevant interference by other exchangeable cations present in the tuffaceous material, NYT was Ca-conditioned as it follows. Suitable amounts of NYT were suspended in $CaCl_2$ 1 M at the ratio w/v = 1:10 in stoppered polyethylene bottles; the suspension was end-over-end shaken at 120 opm for 3 h, centrifuged at 3500 rpm; the supernatant was discharged and replaced by an equivalent volume of $CaCl_2$ 1 M. The conditioning treatment was repeated three times a day for three consecutive days, allowing NYT to react with $CaCl_2$ overnight. After the last treatment, NYT was washed with bi-distilled water (w/v = 1:10) until Cl free, centrifuged and dried in an air-stream oven at 40 °C.

2.1.2. AT solution

5, 10 or 20 g of AT were dissolved in 50, 100, or 200 ml of a H_2O/CH_3OH solution (v:v = 5:1); these solutions will be indicated as AT5, AT10, or AT20, respectively.

2.2. Aggregate models

Three aggregate models, with different NYT/AT weight ratio, i.e., 0.5, 1, or 2 w/w, but with a constant Ca/AT molar ratio, i.e., 3.2 mol/mol to obtain a neutral-subalkaline reaction environment (7.2 ≤ pH ≤ 7.5), were prepared as it follows. 10 g NYT were placed in a 500 ml large-neck polyethylene bottle containing 50 ml bi-distilled water under vigorous continuous stirring. AT5, AT10, or AT20 solutions were slowly added with a simultaneous drop by drop addition of 18.75, 37.50, or 75.00 ml $Ca(OH)_2$ 0.5 M suspensions. The obtained suspensions were then stoppered, end-over-end shaken at 120 opm for 24 h, measured for pH and electrical conductiviy (EC), quantitatively transferred in glass crystallizers, and then dried in an air-stream oven at 40°C. The following aggregates, with the relevant initial compositions, were obtained:
(i) NYT-Ca-AT5, containing 10 g NYT, 5 g ≡ 2.94 mmol AT, 0.94 g ≡ 9.37 mmol Ca ≡ 18.75 ml $Ca(OH)_2$ 0.5 M;

(ii) NYT-Ca-AT10, containing 10 g NYT, 10 g $\equiv$ 5.88 mmol AT, 1.88 g $\equiv$ 18.75 mmol
   Ca $\equiv$ 37.50 ml Ca(OH)$_2$ 0.5 M;
(iii) NYT-Ca-AT20, containing 10 g NYT, 20 g $\equiv$ 11.76 mmol AT, 3.86 g $\equiv$ 37.50 mmol
   Ca $\equiv$ 75.00 ml Ca(OH)$_2$ 0.5 M.

2.3. Reference standards

Reference standards represented by single component, as AT-a or Ca-a, or by binary models, as NYT-Ca, NYT-AT, or Ca-AT were prepared according to the above described procedure utilized to prepare the NYT-Ca-AT10 (see sub-section 2.2.). The suffix –a indicates that such standards underwent the aggregation procedure, being distinguished by the starting materials, NYT-p, AT-p and Ca-p (as Ca(OH)$_2$ pure chemical solid phase). Also pure calcium carbonate, CaCO$_3$-p (Sigma), was considered as a reference standard, since a part of calcium initially added as hydroxide transformed into carbonate.

2.4. Characterization of aggregates

Oxidable organic carbon was determined according to Walkley-Black method [36], oxidizing C by an excess of K$_2$Cr$_2$O$_7$ in solution in the presence of concentrated H$_2$SO$_4$; after 30 min reaction and rapid cooling at room temperature, residual dichromate was back-titrated with a FeSO$_4$ solution.

Water-extractable polyphenol (WPP) were determined according to Lowe procedure [37], by measuring the absorbance at 750 nm after 1 h reaction of water-extract with Folin-Ciocalteau reactant. A Shimadzu 1601 spectrophotometer was utilized.

Color was determined on dry, solid samples by analyzing the bi-directional reflectance response [38] in the visible range 380-770 nm through a spectro-radiometer ASD FieldSpec Pro 350-2500. After data processing, color was expressed as conventional Munsell [39] attributes "hue", "value" and "chroma", on the basis of a standardized YR hue number.

FT-IR spectroscopy analyses were carried out on 100 mg KBr disk containing 0.5 mg sample by a Perkin Elmer 1720 apparatus.

Each model or reference standard was prepared in triplicate. All data are expressed on a 105°C dry weight basis.

3. RESULTS AND DISCUSSION

3.1. Organic carbon (OC) and water-soluble polyphenol (WPP)

Table 1 shows the amount of oxidable organic carbon (OC) and water soluble polyphenol determined in the aggregates and in reference standard models.

Table 1
Organic carbon (OC) and water-soluble polyphenol (WPP) in the aggregates and in reference standard models (g kg^{-1})

Model	OC	WPP
AT-a	450.75	924.23
NYT-AT	191.08	146.35
AT-Ca	368.33	273.01
NYT-Ca-AT5	107.00	12.65
NYT-Ca-AT10	157.10	17.36
NYT-Ca-AT20	224.66	66.75

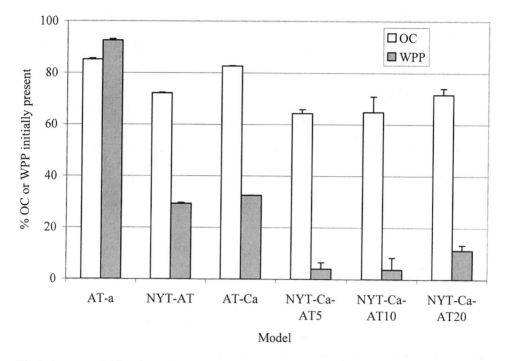

Fig. 1. Amounts of OC and WPP in the reference and aggregate models expressed as percentage of the respective amount initially present in the model. Bar on the histogram indicates standard error.

It is evident that the larger amounts of both OC and WPP were determined for the single AT-a phase and for the binary model AT-Ca and NYT-AT, whereas the opposite was true for the ternary aggregate models, especially with reference to the WPP. This could be a clue for an effective aggregation between NYT and polyphenol via Ca ions-bridge, but it could also depend on the different amount of organic carbon, viz polyphenol, initially present in the various models. Thus, the experimental data are more appropriately read as the percentage of the respective amount of OC or polyphenol initially present in each model (Fig. 1). For calculation, we took into account that the OC and soluble polyphenol content as experimentally determined in AT-p is 529.78 and 999.40 g kg^{-1}, respectively; this data feel the effect of the ordinary impurity present in the tannic acid, which is a natural gallotannin.

The comparative examination of the histograms in Fig. 1 clearly suggests that the experimental aggregation process resulted in an effective and significant reduction of both oxidable carbon and extractable polyphenol in all models, including the "aggregate" tannic acid, likely because a spontaneous polymerization with re-arrangement of phenolic –OH partially occurred. As a general tendency, a significant decrease of the fraction of still oxidable carbon was observed from binary models NYT-AT and AT-Ca (72 - 82 %) to the ternary aggregate models (64 - 72 %). Indeed, the lowest % OC was observed for NYT-Ca-AT5 and –AT10 samples (64 - 65 %). This trend is confirmed by the variability of % WPP values, which showed much more wide differences among samples: from 93% for AT-a, to 29 - 32 % for the binary models, down to 4 - 11% for the ternary aggregates. Also in this case, the larger reduction was observed for –AT5 and –AT10 samples (~ 4%). In previous studies, the reduction of susceptibility of organic matter to be oxidized was assumed as a hint of aggregation [31, 32, 40].

On the whole, the present data show that the organic matter is significantly and substantially protected against both chemical oxidation and water elution: therefore, this can be regarded as a result of a quite stable re-arrangement and organization between discrete tuff particles and colloidal organic phase.

3.2. Color

It is well known that soil color is mainly determined by the dyed component with the highest colloidal behavior, e.g., organic matter, iron or manganese oxi-hydroxides, clay, etc., even if present in minor amounts. Soil color has long been the best-known feature of soil in early pedological studies, but it is still one of the indispensable properties of soil for its classification or composition study.

Soil color is traditionally evaluated by visual comparison of soil sample with standard soil color charts according to Munsell system [39]. Such classification takes into account three main attributes which determine the global optical perception of the color: "hue", "value" and "chroma".

The first refers to the dominant dye in therm of visible optical spectrum, and it is defined by the initial of the fundamental colors red, yellow, green, blue, etc. The five major dyes, R, Y, G, B, P, and the respective in-between combinations YR, GY, BG, PB, and RP are arranged on a loop, and each dye is divided by decimal system from 0 to 10. For instance, 5R represents pure red, 5YR a 1:1 yellow and red mixture, and 10YR a 1:1 mixture red and yellow-red. The "value" identifies the "neutral" component of color, and consists of numbers from 0 for absolute black, to 10 for absolute white. The "chroma" represents the relative purity or strength of the spectral color, i.e., the relative dominance of a hue over a given value. Chroma is numbered starting from 0, so that it gradually increases as the color vividness increases.

According to the Munsell system, the complete description of color of a given material is provided by:

x-"hue" "y-value"/"z-chroma",

where x, y, and z represents the specific number of hue, value, and chroma, respectively; so, 5YR 7/8 is read "five YR seven-slash-eight", and indicates "reddish yellow", while 5YR 3/2 is read "five YR three-slash-two", and indicates dark reddish brown.

Indeed, the visual estimate of color suffers for some approximation, prevailingly due to the source and incidence angle of light, to the poor discriminatory power of human eye, and to observer subjectivity. The recent application of spectro-radiometry for soil color evaluation hugely improved the accuracy and sensibility of measurements [41].

Table 2 reports the Munsell color attributes of reference standard and aggregate models. The Y hue was standardized to YR as $YR_{std} = 10+nY$, where n is the hue number. Each color spectro-radiometrically determined was conveniently classified according to Munsell's conventional color names by approximating the description to the nearest dip in the standard Munsell's color charts.

The color of the starting materials utilized could be classified as follows (Table 2). Light yellowish brown for NYT and bright yellowish brown for AT. Both color have a pale gray neutral component, with a value > 6.0, and are almost vivid with a well-saturated chroma > 4.0. The combination of hue, value and chroma for Ca-p is not considered in the Munsell classification, since value number > 8.0 are not detected in natural soils. Furthermore, the low chroma number, less then 2.00, indicates that the spectral dominant is very scarcely represented in the final color. A convenient description for Ca-p color could be light grayish white; indeed, Ca-p appears visually as sleet-white. On the whole, starting materials exhibit bright colors with a neutral component ranging from light gray to near-absolute white.

Table 2
Munsell color attributes of aggregates and standard reference models

Model*	Hue	YR$_{std}$ hue	Value	Chroma	Color name
NYT-s (*e*)	1.26Y	11.26YR	6.46	4.22	light yellowish brown
Ca-s (*e*)	7.79YR	7.79YR	9.56	1.88	light grayish white
AT-s (*e*)	0.22Y	10.22YR	6.74	5.34	bright yellowish brown
NYT-Ca (*e*)	1.02Y	11.02YR	6.91	3.90	very pale brown
NYT-AT (*e*)	3.51YR	3.51YR	2.08	0.58	dark grayish brown
AT-Ca (*e*)	0.41Y	10.41YR	2.18	1.26	dark grayish brown
NYT-Ca-AT5 (*v*)	10.72YR	10.72YR	6.76	4.43	very pale brown
NYT-Ca-AT10 (*v*)	10.48YR	10.48YR	6.85	4.53	very pale brown
NYT-Ca-AT20 (*v*)	10.24YR	10.24YR	6.96	4.63	very pale brown
NYT-Ca-AT5 (*e*)	9.46YR	9.46YR	1.87	1.73	dark grayish brown
NYT-Ca-AT10 (*e*)	9.67YR	9.67YR	1.80	1.67	dark grayish brown
NYT-Ca-AT20 (*e*)	10.21YR	10.21YR	1.67	1.17	dark grayish brown

* (*e*): experimentally determined; (*v*): "virtual" color expected as a weighed mean of the respective hue, value and chroma numbers of individual components

With respect to binary models, NYT-Ca exhibits a very pale brown color, quite similar to the starting materials for this model. On the contrary, NYT-AT and AT-Ca models are characterized by a much less bright and rich color, and are both classified as dark grayish brown. This could be not surprising, since it is well known that natural tannin in solution spontaneously undergo to browning polymerization; furthermore, the high colloidal activity of tannic acid bestows its own color to the models. This is likely accounted for an intimate surface interaction between starting materials, which should forecast an effective aggregation in ternary models. With reference to these last, let us consider that if the starting materials were not able to undergo aggregation, the ternary models should behave as simple physical mixtures, and then their final color should merely result as a weighed mean of the respective hue, value and chroma numbers of starting materials. Therefore, if aggregation did not eventually occur, the virtual color ("*v*" in Table 2) of the ternary models should be very pale brown, *i.e.* again a bright, chroma-saturated color with light gray neutral component. On the contrary, the color of ternary models experimentally determined by spectro-radiometry ("*e*" in Table 2) is dark grayish brown, practically black, for all of them. In fact, with respect to the virtual calculated color, the spectro-radiometrically determined color is characterized by a hue number similar to the virtual one, i.e., around 10YR, whereas value and chroma are much more lower, less than 2.00. This means that the real color of the ternary models has a near-black neutral component, which prevails on the spectral dominant. It is interesting to remark that, in virtual color, both value and chroma tend to increase along with the organic matter content, whereas the opposite is true for the real color numbers. Also is worthy of note that the absence of Ca in the NYT-AT model dramatically shifted the hue of the starting component from around 10YR to 3.5YR; on the contrary, in the ternary models, the calcium ions, albeit quantitatively much less abundant with respect to both NYT and AT, significantly contributes to "buffer" the real hue number again at a value around 10YR. Thus, on the basis of the spectral radiometric behavior of the ternary models with respect to the starting materials and the binary models, we could infer that an effective aggregating interaction between components occurred.

3.3. FTIR Spectra

Figs. 2 and 3 report the FTIR spectra of starting and reference materials, and of experimental models. In order to evidence distinction among samples, FTIR patterns were zoomed between 2000 and 370 cm^{-1}. Peaks around 3400-3700 (not shown), corresponding to vibration of water –OH, were detected in all spectra.

Diagnostic peaks for pure NYT (Fig. 2a) were at 1031, 720 and 435 cm^{-1}, respectively corresponding to anti-symmetrical and symmetrical stretching of Si-O-Si and vibration of Si-O bonds. Peak at 589.5 cm^{-1} could be diagnostic of albite, likely present as an ancillary phase in the tuff matrix. Pure CaCO$_3$ (Fig. 2a) showed distinct peaks at 1468, 873, and 713 cm^{-1}; such signals are also evident in Ca-a reference standard, clearly suggesting the formation of secondary calcite, and are still present, albeit slightly modified and less intense, in the NYT-Ca model.

The spectrum of pure tannic acid (Fig. 2b) is characterized by an intense absorption band at 3396 cm^{-1} (not shown), corresponding to –OH stretching, followed by a family of distinct, almost sharp peaks at 1715 cm^{-1} (carbonyl –C=O), 1614 cm^{-1} (aromatic nuclei vibrations), 1448 cm^{-1} (aliphatic C-H stretching). Other peaks more specifically diagnostic for tannic acid were at 1321 cm^{-1} (phenolic C–OH stretching), 1202 cm^{-1} (C-O stretching and carboxyl –OH deformation) and 1087 and 1031 cm^{-1} (C–O stretching of glucose present in the tannic acid). Other accessory peaks were observed at 871 and 759 cm^{-1}. All peaks of pure tannic acid, although with some frequency shift and variation in intensity, were detected in the reference binary models At-Ca and NYT-AT. The former showed a shift of phenolic 1321 cm^{-1} peak to 1354 cm^{-1}, and a weakening of glucose 1087 cm^{-1} peak, also shifted to 1098 cm^{-1}.

Such variation could be a clue for a partial interaction between tannic acid and Ca ions. Analogously, NYT-AT (Fig. 2b) showed a weakening of 1321 cm^{-1} peak with a broad shift to 1350 cm^{-1}, whereas the 1087 cm^{-1} peak practically disappeared. Also in this case, the peak modification is likely due to an interaction between NYT and tannic acid; in fact, as recently showed by Capasso and co-workers [42], zeolitized tuff are able to bind humic acids, of which tannic acid is a representative molecule.

Fig. 3a shows the "virtual" ("v") patterns of the ternary models, obtained as a weighed mean of signals of the individual components. As previously observed with reference to color, the "virtual" FTIR spectra should represent those expected in the absence of any actual interaction among the components. The "v" spectra visibly present most of peaks diagnostic of each component, particularly those of tannic acid and, in a lesser extent, those of NYT. Peaks of calcitic phase, which contributed with the minimum weight to the averaged virtual pattern, were less distinct.

Fig. 3b shows the experimental FTIR spectra of ternary models. By comparing such spectra with those of the individual reference standards, the various binary models, and the "virtual" ternary models (Figs. 2 a, b, 3a), several significant modifications are noted. All spectra lost the characteristic peak of tannic acid at 1715 cm^{-1}. In the NYT-Ca-AT20, bands at 1535 cm^{-1} and 1321 cm^{-1} are shifted to 1512 cm^{-1} and 1368 cm^{-1}, respectively, whereas the peak at 1087 cm^{-1}, corresponding to the stretching of the C-O of glucose present in tannic acid, is almost totally smoothed away.

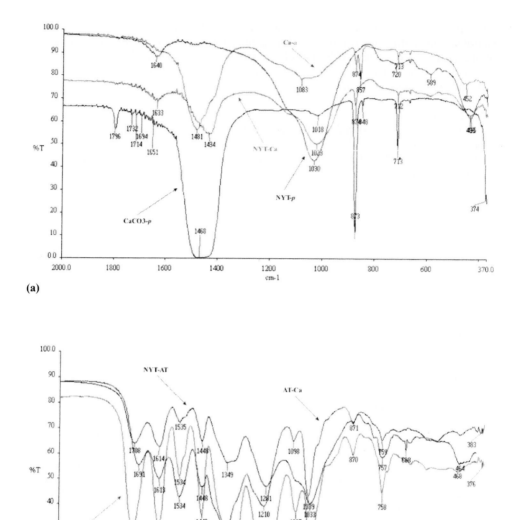

(a)

(b)

Fig. 2. (a) FTIR spectra for NYT-p, Ca-a, CaCO$_3$-p, NYT-Ca, and, (b), FTIR spectra for AT-p, AT-Ca, NYT-AT.

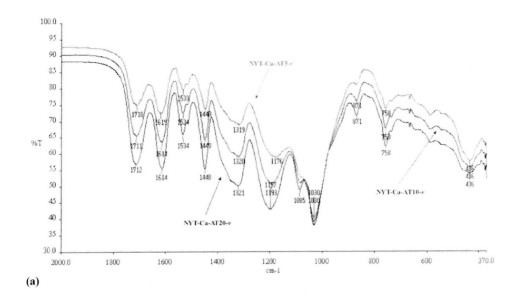

(a)

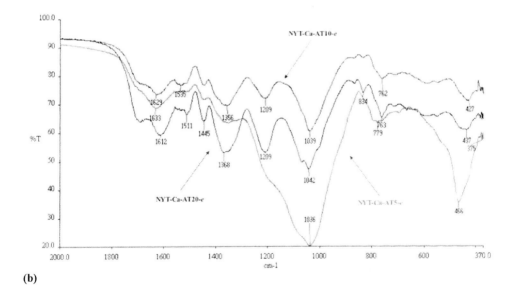

(b)

Fig. 3. (a) FTIR spectra for "virtual" ("*v*") ternary models, and, (b), FTIR spectra for experimental ternary models ("*e*").

114

Also the –OH vibration at 871 cm^{-1} disappears in the NYT-Ca-AT10 model. In the NYT-Ca-AT5 model, peaks at 1535, 1448, 1321, 1203, and 1087 cm^{-1} are completely lost, whereas the aromatic C=C band is shifted to 1633 cm^{-1}. The polysaccharide-C-O stretching moves from 1043 cm^{-1} in the NYT-Ca-AT20 model to 1037 cm^{-1} in the NYT-Ca-AT5 one. On the whole, NYT-Ca-AT5 and NYT-Ca-AT10 showed the FTIR patterns the most dissimilar from the respective "virtual". This suggests that these models underwent a significant chemical-physical reorganization.

4. CONCLUSIONS

The present investigation allowed to produce soil ped models with peculiar chemical and physical-chemical properties. The collected data showed that the interaction between zeolitized tuff, Ca ions and tannic acid led to the formation of aggregates which were substantially stable against chemical degradative treatments. Besides, both infrared and radiometric spectra clearly indicate that solid (NYT), soluble (Ca ions) and colloidal (AT) phases intimately interacted involving chemical and physical-chemical reactions. In particular, ternary models with NYT/AT ratio = 2 or 1 g g^{-1} showed the more effective aggregation. Therefore, organo-zeolite mixtures appear as a promising material suitable to set-up stable proto-horizons for soil rebuilding in pedotechnique strategies.

Taking into account the complexity of factors and processes leading to the soil ped formation, future research steps will be devoted to the comparative analysis of mineralogical and morphological features of NYT-Ca-AT aggregates, also investigating their structural stability and properties of aggregates formed in acidic *milieu* in the presence of Al or Fe as bridge cations.

REFERENCES

[1] D.W. Breck, Zeolite Molecular Sieves, John Wiley & Sons, New York, 1974.
[2] A. Dyer, An introduction to zeolite molecular sieves, John Wiley & Sons, Chichester, U.K., 1988.
[3] C. Colella, in Porous Materials in Environmental Friendly Processes, I. Kiricsi, G. Pál-Borbély, J.B. Nagy and H.G. Karge (eds.), Studies in Surface Science and Catalysis, Vol. 125, Elsevier, Amsterdam, 1999, p. 641.
[4] L.B. Sand and F.A. Mumpton (eds.), Natural Zeolites. Occurence, Properties and Use, Pergamon Press, Elmsford, N.Y., 1978.
[5] W.G. Pond and F.A. Mumpton (eds.), Zeo-Agriculture. Use of Natural Zeolites in Agriculture and Aquaculture, Westview Press, Boulder, Colorado, 1984.
[6] D. Kallò and H.S. Sherry (eds.), Occurrence, Properties and Utilization of Natural Zeolites, Akadémiai Kiadò, Budapest, 1988.
[7] D.W. Ming and F.A. Mumpton (eds.), Natural Zeolites '93. Occurrence, Properties, Use, Int. Committee on Natural Zeolites, Brockport, N.Y., 1995.
[8] G. Kirov, L. Filizova and O. Petrov (eds.), Natural Zeolites-Sofia '95, Pensoft Publ., Sofia, Bulgaria, 1997.
[9] C. Colella and F.A. Mumpton (eds.), Natural Zeolites for the Third Millennium, De Frede Editore, Napoli, Italy, 2000.
[10] P. J. Leggo and B. Ledesert, Miner. Mag., 65 (2001) 563.
[11] E. Coppola, G. Battaglia, M. Bucci, D. Ceglie, A. Colella, A. Langella, A. Buondonno, C. Colella, in Impact of Zeolites and Other Porous Materials on the New Technologies at the Beginning of the New Millennium, R. Aiello, G. Giordano and F. Testa (eds.), Studies in Surface Science and Catalysis, Vol. 142B, Elsevier, Amsterdam, 2002, p. 1759.

[12] E. Coppola, G. Battaglia, M. Bucci, D. Ceglie, A. Colella, A. Langella, A. Buondonno and C. Colella, Clays and Clay Minerals, 51/6 (2003) 609.

[13] C.J. Barrow, Land degradation, Cambridge University Press, Cambridge, U.K., 1991.

[14] K.G. Steiner, Causes of soil degradation and development approaches to sustainable soil management, Margraf Verlag Pub., Weikersheim, Germany, 1996.

[15] J. Boardman, I.D.L. Foster and J.A. Dearing, Soil erosion on agricultural land, John Wiley and Sons, New York, 1996.

[16] R. Horn and T. Baumgartl, in Handbook of Soil Science, M. E. Sumner (ed.), CRC Press, Boca Raton, FL, USA, 2000, p. A/19.

[17] CEC – Commission of the European Communities, Towards a thematic strategy for soil protection, Communication from the Commission to the Council, the European Parliament, the Economic and Social Committee and the Committee of the Regions, COM 179, 16.04.2002, 2002.

[18] P. Mairota, J.B. Thornes and N. Geeson (eds.), Atlas of Mediterranean Environments in Europe. The desertification context, Wiley & Sons, Chichester, U.K., 1998.

[19] A.J. Koolen and J.P. Rossignol, Soil Tillage Res., 47 (1998) 151.

[20] S. Goldberg, I. Lebron and D.L. Suarez, in Handbook of Soil Science, M. E. Sumner (ed.), CRC Press, Boca Raton, FL, USA, 2000, p. C/195.

[21] D.L. Sparks, in Soil physical chemistry, D.L. Sparks (ed.), 2nd ed., CRC Press, Boca Raton, FL, USA, 1998, 135.

[22] J.M. Skopp, in Handbook of Soil Science, M. E. Sumner (ed.), CRC Press, Boca Raton, FL, USA, 2000, p. A/3.

[23] S. Brown, J. Scott Angle and L. Jacobs Beneficial, Co-Utilization of agricultural, municipal and industrial by-products, Kluwer Academic Pub., Dordrecht, TheNetherlands, 1998.

[24] FAO-ISRIC-ISSS, World Reference Base for Soil Resources, World Soil Resources Report 84, Food and Agriculture Organization, Rome, Italy,1998.

[25] A. Buondonno, E. Coppola, M. Bucci, G. Battaglia, A. Colella, A. Langella, C. Colella, Studies in Impact of Zeolites and Other Porous Materials on the New Technologies at the Beginning of the New Millennium, R. Aiello, G. Giordano and F. Testa (eds.), Studies in Surface Science and Catalysis, Vol. 142B, Elsevier, Amsterdam, 2002, p. 1751.

[26] E.A. FitzPatrick, Soils. Their formation, classification and distribution, Longman, New York, 1983.

[27] O.A. Chadwick and R.C. Graham, in Handbook of Soil Science, M. E. Sumner (ed.), CRC Press, Boca Raton, FL, USA, 2000, p. A/41.

[28] M.H.B. Hayes and F.L. Himes, in Interactions of soil minerals with natural organics and microbes, P.M. Huang and M. Schnitzer (eds.), SSSA Sp. Publ. No. 17, Soil Sci. Soc. Am. Ed., Madison, WI, USA, 1986, p. 103.

[29] J.A. McKeague, M.V. Cheshire, F. Andreux and J. Berthelin, in Interactions of soil minerals with natural organics and microbes, P.M. Huang and M. Schnitzer (eds.), SSSA Sp. Publ. No. 17, Soil Sci. Soc. Am. Ed., Madison, WI, USA, 1986, p. 549.

[30] F.J. Stevenson, Humus chemistry. Genesis, composition, reactions, J. Wiley & Sons, New York, 1982.

[31] A. Buondonno, M.L. Ambrosino, E. Coppola, D. Felleca, F. Palmieri, P. Piazzolla and A. De Stradis, in Humic substances in the global environment and implications on human health, N. Senesi and T.M. Miano (eds.), Elsevier, Amsterdam, 1994, p. 1101.

[32] A. Buondonno and E. Coppola, in Oxide-Based Systems at the Crossroads of Chemistry, A. Gamba, C. Colella and S. Coluccia (eds.), Studies in Surface Science and Catalysis, Vol. 140, Elsevier, Amsterdam, 2001, p. 87.

[33] A. Buondonno, C. Colella, E. Coppola, M. de' Gennaro and A. Langella, in: Natural Zeolites for the Third Millennium, C. Colella and F.A. Mumpton (eds.), De Frede, Napoli, Italy, 2000, p. 449.

[34] H.D. Chapman, in Methods of Soil Analysis, Agronomy Series No. 9, Part 2, A. Black (ed.), Am. Inst. of Agronomy , Madison, Wisconsin, 1965, p. 891.

[35] S.J. Chipera and D.L. Bish, Powder Diffr., 10(1) (1995) 47.

116

[36] Mi.P.A.F. (Ministero per le Politiche Agricole e Forestali, Osservatorio Nazionale Pedologico e per la Qualità del Suolo) Metodi di Analisi Chimica del Suolo, No. 1124.2, FrancoAngeli Ed., Milano, Italy, 2000.

[37] L.E. Lowe, in Soil Sampling and Methods of Analysis, Canadian Society of Soil Science, Lewis Publishers, Boca Raton, FL, USA, 1993, p. 409.

[38] R.N. Fernandez and D.G. Schulze, Soil Science Soc. Amer. J., 51 (1987) 1277

[39] Munsell Color Company, Munsell Soil Color Charts, Macbeth Division of Kollmorgen, Baltimore, MD, 1975.

[40] A. Buondonno, D. Felleca and A. Violante, Clays and Clay Minerals, 37/3 (1989) 235.

[41] A.P. Leone, A. Ajmar, R. Escadafal and S. Sommer, Boll. Soc. Ital. Scienza Suolo, 8 (1996) 135.

[42] S. Capasso, S. Salvestrini, E. Coppola, A. Buondonno and C. Colella, Appl. Clay Sci., 28 (2005) 159.

Studies in Surface Science and Catalysis 155
A. Gamba, C. Colella and S. Coluccia (Editors)

Zeolites occurrence in Miocene pyroclastites of Slănic-Prahova (Romania)

L.G. Calotescu[a], V. Boero[a] and M. Angela-Franchini[b]

[a]Università di Torino, DIVAPRA, Chimica Agraria,
Via Leonardo da Vinci 44, 10095 Torino, Italy

[b]Dipartimento di Scienze Mineralogiche e Metrologiche, Università di Torino,
Via Velperga Caluso 35-37, 10125 Torino, Italy

The pyroclastic deposit of Slănic-Prahova (Eastern Carpathes, Romania), having a rhyolitic-rhyodacitic composition, has been analyzed using XRD, XRF, SEM and EDS techniques. The volcanic glass appeared entirely transformed. The secondary mineral assemblage including Ca-clinoptilolite (dominant authigenic phase), smectite, illite, kaolinite, opal-CT, authigenic feldspar, chabazite might indicate a transformation of the volcanic glass in a closed, Ca-rich system.

1. INTRODUCTION

Volcanic glass, a very unstable phase at surface temperature and pressure reacts with water of the depositional systems. The transformation assemblage is easily recognizable and commonly includes zeolites, clay minerals, silica phases and carbonates. The mineral species of zeolites and clay minerals depend on the system conditions (pressure, temperature, silica and cation activities, solid/liquid ratio, etc.).

Clinoptilolite is a widespread zeolite, commonly related to acidic volcanic glass transformed into different hydrological systems. It may be associated with other zeolites, which indicate particular conditions of the transformation process.

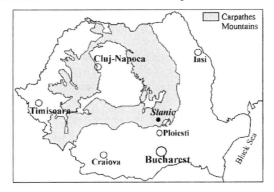

Fig. 1. Location of Slănic-Prahova zeolite deposits in Romania

The aim of this study is to characterize the mineral assemblage of the volcanic tuff from Slănic (Romania) and to propose a genetic model of transformation.

2. LOCATION, GEOLOGY AND PETROGRAPHIC FEATURES

The pyroclastic deposit from Slănic-Prahova (Romania), known as "the tuff with Globigerines" is located in the curvature zone of Eastern Carpathes, within the Tarcău Unit (Fig. 1). The Slănic tuff lies in a synclinal structure (10.5 km long and 1.5-4.5 km large), the stratum being 50-150 m thick, between 1400 m and 1550 m depth (Fig. 2.).

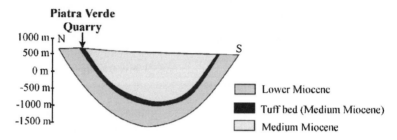

Fig. 2. Transversal section of the Slănic-Prahova deposits

The actual synclinal filling consists of Medium Miocene deposits (salts and breccia, shales with Radiolars and marls with Spirialis) with a maximum thickness of 1400 m. This region has undergone a complex tectonical evolution, which resulted in an almost vertical position of the synclinal borders (inclination ~60°).

The greenish, arenitic to lutitic pyroclastites have a rhyolitic to rhyodacitic composition according the Zr/TiO_2-Nb/Y diagram [1]. No potassium feldspar fenocrystals occur. Biotites have Mg/(Mg+Fe) ratios <0.42, suggesting an acid magma [2]. Plagioclases show a range from 22.5 to 39.4% An, typical for an acid magma. Only a spare crystal from one sample turned out to be a more basic plagioclase with 49% An. The vitroclastic texture is preserved in some samples and appears entirely transformed at microscopic scale. In other samples the glass is re-crystallized and the volcanic texture is cancelled. A minor epiclastic material is associated with the volcanic one either in the same layer or in alternation. Some observations on the glass shards morphology might indicate a subaerian explosion [3]. As concerns the transport, the pyroclastic material appears to have undergone a mass transport, combined with a transport in suspension or by traction, even if only for short distances. The depositional rate would have been generally high, in spite of the presence of some lutitic layers with high Globigerines content, typical of a slow sedimentation in shelf region, under the tempest waves level. As concerns the sedimentary structures, the diagenetic ones are prevalent.

3. MATERIALS AND METHODS

The samples for this study were selected from a stratigraphical succession observable in the northern part of the synclinal, at the surface, in the Piatra Verde quarry, north-south direction (Fig. 2). Preliminary analyses concerning the petrographic features, bulk mineralogical and chemical composition (optical microscopy and instrumental determinations) have been performed on 25 samples. Three samples have been selected for more detailed investigations on the basis of their mineralogical composition (Table 1); they were labeled as 49 bs, 37 bis bs, and 38 bs.

Table 1
Samples position and description

Samples	Depth (m)		Layer thickness (m)
	Within the pyroclastites stratum	Within the actual sedimentary column*	
49 bs	71	1471	>1.50
37 bis bs	110	1510	4.00
38 bs	115	1515	5.00

*Calculated considering the maximum thickness of the synclinal filling deposits (1400 m)

Bulk samples were examined by X-ray powder diffraction (XRD) and X-ray fluorescence (XRF) at SPIN Center, École Nationale des Mines, Saint Étienne (France) using a Siemens D-5000 Kristalloflex diffractometer (CuKα radiation) and a Philips PW1404 spectrometer, respectively.

Mineralogical composition of the <0.2 μm fraction for the three selected samples has been performed by XRD on oriented samples at DIVAPRA, Chimica Agraria, Università degli Studi di Torino, using a PW1710 Philips diffractometer (CoKα radiation).

Polished thin sections and rock fragments were investigated by scanning electron microscope (SEM) using a Cambridge Stereoscan S360 microscope. The chemical composition was determined by energy-dispersive spectroscopy (EDS), using an Oxford instrument installed on the Cambridge Stereoscan S360 microscope.

4. RESULTS AND DISCUSSION

XRD analysis on all 25 samples have shown that clinoptilolite, alkali feldspars, micas, clay minerals, SiO_2 phases (quartz and opal-CT) and carbonates (calcite and/or dolomite) are present in different amounts. Analysis on <0.2 μm fraction have indicated the presence of smectite, illite and kaolinite as secondary clay minerals. In many samples clinoptilolite is the prevalent phase and the reflections of alkali feldspars and carbonates in low amounts are overlapped by the clinoptilolite ones. The mineral composition of the selected samples (Table 2) is dominated by clinoptilolite as secondary mineral, well represented by many reflections (Fig. 3.)

Table 2
Mineralogical composition of the selected samples by XRD

Sample	Minerals (in order of abundance)
49 bs	quartz, clinoptilolite, opal-CT, alkali feldspars (very low), smectite, illite, kaolinite
37 bis bs	quartz, clinoptilolite, mica, kaolinite, opal-CT, alkali feldspars (very low), smectite, illite
38 bs	clinoptilolite, opal-CT, alkali feldspars, smectite, illite

Morphological observations revealed the preserved vitroclastic texture for the samples 49 bs and 37 bis bs (Figs. 4a and 4d) with typical shape of glass shards. The glass appears

120

transformed because its surface is not smooth, but affected by cracks caused by cleavage and authigenic crystal boundaries [4]. In the sample 38 bs (Fig. 4e) the volcanic matrix and glass shards have undergone an extensive diagenetic alteration, being completely recrystallized and replaced by secondary minerals.

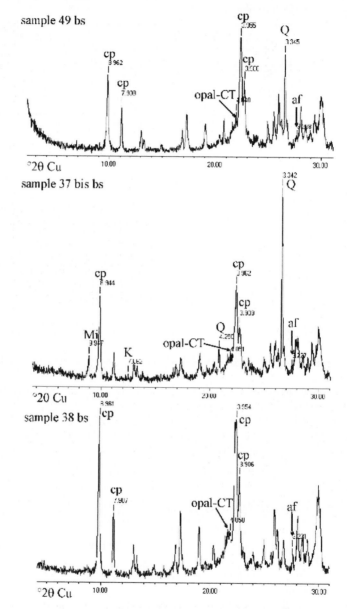

Fig. 3. XRD main reflections of the mineral components. Q = quartz, cp = clinoptilolite, af = alkali feldspars, Mi = mica, K = kaolinite

Clinoptilolite occurs as microcrystalline aggregates inside the matrix or pseudomorphs after glass shards (Figs. 4c and 4d). It also occurs as a marginal massive layer in cavities, succeeded inwardly by crystals perpendicular to the shard-wall (Figs. 4b and 4c).

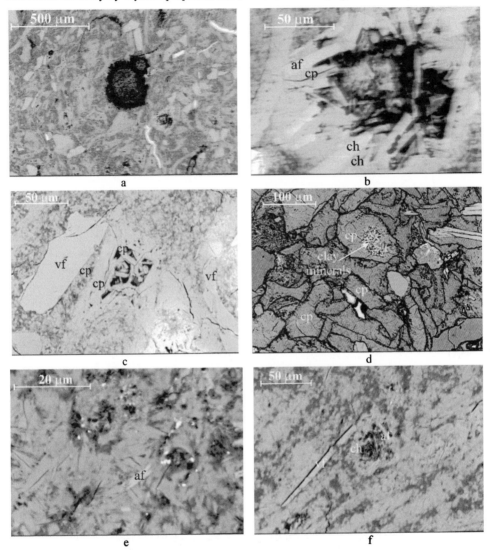

Fig. 4. SEM images of polished thin sections of pyroclastites: (a-c): sample 49 bs; (d): sample 37 bis bs; (e-f): sample 38 bs. Legend: af = authigenic feldspar, cp = clinoptilolite, ch = chabazite, vf = volcanic feldspar

Authigenic feldspar occurs both as neoformation crystals in the recrystallized matrix (Fig. 4e) and as euhedral crystals within cavities (Figs. 4b and 4f); in this last case it is associated with chabazite.

Clay minerals (smectite, illite and kaolinite) occur as replacing phases (Fig. 4d) or as interstitial cements of glass shards.

Representative EDS analysis are given in Tables 3-7. Fe was assumed to be in ferric oxidation state.

Table 3 shows the chemical composition of clinoptilolite from cavities and volcanic matrix. Conventionally, according to Ref. [5], a zeolite may be classified as clinoptilolite if T_{Si} value ($T_{Si} = Si/(Si+Al)$) ranges from 0.80 to 0.84. The cation content is highly variable. Ca-, Na- and K-dominant compositions are known. According to Ref. [6], a mineral may be classified as clinoptilolite if Si/Al >4.0 and (Na+K) >(Ca+Mg+Sr+Ba). The first condition has the same meaning of T_{Si} >0.80. The second one does not appear in Ref. [5]. For the clinoptilolite from Slănic, T_{Si} ranges from 0.81 to 0.84 and Si/Al >4 for all samples. Ca is the dominant cation (1.72 to 2.40 atoms per unit cell), followed by K (1.03 to 1.25 cations per unit cell). Na does not show an important participation, ranging from 0.00 to 0.60 cations per unit cell. The (Na+K) >(Ca+Mg+Sr+Ba) condition is not accomplished for any analysis, but we consider that all the compositions indicate Ca-clinoptilolite according to Ref. [5].

Table 3
EDS analyses of clinoptilolite

sample	37bis bs		49bs				
analysis number	1	3	11	17	20	21	22
location	pore	glass	glass	pore	glass	pore	pore
SiO_2	73.44	70.23	61.59	60.82	71.22	73.13	71.93
Al_2O_3	13.02	12.75	10.01	12.45	12.36	12.62	13.63
MgO	0.43	0.31	0.40		0.54		0.34
CaO	5.53	5.04	3.27	4.65	4.28	4.89	5.05
Na_2O			0.63	0.42	0.22		0.29
K_2O	2.00	2.01	2.01	2.01	2.33	2.54	2.03
Total	94.43	90.35	77.90	80.35	90.96	93.19	93.26
E%	-2.24	5.04	-1.54	9.89	4.22	8.30	9.20
Structural formulae calculated in basis of 72 [O]							
Si	29.75	29.72	30.20	29.13	29.92	30.01	29.52
Al	6.22	6.36	5.78	7.03	6.12	6.11	6.59
Mg	0.26	0.20	0.29		0.34		0.21
Ca	2.40	2.29	1.72	2.39	1.93	2.15	2.22
Na			0.60	0.39	0.09		0.11
K	1.03	1.08	1.25	1.23	1.25	1.33	1.06
Tsi	0.83	0.82	0.84	0.81	0.83	0.83	0.82
Si/Al	4.79	4.67	5.22	4.15	4.89	4.92	4.48
(Na+K)/(Ca+Mg)	0.39	0.44	0.93	0.68	0.59	0.62	0.48
Na/(Na+Ca)	0.00	0.00	0.26	0.14	0.04	0.00	0.05
Ca/k	2.32	2.11	1.37	1.95	1.54	1.62	2.09
Ca/Na			2.85	6.12	21.41		19.39
Na/K	0.00	0.00	0.48	0.32	0.07	0.00	0.11

Table 4 shows the chemical composition of the transformed glass. Except analysis 15 (with Si/Al <4), all the others have the chemical parameters typical for clinoptilolite. However, we did not consider them as clinoptilolite because the condition:
$E\% = 100 \times [(Al + Fe^{3+}) - (Li + Na + K) - 2(Mg + Ca + Sr + Ba)]/[(Li + Na + K) + 2(Mg + Ca + Sr + Ba)] < 10\%$
is not accomplished (E% is the balance error function [7]).

Chabazite was found only as euhedral crystals in cavities (Figs. 4b and 4f) with monoclinic habit. Its chemical composition is showed in Table 5.

Table 4
EDS analyses of the transformed glass

sample	37bis bs	38 bs				49bs			
analysis number	2	5	6	7	8	12	13	14	15
SiO_2	52.15	70.87	76.44	78.75	51.63	66.00	62.36	65.34	62.37
Al_2O3	9.00	12.36	13.47	14.29	9.16	12.02	11.49	11.89	13.32
Fe_2O_3					3.64				
MgO					0.30				
CaO	4.56	3.28	3.55	3.49	2.06	4.13	4.15	3.68	3.86
Na_2O		0.75	0.56	0.77	0.67	0.36	0.31	0.37	1.60
K_2O	1.71	2.40	1.93	2.22	3.42	2.03	2.10	2.17	1.74
Total	67.43	89.66	95.95	99.52	70.87	84.54	80.39	83.44	82.88
cations calculated on the basis of 72 [O]*									
Si	29.76	30.13	30.20	30.05	28.81	29.83	29.71	29.89	28.96
Al	6.05	6.19	6.27	6.43	6.02	6.40	6.45	6.41	7.29
Fe^{3+}					1.53				
Mg					0.25				
Ca	2.79	1.50	1.50	1.43	1.23	2.00	2.12	1.80	1.92
Na	0.00	0.62	0.43	0.57	0.72	0.32	0.28	0.33	1.44
K	1.25	1.30	0.97	1.08	2.44	1.17	1.27	1.26	1.03
Si/Al	4.92	4.87	4.81	4.67	4.78	4.66	4.61	4.66	3.97
(Na+K)/(Ca+Mg)	0.45	1.28	0.93	1.16	2.14	0.74	0.74	0.88	1.29
Na/(Na+Ca)	0.00	0.29	0.22	0.29	0.37	0.14	0.12	0.15	0.43
Ca/K	2.24	1.15	1.55	1.32	0.51	1.71	1.66	1.43	1.86
Ca/Na		2.42	3.51	2.50	1.71	6.31	7.44	5.54	1.34
Na/K	0.00	0.47	0.44	0.53	0.30	0.27	0.22	0.26	1.39

*The calculation on the basis of 72 [O] was made for comparation with clinoptilolite.

Authigenic feldspars commonly occur as euhedral crystals in cavities (Figs. 4b and 4f) or within the volcanic matrix (Fig. 4e). Their composition indicates plagioclases with 36.2 to 54.3% An (Table 5). Potassium authigenic feldspar was not be observed.

Calcite occurs in the Globigerines skeleton and no relationship with zeolites is observed.

Our investigation on the pyroclastites of Slănic showed that the volcanic glass altered to clinoptilolite as dominant secondary mineral, associated with clay minerals (smectite, illite and kaolinite), opal-CT and locally with chabazite and authigenic plagioclase. The textural relationships suggest that the zeolitization processes were very intense, clinoptilolite widely replacing the volcanic matrix. Ca-rich clinoptilolite with Ca>K>Na and Mg does not show large composition variability (Fig. 5 and Table 3). There is no difference between the clinoptilolite formed in the cavities and that formed within the volcanic matrix, but a different cation distribution for the clinoptilolites from different samples is obvious.

The transformed glass composition is similar to that of the clinoptilolite (Table 4 and Fig. 5), but there are some differences between different samples. For a given sample, the transformed glass is richer in alkali than clinoptilolite. Thus, clinoptilolite selectively concentrates more Ca.

Table 5
EDS analyses of chabazite and authigenic feldspar

sample	chabazite			feldspar		
	38 bs	49bs		38 bs		49bs
analysis number	10	18	19	4	9	16
location	pore	pore	pore	glass	pore	pore
SiO_2	67.08	55.77	53.74	63.61	49.32	47.88
Al_2O_3	17.71	16.34	17.93	28.74	19.39	24.21
TiO_2		0.57	0.54			
Fe_2O_3		2.01		1.10	0.60	
MgO		0.61				0.32
CaO	5.35	5.00	5.44	11.43	5.68	7.81
Na_2O	2.78	2.28	2.85	4.82	4.73	3.77
K_2O	1.42	1.62	1.18	0.69	1.30	1.21
Total	94.35	84.19	81.68	110.39	81.02	85.20
E%	11.71	9.27	13.12			
% Or				4.30	9.60	8.80
% Ab				41.30	54.30	42.90
% An				54.30	36.20	48.40
Structural formulae calculated in basis of:						
	24 [O]			8 [O]		
Si	9.21	8.74	8.64	2.59	2.72	2.52
Al	2.87	3.02	3.40	1.38	1.26	1.50
Ti		0.07	0.07			
Fe^{3+}		0.24		0.03	0.02	
Mg		0.14				0.02
Ca	0.79	0.84	0.94	0.50	0.34	0.44
Na	0.74	0.69	0.89	0.38	0.51	0.39
K	0.25	0.32	0.24	0.04	0.09	0.08
Tsi	0.76	0.74	0.72	0.65	0.68	0.63
Si/Al	3.21	2.90	2.54	1.88	2.16	1.68
(Na+K)/(Ca+Mg)	1.25	1.04	1.21	0.83	1.78	1.00
Na/(Na+Ca)	0.48	0.45	0.49	0.43	0.60	0.47
Ca/k	3.16	2.59	3.87	13.88	3.67	5.43
Ca/Na	1.07	1.21	1.05	1.31	0.66	1.14
Na/K	2.96	2.14	3.68	10.58	5.53	4.74

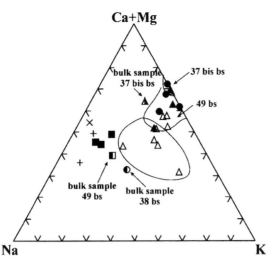

Fig. 5. (Ca+Mg)-Na-K relationships for the Slănic pyroclastites: Δ-transformed glass, ▲ - clinoptilolite formed within the volcanic matrix, ●-cavities clinoptilolite, ■-chabazite, ×-authigenic plagioclase formed within the volcanic matrix, +-cavities authigenic plagioclase. The bulk sample composition was analyzed by XRF.

Chabazite and authigenic plagioclase (Table 5 and Fig. 5) are richer in Na than clinoptilolite and transformed glass. Chabazite with T_{Si} value ranging from 0.72 to 0.76 has Ca as dominant cation. The composition of authigenic feldspar does not vary with sample or occurrence (cavities or volcanic matrix). Its T_{Si} value ranges from 0.63 to 0.68 and the anorthite content varies from 36.2 to 54.3%.

Chemical XRF analyses of bulk samples (Table 6) are plotted in Fig. 5. Important and significative differences can be observed between the three samples: sample 37 bis bs shows the highest Ca+Mg content, sample 38 bs the lowest one. Comparing the composition of bulk sample, clinoptilolite and transformed glass for each sample, the latter two result poorer in Na and K.

Also the volcanic plagioclases (Table 7) show some compositional differences. The plagioclases from the sample 38 bs (25.3-32.0% An) are more acidic than those from the sample 49 bs (22.5-49.0% An).

All the above observations allow some general considerations:

(1) The precursor volcanic glass was probably different for the three samples, statement supported by the volcanic plagioclase composition. The plagioclase of the sample 37 bis bs is the most basic and that of the sample 38 bs the most acidic.

(2) The bulk sample compositions respects the same relationship: the sample 37 bis bs is the richest in Ca+Mg and the sample 38 bs the poorest.

(3) The composition of clinoptilolite and that of the transformed glass respect the same relationship.

(4) No relationship is observed between the composition of authigenic phases and depth.

(5) Clinoptilolite composition does not depend on its occurrence (cavities or volcanic matrix).

Table 6
Chemical analyses of bulk samples (XRF)

sample	37bis bs	38 bs	49 bs
SiO$_2$	65.94	67.48	66.28
Al$_2$O$_3$	11.38	10.67	11.43
TiO$_2$	0.19	0.11	0.17
Fe$_2$O$_3$	1.26	1.12	1.03
MgO	0.81	0.39	0.66
CaO	5.14	2.92	3.64
Na$_2$O	0.72	2.12	2.51
K$_2$O	1.84	2.71	2.09
Total	87.28	87.52	87.81
cations calculated on the basis of 72 [O]*			
Si	29.25	29.85	29.31
Al	5.95	5.56	5.96
Ti	0.06	0.04	0.06
Fe^{3+}	0.42	0.37	0.34
Mg	0.54	0.26	0.44
Ca	2.44	1.38	1.72
Na	0.62	1.82	2.15
K	1.04	1.53	1.18
Si/Al	4.92	5.37	4.92
(Na+K)/(Ca+Mg)	0.56	2.04	1.54
Na/(Na+Ca)	0.20	0.57	0.56
Ca/k	2.35	0.90	1.46
Ca/Na	3.95	0.76	0.80
Na/K	0.59	1.19	1.83

*The calculation on the basis of 72 [O] was made for comparison with clinoptilolite.

Table 7
EDS analyses of volcanic plagioclases

sample	38 bs			49 bs				
SiO$_2$	66.92	68.48	66.83	58.06	57.41	54.38	66.33	66.67
Al$_2$O$_3$	26.84	26.68	25.11	29.15	22.27	24.42	29.15	27.63
CaO	7.10	5.98	5.50	10.47	4.43	6.38	9.14	8.14
Na$_2$O	8.06	8.80	8.35	5.79	8.13	6.64	7.49	7.50
K$_2$O	0.61	0.62	0.77	0.30	0.72	0.22	0.40	0.49
Total	109.53	110.57	106.56	103.78	92.96	92.04	112.50	110.43
%Or	3.10	3.10	4.20	2.00	3.90	1.00	2.00	3.10
%Ab	64.90	70.10	70.50	49.00	73.50	64.60	58.60	60.80
%An	32.0	26.80	25.30	49.00	22.50	34.40	39.40	36.10

(6) Clinoptilolite and transformed glass are poor in Na. Bulk sample has a higher Na+K content. Thus, the alkali are concentrated in other mineral phases: volcanic plagioclases not transformed, clay minerals, authigenic plagioclases, chabazite.

(7) Chabazite and authigenic plagioclases occur in cavities, related probably on the last phase of the transformation.

The transformation of the volcanic glass from Slănic appears to have started in an open system because smectite occurs in all the samples. Then, the system became probably closed. The closed system conditions are supported by the occurrence of Ca-clinoptilolite [8], illite (e.g., [9]), opal-CT [10], and chabazite [11].

Ca-clinoptilolite occurrence could indicate a reduced activity of $(Na^+ + K^+)$ in the pore water. Probably in the early phase of the transformation Ca was provided by volcanic glass, seawater and Globigerines skeletons, being consumed by smectite crystallization. The closing of the system leaded to the smectite transformation, Ca, Na and Mg being released in solution. The Mg presence favored the Al octahedral coordination [12, 13] and illite authigenesis with a K depletion from solution. The continue smectite consumption and silica releasing by glass dissolution allowed the Ca-clinoptilolite crystallization because its cation selectivity suggested by [14] is K >Ca >Na.

Local occurrence of chabazite and authigenic plagioclase with lower Si/Al ratios could indicate some local Ca availability in conditions of silica depletion from solution. This situation may be seen as a competition for available chemical elements in the last period of alteration, in the conditions of a closed system.

Clinoptilolite occurrence in all the samples of volcanic pyroclastites would indicate similar overall conditions of the volcanic glass transformation in a closed system. Some differences have been observed as concerns composition of bulk samples, volcanic plagioclases, transformed glass and clinoptilolite. These differences could indicate a different volcanic glass composition, as a result of the evolution of the explosive process. We do not exclude some important contribution of the chemical conditions of the marine sedimentary basin (depositional rate, burial depth of sediments, porosity etc.). Further volcanological and sedimentological investigations and analysis of more samples are necessary to support the mineralogical relationships.

ACKNOWLEDGEMENTS

We would like to thank Dr. Essaïd Bilal, Dr. Bernard Guy, Dr. Daniel Garcia and Dr. Jean-Jacques Truffat from SPIN Center, École Nationale des Mines, Saint Étienne (France) for the opportunity of the TEMPRA-PECO project, powder XRD and XRF analysis and their scientific support. We are also indebted with Dr. Raffaella Ruffini and Dr. Emanuele Costa from the Department of Mineralogy and Petrology, Università di Torino, for their support in SEM and EDS analyses, and with Dr. Cristina Panaiotu and Barbara Soare, from the Department of Mineralogy, Universitatea București, for petrological indications.

REFERENCES

[1] J.A. Winchester and P.A. Floyd, Chem. Geol., 20 (1977) 325.

[2] A. Kassoli-Fournaraki, M. Stamatakis, A. Hall, A. Filippidis, K. Michailidis, A. Tsirambides, and Th. Koutles, in Natural Zeolites for the Third Millenium, C. Colella and F.A. Mumpton (eds.), De Frede Editore, Napoli, Italy, 2000, p.193.

[3] J. McPhie, M. Doyle and R. Allen, Volcanic Textures. A Guide to the Interpretation of Textures in Volcanic Rocks, University of Tasmania, Center for Ore Deposits and Exploration Studies, 1993, 196 pp.

[4] S. Ogihara, Clays Clay Miner., 48 (2000) 106.

128

[5] D.S. Coombs, A. Alberti, Th. Armbruster, G. Artioli, C. Colella, E. Galli, J.D. Grice, F. Liebau, J.A. Mandarino, H. Minato, E.H. Nickel, E. Passaglia, D.R. Peacor, S. Quartieri, R. Rinaldi, M. Ross, R.A. Sheppard, E. Tillmanns, and G. Vezzalini, Eur. J. Mineral., 10 (1998) 1037.

[6] J.R. Boles, Am. Miner., 57 (1972) 1563.

[7] E. Passaglia, Am. Miner., 55 (1970) 1278.

[8] S.P. Altaner and R.E. Grim, Clays Clay Miner., 38 (1990) 561.

[9] M. de' Gennaro, C. Colella, E. Franco and D. Stanzione, N. Jahrb.Miner., 4 (1988) 149.

[10] R.A. Sheppard, U.S. Geol. Surv. Bull., 2108, 1994, 28 pp.

[11] R.C. Surdam, in Mineralogy and Geology of Natural Zeolites, F.A. Mumpton, (ed.), Reviews in Mineralogy, No. 4, Mineral. Soc. Amer., Washington, 1977, p. 65.

[12] M. de' Gennaro and C. Colella, Miner. Petr. Acta, 35 A (1992), 275.

[13] M. de' Gennaro, C. Colella and M. Pansini, N. Jahrb.Miner., 3 (1993) 97.

[14] L.L. Ames, Jr., Am. Miner., 45 (1960), 689.

Studies in Surface Science and Catalysis 155
A. Gamba, C. Colella and S. Coluccia (Editors)

Data processing of cation exchange equilibria in zeolites: a modified approach

D. Caputo[a], B. de Gennaro[a], P. Aprea[a], C. Ferone[b], M. Pansini[b] and C. Colella[a]

[a]Dipartimento di Ingegneria dei Materiali e della Produzione, Università Federico II, Piazzale V. Tecchio 80, 80125 Napoli, Italy

[b]Dipartimento di Meccanica, Strutture, Ambiente e Territorio dell'Università, Via G. Di Biasio 43, 03043 Cassino (FR), Italy

After reviewing the fundamental literature concerning the equilibrium of binary cation exchange reactions on zeolites, this paper provides a straightforward procedure of computation of their thermodynamic parameters, K_a and $\Delta G°$, based on a method of evaluation of cations activity coefficients in solution, which is simple and accurate and is equipped with a sufficiently large data-base. The computation procedure was tested on isothermal equilibrium data of Cd^{2+} and Zn^{2+} exchange for Na^+ in zeolite A at 25.0°C and 0.05, 0.10, and 0.50 total normality, by calculating their K_a and $\Delta G°$ and by comparing these results to those obtained evaluating the cations activity coefficients through a reference largely used method and to previous literature data. The results obtained show that the proposed methods give results in good agreement with those obtained with the reference method so that it appears to be a good alternative for the determination of cation activity coefficients in solution.

1. INTRODUCTION

Cation exchange properties of zeolites have been known for over 100 years. The first devised application of zeolite minerals was in fact in processes of water softening [1]. More recently the use of these materials as ion exchangers has greatly been enlarged. Apart from the wide utilization of synthetic sodium zeolites as builders and water softeners in detergent formulation, the application of either natural or synthetic zeolites in environment protection is becoming more and more frequent. Processes have been reported to remove noxious and toxic cations from wastewaters, for instance, (a) heavy metal cations from industrial wastewaters, (b) NH_4^+ from municipal sewages, and (c) Cs^+ and Sr^{2+} radionuclides from nuclear power plant wastewaters [2, 3]. On the other hand, cation exchange is essential to modify zeolites for attaining their best performances in adsorption, molecular sieving and catalysis [4], and, besides, it appears to be a step of crucial importance in the preparation of zeolite precursors for obtaining advanced ceramics [5-9].

The evaluation of the ion exchange behavior of a zeolite (and every ion exchanger) involves to study its relevant equilibrium reactions in solution with couples of competing cations. Usually these equilibria are described by exchange isotherms in which the equivalent fraction of the cation entering the solid phase is reported as a function of the equivalent fraction of the same cation in the liquid phase [1]. The isothermal data can be processed on the basis of the proper theories to obtain the relevant thermodynamic quantities, essentially the equilibrium constant K_a and the standard free energy of exchange $\Delta G°$. From a practical point of view these parameters represent an overall estimation of the ion exchange selectivity, therefore they are very important to forecast the zeolite behavior in an exchange process.

In the normally used procedure to estimate the thermodynamic quantities, due to Pitzer [10], one critical step is the knowledge of the mean ionic activity coefficients of the specific electrolytes, not always available in the literature. Some years ago a novel method of evaluation of the activity coefficients of cations in solution, which is rather simple and, in addition, allows the use of a sufficiently large data-base [11], was proposed by Ciavatta [12]. The Ciavatta's procedure has occasionally been utilized [13-17], but a full evaluation of its reliability in comparison with the Pitzer's procedure is still missing. This paper aims to fill this gap, comparing the results of both procedures in binary cation exchange reactions involving the synthetic zeolite A [LTA] [18] and the Cd^{2+}/Na^{+} and Zn^{2+}/Na^{+} cation pairs at three different total normalities.

2. THEORY

Based on previous theoretical treatments of ion exchange equilibria [19-21], Gaines and Thomas [22] introduced the thermodynamic formulation for the study of exchange equilibria in zeolites. The theory was later re-analyzed and confirmed by Barrer and Townsend [23] and Townsend [24] and furthermore discussed in a number of valuable publications [25-36].

Accordingly, if we consider the generic cation exchange reaction:

$$z_B A_{(s)}^{z_A+} + z_A B_{(z)}^{z_B+} \leftrightarrows z_B A_{(z)}^{z_A+} + z_A B_{(s)}^{z_B+}, \tag{1}$$

in which z_A and z_B are the cation valences and the subscripts s and z denote solution and zeolite phase, respectively, the mass action law may be written as:

$$K_a = \frac{a_{A_{(z)}}^{z_B} a_{B_{(s)}}^{z_A}}{a_{B_{(z)}}^{z_A} a_{A_{(s)}}^{z_B}}, \tag{2}$$

where K_a is the equilibrium constant and a represents the cation activity both in solid and in liquid phase. The application of the Gibbs-Duhem equations leads to the following expressions:

$$\log f_{A_{(z)}}^{z_B} = 0.4343(z_B - z_A)E_{B_{(z)}} - E_{B_{(z)}} \log K_c + \int_0^{E_{B(Z)}} \log K_c dE_{B_{(z)}}, \tag{3}$$

$$\log f_{B_{(z)}}^{z_A} = -0.4343(z_B - z_A)E_{A_{(z)}} + E_{A_{(z)}} \log K_c - \int_0^{E_{A(Z)}} \log K_c dE_{A_{(z)}}, \tag{4}$$

$$\log K_a = 0.4343(z_B - z_A) + \int_0^1 \log K_c dE_{A_{(z)}}, \tag{5}$$

where E_A and E_B are the equivalent fractions of cations A^{z_A+} and B^{z_B+} in zeolite phase, respectively, f_A and f_B their rational activity coefficient in zeolite and K_c is the corrected selectivity defined as:

$$K_c = \frac{E_{A_{(z)}}^{z_B} \, m_{B_{(s)}}^{z_A} \, \gamma_{B_{(s)}}^{z_A}}{E_{B_{(z)}}^{z_A} \, m_{A_{(s)}}^{z_B} \, \gamma_{A_{(s)}}^{z_B}} . \tag{6}$$

Here m_A and m_B are the molalities of the cations A^{z_A+} and B^{z_B+} and γ_A and γ_B their molal ionic activity coefficients. K_c may be related to K_a and Eq. (2) becomes:

$$K_a = \frac{E_{A_{(z)}}^{z_B} \, f_{A_{(z)}}^{z_B} \, m_{B_{(s)}}^{z_A} \, \gamma_{B_{(s)}}^{z_A}}{E_{B_{(z)}}^{z_A} \, f_{B_{(z)}}^{z_A} \, m_{A_{(s)}}^{z_B} \, \gamma_{A_{(s)}}^{z_B}} = K_c \, \frac{f_{A_{(z)}}^{z_B}}{f_{B_{(z)}}^{z_A}} . \tag{7}$$

In Eqs. (3), (4), and (5) the terms that take into account salt imbibition and possible change in water content and water activity within the zeolite framework were omitted. This approximation is allowed, as Barrer and Walker [37] reported salt imbibition to be negligible when salt concentration in solution is lower than $1\,m$, and Barrer and Klinowski [29] demonstrated that change in water content and water activity within the zeolite framework is insignificant.

The evaluation of Eqs. (3), (4), and (5) requires the knowledge of K_c as a function of $E_{A_{(z)}}$. In Eq. (6), which defines K_c, instead of determining the individual activity coefficients $\gamma_{A_{(s)}}$ and $\gamma_{B_{(s)}}$ of cations, it is possible to estimate the ratio $\Gamma = \gamma_{B_{(s)}}^{z_A} / \gamma_{A_{(s)}}^{z_B}$.

Actually, according to Munday [38]:

$$\Gamma = \frac{\gamma_{B_{(s)}}^{z_A}}{\gamma_{A_{(s)}}^{z_B}} = \frac{\gamma_{\pm BX}^{[z_A(z_B+z_X)/z_X]}}{\gamma_{\pm AX}^{[z_B(z_A+z_X)/z_X]}} , \tag{8}$$

$\gamma_{\pm AX}, \gamma_{\pm BX}$ being the mean ionic activity coefficients of the salts $A_{z_X}X_{z_A}$ and $B_{z_X}X_{z_B}$.

Since such coefficients are not always available in the literature, a generally valid computation procedure, which was proposed by Ciavatta [12], may be used. Accordingly, the mean ionic activity coefficient of a generic compound M_iX_j at the ionic strength $I = \frac{1}{2}\Sigma\, m_i z_i^2$ is given by:

$$\log \gamma_{\pm M_i X_j} = \frac{b(M_iX_j)4I}{(z_M + z_X)^2} - z_M z_X D . \tag{9}$$

In this expression b is a parameter characteristic of each salt, which may be obtained from the relation:

$$b(M_iX_j) = \underline{b} + \underline{c} \, \log I, \tag{10}$$

where $\underline{b}$ and $\underline{c}$ are constants characteristic of each salt. For some salts the constant $\underline{c}$ is zero: in this case b is not considered to appreciably vary with I. The values of the constants b, $\underline{b}$ and $\underline{c}$ are reported in Table 1 which was taken from Ciavatta's paper [12].

The reported values were estimated from the isopiestic data of Robinson and Stokes [11] and from other literature data [12].

Table 1
Survey of b values for various compounds M_iX_j at 25.0°C, see Eq. (10)[a]

Species	Cl⁻	ClO₄⁻	NO₃⁻	Species	Li⁺	Na⁺	K⁺
H^+	0.12	0.14	0.07	OH^-	-0.04 (0.07)	0.04	0.08
NH_4^+	-0.01	-0.10 (0.14)	-0.08 (0.06)	CrO_4^{2-}		-0.09 (0.06)	-0.12 (0.10)
Gly^+	-0.05			$B(OH)_4^-$		-0.09 (0.11)	
Cu^+		0.11		SCN^-		0.05	-0.01
Ag^+		0.0	-0.14 (0.10)	HCO_3^-		0.0	
Tl^+		-0.18 (0.09)	-0.27 (0.18)	CO_3^{2-}		-0.08 (0.05)	0.02
$CdCl^+$		0.25		NO_2^-	0.02 (0.11)	0.0	-0.03
CdI^+		0.27		NO_3^-	0.08	-0.05 (0.04)	-0.13 (0.09)
$CdSCN^+$		0.31		$H_2PO_4^-$		-0.11 (0.09)	-0.15 (0.12)
$HgCl^+$		0.20		HPO_4^{2-}		-0.19 (0.06)	-0.15 (0.09)
Mg^{2+}	0.19	0.33	0.17	$PO4^{3-}$		-0.29 (0.10)	-0.09
Ca^{2+}	0.14	0.27	0.02	$P_2O_7^{4-}$		-0.3	~ -0.15
Ba^{2+}	0.07	0.15	-0.28	SO_3^{2-}		-0.12 (0.10)	
Mn^{2+}	0.13			$S_2O_3^{2-}$		-0.12 (0.10)	
Co^{2+}	0.16	0.34	0.14	SO_4^{2-}	-0.06 (0.07)	-0.18 (0.13)	-0.20 (0.22)
Ni^{2+}	0.17			HSO_4^-		-0.01	
Cu^{2+}	0.08	0.32	0.11	F^-		-0.04	0.02
Zn^{2+}		0.33	0.16	Cl^-	0.10	0.03	0.00
Cd^{2+}			0.09	ClO_3^-		-0.01	
Hg_2^{2+}		0.09	-0.23 (0.20)	ClO_4^-	0.15	0.01	
Hg^{2+}		0.34	-0.15 (0.20)	Br^-	0.13	0.05	0.01
Pb^{2+}		0.15	-0.34 (0.32)	BrO_3^-		-0.06 (0.06)	
UO_2^{2+}	0.21	0.46	0.24	I^-	0.16	0.08	0.02
Al^{3+}	0.33			For^-		0.03	
Cr^{3+}	0.30		0.27	AcO^-	0.05	0.08	0.09
Fe^{3+}		0.56	~ 0.42				
La^{3+}	0.22	0.47					
Th^{3+}	0.25		0.11				

[a] The values are given as the weighed average, $\Sigma I \cdot b(M_iX_j)/\Sigma I$, when a trend not exceeding 0.03 was observed in the I range from 0.5 to 3.5 mol kg⁻¹. Otherwise, the data were fitted with the linear function $b(M_iX_j) = \underline{b} + \underline{c} \log I$, where $\underline{c}$ is a constant for a specific electrolyte; in this case $\underline{b}$ and $(\underline{c})$ values are listed. Gly^+=glycinium, For^-=formate, AcO^-=acetate.

In Eq. (9), D is the Debye-Huckel term, which is the resultant of electrostatic, non-specific long range forces. It may be computed as follows:

$$D = \frac{0.5109\sqrt{I}}{(1 + \rho\sqrt{I})}. \tag{11}$$

Guggenheim [39], Pitzer and Mayorga [40], and Scatchard [41], on the basis of different considerations, suggested for ρ the values of 1.0, 1.2, and 1.5, respectively. This uncertainty is overcome, choosing for ρ the value that gives, at the ionic strength corresponding to 0.1 m, the calculated mean ionic activity coefficient closest to the value reported in the literature [11].

In a stricter thermodynamic treatment, the mean ionic activity coefficients that should be used for a mixed (A,B)X solution, are those calculated with the following expressions, proposed by Glueckauf [42] for the salts $A_{zX}X_{zA}$ and $B_{zX}X_{zB}$:

$$\log \gamma_{\pm AX}^{(BX)} = \log \gamma_{\pm AX} - \frac{m_{B(s)}}{4I}\left[K_1 \log \gamma_{\pm AX} - K_2 \log \gamma_{\pm BX} - \frac{K_3}{\left(1 + I^{-\frac{1}{2}}\right)} \right], \tag{12}$$

$$\log \gamma_{\pm BX}^{(AX)} = \log \gamma_{\pm BX} - \frac{m_{A(s)}}{4I}\left[K_1' \log \gamma_{\pm BX} - K_2' \log \gamma_{\pm AX} - \frac{K_3'}{\left(1 + I^{-\frac{1}{2}}\right)} \right], \tag{13}$$

where:

$$K_1 = z_B(2z_B - z_A + z_X), \tag{14}$$
$$K_2 = z_A(z_B + z_X)^2(z_A + z_X)^{-1}, \tag{15}$$
$$K_3 = 0.5\, z_A z_B z_X (z_A - z_B)^2(z_A + z_X)^{-1}, \tag{16}$$
$$K_1' = z_A(2z_A - z_B + z_X), \tag{17}$$
$$K_2' = z_B(z_A + z_X)^2(z_B + z_X)^{-1}, \tag{18}$$
$$K_3' = 0.5\, z_A z_B z_X (z_B - z_A)^2(z_B + z_X)^{-1}. \tag{19}$$

Γ, Eq. (8), can be calculated substituting these expressions in Eqs. (12) and (13). The values of $\gamma_{\pm AX}$ and $\gamma_{\pm BX}$ are the mean ionic activity coefficients of salts $A_{zX}X_{zA}$ and $B_{zX}X_{zB}$ at the actual ionic strength I of the solution.

Finally the standard free energy of the exchange per equivalent of cation exchanger can be computed as:

$$\Delta G° = -\frac{RT}{z_A z_B}\ln K_a, \tag{20}$$

where R is the gas constant, T is the absolute temperature and ln represent natural logarithm.

The computation procedure here proposed allows also the prediction of the exchange isotherms at a total normality, N_{cal}, starting from experimental data at a given normality N_{exp}. Such prediction is based, besides the assumption that salt imbibition can be neglected [37] and change in water content and water activity within the zeolite framework is insignificant [29], also on the hypothesis that the ratio of the activity coefficient in zeolite is not sensibly affected by varying the total concentration of aqueous solution [30]. From this third assumption one may infer that this ratio, $f_A^{z_B}/f_B^{z_A}$, depends only on the cation concentration in zeolite. Thus, once that the composition of the zeolite phase is fixed by giving a value of the equivalent fraction of the ingoing cation, E_A^*, the above value of the ratio of the activity coefficients in zeolite can be computed starting from the mass action law, Eq. (2), using the value of K_a previously determined, the experimental equilibrium data obtained at the total normality N_{exp}, and the related values of cation activity coefficient in solution:

$$\frac{f_A^{z_B}}{f_B^{z_A}} = K_a \cdot \frac{(E_B^*)^{z_A}}{(E_A^*)^{z_B}} \cdot \frac{\gamma_{\pm AX}^{[z_B(z_A+z_X)/z_X]}}{\gamma_{\pm BX}^{[z_A(z_B+z_X)/z_X]}} \cdot \frac{m_A^{z_B}}{m_B^{z_A}}. \tag{21}$$

On the assumption that such ratio is not affected by varying the total normality of solution, it is possible to determine the value of cation molality, in equilibrium with the given composition of zeolite phase at a total normality N_{cal} different from N_{exp}:

$$m_A = \left(\frac{E_A^{z_B}}{E_B^{z_A}} \cdot \frac{\gamma_{\pm BX}^{|z_A(z_B+z_X)/z_X|}}{\gamma_{\pm AX}^{|z_B(z_A+z_X)/z_X|}} \cdot \frac{f_A^{z_B}}{f_B^{z_A}} \cdot \frac{\left(\dfrac{N_{cal} - z_A m_A}{z_B} \right)^{z_A}}{K_a} \right)^{\frac{1}{z_B}}. \tag{22}$$

This equation, which is obtained from the mass action law, is implicit in m_A and, thus, its solution requires fixing an initial value for m_A and applying a trial and error procedure until the following condition is satisfied:

$$\left| m_A^n - m_A^{n-1} \right| \le 0.0001. \tag{23}$$

3. EXPERIMENTAL

Carlo Erba reagent-grade zeolite A (marketed as zeolite 4A) ($Na_{12}Al_{12}Si_{12}O_{48} \cdot 27H_2O$) was used in cation exchange operations. Its cation exchange capacity was determined by the cross-exchange method [43]. Accordingly, a 1-g sample of zeolite A, placed on a Gooch filter, was eluted at about 60°C with a 0.5 M K^+ solution up to exhaustion (assumed as the condition when eluted Na^+ concentration attained the value present as impurity in the ingoing solution). The obtained zeolite A in its K form was then re-exchanged with Na^+ under the same conditions. To obtain monocationic zeolite A samples as more pure as possible, elution was made first with K^+ and Na^+ solutions prepared from Carlo Erba reagent-grade KCl and NaCl (purity 99.5%), and then, in the final stage of the elution, with solutions prepared from high purity Aldrich KCl and NaCl chemicals (purity>99.999%).

Collected eluates of both exchange operations were analyzed by atomic absorption spectrophotometry (AAS), using a Perkin-Elmer 2100 apparatus, and their K^+ and Na^+ concentrations were used to calculate two values of cation exchange capacity. The mean value of these two determinations turned out equal to 5.38 meq·g^{-1}, which is in good agreement with the calculated theoretical value of 5.48 meq·g^{-1} from the above reported formula.

Zeolite A in its original Na form was allowed to react at 25.0°C (±0.1°C) in rotating Teflon-lined stainless steel vessels with solutions containing different amounts of Cd^{2+} or Zn^{2+} and Na^+ at 0.05, 0.10, and 0.50 total normality. The relevant solutions were prepared from reagent-grade Carlo Erba $Cd(NO_3)_2 \cdot 4H_2O$, $Zn(NO_3)_2 \cdot 6H_2O$, and $NaNO_3$. The solid-to-liquid ratio was allowed to range between 1/40 and 1/1000; the reaction time was fixed at seven days, which was proved in advance to be sufficient to attain equilibrium. The pH of each reacting system, recorded at the end of the reaction time, ranged between 4.3 and 5.5. Cd^{2+} or Zn^{2+} and Na^+ concentrations or contents were measured either in the liquid or in the solid phase, respectively. In the former case Na^+ was determined by AAS, whereas Cd^{2+} and Zn^{2+} concentrations were measured by AAS or by titrations with EDTA (xylenol orange was used as indicator and urotropine as buffer for Cd^{2+}, and Erio-T as indicator and a solution of NH_3 and NH_4^+ at pH 10 as buffer for Zn^{2+} [44]). In the latter case the zeolite was chemically dissolved in a hydrofluoric and perchloric acid solution and its Cd^{2+} or Zn^{2+} and Na^+ concentrations were determined by AAS.

Reversibility tests of the cation exchange was performed following the recommendations of Fletcher and Townsend [45], i.e., (a) the vessels, in which the reacting systems were kept, were centrifuged after the reaction time was elapsed, (b) a known volume of the solution in equilibrium with the zeolite was removed and substituted with an equal volume of Na^+ solution of the same total normality, (c) the vessel was rotated and the system was allowed to react for further seven days in order to attain a new equilibrium condition.

To determine to which extent zeolite A exchange capacity is available for Cd^{2+} or Zn^{2+} exchange, a sample of zeolite A in its original Na form was contacted with a 0.05, 0.10, and 0.50 N Cd^{2+} or Zn^{2+} solution for a week, replacing the exhausted solution for a fresh one every 12 hours. At the end of the test, the mother solution was analyzed by AAS.

4. RESULTS AND DISCUSSION

The isotherms of Cd^{2+} and Zn^{2+} exchange for Na^+ in zeolite A at 25.0°C and 0.05, 0.10, and 0.50 total normality are reported in Fig. 1. Circles refer to the experimental points, solid lines to the model described in the Theory section (Sec. 2). Accordingly, assuming the 0.10 N total cation concentration as N_{exp}, the $E_{Cd(z)}$ and $E_{Cd(s)}$ (or $E_{Zn(z)}$ and $E_{Zn(s)}$) equilibrium values for N_{cal} equal to 0.05 and 0.50 N were calculated through the Eqs. (21-23) and compared to the relevant experimental data.

The most striking features of these exchange isotherms are the following:

-) The two exchange reactions are reversible at all the considered total normalities inasmuch as direct- and back-exchange runs give results that can be fitted by single curves.
-) The total cation exchange capacity of zeolite A is available for either Cd^{2+} or Zn^{2+} exchange at all the three selected total normalities of the contact solutions.
-) The isotherms of either Cd^{2+} or Zn^{2+} exchange for Na^+ show, at the lower equivalent cation concentration in solution, as steeper upward trends as the total normality of the solution decreases. This finding confirms that, as one may have expected, the selectivity for the cation of higher valence increases with increasing the dilution of the solution [29].

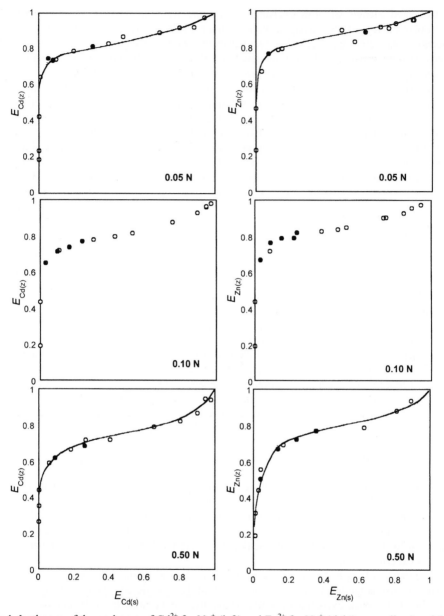

Fig.1. Isotherms of the exchange of Cd^{2+} for Na^+ (left) and Zn^{2+} for Na^+ (right) on zeolite A at 25.0°C and 0.05 N, 0.10 N, and 0.50 N. Empty circles = forward points, full circles = reverse points. $E_{Cd(s)/Zn(s)}$ = cadmium/zinc equivalent fraction in solution, $E_{Cd(z)/Zn(z)}$ = cadmium/zinc equivalent fraction in zeolite. Solid lines obtained from the model, Eqs. (21-23).

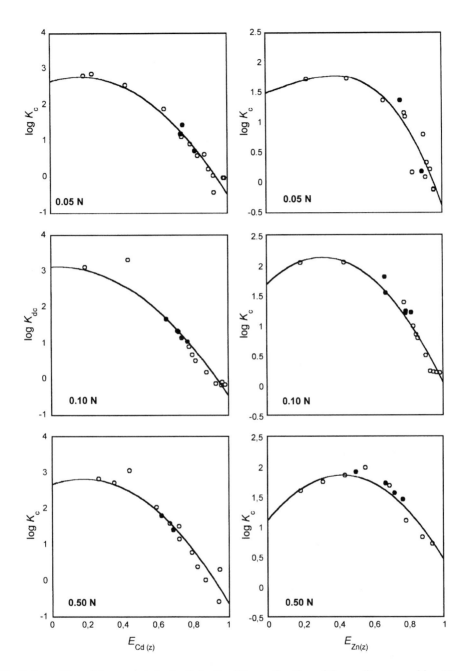

Fig. 2. Logarithm of the corrected selectivity ($\log K_c$) as a function of the zeolite composition ($E_{Cd(z)}$) for the exchange of Cd^{2+} for Na^+ (left) and the zeolite composition ($E_{Zn(z)}$) for the exchange of Zn^{2+} for Na^+ (right) on zeolite A at 25.0 °C and 0.05 N, 0.10 N, and 0.50 N. Empty circles = forward points, full circles = reverse points.

-) The isotherms of Cd^{2+} exchange for Na^+ at 0.05 and 0.10 total normality lie abundantly above the diagonal, thus denoting good selectivity for Cd^{2+} over the whole composition range. As far as the isotherm of Cd^{2+} exchange for Na^+ at 0.50 total normality is concerned, a selectivity reversal at about 0.90 Cd equivalent fraction is recorded.

-) The isotherms of Zn^{2+} exchange for Na^+ at all the considered total normalities lie abundantly above the diagonal, thus denoting good selectivity for Zn^{2+} over the whole composition range.

-) The calculated curves for the 0.05 and 0.50 N total cation concentrations appear to well fit the experimental points, confirming the validity of the assumptions made in Sec. 2.

The equilibrium data reported in the exchange isotherms were subjected to the computation procedure previously described. Accordingly, the corrected selectivity K_c was computed over the whole composition range and the decimal logarithm of their values plotted as a function of cation composition of zeolite in Figs. 2 (a, b) (Kielland plots). An accurate evaluation of the area delimited by the coordinate axes and the relevant curves over the whole composition range is of crucial importance in the determination of the thermodynamic quantities K_a and $\Delta G°$. It must be evidenced that the extrapolation of the curves at values of Cd or Zn equivalent fraction in zeolite very close to zero may result in very serious errors in the determination of the integrals, as it is often found that, in this area of the plot, very small changes in the equivalent fraction may give rise to dramatic variations of K_c. This drawback may be avoided either by properly adjusting the solid-to-liquid ratio so as to cover the whole composition range or by carefully recording fictitious experimental points from the fitting curve of the exchange isotherm and using them in the Kielland plot [46].

The values of the thermodynamic quantities K_a and $\Delta G°$, computed through the described computation procedure, after the Ciavatta's method of the activity coefficients evaluation [12] are reported in Tab. 2, in comparison with those computed evaluating the activity coefficients by the Pitzer's method [10]. Reference literature data [47], obtained by computing the activity coefficients by interpolation from tabulated values [48, 49], are also reported.

Inspecting the data in Tab. 2 allows the following observations to be made:

(a) The equilibrium data obtained at the three different total normalities are on the whole reasonably consistent with each other. Actually, differences in K_a values not larger than +21.2% are recorded using Ciavatta's and Pitzer's methods, respectively. These differences are acceptable if one considers that K_a is computed as exponential of the integral of $\log K_c$, see Eq. 5. This statement is supported by the fact that differences in $\Delta G°$ values, which do not depend on an exponential factor, are not larger than +5.7%, using Ciavatta's and Pitzer's methods, respectively.

(b) The equilibrium data obtained at the three different total normalities appear sufficiently consistent also with the literature data [47].

(c) The values of K_a and $\Delta G°$ computed starting from equilibrium data at 0.05 and 0.10 N using Ciavatta's method and Pitzer's method are very close to each other. On the contrary, a greater discrepancy has been found on the values of the thermodynamic parameters computed for the data at 0.50 N.

(d) The values of K_a and $\Delta G°$ computed at the three different total normalities using the Ciavatta's method are closer to each other, than the values calculated using the Pitzer's method (referring in particular to K_a, the maximum recorded deviation from the mean values is equal to -2.5 and -4.7, respectively, for the couple Cd/Na and -0.5 and -1.4, respectively, for the couple Zn/Na).

Table 2. K_a and $\Delta G°$ values of Cd^{2+} and Zn^{2+} exchange for Na^+ on zeolite A at 25.0°C for different total normalities

		$2Na^+ \leftrightarrows Cd^{2+}$			$2Na^+ \leftrightarrows Zn^{2+}$		
		Ciavatta's method	Pitzer's method	Gal's data	Ciavatta's method	Pitzer's method	Gal's data
0.05 N	K_a	36.1	35.1	-	19.1	18.5	-
	$\Delta G°$ (kJ/eq)	-4.4	-4.4	-	-3.6	-3.6	-
0.1 N	K_a	34.7	32.6	38.5	19.8	18.8	18.6
	$\Delta G°$ (kJ/eq)	-4.4	-4.3	- 4.6	-3.7	-3.6	- 3.6
0.5 N	K_a	31.6	26.8	-	20.0	16.5	-
	$\Delta G°$ (kJ/eq)	-4.3	-4.1	-	-3.7	-3.5	-

5. CONCLUSIONS

The findings of the present study seem to suggest that very similar final results may obtained by either using Ciavatta's or Pitzer's method. However a point that should be stressed is that the values of the thermodynamic quantities K_a and $\Delta G°$ should not depend on the total normality of the solution within the validity of the assumptions that salt imbibition can be neglected [30] and that change in water content and water activity within the zeolite framework is insignificant [29], what, as just mentioned in the above point (d), is better assured by the Ciavatta's method than by the Pitzer's method. On the whole it could be said that the method of evaluation of cation activity coefficients in solution proposed by Ciavatta [12] is not only simple and accurate, but appears to give more reliable results than the reference method.

Eventually, the computation procedure proposed to predict the exchange isotherms at a total normality, N_{cal}, starting from experimental data at a given normality, N_{exp}, appears to give reliable results, as the calculated curves well fit the experimental points.

AKNOWLEDGEMENTS
The authors are indebted to Prof. L. Ciavatta for the useful discussions.

REFERENCES
[1] D.W. Breck, Zeolite Molecular Sieves: Structure, Chemistry and Use, Wiley, New York, 1974.
[2] M. Pansini, Miner. Deposita, 31 (1996) 563.
[3] C. Colella, in Porous Materials in Environmentally Friendly Processes, I. Kiricsi, G. Pal-Borbely, J.B. Nagy and H.G. Karge (eds.), Studies in Surface Science and Catalysis, vol. 125, Elsevier Science B.V., Amsterdam, 1999, 641.
[4] A. Dyer, An Introduction to Zeolite Molecular Sieves, J. Wiley and Sons, Chichester, U.K., 1988.
[5] M.A. Subramanian, D.R. Corbin and R.D. Farlan, Mat. Res. Bull., 21 (1986) 1525.
[6] U.V. Chowdry, D.R. Corbin and M.A. Subramanian, US Patent No. 4 813 303 (1989).
[7] M.A. Subramanian, D.R. Corbin and U.V. Chowdry, Adv. Ceram., 26 (1989) 239.
[8] D.R. Corbin, J.B. Parise, U.V. Chowdry and M.A. Subramanian, Mater. Res. Symp. Proc., 233 (1991) 213.
[9] G. Dell'Agli, C. Ferone, M.C. Mascolo and M. Pansini, Solid State Ionics, 127 (2000) 309.
[10] K.S. Pitzer, Activity Coefficients in Electrolyte Solution, CRC Press, Boca Raton, FL, USA, 1991.
[11] R.A. Robinson and R.H. Stokes, Electrolyte Solutions, Butterworths, London, 1959.
[12] L. Ciavatta, Annali di Chimica, 70 (1980) 551.

140

[13] M. Pansini, C. Colella, D. Caputo, M. de' Gennaro and A. Langella, Microp. Mat., 5, (1996) 357
[14] E. Torracca, P. Galli, M. Pansini and C. Colella, Microp. Mesop. Mat., 20 (1998) 119.
[15] C. Colella, M. de' Gennaro, A. Langella and M. Pansini, Sep. Sci. Technol., 33 (4) (1998) 467.
[16] M. Adabbo, D. Caputo, B. de Gennaro, M. Pansini and C. Colella, Microp. Mesop. Mat., 28 (1999) 315.
[17] A. Langella, M. Pansini, P. Cappelletti, B. de Gennaro, M. de' Gennaro and C. Colella, Microp. Mesop. Mat., 37, (2000) 337.
[18] Ch. Baerlocher, W.M. Meier and D.H. Olson, Atlas of Zeolite Framework Types, Elsevier, Amsterdam, 2001, 168.
[19] G.E. Boyd, J. Schubert and A.W. Adamsom, J. Amer. Chem. Soc., 69 (1947) 2818.
[20] H.P. Gregor, J. Amer. Chem. Soc., 73 (1951) 642.
[21] O.D. Bonner, W.J. Argensinger, Jr., and A.W. Davidson, J. Amer. Chem. Soc., 74 (1952) 1044.
[22] G.L. Gaines and H.C. Thomas, J. Chem. Phys., 21 (1953) 714.
[23] R.M. Barrer and R.P. Townsend, Zeolites, 5 (1985) 287.
[24] R.P. Townsend, in New Development in Zeolite Science and Technology, Y: Murakami, A. Iijima and J.W: Ward (eds.), Kodansha-Elsevier, Tokyo-Amsterdam, 1986, 273.
[25] F. Helferrich, Ion Exchange, Mc Graw-Hill, New York, 1962.
[26] D.G. Howery and H.C. Thomas J. Phys. Chem., 69 (1965) 531.
[27] R.M. Barrer and J. Klinowski, J. Chem. Soc. Farad. Trans. I, 68 (1972) 73.
[28] R.M Barrer, J. Klinowski and H.S. Sherry, J. Chem. Soc. Farad. Trans. II, 69 (1973) 1669.
[29] R.M. Barrer and J. Klinowski, J. Chem. Soc. Farad. Trans. I, 70 (1974) 2080.
[30] R.M. Barrer, in Natural Zeolites. Occurrence, Properties, Uses, L.B. Sand and F.A. Mumpton (eds.), Pergamon Press, Elmsford, USA, 1976, 385.
[31] R.M. Barrer and J. Klinowski, J. Chem. Soc. Farad. Trans. I, 75 (1979) 637.
[32] A. Dyer, H. Enamy and R. P. Townsend, Sep. Sci. Technol., 16 (1981) 173.
[33] P. Fletcher, K.R. Franklin and R.P. Townsend, Phil. Trans. Roy. Soc., A312 (1984) 141.
[34] R.P. Townsend, Pure Appl. Chem., 58 (1986) 1359.
[35] R.P. Townsend, in Introduction to Zeolite Science and Practice, H. van Bekkum, E.M. Flanigen, and J.C. Jensen (eds.), Studies in Surface Science and Catalysis, vol. 58, Elsevier Science B.V., Amsterdam, 1991, 359.
[36] R.T. Pabalan, Geochim. Cosmochim. Acta, 58 (1994) 4573.
[37] R.M. Barrer and A.J. Walker, J. Chem. Soc. Farad. Trans., 60 (1964) 171.
[38] B.M. Munday, Ph.D. Thesis, Department of Chemistry, Imperial College of Science and Technology, London, 1969.
[39] E.A. Guggenheim, Phil. Mag., 19 (1935) 588.
[40] H.S. Pitzer and G.J. Mayorga, Phys. Chem., 77 (1973) 2300.
[41] G. Scatchard, Equilibrium in Solution. Surface and Colloid Chemistry, Harvard University Press, Cambridge, Massachussets, 1976.
[42] E. Glueckauf, Nature, 163 (1948) 414.
[43] C. Colella, M. de' Gennaro, E. Franco and R. Aiello, Rendiconti della Società Italiana di Mineralogia e Petrografia (Milan), 38 (1982-83) 1423.
[44] G. Schwarzenbach and H. Flaschka, Complexometric Titration, Methuen, London, 1969.
[45] P. Fletcher and R.P. Townsend, J. Chem. Soc. Faraday Trans., 77 (1981) 497.
[46] R.M. Barrer and B.M. Munday, J. Chem. Soc. (A), (1971) 2904.
[47] I.J. Gal, O. Jankovic, S. Malcic, P. Radovanov and M. Todorovic, J. Chem. Soc. Farad. Trans. I, 67 (1971) 999.
[48] H.S. Harned and B.B. Owen, The physical chemistry of electrolyte solution, Reinhold, New York, 1950.
[49] R.A. Robinson and H.S. Harned, Chem. Rev., 28 (1941) 419.

Studies in Surface Science and Catalysis 155
A. Gamba, C. Colella and S. Coluccia (Editors)
© 2005 Published by Elsevier B.V.

Synthesis and characterization of TAAET(Fe)S-10, a microporous titano silicate obtained in the presence of iron and TAABr salts

L. Catanzaro[a], P. De Luca [a], D. Vuono [a], J. B.Nagy[b] and A. Nastro[a]

[a] Dipartimento di Pianificazione Territoriale, Università della Calabria, 87030 Rende (CS), Italy

[b] Laboratoire de R.M.N., Facultès Universitaires Notre Dame de la Paix, B-5000 Namur, Belgium.

The present paper deals with the synthesis and characterization of ET(Fe)S-10 containing organic compounds. The synthesis of the ET(Fe)S-10 molecular sieves was obtained from systems having the following general mole composition: a Na_2O-0.6 KF-b TiO_2-0.03 Fe_2O_3-1.28a HCl-c TAABr-1.49 SiO_2-39.5 H_2O, where TAA (tetralkylammonium) is one of the following compounds: tetramethyl- (TMA), tetraethyl- (TEA), tetrapropyl- (TPA) or tetrabutylammonium (TBA) and a, b, and c are three coefficients ranging as follows: $1.0 < a < 1.5$; $0.0 < b < 0.25$; $c = 0$; 0.03. The effect of composition of the reaction mixture on the synthesis of ETS-10, the kinetic parameters, the morphology, size and composition of the crystals was thoroughly studied. The role played by the organic compounds in the reaction mechanism is at last suggested.

1. INTRODUCTION

While novel structures and compositions of molecular sieves are incessantly being obtained, there are indications that a turn point is made toward a potentially large class of molecular sieve materials composed of interconnected octahedral and tetrahedral oxide polyhedra. Indeed, the incorporation of octahedra into molecular sieve frameworks should allow for a rich diversity of new compositions because of the large number of elements that present octahedral co-ordination.

ETS (Engelhard Titanium Silicate) materials are a new class of compounds containing both octahedral and tetrahedral framework atoms. Two titano-silicate members of this family, ETS-4 and ETS-10, show adsorption characteristics typical of microporous materials [1-4]. A new family of stable molecular sieves results from the replacement of a portion of the octahedral titanium in ETS-10 with at least another octahedral metal atom and/or replacement of a portion of the tetrahedral silicon [5-9].

The aim of this paper is to study the synthesis of TAAET(Fe)S-10 materials, i.e., phases ETS-10 and ET(Fe)S-10 obtained from systems containing TAABr salts. Crystallization fields and kinetic parameters of the reaction will be investigated and the obtained products fully characterized, studying also their morphology as a function of TiO_2, Na_2O and TAABr contents.

2. EXPERIMENTAL

The original gels were prepared from systems, having the following general mole composition:

a Na$_2$O-0.6 KF-b TiO$_2$-0.03 Fe$_2$O$_3$-1.28° HCl- c TAABr-1.49 SiO$_2$-39.5 H$_2$O,

where TAA (tetralkylammonium) is one of the following compounds: tetramethyl- (TMA), tetraethyl- (TEA), tetrapropyl- (TPA) or tetrabutylammonium- (TBA) and a, b, and c are three coefficients ranging as follows: 1.0 < a < 1.5; 0.0 < b < 0.25; c = 0; 0.03.

The hydrothermal syntheses were carried out in static conditions under autogenous atmosphere at 190°C in PTFE lined autoclaves for a prefixed time. The reaction mixtures were prepared by mixing together an alkaline aqueous solution, containing sodium silicate solution, NaOH solution and the selected organic salts (TMABr, TEABr, TPABr, TBABr) with an acidic aqueous solution prepared from suitable KF, TiCl$_4$ and HCl solutions and from FeCl$_3$. At the end of the reaction the products were filtered, washed with distilled water and dried overnight at 105°C.

The identity of the solid phase and the degree of crystallinity were investigated using X-ray powder diffraction. The amount of TAA$^+$ cations and water trapped into the crystals was determined using TG analysis. DSC curves were recorded to evaluate the decomposition path of the organic molecules. DTG, TG and DSC analyses were performed on the basis of 20 mg samples, under constant nitrogen flow (flow rate, 15 ml·min^{-1}) with a heating rate of 10°C·min^{-1} using a NETZSCH, STA 429 instrument.

Morphology and crystal size were determined by SEM microscopy. Atomic absorption spectrophotometry (AAS) was used to determine the M·u.c.$^{-1}$ values (M = alkali cation).

Magic angle spinning NMR spectra of ^{13}C and ^{29}Si were carried out on a Bruker MSL 400 spectrometer. The ^{29}Si-NMR (79.47 MHz) spectra were measured with a 4.0 µs pulse (π/4) and a repetition time of 6.0s, while the ^{13}C-NMR (100.61MHz) spectra were taken with a 4.5 0 µs (π/4) pulse and a repetition time of 4.0 s. The number of scans varied between 1000 and 2000.

3. RESULTS AND DISCUSSION

Table 1 reports the crystalline phases obtained at 190°C after 3 days of treatment from systems without organic cations, having the following mole composition: a Na$_2$O-0.6 KF-b TiO$_2$-0.03 Fe$_2$O$_3$-1.28a HCl-1.49 SiO$_2$-39.5 H$_2$O, with a ranging between 1.0 and 1.5 and b from 0 to 0.25 (system S1).

The ET(Fe)S-10 compound was obtained, without co-crystallizing phases from systems of the above composition with 1.0 < a < 1.26 and 0.1 < b < 0.25. The Na$_2$O and TiO$_2$ amounts appear to be critical in obtaining ET(Fe)S-10. Actually, it was not observed from systems with Na$_2$O content greater than 1.5 moles and TiO$_2$ greater than 0.25 moles. The pH value of the initial gel affected the reactivity and selectivity of reaction batch, in fact, when 10.5 ≤ pH ≤ 11.5, ET(Fe)S-10 was obtained without co-crystallizing phases. The patterns of ET(Fe)S-10 were substantially similar to ETS-10 patterns but showed a shift of all reflections, likely due to a small variation of the unit cell. Note that most of the ET(Fe)S-10 samples were white, which suggests Fe^{3+} was incorporated in tetrahedral framework positions [10].

Table 1
Results of the syntheses at 190°C after 3 days from the system S1*

Moles Na$_2$O	0.0 moles TiO$_2$	0.1 moles TiO$_2$	0.2 moles TiO$_2$	0.25 moles TiO$_2$
1.50	Quartz pH gel = 13	Unreacted pH gel = 12.5	Unreacted pH gel= 12.5	Unreacted pH gel= 12.5
1.26	Quartz pH gel=12	ET(Fe)S-10^{+++} pH gel=10.5	ET(Fe)S-10^{++} pH gel=11	Unreacted pH gel=11
1.00	Quartz pH gel=12	ET(Fe)S-10+ pH gel=10.5	ET(Fe)S-10+ pH gel=10.5	ET(Fe)S-10+ pH gel=11.5

*+ = low crystallinity; ++ = medium crystallinity; +++ = high crystallinity

If 0.2 moles of TPABr were added to the systems leading to ET(Fe)S-10 crystallization, i.e., starting from the following mole composition: 1.26 Na$_2$O-0.6 KF-0.1 TiO$_2$-0.03 Fe$_2$O$_3$-0.2 TPABr-1.61 HCl-1.49 SiO$_2$-39.5 H$_2$O (system S2), TPAETS-10 and TPAET(Fe)S-10 were obtained (see X ray diffractogams in Fig. 1).

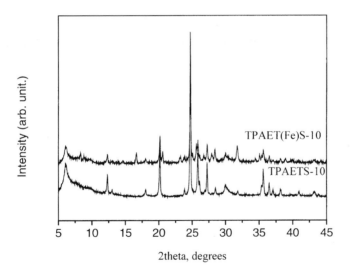

Fig. 1. X-Ray diffractograms of the phases TPAETS-10 and TPAET(Fe)S-10 obtained from system S2

TPAETS-10 sample appears to be similar to the phase obtained without organic ions [4]. Note that the TPAET(Fe)S-10 sample contains as impurity a novel phase named NTS [11].

Fig. 2 reports the crystallization curves of the phases ETS-10 and ET(Fe)S-10 obtained starting from compositions with organic cations. Differences in reaction kinetics are evident.

144

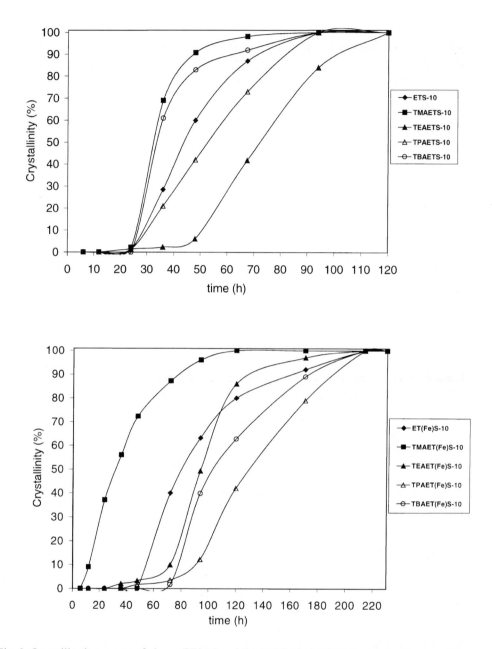

Fig. 2. Crystallization curves of phases ETS-10 and TAAETS-10 at 190°C from gels of composition: 1 Na_2O-0.6 KF-0.2 TiO_2-0.2 TAABr-1.61 HCl-1.49 SiO_2-39.5 H_2O (top diagram) and of ET(Fe)S-10 and TAAET(Fe)S-10 at 190°C from gels of composition: 1.26 Na_2O-0.6 KF-0.1 TiO_2-0.03 Fe_2O_3-0.2 TAABr-1.61 HCl-1.49 SiO_2-39.5 H_2O (bottom diagram).

Table 2

Induction time I (h) and crystallization rate of R (w-%·h⁻¹) of the TAAETS-10 and TAAET(Fe)S-10 samples obtained from the systems referred to in Fig. 2

	ETS-10	TMAETS-10	TEAETS-10	TPAETS-10	TBAETS-10
I	26.6	24	46	26.7	26
R	2.5	6.0	1.8	1.6	5.4

	ET(Fe)S-10	TMAET(Fe)S-10	TEAET(Fe)S-10	TPAET(Fe)S-10	TBAET(Fe)S-10
I	50	2.5	50	75	75
R	1.4	1.5	1.6	0.8	1.2

Table 2 reports the values of induction time, defined as the time of appearance of ca. 4% crystallinity, and crystallization rates, taken from the data of Fig. 2. The samples of highest crystallinity, lacking of any remaining amorphous phase, are taken as 100% crystalline samples.

The induction times for the formation of ETS-10 are not greatly affected by the presence of organic salts and are close to 25 h. Only the induction time for the formation of TEAETS-10 is longer, 46 h. The crystallization rate increases from ETS-10 to TMAETS-10. The incorporation of TMA^+ ions within the ETS-10 channels is possibly favored by their size, which fit easily the micropores in the structure. On the other hand, the crystallization rates decrease for TEA^+ and TPA^+ ions. These large cations can not be easily accommodated within the channels and this could determine a decrease of the crystallization rate. Finally, although TBA^+ ions are rather big, they seem to be responsible for the increase in the crystallization rate. However, the final amount of these ions is rather small in the TBAETS-10 sample, 0.08·u.c⁻¹ (Table 3). The role of the TAA cations is thus not only linked to their size, but also to their capacity of stabilization of crystals in formation.

In presence of Fe^{3+} ions, the induction periods are generally twice (for ET(Fe)S-10 and TEAET(Fe)S-10) or threefold (for TPA- and TBAET(Fe)S-10) longer than for the corresponding ETS-10 samples. Only the synthesis of TMAET(Fe)S-10 sample shows a ten times shorter induction time. Also the crystallization rates are systematically smaller in the presence than in the absence of $FeCl_3$. Note that a leveling of values is evident in the presence of iron, i.e., the crystallization rates are all quite close to each other. This stems from the difficulty to introduce other trivalent ions in the tetrahedral structure.

Table 3 shows the chemical compositions of the ETS-10 and the ET(Fe)S-10 samples, respectively, as determined by AAS (Na, K, Fe and Ti) and thermal analysis (H_2O and TAA). The TPA and TBA ions could accommodate only in large micropores (14.3Å × 7.6Å), produced by a partial dislocation line [4]. This is suggested by the low content of TPA·u.c⁻¹ (0.17) and of TBA·u.c⁻¹ (0.08) and these ions could be hydrated, as it is suggested by the ¹³C-NMR chemical shift (see below).

The amount of H_2O·u.c⁻¹ decreases from the ETS-10 to the TAAETS-10 samples, as the amount of alkali ions greatly decreases too. The H_2O/M ratio (with M= alkali cation) varies from 2.2 in ETS-10 sample to 9-14, showing that the amount of water is greater than the coordination number reported in the literature, i.e., 9 for Na^+ and 12 for K^+[12].

Table 3
Chemical characterization of the TAAETS-10 and TAAET(Fe)S-10 samples obtained from the systems referred to in Fig. 2*

Sample	H_2O per u.c.	TAA per u.c.	Na per u.c.	K per u.c.	Fe per u.c.	Ti per u.c.	(TAA+Na+ K-2Ti)/u.c.	H_2O/M
ETS-10	2.7	--	0.84	0.37	----	1.0	-0.8	2.2
TMAETS-10	1.8	0.32	0.11	0.037	----	0.92	-1.37	12
TEAETS-10	1.1	0.30	0.09	0.033	----	0.85	-1.28	9
TPAETS-10	1.1	0.17	0.07	0.027	----	0.77	-1.27	11
TBAETS-10	1.8	0.08	0.1	0.030	----	0.85	-1.49	14
ET(Fe)S-10	3.1	----	0.86	0.28	0.28	0.69	-0.52	2.7
TMAET(Fe)S-10	1.7	0.28	0.79	0.067	0.17	0.69	-0.41	2.0
TEAET(Fe)S-10	2.4	0.1	0.75	0.067	0.15	0.62	-0.37	3.0
TPAET(Fe)S-10	2.0	0.1	0.66	0.050	0.11	0.55	-0.40	2.8
TBAET(Fe)S-10	2.6	0.1	0.56	0.048	0.11	0.59	-0.58	4.3

* H_2O and TAA obtained by TG analysis (0-200°C and 200-700°C ranges, respectively; Na, K, Fe and Ti obtained by AAS

The difference between the positive charge (TAA + Na + K) and the framework negative charge (2Ti) is also very revealing. There is an excess negative charge of –0.8 for the ETS-10 sample which increases to ca. –1.3 for the TAAETS-10 samples. This negative charge in excess has to be neutralized with protons, which introduce therefore a Brønsted acidity in the samples [9, 16].

The chemical compositions of the ETS(Fe)-10 samples are quite different as it is shown in Table 3. The Ti u.c.$^{-1}$ values do not vary much, they decrease only from 0.7 to 0.55 u.c.$^{-1}$ but they remain always smaller than the corresponding ETS-10 values. The Fe u.c.$^{-1}$ values decrease from ET(Fe)S-10 to TBAET(Fe)S-10 samples. The Na u.c.$^{-1}$ are also decreasing, but the extent of the decrease is much smaller than in the ETS-10 samples. This means that part of Na^+ ions might neutralize the negative charges connected to the presence of Fe^{3+} ions in the tetrahedral framework. The K u.c.$^{-1}$ values also decrease from the ET(Fe)S-10 to the TAAET(Fe)S-10 samples, but this decrease is only a fivefold decrease. The amount of TMA·u.c.$^{-1}$ is similar to that of the corresponding TMAETS-10 sample (0.3) but for the other TAAET(Fe)S-10 samples this value drops to 0.1. This means that, due to the introduction of Fe^{3+} in the tetrahedral framework, the presence of counter-cations impedes the incorporation of TAA ions. The H_2O·u.c.$^{-1}$ values are slightly decreasing from the ET(Fe)S-10 sample. Interestingly, the H_2O/M (M= alkali cation) ratio does not change much and remains in most cases close to 3.

The difference between the positive charge (TAA + Na + K) and the negative charge (Fe + 2Ti) is much smaller in the ET(Fe)S-10 samples, being close to –0.45, and does not change much in presence of TAA ions. This suggests a lower amount of Brønsted acid sites in the ET(Fe)S-10 samples.

ETS-10

TMA ETS-10

TEA ETS-10

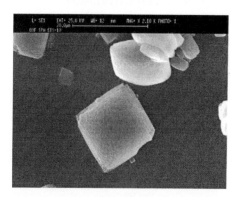

TPA ETS-10

TBA ETS-10

Fig. 3. SEM micrographs of ETS-10 crystals synthesized from gels of composition: 1 Na$_2$O-0.6 KF-0.2 TiO$_2$-0.2 TAABr-1.28 HCl-1.49 SiO$_2$-39.5 H$_2$O at 190°C.

148

ET(Fe)S-10

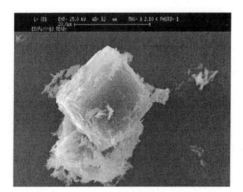

TMA ET(Fe)S-10

TEA ET(Fe)S-10

TPA ET(Fe)S-10

TBAET(Fe)S-10

Fig. 4. SEM micrographs of ET(Fe)S-10 crystals synthesized from gels of composition: 1.26 Na$_2$O-0.6 KF-0.1 TiO$_2$-0.2 TAABr-0.03 Fe$_2$O$_3$-1.61 HCl-1.49 SiO$_2$-39.5 H$_2$O at 190°C.

The templating action of the tetralkylammonium salts was evaluated by thermal analysis (DSC, TG, DTG). The trend of DTG curve of the TAAET(Fe)S-10 and TAAETS-10 samples suggests the presence of one type of organic cation incorporated inside the ET(Fe)S-10 and ETS-10 channels. In fact, the thermal decomposition of TAA is connected with a single endothermal effect, its temperature being different for each type of organics.

The TG data show that the water adsorbed into the pores of ET(Fe)S-10, synthesized from systems without organic cation, is partially substituted by the TAA added. Water is desorbed between 20 and 200°C. The amount of $TAA \cdot u.c.^{-1}$ is a function of the molecular weight and of the molecular size of the organic cation.

The temperature of decomposition of the TAA cations increases from TMA to TBA in TAAETS-10 samples: TMA, 410°C; TEA, 430°C; TPA, 450°C and TBA, 460°C.

The DTG curves are rather broad for TMA and TEA ions. It is possible that these smaller cations could be found in states where they are more or less hydrated as it was previously suggested in various gel phases [14].

The temperature of decomposition of TAA cations occluded in the TAAET(Fe)S-10 samples is similar to the corresponding TAAETS-10 samples. The only difference stems from the DTG curves for TMA and TEA ions, which are better defined with a single peak in this case. In addition, a small DTG peak is also present at ca 620°C in the TMAET(Fe)S-10 sample, which could be due to the decomposition of the products of decomposition of TMA^+ ions. The crystals of ET(Fe)S-10 are bigger and more regular than the crystals of ETS-10.

The differences could stem from both the nucleation and crystal growth rate. The ETS-10 and TMAETS-10 crystals are both rod –like (Figs. 3 and 4).

When the induction time (linked to the nucleation time) increases, and the crystal growth decreases, cubic crystals are essentially obtained. The biggest crystals are obtained for TPAETS-10 samples. The faster crystallization rate of TBAETS-10 samples starting from a larger number of initial nuclei could explain the smaller crystal size in this case. The TAAET(Fe)S-10 crystals are all cubic and their size is generally bigger than the corresponding TAAETS-10 samples. Sometimes some secondary crystallization can also be seen on the nice cubic crystallites.

A typical ^{29}Si-NMR spectrum of ETS-10, obtained with TMABr, is shown in Fig. 5 (top diagram). All the other spectra taken without organics or with TEA, TPA and TBA are similar to that of the above figure.

The NMR lines at –94.6 and –97.0 ppm stem from the Si(3Si,1Ti) configuration, while the –104.0 ppm line stems from Si(4Si,0Ti) configuration as it was proposed previously [4]. The relative intensities of the lines are 2:2:1. Note that the –97.0 ppm line is not resolved. The $\{^1H\}$-^{29}Si cross polarized spectra shows that the – 94.6 ppm line is less influenced than the other two NMR lines.

Figure 5 (bottom diagram) shows the ^{29}Si-NMR spectrum of TMAETS(Fe)-10. All the other spectra obtained without organics or with TEA, TPA and TBA are similar. The lines are rather broad due to the presence of Fe^{3+} ions. The NMR lines at –94.6 and –97 ppm overlap and only the line at –104 ppm is shown distinctly. The relative intensities of the lines are 4:1. In addition, a new NMR line appears in the spectra of ETS(Fe)-10 at –107 ppm due to the presence of a new crystalline titanosilicate phase called NTS [11].

The ^{13}C-NMR spectra of the ETS-10 samples show that the organic ions have been incorporated intact in the channels of ETS-10. The corresponding spectra of the ETS(Fe)-10 could not be taken in normal conditions because of the fast relaxation due to the presence of paramagnetic Fe^{3+} ions.

The chemical shifts of the various organic ions are: TMA, 56.8 ppm; TEA, 54.6 and 9.4 ppm; TPA, 61.0, 16.1 and 11.0 ppm; TBA, 58.5, 23.8, 19.8 and 14.4 ppm. These values are close to the ones obtained in water [13-15]. This means that the TAA ions do not interact specifically with the channel walls as it was the case of the ZSM-5 structure [13]. The larger amount of water in the TAAETS-10 samples already suggested the possible hydration of TAA$^+$ cations. This is confirmed by the ^{13}C-NMR spectra.

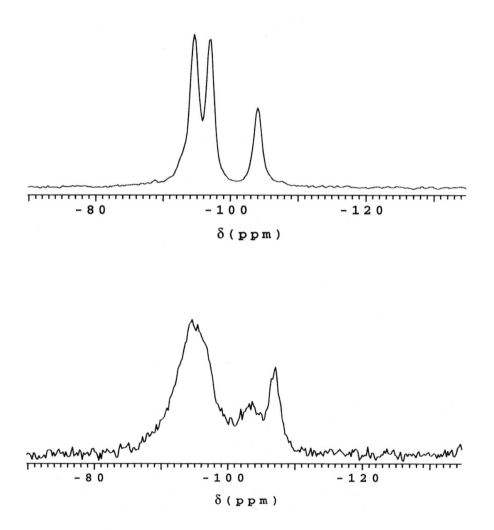

Fig. 5. Magic angle spinning ^{29} Si NMR spectra of TMAETS-10 (top diagram) and of TMAET(Fe)S-10 (bottom diagram) samples.

4. CONCLUSIONS

The presence of tetralkylammonium salts in a reaction mixtures leading to the formation of the ETS-10 phase, confirms the synthesis of such type of compound either in the presence or in the absence of a iron source. The crystals have a cubic shape and are bigger when TAA salts are added to the reaction mixture. The presence of organics modifies both the nucleation and crystallization rates and hence influences the crystal morphology and size. The white color of the samples and the broadening of the ^{29}Si-NMR lines accounts for the incorporation of iron in the framework. Others studies are, however, in progress to confirm the incorporation of iron atoms.

ACKNOWLEDGMENTS
The authors are indebted to Mr. Guy Daelen for taking the NMR spectra. The work was partly supported by the Belgian Programme on Inter University Poles of attraction initiated by the Belgian State, Prime Minister's Office for Scientific, Technical and Cultural Affair (OSTC-PAI-IUAP N°. 5/10 on Quantum size effects in nanostructured materials) and partly by the Italian Research Council, Progetto Finalizzato Materiali 2.

REFERENCES
[1]   S.M. Kuznicki, U.S. Patent No.4,938,939, assigned to Engelhard Corporation, 1990.
[2]   S.M. Kuznicki, U.S. Patent No. 4,853,202, assigned to Engelhard Corporation, 1989.
[3]   M.W. Anderson, O. Terasaki, T. Ohsuna, P.J. O'Malley, A. Philippou, S.P. Mackay, A. Ferreira, J. Rocha and S. Lidin, Phil. Mag. B71, (1995) 813.
[4]   M.W. Anderson, O. Terasaki, A. Philippou, S.P. Mackay, A. Ferreira, J. Rocha and S. Lidin, Nature, 367 (1994) 347.
[5]   S.M. Kuznicki, U.S. Patent No. 5,208,006, assigned to Engelhard Corporation, 1993.
[6]   J. Rocha, P. Brandao, JDP de Jesus, A. Philippou and M.W. Anderson, Chem. Commun. (1999) 471.
[7]   S.S. Lin and H.S. Weng, Appl. Catal. A: General, 105 (1993) 289.
[8]   M.W. Anderson, A. Philippou, Z. Lin, A. Ferreira and J. Rocha, Angew. Chem. Int. Ed., 34 (1995) 1003.
[9]   M.W. Anderson, J. Rocha, Z. Lin, A. Philippou, I. Orion and A. Ferreira, Microporous Mater., 6 (1996) 195.
[10] P. Fejes, J. B.Nagy, K. Làzàr and J. Halàsz, Appl. Catal., A: General, 190 (2000) 117.
[11] A. Nastro, P. De Luca, M. Turco, G. Bagnasco, G. Busca; Italian Patent Application No. RM2001A000172, 2001.
[12] J. E.Huheey, Chimica Inorganica, Piccin Editore, Padova, 1977, p 75-76.
[13] J. B.Nagy, Z. Gabelica and E.G. Derouane, Zeolites, 3 (1983) 43.
[14] Q. Chen, J. B.Nagy, J. Fraissard, J. El Hage-Al Asswad, Z. Gabelica, E.G. Deroune, R. Aiello, F. Crea, G. Giordano and A. Nastro, in Guidelines for mastering the properties of molecular sieves. Relationship between the physico-chemical properties of zeolitic systems and their low dimensionality, NATO ASI Series, Plenum, New York, 1990, p 87.
[15] E. Breitmeier and W. Voelter, Carbon-13 NMR Spectroscopy, 3rd Ed., Wiley-VCH, Weiheim, 1984, p 237.
[16] P. De Luca and A. Nastro, Progress in Zeolites and Microporous Materials, U. Chon, S.-K. Ihm and Y.S. Uh (eds.), Studies in Surface Science and Catalysis Elsevier, vol. 105, Amsterdam, 1997, p 221.

Studies in Surface Science and Catalysis 155
A. Gamba, C. Colella and S. Coluccia (Editors)

A new approach to the characterization of V species in doped-V/Ti/O catalysts by means of TPR and TPO measurements: a study of the effect of promoters in the oxidation of o-xylene

F. Cavani[a]*, C. Cortelli[a], A. Frattini[a], B. Panzacchi[a], V. Ravaglia[a], F. Trifirò[a], C. Fumagalli[b], R. Leanza[b], G. Mazzoni[b]

[a]Dipartimento di Chimica Industriale e dei Materiali, Viale Risorgimento 4, 40136 Bologna, Italy. *cavani@ms.fci.unibo.it, INSTM, Research Unit of Bologna.
[b]Lonza SpA, Via E. Fermi 51, 24020 Scanzorosciate (BG), Italy.

Titania-supported vanadium oxide systems, catalysts for the oxidation of o-xylene to phthalic anhydride, were characterized by means of Raman spectroscopy, Thermal-Programmed Reduction and Oxidation, and adsorption/TP Desorption of methanol, with the aim of defining a method for the quantification of the different V species. It was found that vanadium oxide, either as polyvanadate dispersed over titania, or in the form of bulk vanadia, spontaneously releases molecular oxygen at 600-650°C, whereas isolated V species, chemically interacting with the support, is not susceptible of self-reduction. The latter species is that predominant in samples having low vanadium oxide loading ($\leq$ 2 wt.% V_2O_5, with TiO_2 surface area 22.5 m^2/g), and possesses the highest intrinsic activity in o-xylene conversion. In samples having higher vanadia loading, instead, the activity is determined by the amount of dispersed polyvanadate and of bulk vanadia. The effect of Sb, promoter of activity for V/Ti/O catalysts, was explained in terms of an increase of the dispersion of the most active species, and of stabilization of the latter towards segregation. These promotional effects are more pronounced in the co-presence of Cs and Sb.

1. INTRODUCTION

Supported vanadium oxide is the catalyst for several oxidation reactions; of great industrial importance is the V_2O_5/TiO_2 system (V/Ti/O), for the selective oxidation of o-xylene to phthalic anhydride [1-3]. Indeed, the industrial V/Ti/O catalyst is very complex, since it contains several dopants, which are fundamental in order to increase the activity and the selectivity to the desired product, as compared to the undoped catalyst [4].

The V/Ti/O system has been the object of several investigations in the past [1-3], and in most cases the aim of the work was to characterize the different vanadium species which develop at the titania surface, for an increasing loading of active phase. It is known that the interaction between vanadium oxide and titania generates: (i) a V species which is chemically bound to the support via oxo bridges (isolated V in octahedral or tetrahedral coordination, depending on the hydration degree), (ii) a polymeric species spread over titania, and (iii) bulk vanadium oxide, either amorphous or crystalline. Relationships have been looked for between the amount and properties of each of them, and the catalytic performance [5]. On the contrary, very few works on doped, industrial-like V/Ti/O systems can be found in literature [6-8];

indeed, the effect of dopants on the catalyst characteristics, and the reasons why they promote the catalytic performance, is not well understood, despite they play a fundamental role in the industrial system.

Promoters have been classified as belonging to either the non-interacting-elements class (WO_3, Nb_2O_5, SiO_2; the two former are present in V/Ti/O systems for SCR, while the latter is often present as an impurity in the raw materials rather than being intentionally added), which thus coordinate directly to the support and do not significantly interact with V oxide, or to the interacting-elements class (K_2O and P_2O_5), which coordinate with the V ion and thus may profoundly affect the redox property of the latter [4,6]. Several experimental techniques have been used to investigate the nature and amount of each V species; however, most techniques, but potentiometric titration after dissolution of the samples [9], only allow the qualitative identification, but not the quantitative determination of each of them. Thermal-programmed-reduction and oxidation (TPR, TPO) [10,11] and adsorption of probe molecules followed by desorption [12,13] are used to determine the reducibility or oxidizability of V species, and the number of active sites which may chemically interact with specific reactants. These techniques give reliable quantitative information when a clear and unambiguous assignment of peaks in reduction, oxidation or desorption profiles can be done; this becomes particularly difficult when complex, multi-element catalytic systems are studied.

In a previous work [7], we have characterized Cs-doped V/Ti/O systems by means of TPR and TPR+O (re-oxidation after TPR), together with conventional spectroscopic methods. This allowed us to give an interpretation to the activity-enhancement effect observed when small amounts of Cs are added to a 7 wt.% V_2O_5-containing, V/Ti/O catalyst. In Cs-containing catalysts, a relationship was found between V re-oxidation (the rate-determining step of the redox process for the Cs-doped catalyst), and catalytic activity for samples containing increasing amounts of Cs. The experimentation also led us to conclude that the thermal characterization methods employed can be used to directly discriminate the different V species, while isolating the effects of promoters. In the present work we extend this investigation approach to both the undoped and doped V/Ti/O systems, the latter containing two important promoters for the industrial catalyst, Cs and Sb.

2. EXPERIMENTAL

Catalysts were prepared with the wet impregnation technique: an aqueous solution containing the desired amount of NH_4VO_3, $CsNO_3$ and $Sb(CH_3COO)_3$ was added to the support (TiO_2 anatase, having a specific surface area of 22.5 m^2/g). Water was evaporated under vacuum at 70°C. Then the wet solid was dried at 150°C for 3h, and calcined at 450°C for 5h in static air.

The catalysts prepared were characterized by means of X-Ray Diffraction, Raman Spectroscopy, Thermal Programmed Reduction (TPR) and Oxidation (TPO, and TPR+O for the oxidation after pre-reduction) and Thermal Programmed Desorption (TPD) of methanol. The XRD measurements were carried out using a Philips PW 1710 apparatus, with Cu $K\alpha$ (λ = 1.5406 nm) as radiation source. Raman studies were performed using a Renishaw 1000 instrument, equipped with a Leica DMLM microscope, laser source Argon ion (514 nm). Thermal programmed measurements were performed using a Thermoquest TPDRO1100 instrument; calcined samples were loaded in a quartz reactor and pre-treated in N_2 at 180°C for 30 min to eliminate weakly adsorbed species. After cooling at room temperature, samples were reduced by heating under H_2 atmosphere (5% H_2 in Ar) with a linear increase of temperature (10°C/min) up to 650°C, and with a final isothermal step at 650°C for 30min; the corresponding TPR profile was recorded. After cooling down to room temperature, the samples were re-oxidized by a gas mixture of 5% O_2 in He, using the same temperature program as for the reduction (TPR+O). In TPO tests, calcined samples were directly oxidized, i.e., without pre-reduction, with the same conditions described above. For Thermal Programmed Desorption of methanol (TPD), the samples were pre-treated in He from room temperature to 550°C, with a temperature increment of 10°C/min, and then maintained at 550°C for 1h. The samples were then cooled down to 110°C, and pulses of 1 μl of methanol, with injector at 250°C, were done [13]. Physi-adsorption of methanol did not occur at 110°C. After the last pulse, the samples were heated again at 550°C, while recording the desorption profile of methanol.

Catalytic tests were carried out in a continuous-flow, fixed bed, stainless steel reactor. The feed composition was: 1 mol.% o-xylene in air. The products in the outlet stream were condensed in acetone. The reactants and the products were analyzed with a GC equipped with a HP-5 semicapillary column (FID) for organic compounds and with a Carbosieve S column (TCD) for O_2, CO and CO_2.

3. RESULTS AND DISCUSSION

3.1 Characterization and reactivity of undoped V/Ti/O catalysts

The composition of samples prepared is summarized in Table 1. Two series of samples were prepared: the first one containing only vanadium oxide (Vx: with an amount ranging from 1 to 15 wt.% V_2O_5, with respect to the overall weight of the sample), and the second one with a fixed amount of vanadium (7 wt.% V_2O_5), and a variable amount of the two promoters.

Table 1
Samples prepared and their composition.

Sample, code	Wt. % V_2O_5	Wt. % Cs_2O	Wt. % Sb_2O_3
V1	1	-	-
V2	2	-	-
V3.5	3.5	-	-
V5	5	-	-
V7	7	-	-
V10	10	-	-
V15	15	-	-
Cs0.35	7	0.35	-
Sb0.7	7	-	0.7
Cs0.35Sb0.35	7	0.35	0.35
Cs0.35Sb0.7	7	0.35	0.7
Cs0.35Sb3.5	7	0.35	3.5

Figure 1 shows the Raman spectra of samples Vx. The Raman bands attributable to bulk V_2O_5 (at 998, 705, 483, 305, 285 cm^{-1}) were observed only for samples having an overall vanadium oxide content higher than 2 wt.%. Therefore, in samples V1 and V2, vanadium oxide was mainly present in the form of dispersed V species, while in samples having higher V_2O_5 content, both dispersed and segregated, bulk vanadium oxide were present. XRD evidenced the presence of crystalline V_2O_5 only in samples having more than 2 wt.% vanadia. It is worth mentioning that a mixture prepared by grinding together 1 wt.% V_2O_5 with TiO_2 led to a sample having a Raman spectrum and a XRD pattern in which the bands or reflections relative to bulk vanadia were clearly visible. This means that in samples V1 and V2 the absence of Raman bands and XRD reflections typical of V_2O_5 is due to the complete spreading of vanadium oxide on TiO_2, and not to a detection sensitivity limit of the techniques employed. Since the theoretical monolayer coverage in our samples should correspond to ≈3.4 wt.% V_2O_5 [14], the data evidence that with the preparation adopted, or due to the characteristics of the support employed, it is not possible to obtain a complete spreading of vanadium oxide over the support; in fact sample V3.5 already shows a non-negligible amount of bulk V_2O_5. Nevertheless, our data also agree with indications from literature, that only for very low vanadium oxide loadings (< than 20% monolayer), the isolated monomeric or oligomeric species are the only ones present in V/Ti/O systems [15].

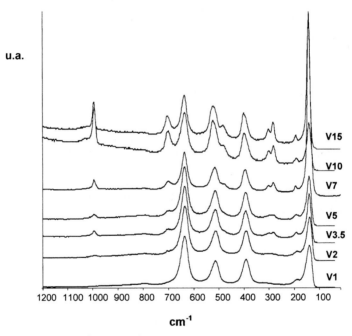

Figure 1. Raman spectra of Vx samples.

The determination of the relative amounts of the different V species can be done with TPR, since each species is characterized by a different reducibility, and hence by a defined reduction peak in the TPR profile [10,11,16]. Figure 2(left) shows the reduction profile of sample V2; in the case of samples having very low vanadium oxide content, e.g., V1 and V2, a single reduction peak was observed in the low temperature range, which clearly indicates the exclusive presence of isolated, easily reducible V species. For samples having higher vanadium oxide content, the reduction profile became more complex: three different peaks were present, which always overlapped, making the de-convolution procedure for assignment of each contribution not unambiguous. An example is given in Figure 2(right), which plots the profile of sample V5, and the result of the de-convolution procedure, when the three different contributions are assigned [7,11]: (i) at ≈ 500°C, for isolated V species, which are more easily reducible, (ii) at 550°C for polyvanadates and (iii) at 600°C for bulk vanadia, either amorphous or crystalline. It is evident that the complexity of the profile does not allow a reliable assignment of the relative contribution of each peak, and that the de-convolution of the experimental peak may suffer from a large degree of uncertainty.

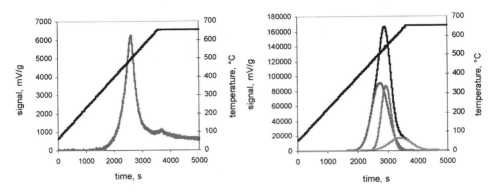

Figure 2. Left: TPR profile of sample V2. Right: TPR profile of V5 sample, and de-convolution into three reduction peaks

An additional problem arises when dopants are present which are themselves susceptible of reduction; for instance, the reduction of Sb in doped V/Ti/O samples occurs at temperature similar to that of isolated V or polyvanadate [17]. This is shown in Figure 3, which reports the TPR profile of a reference sample prepared by grinding 4.5 wt.% Sb_2O_3 and TiO_2, thermally treated in air at 450°C. The same Figure also reports the reduction profile of Sb_2O_3, calcined at 450°C; the latter was non reducible, while the interaction between antimony oxide and titania led to the development of a species characterized by a reduction peak at low temperature. This demonstrates that the interaction with titania is very strong, and leads to a profound modification of Sb oxide properties.

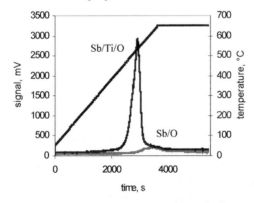

Figure 3. TPR profile of a Sb/O/Ti sample, prepared by grinding and calcination of 4.5 wt.% Sb_2O_3 with TiO_2 (Sb/Ti/O), and of reference Sb_2O_3 (Sb/O).

In order to overcome this problem, we have used an alternative approach, by taking advantage of the fact that in an oxidizing atmosphere (5% O_2 in He), supported vanadium oxide spontaneously releases molecular oxygen at 650°C [7]. A preliminary analyses evidenced that in TPO and in TPR+O tests, the amount of oxygen released was not simply proportional to vanadium oxide loading in catalysts, and that the self-reduction effect was not a peculiarity of all V species present in V/Ti/O catalysts. Therefore, we have investigated more thoroughly this aspect.

Figure 4 shows the TPO profiles of a few samples. Samples exhibited very small oxygen consumption (corresponding to positive peaks), indicating that V was in its highest oxidation state, or that it was eventually stabilized in a partially reduced state [18]. At 650°C, in the isothermal step, samples V3.5-V15 released molecular oxygen, as evident from the negative peak; on the contrary, this did not occur in the case of V1 and V2 samples. The spontaneous release of oxygen is due to the formation of the thermodynamically more stable vanadium sub-oxides, the so-called Magneli phases, or of VO_2. It is known that V_2O_5 spontaneously self-reduces to V_6O_{13} at temperatures higher than 550°C, under an oxidizing environment [19,20]. Data indicate that isolated vanadium, present as the only species in samples V1 and V2, does not release oxygen; therefore, the TPO profile can be used for the quantitative discrimination of this species from polyvanadate and bulk vanadia.

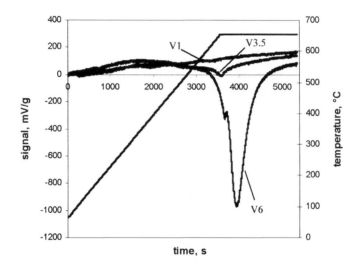

Figure 4. TPO profiles of some Vx samples.

Table 2 reports the amount of (i) isolated V, chemically interacting with the support, and of (ii) vanadium oxide which underwent spontaneous self-reduction (polyvanadate + bulk vanadia), as evaluated from the quantitative determination of the O_2 released. It is evident that samples V3.5 and V5, having intermediate vanadium oxide amount, possessed both V species, while samples having the highest loading (V10 and V15), did not possess isolated V at all. This agrees with literature indications, that for increasing vanadium oxide contents, the formation of dispersed polyvanadates is preferred over the formation of isolated V species [1,4,16].

Also when pre-reduced samples were re-oxidized (TPR+O tests), the oxidation profile showed a negative peak; in this case, however, the amount of released oxygen was different with respect to that one calculated from TPO profiles. Figure 5 compares the molar ratio between the amount of O_2 released and V, as functions of the vanadium oxide content in catalysts, for both TPO and TPR+O tests. When this ratio is equal to 0.167, all the V present in the sample self-reduces to V_6O_{13}, and this means that the sample does not contain isolated V species. It is shown that the largest difference between TPO and TPR+O was for samples V2, V3.5 and V5, while with samples having either less (V1) or more (V7, V10 and V15) vanadium oxide, the amount of oxygen released was similar in the two cases. This suggests that the pre-reduction in TPR+O tests induced changes in the distribution of the different V species; the isolated V was not stable, and the reduction cycle led to a segregation of part of it into polyvanadate or bulk vanadium oxide.

Table 2
Relative amounts of vanadium species in samples Vx, as evaluated from TPO measurements.

Sample	Relative amount, %[a]	Absolute amount, mmol V	Relative amount, %[a]	Absolute amount, mmol V
	Isolated V species		Polyvanadate + bulk vanadia	
V1	100	0.11	0	0
V2	100	0.22	0	0
V3.5	66	0.25	34	0.13
V5	55	0.30	45	0.25
V7	4	0.03	96	0.73
V10	0	0.00	100	1.07
V15	0	0.00	100	1.58

[a] relative amount, with respect to the overall V content.

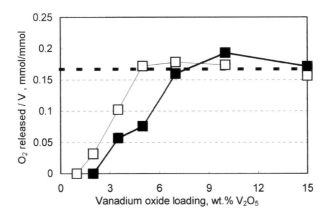

Figure 5. Molar ratio between the amount of O_2 released and the V content in samples V1-V15, as a function of vanadium oxide loading, calculated from TPO (■) and TPR+O (□) tests. Dotted line corresponds to the stoichiometric reduction of V_2O_5 to V_6O_{13}.

Samples were tested as catalysts for the oxidation of o-xylene to phthalic anhydride. Figure 6 summarizes the catalytic activity: the temperature at which 60% o-xylene conversion was reached is plotted as a function of vanadium oxide loading. The more active was the catalyst, the lower was the temperature necessary to reach a defined level of conversion.

160

Two different catalytic behaviours can be observed:
1. Samples V1 and V2 were the most active ones, despite the very low vanadium oxide content.
2. Sample V3.5 was the least active; therefore, a drastic fall of activity occurs when the vanadium oxide content is increased from 2 to 3.5 wt.%. Thereafter, for a loading higher than 3.5 wt.%, the activity increased again, but nevertheless remained lower than that of sample V2.

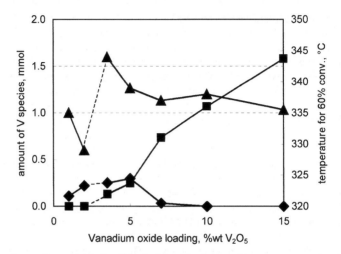

Figure 6. Effect of vanadium oxide loading in Vx samples on the temperature at which 60% o-xylene conversion is reached (residence time 0.25 s) (▲), on the absolute amount of isolated V species (♦), and on the absolute amount of polyvanadate + bulk vanadia (■).

The same Figure also reports the absolute amount of the isolated V species, and that of polyvanadate + bulk vanadia (the species which spontaneously released O_2 on heating), as evaluated from TPO profiles (Table 2). In the case of samples V1 and V2 (those having exclusively this V species), a relationship exists between the amount of the former species and the activity, while in the case of Vx samples having x ≥ 3.5, the activity increased for 3.5 ≤ x ≤ 7, and then remained approximately constant for x > 7. These data suggest that: (i) the isolated V species was that one characterized by the highest specific activity in the activation of o-xylene; (ii) the development of polyvanadate and bulk vanadia (as inferred from Raman spectra and XRD patterns), starting from sample V3.5, caused the activity of isolated V to be surpassed, probably because the latter species was covered by the former ones. From this catalyst onwards (3.5 ≤ x ≤ 15), the catalytic performance was determined by the amount of polyvanadate and bulk vanadia. Polyvanadate is known to be more active than V in bulk vanadia [1], but the latter species was likely the predominant one in samples V10 and V15; this led to the levelling off of activity in these samples.

In summary, the following are the main conclusions concerning undoped V/Ti/O systems:
1. By TPO, it is possible to calculate the amount of isolated V species and of polyvanadate + bulk vanadia;

2. The isolated V species is the most active one in the oxidation of o-xylene to phthalic anhydride;
3. Polyvanadates and bulk vanadia, if present, cover the isolated V and lead to a less active catalyst;
4. The isolated V species is not stable, and during the reduction step in TPR measurements a great part of it segregates to polyvanadates or bulk vanadia, especially in those samples having higher vanadium oxide content (> 3.5 wt.% V_2O_5).

3.2 Characterization and reactivity of Sb-, Cs- and Cs/Sb-doped V/Ti/O catalysts

The addition of Sb, or the combination of Cs and Sb, greatly affected the catalytic performance of V/Ti/O catalysts. As an example, Figure 7(left) compares the o-xylene conversion as a function of temperature for samples V7, Sb0.7 and Cs0.35Sb0.7 (all doped catalysts contained 7 wt.% V_2O_5). The presence of only Sb promoted the catalytic activity, but when both Sb and Cs were present, the activity increase was even more pronounced. It is worth reminding that Cs contents higher than 0.35 wt.% Cs_2O in Cs-doped samples [7], or Sb contents higher than 0.7 wt.% Sb_2O_3 in Sb-doped samples, had a negative effect on catalytic performance. Therefore, a sinergic promotional effect of the two elements occurred in sample Cs0.35Sb0.7.

The promoting effect of Cs on catalytic activity was studied in a previous work [7]. It was found that Cs has a positive effect on conversion even when amounts as low as 0.1 wt.% Cs_2O were added. Moreover, Cs profoundly affects the redox properties of V; in undoped V/Ti/O samples, the overall rate of o-xylene transformation was a function of the hydrocarbon concentration in feed, and was only slightly affected by oxygen concentration, while in the Cs-doped samples the rate was unaffected by the hydrocarbon concentration, and had a clear positive dependence on oxygen concentration. Thus, the presence of Cs modified the rate-determining step of the redox process. It was also found that the addition of Cs decreases V reducibility, but improves V re-oxidizability. Therefore, one effect of Cs on activity was related to the acceleration of V re-oxidation rate.

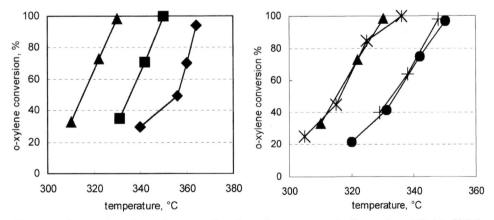

Figure 7. Conversion of o-xylene as a function of temperature. Left: catalysts V7 (♦), Sb0.7 (■), and Cs0.35Sb0.7 (▲). Right: catalysts Cs0.35 (●), Cs0.35Sb0.35 (✳), Cs0.35Sb0.7 (▲) and Cs0.35Sb3.5 (+). Residence time 0.13 s.

The effect of variations of Sb contents on activity, for samples containing 7 wt.% V_2O_5 and 0.35 wt.% Cs_2O, is shown in Figure 7(right); a small amount of Sb ($\leq$ 0.7 wt.% Sb_2O_3) improved the catalytic activity with respect to sample Cs0.35 (that is, without Sb). When an higher amount of Sb was added (sample Cs0.35Sb3.5), the activity became comparable to that of sample Cs0.35.

Table 3 reports the value of the molar ratio between released O_2 and V, as evaluated from negative peaks in TPO and TPR+O profiles, for the doped samples.

Table 3
Ratio between released O_2 and V, and amount of isolated V species in doped samples, as evaluated by means of TPR+O and TPO.

Sample	O_2 released/V, mmol/mmol	Isolated V, by TPR+O, %[a]	Isolated V, by TPR+O, mmol	Isolated V, by TPO, %[a]
V7	0.1648	0	0	4
Sb0.7	0.1521	12	0.09	nd
Cs0.35Sb3.5	0.1508	12	0.09	nd
Cs0.35Sb0.35	0.1370	20	0.15	nd
Cs0.35Sb0.7	0.1361	21	0.15	28

[a] relative amount, with respect to the overall V content.

The presence of the promoters decreased this ratio, and hence decreased the amount of polyvanadate + bulk vanadia, and correspondingly increased that of the isolated V species. The greater effect was obtained for the sample containing both Cs and Sb, when Sb was present in low quantity (samples Cs0.35Sb0.35 and Cs0.35Sb0.7). In these catalysts, the amount of isolated V species was comparable to the maximum one exhibited by Vx samples, that is the amount obtained with the V3.5 catalyst. Therefore, Sb favoured the development of a strong chemical interaction between V and the underlying support, with generation of a higher amount of isolated V species, and a correspondingly lower amount of polyvanadate and bulk vanadia. This is particularly important, because the latter species not only were the least active ones, but also covered the isolated, more active species, already when a very small amount of them was present (e.g., in sample V3.5). In the presence of Sb, instead, the isolated V species controlled the catalytic activity, even for relatively high amounts of vanadium oxide content. The effect was even more relevant in the co-presence of Sb and Cs. An excessive amount of Sb (sample Cs0.35Sb3.5) was instead detrimental, because the number of isolated active sites decreased with respect to samples having lower dopant quantity. Furthermore, it is likely that in this case the development of bulk antimony oxide led to the coverage of the active sites [21], also contributing to the decrease of activity.

One additional important consideration is that the presence of the dopant stabilizes the isolated V species, and avoids its segregation during the pre-reduction treatment in TPR+O tests. This is evident from the comparison of the amount of isolated V species for sample Cs0.35Sb0.7, as calculated from TPR+O and TPO analysis (Table 4); only a slight decrease of it occurred during the pre-reduction in TPR+O tests (from 28% to 21%). In the case of Vx samples, instead, the pre-reduction treatment led to the segregation of a great part of the isolated V species (see also Figure 5).

A confirmation of the effect of Cs and Sb on the formation of the different V species, has been obtained through a direct titration of active sites, following the method proposed by Wachs et al. [13]. Specifically, the adsorption of controlled amounts of methanol is a method to quantitatively evaluate the number of active sites in bulk or dispersed vanadium or molybdenum oxide. Figure 8 plots the amount of methanol adsorbed per unit weight of sample, as a function of the Sb content in samples Cs0.35Sbx; also, the temperature for 70% o-xylene conversion is reported. The number of sites able to adsorb methanol was the highest in samples Cs0.35Sb0.35 and Cs0.35Sb0.7, which also were the most active ones.

Table 4
Isolated V species, as evaluated by TPO and TPR+O tests.

Sample	Isolated V, by TPO, % [a]	Isolated V, by TPR+O, % [a]
V5	55	0
V7	4	0
Cs0.35Sb0.7	28	21

[a] relative amount, with respect to the overall V content.

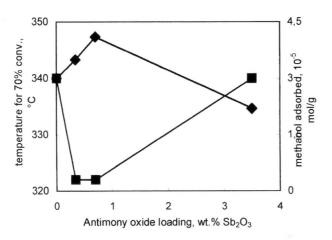

Figure 8. Amount of methanol adsorbed per unit weight of catalyst (♦), and temperature for 70% o-xylene conversion (■), as functions of the Sb content in Cs0.35Sbx samples. Residence time 0.13 s.

Figure 9 summarizes the results of catalytic tests; the temperature for 70% o-xylene conversion is plotted as a function of the absolute amount of the isolated V species, for both undoped and doped V/Ti/O samples containing 7 wt.% V_2O_5. A relationship is present, which confirms that this V species is that one which mainly determines the conversion of o-xylene, when suitable amounts of Sb and Cs are added. Sample V7 was the least active, because the amount of isolated V species was very low, and the activity was controlled by the polyvanadate species (Figure 6).

3.3 Conclusions: a model for Sb-doped V/Ti/O catalysts

The addition of a suitable amount of Sb and Cs in V/Ti/O makes possible to develop and stabilize an higher amount of isolated V species, with a correspondingly lower amount of polyvanadate and bulk vanadia, as compared to the undoped samples.

Grasselli et al. [8,21] proposed the development of an amorphous V/Sb/O compound over titania, in which an improved isolation of V species occurs as a consequence of the interruption of the monolayer continuity by Sb oxide. A slightly different picture was

164

proposed by Bruckner et al. [22], in which V species are spread over antimony oxide; the formation of amorphous VSbO$_4$ was also hypothesized. These models explain the effect of Sb when it is added to V/Ti/O, in amounts which are greater than those employed in the present work. In our case, it is possible to hypothesize that Sb ions develop a strong chemical interaction with titania surface, through formation of oxo bridges; this occurs for low Sb oxide amounts. Vanadium then interacts with Sb, and this interaction represents a more favourable situation for the generation of isolated V species, highly active in the oxidation of o-xylene. Indeed, the absolute amount of additional isolated V species which developed in the presence of Sb did approximately correspond to the amount of Sb added; this suggests the formation of specific Sb-O-V bonds. Even more favourable was the contemporaneous presence of Cs and Sb, with development of a higher relative amount of isolated V species.

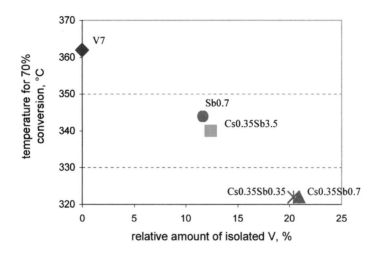

Figure 9. Temperature for 70% o-xylene conversion as a function of the relative amount of isolated V species, in different undoped (V7) and doped samples containing 7 wt.% V$_2$O$_5$.

REFERENCES
1. G. Centi, Appl. Catal. A, 147 (1996) 267.
2. P. Courtine and E. Bordes, Appl. Catal. A, 157 (1997) 45.
3. C.R. Dias, M.F. Portela and G.C. Bond, Catal. Rev.-Sci. Eng., 39(3) (1997) 169.
4. B. Grzybowska, Topics Catal., 21 (2002) 35.
5. J.C. Vedrine (Ed.), Catal. Today, 20 (1994) (Eurocat Oxide).
6. G. Deo and I.E. Wachs, J. Catal., 146 (1994) 335.
7. S. Anniballi, F. Cavani, A. Guerrini, B. Panzacchi, F. Trifirò, C. Fumagalli, R. Leanza and G. Mazzoni, Catal. Today, 78 (2003) 117.
8. U.A. Schubert, F. Anderle, J. Spengler, J. Zuehlke, H.J. Eberle, R.K. Grasselli and H. Knoezinger, Topics Catal., 15 (2001) 195.
9. J.Ph. Nogier, Catal. Today, 20 (1994) 23.
10. S. Besselmann, C. Freitag, O. Hinrichsen and M. Muhler, Phys. Chem. Chem. Phys., 3 (2001) 4633.
11. D.A. Bulushev, L. Kiwi-Minsker, F. Rainone and A. Renken, J. Catal., 205 (2002) 115.

12. M. Badlani and I.E. Wachs, Catal. Lett., 75 (2001) 137.
13. L.J. Burcham, L.E. Briand and I.E. Wachs, Langmuir, 17 (2001) 6164.
14. G.C. Bond and S.F. Tahir, Appl. Catal., 71 (1991) 1.
15. B. Grzybowska, Appl. Catal. A, 157 (1997) 263.
16. G.T. Went, L.J. Leu and A.T. Bell, J. Catal., 134 (1992) 479.
17. B. Pillep, P. Behrens, U.A. Schubert, J. Spengler and H. Knoezinger, J. Phys. Chem. B, 103 (1999) 9595.
18. F. Cavani, E. Foresti, F. Trifirò and G. Busca, J. Catal., 106 (1987) 251.
19. K. Devriendt, H. Poelman and L. Fiermans, Surf. Sci., 433 (1999) 734.
20. J. Haber, M. Witko and R. Tokarz, Appl. Catal. A, 157 (1997) 3.
21. J. Spengler, F. Anderle, E. Bosch, R.K. Grasselli, B. Pillep, P. Beherens, O.B. Lapina, A.A. Shubin, H.J. Eberle and H. Knoeziger, J. Phys. Chem. B, 105 (2001) 10772.
22. U. Bentrup, A. Brueckner, C. Ruedinger and H.J. Eberle, Appl. Catal. A., 269 (2004) 237.

Studies in Surface Science and Catalysis 155
A. Gamba, C. Colella and S. Coluccia (Editors)

How basic properties of MgO-based mixed oxides affect the catalytic performance in gas-phase and liquid-phase methylation of m-cresol

F. Cavani[1*], L. Maselli[1*], D. Scagliarini[1], C. Flego[2], C. Perego[2]

[1]*Dipartimento di Chimica Industriale e dei Materiali, Viale Risorgimento 4, 40136 Bologna, Italy. cavani@ms.fci.unibo.it. *INSTM, Research Unit of Bologna.*
[2]*EniTecnologie SpA, Via Maritano 26, 20097 S. Donato MI, Italy*

The reactivity of MgO-based mixed oxides in the gas-phase alkylation of m-cresol with methanol was studied, with the aim of finding relationships between basic surface features and catalytic performance. Conversion and products distribution were compared with those obtained in liquid-phase methylation. Catalysts investigated included: MgO, Mg/Al/O, Mg/Fe/O, Mg/Al/Zr/O, Mg/Al/Ce/O and H-Y zeolite, the latter used as reference for Brønsted-type activity. It was found that the regio-selectivity of the reaction is greatly affected by the presence of (i) medium-strength and strong basic sites, the latter playing a relevant role only in gas-phase reaction, and (ii) metal cations having acid, Lewis-type, coordination properties.

1. INTRODUCTION

The basic properties of transition, alkali and alkaline-earth metal oxides are affected by the physico-chemical characteristics of the metal ion, and by the nature of its bond with the neighbour oxygen anion [1-6]. When metal oxides are used as catalysts in transformations requiring the basic-catalyzed activation of organic substrates, high nucleophilicity of the O^{2-} species may over-increase the reactivity of the oxide. This implies a too strong interaction with the reactants and with all molecules which exhibit acid character, including CO_2 and H_2O. Strong interactions cause easy deactivation by adsorbed molecules, as reported for alkali metal oxides, like MgO or Cs_2O, which suffer from quick poisoning even by simple storing at room conditions. Preferred basic materials are some transition metal-based oxides, which exhibit medium-strength basic properties, and are indeed claimed to be the preferred catalysts for specific industrial basic-catalyzed processes [7]. An interest exists, however, for the use of alkali- and alkaline earth metal-based systems, due to their low toxicity, high structural stability, and lack of undesired redox reactivity. A wide literature exists on the control of the basic features of these materials through the formation of mixed oxides, realized by dissolution of guest elements, including transition metal ions, in MgO or in other oxides. These systems can be prepared by decomposition of hydrotalcite-like precursors, or by the sol-gel method [8-15]. Well-known examples are Mg/Al and Mg/Fe mixed oxides that represent a heterogeneous alternative to homogeneous catalysts (e.g. NaOH or KOH), the most widely used in liquid-phase applications.

In the present work we describe the preparation, characterization and reactivity of several mixed metal oxides, with the aim to achieve a control of the basic features of potential catalysts. In order to develop solid solutions of the different components, the preparation of the materials was carried out by precipitation of corresponding hydrotalcite-like precursors. The catalytic performance of the mixed oxides was evaluated in the gas-phase methylation of m-cresol. The methylation of phenols and of its derivatives is industrially applied for obtaining intermediates for the synthesis of antiseptics, dyes and antioxidants, and for the manufacture of polyphenylenoxide resins [16-20]. Moreover, the reaction between phenol derivatives and methanol represents a probe test of the surface basic properties of the materials, since basic features greatly influence the distribution of products. In previous works we examined the reactivity in the liquid-phase methylation of m-cresol of Mg/Al and Mg/Fe mixed oxides prepared from hydrotalcite precursors or synthesized via the sol-gel procedure [21-24]. This reaction, when carried out under mild conditions (i.e., in liquid phase), gives indication about the occurrence of different parallel reactions of methylation, either at the aromatic ring (*C*-methylation) or at the O atom (*O*-methylation), but suffers from low conversion and accumulation of heavy compounds, responsible for rapid catalyst deactivation.

2. EXPERIMENTAL

Catalysts were prepared from the corresponding hydrotalcite-like precursor [21,24]. The precipitate was then dried and calcined in air at 450°C for 8 h. The materials were characterised by means of: (i) X-ray diffraction analysis (Philips PW 1050/81; CuKα, λ 1.5406 nm, as the radiation source), (ii) surface area measurements (single point BET, Sorpty 1700 Carlo Erba), (iii) adsorption and thermal-programmed-desorption of CO_2 (PulseChemisorb 2705, Micromeritics), and (iv) atomic absorption spectroscopy (Philips PU 9100).

Adsorption of CO_2 was carried out on the pre-activated material (450°C, 2 hours, He flow) under He flow to avoid physisorption. Profiles of desorption of CO_2 were recorded up to 450°C, and then analysed with a deconvolution procedure. Curve fit quality was measured by the correlation coefficient (R^2), standard deviation (σ) and the Levenberg-Marquardt algorithm (χ^2). The quality of the deconvolution was high in all experiments: $R^2 > 0.996$, $\delta < 0.004$ and $\chi^2 < 1.133$.

The liquid-phase catalytic tests of m-cresol methylation were carried out in batch reactor, as described elsewhere [21-24]. The gas-phase catalytic tests were carried out in a continuous-flow reactor operating at atmospheric pressure in a temperature range between 250 and 450°C. A mixture of m-cresol and methanol (1/5 molar ratio) with a molar flow of 0.03 mol/h (F) was injected using a syringe pump in a N_2 flow of 60 mL/min (N_2/reactants = 5/1 molar ratio). The catalyst bed contained 1.5 g catalyst (30-60 mesh), which results in a W/F ratio of 50 g·h·mol^{-1}. For each test, the reaction products were collected during 60 min time-on-stream. The reaction mixture (liquid and gas phases) was analyzed using a gas chromatograph equipped with a HP-5 capillary column. The activation of the catalysts before reaction was carried out at 450°C for 3 hours in N_2 flow (gas-chromatographic grade).

For comparison acidic Y type zeolites (H-Y and USY) were used as supplied by Tosoh (HSZ-320HOA and HSZ-330HUA), without any other pre-treatment but the activation above reported.

3. RESULTS AND DISCUSSION

3.1 Characterization of catalysts

Table 1 summarizes the main features of the catalysts prepared. Figure 1 compares the XRD patterns for calcined catalysts with patterns of reference single metal oxides.

Table 1
Main features of the catalysts after thermal treatment.

Sample	Atomic ratios	Specific surface area, m^2/g
Mg	-	206
MgAl	Mg/Al 1.8/1.0	185
MgFe	Mg/Fe 2.0/1.0	149
MgAlCe	Mg/Al/Ce 5.3/1.8/1.0	120
MgAlZr	Mg/Al/Zr 4.7/0.5/1.7	126

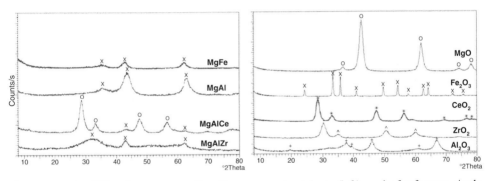

Figure 1. X-ray diffraction patterns of multi-component catalysts (left), and of reference single metal oxides (right). Left: MgO-like phase (✖), cerianite CeO_2 (O), $Mg_5Al_{2.4}Zr_{1.7}O_{12}$ (✳). Right: MgO periclase (O), Fe_2O_3 hematite (✖), cerianite CeO_2 (✳), ZrO_2 baddeleyte (^), γ-Al_2O_3 (+).

The precursors of bi-component systems (MgAl, MgFe) had XRD patterns corresponding to that one of the hydrotalcite-like structure ($Mg_6Me_2(OH)_{16}CO_3nH_2O$). In the case of MgAlCe, evidence was obtained for the additional presence of a Ce hydroxycarbonate phase. The MgAlZr precursor had a pattern similar to that obtained in the sole presence of Zr; therefore, the presence of Zr hindered the development of the crystalline hydrotalcite-like precursor.

After thermal treatment, XRD patterns evidenced the development of a MgO-like phase for MgAl and MgFe [21,23] (Figure 1). In the former, the shift towards higher 2θ values with respect to the reflections of MgO was due to a decrease in the crystallographic cell volume of periclase, in agreement with the progressive isomorphic replacement of Mg^{2+} cations with the smaller Al^{3+} cations (Mg^{2+} = 0.65 Å; Al^{3+} = 0.50 Å). This shift was not observed in the case of MgFe, due to the similar size of metal cations (Fe^{3+} = 0.69 Å). In the tri-component systems, the characteristics of the third component influenced the pattern [22]. In the case of MgAlCe, the dominant reflections were those of CeO_2 (cerianite), but reflections of the Mg/Al mixed oxide were also present. In MgAlZr pattern, the main reflections were relative to the Mg/Al mixed oxide, and only very weak reflections attributable to ZrO_2 were detected. The intense broad reflection centred at $2\theta \approx 32°$ did not correspond neither to zirconia nor to Mg/Al mixed oxide, while it corresponded to the most intense reflection of the compound $Mg_5Al_{2.4}Zr_{1.7}O_{12}$.

Basic properties of solids can be checked by means of adsorption and TPD of CO_2 [25]. The complex desorption profiles usually observed are related to the presence of basic sites with different strengths. For all samples, the desorption profiles consisted of: i) one low-temperature peak, with maximum at T < 100°C, and attributed to the interaction with sites having weak basic strength, ii) one peak with a maximum in the range 140-170°C, related to desorption of CO_2 from sites of medium basic strength, and iii) one broad desorption area which covered the temperature range from 250 to 450°C, attributed to CO_2 desorption from sites with strong basicity. In MgO-based mixed oxides, the replacement of octahedral Mg^{2+} by octahedral Al^{3+} generates an excess of positive charges which is compensated by cationic vacancies [26,27]. The oxygen anions adjacent to the defects are coordinatively unsaturated, and provide the strongest basic sites. On the other hand, fully-coordinated O^{2-} anions are less basic that the corresponding anions in MgO, due to the characteristics of the guest cation. Increasing amounts of Al^{3+} in the Mg/Al/O lattice cause an increase of the overall electronegativity of the solid, with a decrease of the electronic density of the unsaturated O species.

The results of the deconvolution procedure of the CO_2-desorption profile are reported in Table 2, together with the amount of CO_2 adsorbed at room temperature, and the overall amount of CO_2 desorbed. MgO and MgAlZr possessed the highest fraction of strong basic sites, while the opposite was true for MgAl and MgFe; MgAlCe had an intermediate behavior.

Table 2
Results of the CO_2 adsorption and of the deconvolution procedure of TPD profiles.

Sample (oxides)	Total amount of adsorbed CO_2		Total amount of desorbed CO_2,	Weak basic sites,	Medium basic sites,	Strong basic sites,
	μmol/g	μmol/m^2	μmol/g	μmolCO$_2$/g	μmolCO$_2$/g	μmolCO$_2$/g
Mg	356	1.73	346	56	25	265
MgAl	263	1.42	274	50	148	77
MgFe	159	1.07	143	33	64	46
MgAlCe	350	2.92	350	27	150	173
MgAlZr	429	3.40	441	18	76	347

3.2 Reactivity in the gas-phase m-cresol methylation

The catalytic performance of selected catalysts in gas-phase methylation is reported in Figure 2, where the conversion of m-cresol and the selectivity to the products are plotted as functions of the reaction temperature. Deactivation phenomena were evident for the H-Y zeolite (not reported), less relevant for MgAl and MgFe (with an initial activity decay followed by a stable catalytic performance), and substantially absent for MgAlCe and MgAlZr.

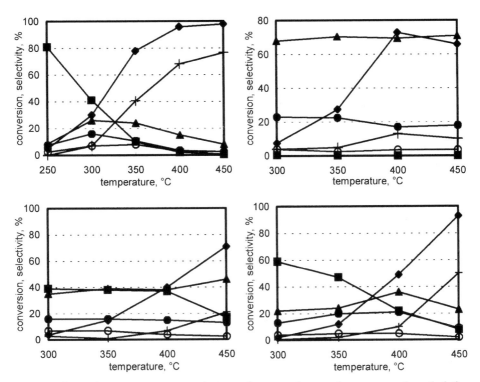

Figure 2. Effect of temperature on catalytic performance in gas-phase m-cresol methylation. Catalysts: MgAl (top, left); MgFe (top right); MgAlCe (bottom left); MgAlZr (bottom right). Symbols: m-cresol conversion (◆), selectivity to 3-methylanisole (■), 2,3-dimethylphenol (●), 2,5-dimethylphenol (▲), 3,4-dimethylphenol (○), and polyalkylates + dimethylanisoles (✚).

The rank of activity, expressed as overall conversion of m-cresol, was as follows: H-Y (HSZ-330) > MgAl > MgFe > MgAlZr ≥ MgAlCe > MgO. In the case of liquid-phase tests [22] the ranking was different: H-Y > MgAl > MgO ≥ MgAlCe ≈ MgAlZr ≥ MgFe. In gas-phase operation, differences of activity among samples were greater than in the liquid-phase, since the results were less affected by phenomena of accumulation of heavy compounds on catalytic surface. For example, in the case of MgFe, operation in liquid-phase evidenced the formation of heavy, diarylic compounds, originated from the dehydrogenation of methanol to yield formaldehyde, followed by hydroxymethylation of m-cresol (with the substitution of a –CH₂OH group on the aromatic ring of m-cresol), and by condensation of two molecules of this product to yield the diaryl compounds. The latter were responsible for rapid catalyst deactivation, and hence of the very low conversion achieved. This side reaction was not relevant in the case of the other catalysts, neither occurred with MgFe under gas-phase conditions.

Concerning the selectivity, the following indications are drawn from data shown in Figure 2:

1) MgAl gave 3-methylanisole as the main product at low temperature and low conversion. Its selectivity decreased with increasing the temperature, with a consequent increase in the formation of 2,5- and 2,3-dimethylphenols (the products of ortho-*C*-alkylation), and of polyalkylates (including small amounts of dimethylanisoles). The selectivity to products of mono-*C*-alkylation (2,5-, 2,3- and 3,4-dimethylphenols) reached a maximum, due to the relevant formation of polyalkylates at high temperature; the latter became the prevailing products at high m-cresol conversion. The selectivity to the product of para-*C*-alkylation (3,4-dimethylphenol) was lower than that to ortho-*C*-alkylated products. MgAl and MgAlZr had comparable behaviors, although the presence of a considerable fraction of strong basic sites in the former and of basic sites with medium-strength in the latter. The behavior of MgAlCe showed differences with respect to the other samples concerning (i) the higher stability of the selectivity to 3-methylanisole at increasing temperature, (ii) the higher formation of *C*-methylated compounds (especially 2,5-dimethylphenol), and (iii) the lower formation of polymethylated compounds (accordingly, the selectivity to the products of primary *C*-methylation did not show a maximum).

2) MgFe yielded almost exclusively products of ortho-*C*-methylation, with negligible formation of 3-methylanisole and very low formation of 3,4-dimethylanisole. The selectivity to polyalkylates was very low. A similar behavior was also observed with MgO (not shown in Figure 2), but in this case the conversion achieved was much lower (30% at 450°C), and the selectivity to 3,4-dimethylphenol was comparable to that of 2,3-dimethylphenol. Differences between MgFe and the other catalysts can not be ascribed exclusively to the different basicity of the materials, since the latter was not much different for MgFe and MgAl. Therefore, the presence of Fe plays a fundamental role in the reaction (see discussion below).

3) H-Y zeolite (not reported in Figure 2) was extremely active, since already at 250°C, 80% m-cresol conversion was reached; however, the latter remained constant up to 450°C. This was due to deactivation phenomena which overlapped to the activity-enhancement effect due to the increased temperature. The distribution of products was different from that obtained with the basic catalysts, since over the entire range of temperature examined (250-450°C) the prevailing compounds were the polyalkylates. Among the mono-alkylated compounds, the prevailing one was 2,5-dimethylphenol (selectivity around 20-25%), while the selectivities to 3-methylanisole, 2,3-dimethylanisole and 3,4-dimethylanisole were similar (around 10% each).

3.3 Comparison with the liquid-phase m-cresol methylation

Analogies and differences between liquid-phase and gas-phase methylation can be inferred from the main selectivity parameters obtained under the two different conditions: (i) the *O*/*C*-alkylation selectivity ratio, that is the ratio of selectivity between 3-methylanisole and the sum of 2,3-, 2,5- and 3,4-methylphenols; (ii) the ortho/para-*C*-methylation selectivity ratio, that is the ratio between the sum of selectivity to 2,3- and 2,5-dimethylphenols, and the selectivity to 3,4-dimethylphenol. The selectivity parameters were compared at similar conversion degree for all the catalysts: ≈30% for gas-phase and <15% for liquid-phase operation. These two conversion levels were chosen in agreement with the finding that at low m-cresol conversion, the selectivity parameters give an indication of the ratio between the rates of the various parallel reactions, without the interference of contributions deriving from consecutive reactions.

The main results obtained in the liquid-phase methylation of m-cresol [21-24] can be summarized as below:

1) The medium-strength sites, the prevailing ones in MgAl as compared to alumina or magnesia, played the major catalytic role in the reaction. Stronger sites were probably very soon deactivated under the reaction conditions employed (condensed phase, batch operation). An increase of the Al content in these systems (i.e., a decrease of the Mg/Al ratio in the 4 to 2 range), increased the number and relative amount of these sites, with an increase of the activity. The selectivity parameters were little affected by variations of the Mg/Al ratio in the mentioned range [21]. The same behaviour was observed, in both basicity and reactivity, with tri-component systems. In MgAlLa, the number of basic sites was lower than in MgAl, and in MgAlZr, the number of medium-strength sites was comparable to that of MgAl, but strong basic sites were also present [22]. An anomalous behavior was observed with MgAlCe, in which the low ortho/para-C-alkylation ratio (close to 2, as expected in the absence of specific adsorptive effects) was attributed to the coordinative properties of Ce^{4+}, with a preferred planar coordination of the aromatic ring. The O/C-alkylation ratio also was lower than with the other systems.

2) The importance of Lewis acid sites was confirmed by the comparison of MgAl and MgFe. The contemporaneous presence of medium-strength basic sites and Lewis acid sites in the latter catalyst (which, therefore, exhibited multifunctional properties) decreased the ortho/para-C-alkylation ratio and the O/C-alkylation ratio with respect to those obtained with MgAl having comparable Mg/Me ratio [23]. One major drawback of MgFe was the very low conversion, due to the accumulation of heavy compounds on catalyst under liquid-phase conditions.

3) A comparison of different preparation procedures for MgAl catalysts having different Mg/Al ratios (i.e., preparation by decomposition of an hydrotalcite-like precursor, and preparation via sol-gel route) confirmed that in liquid-phase the major role in m-cresol methylation was played by the medium-strength basic sites [24]. An increase of the number or density of these sites caused (i) an increase of activity, and (ii) a decrease of the O/C-methylation ratio (i.e. the ring alkylation became preferred over the etherification). Another relevant effect was the increase of the ortho/para-C-alkylation selectivity ratio. The presence of residual Brønsted acid sites (from alumina) had the opposite effect of a decrease of the ortho/para-C-alkylation ratio and of an increase of the O/C-alkylation ratio.

In liquid-phase conditions, the activity ranking was in agreement with the overall number of medium-strength basic sites, affected by both the catalyst surface area and the basic-strength distribution, with stronger sites being substantially inactive in the reaction. In gas-phase operation, the activity is less affected by these catalyst characteristics; at high reaction temperature it may be supposed that also the stronger basic sites contribute to the reaction. However this assumption is valid only under the hypothesis that the basic sites are directly involved in the rate-determining step of the reaction, and that the latter is represented by the H^+-abstraction from phenol (the most acid proton) with generation of the phenolate anion. The very high catalytic activity of the H-Y zeolite, both in liquid-phase and in gas-phase operation, is due to the well-known mechanism of Brønsted-type, acid-catalyzed activation of methanol, to generate strong electrophilic species which may either evolve with formation of olefins and aromatics, or give rise to the aromatic electrophilic substitution on m-cresol. Also the basic catalysts may direct their activity towards both m-cresol and methanol, the aromatic activation not being necessarily the rate-controlling step.

The comparison of the selectivity parameters between liquid-phase and gas-phase operations (Figure 3) allows the following considerations:

1) Gas-phase conditions favoured *C*-alkylation with respect to *O*-alkylation. The lower *O/C*-alkylation ratios may be attributed to the higher reaction temperature employed in gas-phase with respect to liquid-phase, even though the temperature was close to 300-350°C in both cases when most active catalysts were used.

2) MgO gave a low *O/C*-alkylation ratio in gas-phase operation, while gave the highest *O/C*-alkylation selectivity ratio in liquid-phase tests. Therefore, the presence of strong basic sites played a role in the distribution of products, although not in the rate-determining step of the reaction (see discussion above concerning m-cresol conversion). In gas-phase, the regio-selectivity of methylation was not only a function of the catalyst basic strength. In fact, MgAlZr gave a high *O/C*-alkylation ratio, despite a remarkable amount of strong basic sites, and MgFe gave a very low *O/C*-alkylation ratio, with negligible formation of 3-methylanisole, despite the predominance of medium-strength sites.

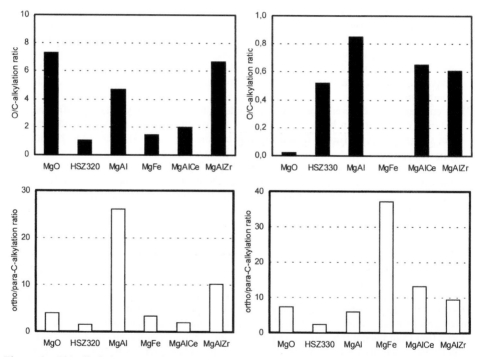

Figure 3. *O/C*-alkylation (top Figures) and ortho/para-*C*-alkylation (bottom Figures) selectivity ratios for m-cresol methylation in liquid phase (left Figures) and in gas phase (right Figures). Liquid-phase: m-cresol conversions < 15%, except HSZ-320HOA (26.6%) (in all cases, reaction temperature was 300°C, and reaction time 6h). Gas-phase: m-cresol conversion ≈ 30% (either experimental or interpolated), except HSZ-330HUA (78%, 250°C), and MgAlCe (40%, 400°C).

3) In gas-phase tests, MgFe gave a very low O/C-alkylation ratio, the highest ortho/para-C-alkylation ratio, and the highest selectivity to 2,5- and 2,3-dimethylphenols. Remarkably, this effect was relevant only in gas-phase conditions. The peculiarity of MgFe was also claimed by other authors [14], and has analogies with the behavior of Fe_2O_3 in 1-naphthol methylation [28]. The Lewis-type, coordinative properties of Fe cations have been claimed to be important in addressing the methylation in the ortho positions with respect to the hydroxy group (which is coordinated to the metal cation through the $-O^-$ moiety). Indeed, according to the Tanabe model for the adsorption of activated aromatic molecules on the surface of basic oxides [29], the presence of Lewis-type acid sites should rather favour the methylation in the para position (so decreasing the ortho/para-C-methylation ratio), due to the increased interaction of the aromatic ring with the metal cation. This effect was claimed to be the reason for the lower ortho/para-C-methylation ratio as compared to MgAl for tests in liquid phase (see Figure 3, left bottom) [23], but can not be applied for tests carried out in gas phase. Therefore, the adsorptive model, which takes into account only geometric factors for the adsorption of the aromatic compound, does not explain the very high ortho/para-C-methylation ratio and the very low O/C-methylation ratio of MgFe. It is likely that electronic factors may control the regioselectivity. One possible interpretation considers the decrease of the mesomeric delocalization effect of the negative charge of the phenolate anion in the aromatic ring, as a consequence of its interaction with the Fe cation. This might enhance the inductive effect of the $-O^-$ moiety, making the ortho positions more prone to undergo electrophilic substitution by activated methanol.

4) A low ortho/para-C-methylation ratio was exhibited by the H-Y zeolite, under both liquid-phase and gas-phase conditions. The relevant formation of the product of para-C-methylation is attributed to the generation of a strong electrophilic species as the consequence of the acid, Brønsted-type activation of methanol [30]. By contrast, with basic catalysts relevant differences were observed between liquid-phase and gas-phase tests. In the former case, the higher ortho/para-C-methylation was obtained with MgAl and MgAlZr. This behaviour was attributed to both the role of medium-strength basic sites, which presence addressed the methylation preferably in the ortho position [21], and the presence of metal cations having Lewis-type, acid characteristics, that according to the Tanabe's model favoured a planar coordination of the aromatic ring on the catalyst [22]. The co-presence of Lewis acid cations made the para position accessible for methylation, and caused the lower ortho/para-C-methylation ratio in MgFe and, at a lower extent, in MgAlCe. In gas phase, all systems but MgFe showed similar selectivity parameters. This suggests that under gas-phase conditions geometric adsorptive factors become less relevant, and that major deviations from a ortho/para-C-methylation ratio of 10 are only obtained with MgFe, a multifunctional system containing basic sites and a metal cation having relevant Lewis-type, coordinative properties.

4. CONCLUSIONS

Several MgO-based mixed oxides were prepared, characterized and tested as catalysts for the methylation of m-cresol. Relationships were evaluated between surface basic characteristics, as inferred from CO_2 adsorption and TPD, and activity and regio-selectivity for tests done in gas-phase and liquid-phase. Main conclusions were:

1) Under gas-phase conditions, which imply higher reaction temperature than in the liquid-

phase, even stronger basic sites played a role in the reaction, especially for what concerns the distribution of products, but the basic strength was not the main parameter which determined the catalyst activity. Lower O/C-alkylation selectivity ratios were obtained in gas-phase than in liquid-phase operation.

2) The ortho/para-C-methylation ratio ranking was different in liquid-phase and gas-phase conditions, especially for the bifunctional MgFe system, in which the role of Lewis-type, Fe^{3+} cations was the opposite under the two conditions. This has been attributed to the prevalence of an electronic effect over a geometric, adsorptive effect under gas-phase conditions, as a consequence of the interaction of the aromatic ring with the metal cation.

ACKNOWLEDGEMENTS

EniTecnologie is acknowledged for the PhD grant of DS. INSTM is acknowledged for the financial support (project PRISMA).

REFERENCES

1. Y. Ono, J. Catal., 216 (2003) 406.
2. G. Busca, Phys. Chem. Chem. Phys., 1 (1999) 723.
3. A. Cimino and F.S. Stone, Adv. Catal., 47 (2002) 141.
4. M. Barteau, Chem. Rev., 96 (1996) 1413.
5. A. Zecchina, D. Scarano, S. Bordiga, G. Spoto and C. Lamberti, Adv. Catal., 46 (2001) 265.
6. G. Spoto, E.N. Gribov, G. Ricchiardi, A. Damin, D. Scarano, S. Bordiga, C. Lamberti and A. Zecchina, Prog. Surf. Sci., 76 (2004) 71.
7. K. Tanabe and W.F. Hölderich, Appl. Catal., A: General, 181 (1999) 399.
8. T. Lopez, P. Bosch, E. Ramos, R. Gomez, O. Novaro, D. Acosta and F. Figueras, Langmuir, 12 (1996) 189.
9. F. Prinetto, G. Ghiotti, P. Graffin and D. Tichit, Microp. Mesop. Mater., 39 (2000) 229.
10. A. Guida, M.-H. Lhouty, D. Tichit, F. Figueras and P. Geneste, Appl. Catal. A, 64 (1997) 251.
11. A. Corma, V. Fornés, R.M. Martin-Aranda and F. Rey, J. Catal., 134 (1992) 58.
12. A. Corma, V. Fornés and F. Rey, J. Catal., 148 (1994) 205.
13. R.J. Davis and E.G. Derouane, J. Catal., 132 (1991) 269.
14. S. Velu and C.S. Swamy, Appl. Catal. A, 162 (1997) 81.
15. J. Sanchez Valente, F. Figueras, M. Gravelle, P. Kumbhar, J. Lopez and J.-P. Besse, J. Catal., 189 (2000) 370.
16. R.F. Parton, J.M. Jacobs, D.R. Huybrechts and P.A. Jacobs, Stud. Surf. Sci. Catal., 46 (1989) 163 and references therein.
17. S. Velu and C.S. Swamy, Appl. Catal. A, 119 (1994) 241.
18. S. Sato, R. Takahashi, T. Sodesawa, K. Matsumoto and Y. Kamimura, J. Catal., 184 (1999) 180.
19. T.M. Jyothi, T. Raja, M.B. Talawar and B.S. Rao, Appl. Catal. A, 211 (2001) 41.
20. V. Durgakumari and S. Narayanan, J. Mol. Catal., 65 (1991) 385.
21. M. Bolognini, F. Cavani, D. Scagliarini, C. Flego, C. Perego and M. Saba, Catal. Today, 75 (2002) 103.
22. F. Cavani, C. Felloni, D. Scagliarini, A. Tubertini, C. Flego and C. Perego, Stud. Surf. Sci.

Catal., 143 (2002) 953.

23. M. Bolognini, F. Cavani, C. Felloni, D. Scagliarini, C. Flego and C. Perego, in "Catalysis of Organic Reactions", D.G. Morrell (Ed), Marcel Dekker, New York, 2002, 115.

24. M. Bolognini, F. Cavani, D. Scagliarini, C. Flego, C. Perego and M. Saba, Microp. Mesop. Mater., 66 (2003) 77.

25. R.J. Davis, J. Catal., 216 (2003) 396.

26. T. Sato, T. Wakahayash and, M. Shimada, Ind. Eng. Chem., Prod. Res. Dev., 25 (1986) 89.

27. D. Tichit, M.-H- Lhouty, A. Guida, B.-H- Chiche, F. Figueras, A. Auroux, D. Bartalini and E. Garrone, J. Catal., 151 (1995) 50.

28. J. Wrzyszcz, H. Grabowska, W. Mista, L. Syper and M. Zawadzki, Appl. Catal. A, 166 (1998) L249.

29. K. Tanabe, Stud. Surf. Sci. Catal., 20 (1985) 1.

30. A. Corma, G. Sastre and P.M. Viruela, J. Mol. Catal. A, 100 (1995) 75.

Studies in Surface Science and Catalysis 155
A. Gamba, C. Colella and S. Coluccia (Editors)

Cyclohexane photocatalytic oxidative dehydrogenation to benzene on sulphated titania supported MoO$_x$

Paolo Ciambelli, Diana Sannino, Vincenzo Palma, Vincenzo Vaiano

Department of Chemical and Food Engineering, University of Salerno, 84084 Fisciano (SA), Italy.

Abstract

The effect of sulphate on the photocatalytic properties of MoO$_x$/TiO$_2$ catalysts for cyclohexane photocatalytic oxydehydrogenation has been investigated with a gas-solid continuous flow reactor. Photocatalytic tests performed on anatase titania at different sulphate content, ranging from 0 to 2 wt%, have shown that only the total oxidation of cyclohexane occurs. Cyclohexane conversion to carbon dioxide at steady state conditions decreases as the sulphate load increases.

Photocatalytic oxidative dehydrogenation of cyclohexane to benzene has been studied on MoO$_3$ 4.7 wt % supported on sulphated titania. The presence of sulphate species on the surface of titania enhances selectivity and yield to benzene as more as higher is the sulphate content. On the catalyst with the highest sulphate content very high selectivity to benzene was obtained, very weakly dependent of cyclohexane conversion.

The selectivity properties of MoO$_x$/TiO$_2$ catalysts are associated to the presence of both sulphate and polymolybdate species on the titania surface.

1. INTRODUCTION

Titanium dioxide is widely used in heterogeneous photocatalysis as a semiconductor photocatalyst because of its long-term stability, no toxicity and good, often the best, photocatalytic activity [1]. Since TiO$_2$ is active only under ultraviolet (UV) light in recent years transition metal ions doping has been widely performed by chemical synthesis and other methods, in order to improve photoactivity [2]. Karakitsou and Verykios [3] showed that doping with cations having a valence higher than +4 can increase the photoactivity, whereas Mu et al. [4] reported that doping with trivalent or pentavalent metal ions was detrimental to the photoactivity even in the UV region. Furthermore, according to a systematic study on the photoactivity and transient absorption spectra of quantum-sized TiO$_2$ doped with 21 different metals, the energy level and d-electron configuration of the dopants were found to govern the photoelectrochemical process in TiO$_2$ [5]. Even though the effects of metal doping on the activity of TiO$_2$ have been a frequent topic of investigation, it remains difficult to make general conclusions.

The presence of sulphate and metal oxides as dopant of titania has been reported to enhance the photooxidation reactivity of several organic compounds. It was shown [6, 7] that TiO$_2$ treatment with sulphuric acid could increase its photoactivity, although the mechanism was not clarified. Kozlov et al. [8] reported that the treatment of TiO$_2$ with sulphuric acid enhances the photocatalytic activity in acetone oxidation by 20–30%. Colon et al. [9], studying the photocatalytic degradation of phenol on sulphated TiO_2 in a slurry reactor, evidenced that the improvement of the photocatalytic activity is related to the optimisation of the redox step in the photocatalytic process instead of the acidity properties that could favour the adsorption of the organic substrate. Fu et al. [10] studied the structure of SO_4^{2-}/TiO_2 and its activity for room temperature photocatalytic oxidation (PCO) of CH_3Br, C_6H_6, and C_2H_4 in air. For catalysts calcined at 723 K, conversion of CH_3Br over SO_4^{2-}/TiO_2 was six times higher than over TiO_2. Moreover, TiO_2 deactivated faster than SO_4^{2-}/TiO_2; after 6 h of PCO, SO_4^{2-}/TiO_2 did not deactivate, whereas on TiO_2 conversion decreased from 88 to 20% for C_6H_6 and from 60 to 12% for CH_3Br. They concluded that the improved rate for SO_4^{2-}/TiO_2 was due to a greater surface area as well as a larger fraction of the anatase phase of TiO_2, which is more active than rutile for PCO.

Doping TiO_2 with metal oxides such as WO_3, MoO_3, and Nb_2O_5 increases surface acidity and PCO activity of TiO_2 [11-14]. Cui et al. [11] studied the activity of Nb_2O_5/TiO_2 during PCO of 1,4-dichlorobenzene. They found that surface acidity increased and photocatalytic activity doubled when niobium oxide was deposited on titania up to a monolayer. However, higher niobium loading did not increase surface acidity and PCO photocatalytic activity due to Nb_2O_5 segregation. They concluded that the same structural feature that enhances surface acidity increases PCO rate. Similarly, the authors found correlation between PCO activity and acidity for TiO_2 doped with WO_3, MoO_3, or Nb_2O_5 [12-14].

Cyclohexane photocatalytic oxidation has been studied on silica supported vanadia or polyoxytungstate catalysts [15-19] in slurry systems, yielding cyclohexanol, cyclohexanone, and polyoxygenates. On titania [20] cyclohexane photo-oxidation in dichlorometane leads to the formation of cyclohexene traces.

In gas-solid reactors the oxidation of cyclohexane and cyclohexene in humidified air at 303 K on P-25 titania powder [21] leads substantially to deep oxidation to CO_2. Very recently we have found that cyclohexane is selectively oxidised to benzene and cyclohexene on Mo-supported titania catalysts in the presence of gaseous oxygen at temperature of 308 K under UV illumination [22]. Higher molybdenum loading up to the monolayer resulted in higher benzene selectivity, whereas titania alone was 100% selective to carbon dioxide. The selective formation of benzene was associated to the presence of polymolybdate species on the titania surface. In order to elucidate the effect of sulphate content on MoO_x/TiO_2 activity and selectivity, in this work cyclohexane photocatalytic oxydehydrogenation has been studied on molybdenum oxide supported on titania at various sulphation extents in a gas-solid continuous flow reactor.

2. EXPERIMENTAL

2.1 Catalysts preparation

Three titania were used as support: two commercial titania samples (DT and DT51, Rhone Poulenc) with different sulphate content (respectively 0.5 wt% and 2 wt%) and an ultrafine sulphate-free titania produced by laser-pyrolisis [23]. The samples are named, respectively, T0, T5 and T20 with reference to the sulphate content. MoO_x-based catalysts were prepared by wet impregnation of titania with aqueous solution of ammonium heptamolybdate $(NH_4)6Mo_7O_{24}\cdot4H_2O$, followed by drying at 393 K and calcination at 673 K for 3 hrs.

2.2 Catalysts characterisation

Physico-chemical characterisation of catalysts was performed by different techniques. Thermogravimetric analysis (TG-DSC-MS) was carried out in air flow on powder samples with a thermoanalyzer (Q600, TA) in the range 293-1273 K with an heating rate of 10 K/min. Sulphate content has been evaluated from the weight loss in the range 623-1073 K, correspondent to SO_3 release [24].

Chemical analysis of molybdenum loading was performed by inductive coupled plasma-mass spectrometry (7500c ICP-MS, Agilent) after sample microwave digestion (Ethos Plus from Milestone) in HNO_3/HCl and HF/HCl mixtures.

Surface area and porosity characteristics were obtained by N_2 adsorption-desorption isotherm at 77 K with a Costech Sorptometer 1040. Powder samples were treated at 423 K for 2 h in He flow (99.9990%) before testing.

Laser Raman spectra of powder samples were obtained with a Dispersive MicroRaman (Invia, Renishaw), equipped with 785 nm diode-laser, in the range 100-2500 cm^{-1} Raman shift. FT-IR spectra of 1 wt% catalyst in KBr disk were obtained with a Bruker IFS 66 spectrophotometer in the range 4000-400 cm^{-1} (2 cm^{-1} resolution).

The zero point charge (ZPC) of supports and catalysts was determined by mass titration according to Noh and Swarz [25].

2.3 Catalytic tests

In Fig.1 a schematic picture of the experimental apparatus is reported. Oxygen and nitrogen were fed from cylinders, nitrogen being the carrier gas for cyclohexane vaporised from a temperature controlled saturator. The gas flow rates were measured and controlled by mass flow controllers (Brooks Instrument). The reactor inlet or outlet gas were fed to an on-line modified quadrupole mass detector (MD800, ThermoFinnigan) and a continuous CO-CO_2 NDIR analyser (Uras 10, Hartmann & Braun).

The annular section of the photocatalytic reactor was realised with two axially mounted 500 mm long quartz tubes of 140 and 40 mm diameter, respectively. The reactor was equipped with seven 40 W UV fluorescent lamps providing photons wavelengths in the range from 300 to 425 nm, with primary peak centered at 365 nm. One lamp (UVA Cleo Performance 40 W, Philips) was centered inside the inner tube while the others (R-UVA TLK 40 W/10R flood lamp, Philips) were located symmetrically around the reactor. Both photoreactor and lamps were covered with reflectant aluminum foils. In order to avoid temperature gradients in the reactor caused by irradiation, the temperature was controlled to 308 $\pm$ 2 K by cooling fans.

The catalytic reactor bed was prepared in situ, by coating quartz flakes previously loaded in the annular section of a quartz continuous flow reactor with an aqueous slurry of catalysts powder. The coated flakes were dried at 393 K for 24 hours in order to remove the excess of physisorbed water. This treatment resulted in uniform coating well adhering to the quartz flakes surface. The amount of deposited catalyst, evaluated by weighing the reactor before and after the coating treatment, was 20 g. Catalytic tests were carried out feeding 830 cm^3/min (STP) N_2 stream containing 1000 ppm cyclohexane, 1500 ppm oxygen and adding 1600 ppm water to minimise catalyst photodeactivation. Lamps were switched on after complete adsorption of cyclohexane on catalyst surface.

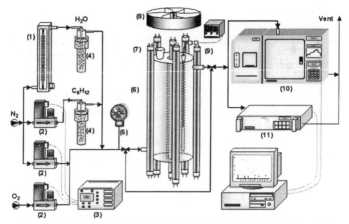

Experimental set up apparatus: (1) rotameter; (2) mass flow controllers; (3) MFC control
unit; (4) cycloexane and water saturators; (5) manometer; (6) anular photoreactor; (7) UV
lamps; (8) cooling system; (9) thermocouple; (10) GC-MS; (11) CO-CO2 analyzer.

Fig. 1. Schematic picture of the experimental apparatus

3. RESULTS AND DISCUSSION

The list of catalysts and their characteristics is reported in Table 1.

Table 1.Catalysts and their characteristics

Sample	Specific surface area (BET), m^2/g	MoO_3 content, wt %	MoO_3 surface density, $\mu mol/m^2$	SO_3^*, wt %	SO_4^{2-} surface density, $\mu mol/m^2$
T0	88	-		-	-
Mo/T0	88	4.5	3.5	-	-
T05	67	-		0.5	0.8
Mo/T05	68	4.2	4.3	0.5	0.8
T20	71	-		1.9	2.8
Mo/T20	68	4.4	4.5	1.7	2.6

*by evaluation of TG weight loss in the range 623-1073 K.

Blank experiments were performed in order to verify that cyclohexane was converted in heterogeneous photocatalytic process. A control test was carried out with the reactor loaded with uncoated quartz flakes. No conversion of cyclohexane was detected during this test, indicating the necessity of the catalyst for the observed reaction. A second test was performed with the catalyst loaded reactor, but switching the lamps off even after cyclohexane adsorption equilibrium was reached. In these conditions the composition of the outlet reactor was identical to that of the reactor inlet, indicating that no reaction occurred in dark conditions.

In Fig.2, cyclohexane apparent conversion on T0, T05 and T20 as a function of illumination time is reported. Photocatalytic test performed on these samples showed that cyclohexane conversion reached a maximum after about 5 min of illumination time and then decreased for all catalysts to reach a steady state conversion of about 2%, 5% and 8% on T2, T05 and T20, respectively.

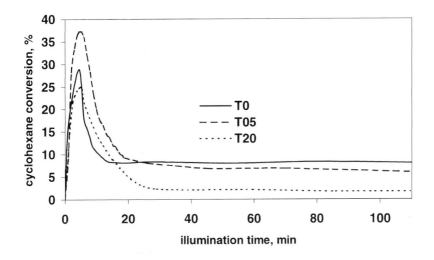

Fig. 2. Cyclohexane apparent conversion on T0, T05 and T20 as a function of illumination time

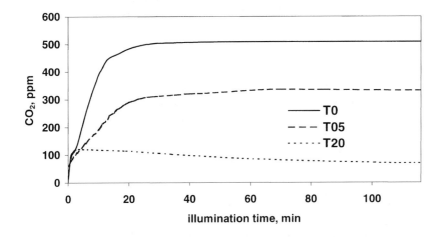

Fig. 3. CO_2 concentration in the outlet reactor on T0, T05 and T20 as a function of illumination time

184

In Fig.3 CO$_2$ production on T0, T05 and T20 as a function of illumination time is reported. For all catalysts, carbon dioxide was the only product detected in the gas phase and started forming immediately after lamp on, reaching steady state values (510 ppm, 335 ppm and 70 ppm on T0, T05 and T20 respectively) after about 110 min. Therefore, cyclohexane conversion and CO$_2$ yield decreased with the sulphate content. It has been reported that the activity of titania increases with sulphate content, likely due to the acidic characteristics of the most intermediates of photocatalytic oxidation [26, 27]. On T0, T05 and T20, experimental results suggest that during photooxidation of cyclohexane the formation of less acidic intermediates should be involved. Investigations on the mechanism reaction is currently in progress.

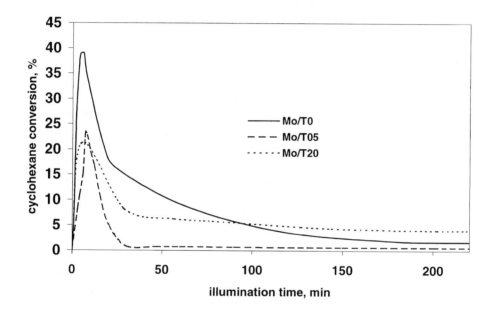

Fig. 4. Cyclohexane apparent conversion on Mo/T0, Mo/T05 and Mo/T20 catalysts as a function of illumination time.

The apparent cyclohexane conversion on Mo/T0, Mo/T05 and Mo/T20 is shown in Fig. 4. The analysis of the outlet stream disclosed the presence of benzene, cyclohexene (less than 1 ppm) and CO$_2$ for all catalysts. A maximum value of conversion was reached after about 5 min, then activity decreased approaching a steady state conversion. On Mo/T0 the maximum cyclohexane conversion was about 40 %, decreasing to 17% in 25 minutes, less quickly with respect to Mo/T05 and Mo/T20. On Mo/T20 the initial maximum conversion was lower (about 21% after 8 min of illumination), reached 7% after 30 min and 4% after 220 min. On Mo/T05 the initial maximum conversion was higher with respect to Mo/T20 (about 23%), while steady state conversion was about 1% after 60 min.

In Fig.5 benzene selectivity as a function of cyclohexane conversion is reported. Low benzene selectivity and high selectivity to CO$_2$ were observed on Mo/T0 in the whole conversion range. On Mo/T05 benzene selectivity ranged from 35 to about 70%. On Mo/T20 benzene selectivity reached values higher than 80 % (86% with a cyclohexane conversion of 6%). The dependence of benzene selectivity on cyclohexane conversion for all catalysts is less strong with respect to the usually found effect in hydrocarbon catalytic partial oxidation processes especially in the case of Mo/T20, for which the selectivity to benzene is very weakly decreasing with conversion.

Therefore, the presence of sulphate species on titania surface enhanced benzene yield as more as higher is the sulphate content.

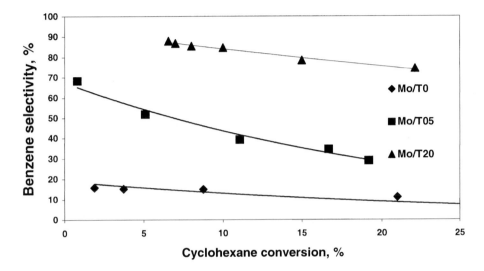

Fig. 5. Benzene selectivity versus cyclohexane conversion on Mo/T0, Mo/T05 and Mo/T20.

Table 2 shows that the value of ZPC of T0 is 6, according to the amphoteric character of anatase titania. The presence of sulphate increases surface acidity, leading to ZPC value of 4.7 on T05. As the sulphate load increases, ZPC decreases to 2.

Table 2. ZPC, FTIR and Raman spectra of support and catalysts.

Catalyst	ZPC pH unit	Mo=O stretching (FTIR) cm^{-1}	Mo=O stretching (Raman) cm^{-1}
T0	6	-	-
Mo/T0	3.8	947	950
T05	4.7	-	-
Mo/T05	1.8	951	956
T20	2	-	-
Mo/T20	1.7	957	958

The comparison of ZPC values of catalysts and the relevant support indicates that the presence of molybdenum confers, in all cases, strongest acidity than that of the relevant support. However, selectivity to benzene doesn't appear directly related to the lower values of ZPC (see values for Mo/T05 and Mo/T20).

FTIR spectra of TiO_2 supported catalysts show MoO_x species bands overimposed to typical absorptions from the support. Sulphate absorption main bands are located at 1050 and 1132 cm^{-1} on T05 and, with enhanced intensity, on T20. For all Mo-based catalysts no evidence for MoO_3 crystallites (sharp bands from Mo=O stretching vibrations at 992 cm^{-1} and bulk vibrations and 820 cm^{-1}) are present [28]. Bands at 947 cm^{-1} on Mo/T0, at 951 cm^{-1} on Mo/T05 and at 957 cm^{-1} on Mo/T20 are observed, assigned to terminal Mo=O stretching of polymeric surface species [29]. In the Raman spectra, these modes are shown at 950 cm^{-1} on Mo/T0, at 956 cm^{-1} on Mo/T05 and at 958 cm^{-1} on Mo/T20, all characteristic of octahedral MoO_x species. The increasing of wavenumber can be attributed to a higher degree of polymerisation of Mo species [30].

In a previous work [22] heterogeneous oxidative dehydrogenation of cyclohexane to benzene has been investigated on MoO_x supported on different oxides, like TiO_2, α- or γ- alumina, or on Mo-exchanged FER zeolites [31]. We found that polymolybdate species change the photoactivity of titania, favouring the formation of benzene as well as high the polymerisation degree is. Nevertheless, when supported on α- or γ- alumina, they promote only the formation of carbon dioxide. Thus, the ability to give oxidative dehydrogenated products seems to be related to both the presence of highly aggregated (not segregated) MoO_x species and a photocatalytic support.

The present results show that highly polymerised MoO_x species are present on Mo/T20 catalyst that gives the best benzene yield. Even if, due to the presence of sulphate, the higher polymerisation degree could be associated to the lower surface area available to molybdenum oxide species deposition, leading to an higher surface density, the whole results suggest an active role of sulphate in promoting the selectivity to benzene. Further investigations to elucidate the reaction mechanism are in progress. Recent photocatalytic tests in a fluidised bed reactor with a polymolybdate richer sulphated catalyst have given cyclohexane conversion up to 9% with selectivity of about 98 % at 393 K [32].

3. CONCLUSIONS

We have found that in photocatalytic tests performed on anatase titania with different sulphate content deep oxidation of cyclohexane occurs. The steady state conversion of cyclohexane to carbon dioxide decreases as the sulphate load increases.

Photocatalytic oxidative dehydrogenation of cyclohexane to benzene has been observed on Mo-supported on sulphated titania. The presence of sulphate species on the surface of titania enhances benzene yield as more as higher is the sulphate content. Therefore, sulphate species together with polymeric octahedral Mo species seem to be necessary to obtain selective cyclohexane photodehydrogenation to benzene.

REFERENCES

1. L. Palmisano, A. Sclafani in "Heterogeneous photocatalysis", M. Schiavello (ed), 1977 John Wiley & Sons.
2. O. Legrini, E. Oliveros, A.M. Braun, Chem.Rev.2 (1993) 671.
3. K.E. Karakitsou, X. E. Verykios, J. Phys. Chem. 97 (1993) 1184.
4. W. Mu, J.M. Herrmann, P. Pichat, Catal. Letters 3 (1989) 73.
5. H. Yamashita, Y. Ichihashi, M. Takeuci, S. Kishiguchi, M. Anpo, J. Synchrotron Rad. 6 (1999) 451.
6. J.C. Yu, J. Yu and J. Zhao, Appl. Catal. B: Environ. 36 (2002) 31.
7. D.S. Muggli and L. Ding, Appl. Catal. B: Environ. 32 (2001) 181.
8. D. Kozlov , D. Bavykin, E. Savinov, Catal. Letters 86 (2003) 169.
9. G. Colón., M.C. Hidalgo, J.A. Navìo, Appl. Catal. B: Environ. 45 (2003) 39.
10. X. Fu, Z. Ding, W. Su, Chin., J. Catal 20 (1999) 321.
11. K. Cui, S. Dwight, A. Soled, Wold, Solid State Chem. 115 (1995) 187.
12. J. Papp, S. Soled, K. Dwight, A. Wold, Chem. Mater. 6 (1994) 496.
13. S. Okasaki, T. Okuyama, Bull. Kor. Chem. Soc. Jpn. 56 (1983) 2159.
14. W. I. Lee, G. J. Choi, Y. R. Do, Bull. Kor. Chem. Soc. 18 (1997) 667.
15. A. Maldotti, R. Amadelli, G. Varani, S. Tollari, F. Porta, Inorg. Chem., 33 (1994) 2968.
16. A. Maldotti, R. Amadelli, V. Carassiti, A. Molinari, Inorg. Chim. Acta, 256 (1997) 309.
17. A. Molinari, A. Maldotti, R. Amadelli, A. Sgobino, V. Carassiti, Inorg. Chim. Acta, 272 (1998) 197.
18. C. Giannotti, C. Richter, Int. J. Photoenergy, 1 (1999) 1.
19. K. Teramura, T. Tanaka, T Yamamoto, T. Funabiki, J. Mol. Catal. A: Chemical, 165 (2001) 299.
20. C. B. Almquist, P. Biswas, Appl. Catal. A 214 (2001) 259.
21. H. Einaga, S. Futamura, T. Ibusuki, Appl. Catal. B: Enviromental 38 (2002) 215.
22. P. Ciambelli, D. Sannino, V. Palma, V. Vaiano, Catal. Today, in press.
23. F. Curcio, M. Musci, M. Notaro, G. Quattroni, Mat. Sci. Mon. 66D (1991) 2569.
24. P. Ciambelli, M. E. Fortuna, D. Sannino, A. Baldacci , Catal. Today 29 (1996) 161.
25. J. Noh, J. Schwarz, J. Colloid Interface Sci. 130 (1989) 157.
26. D.V. Kozlov, E.A. Paukshtis, E.N. Savinov, Appl. Catal. B:Environmental 24 (2000) L7.
27. M.L. Sauer, D.F. Ollis, J. Catal. 158 (1996) 570.
28. H. Matralis, S. Theret, Ph. Sebastians, M. Ruwet, P. Grange, Appl. Catal. B 5 (1995) 271.
29. S. K. Maity, M. S. Rana, S. K. Bej, Appl. Catal. A 205 (2001) 215.
30. C. P. Cheng, G. L. Schrader, J. Catal. 60 (1979) 276.
31. P. Ciambelli, D. Sannino, V. Palma, V. Vaiano, Prepr. 6° Italian Congress on Zeolites, Vietri sul Mare, Italy, 20-23 Sept. 2003, p. 125.
32. P. Ciambelli, D. Sannino, V. Palma, V. Vaiano, S. Vaccaro, 7 th World Congr. Chem. Eng., Glascow, Scotland, 10-14 July 2005, submitted.

Studies in Surface Science and Catalysis 155
A. Gamba, C. Colella and S. Coluccia (Editors)
© 2005 Elsevier B.V. All rights reserved

ZEOLITE-CHROME TANNING: FROM LABORATORY TO PILOT SCALE

P. Ciambelli[a], D. Sannino[a], B. Naviglio[a], A.M. Manna[a], V. Vaiano[a], G. Calvanese[b], G. Vietri[a], S. Gallo[a]

[a]Dipartimento di Ingegneria Chimica e Alimentare, Università di Salerno, 84084 Fisciano (SA), Italy
[b]Stazione Sperimentale per l'Industria delle Pelli e delle Materie Concianti, Napoli, Italy

ABSTRACT

Synthetic Na-zeolites (NaA, NaX, NaAX) were investigated as tanning agents in leather production from sheepskin and calfskin pelts. It was found in laboratory scale testing that the combined use of zeolite and chrome sulphate results in both higher float exhaustion and higher shrinkage temperatures in shorter time than in conventional chrome tannage. The best results were obtained with zeolite NaA. The operating conditions were optimized with respect to chrome and zeolite concentration, tanning bath pH and pelt weight/bath volume ratio. The analysed Si/Al ratio in the leather is about 1.1, similar to that of NaA zeolite.

The promoting effect of zeolite A in chrome tanning has been confirmed on pilot scale: higher tanning rate as well as higher float exhaustion were found. Mechanical properties of the finished leathers, such as tensile strength, elongation, tearing strength and ball bursting are similar or better when compared with chrome usually tanned leather.

1. INTRODUCTION

The tanning process converts unstable raw hides into leather with adequate strength properties and resistance to various biological and physical agents. Skin matrix is primarily composed of collagen, accounting for 60-80% of skin dry matter. Collagen is a fibrillar protein in which each helical macromolecule is constituted by a sequence of amino acids (primary structure) that assume local spatial configuration (secondary structure). Its tertiary structure consists of three peptide chains stranded in a triple helix and interconnected by a system of crosslinking bonds, perpendicular and parallel to the molecules, forming intramolecular and intermolecular bridges. Triple helical macromolecular aggregates form a fibrillar arrangement (quaternary structure), stabilised by various non-covalent and covalent intermolecular crosslinks [1 - 6]. Therefore, the tanning process consists in introducing additional crosslinks into collagen, which bind the active groups of the tanning agents to functional groups of the protein. In particular, every tanning molecule may react with several identical or different polar groups of peptide chains, forming supplementary water-resistant bridges. The introduction of intermolecular covalent crosslinks between collagen molecules is responsible for the mechanical and thermal stability of the fibrous network [7 - 9], that is related to the transition from the triple helix to a randomly coiled form. The bonds which stabilize the superhelix are hydrogen, hydrophobic, van der Waal's bonds and interactions between oppositely charged residues on side chains. All these non covalent bondings break down on heating, the breakdown starting at the weakest points of the helix, between the stabilizing clusters. A small region containing a few linkages of low energy will act as a favorable site to initiate denaturation. The temperature of this phase transition is evaluated either as denaturation temperature, manifested by an abrupt change in viscosity and optical rotation of colloidal collagen upon heating, and as shrinkage temperature, manifested by change in solid sample length and/or stress [10]. Covalent bonds increase the size of cooperating units through inter- and intramolecular cross-links and increase the temperature of denaturation (Td). Therefore tanning, by introducing cross-links, generally increases Td, but its influence on the shrinkage behaviour varies with its nature and the kind of reactive groups which are involved [11 - 13]. Different factors, classified as biological and non biological, influence either the denaturation and shrinkage temperature [14]. Moreover, the content of water in the material strongly affects the value of Td; in particular the lower the water content, the higher the Td [15]. Therefore the thermal stability of collagen is used to evaluate the modifications occurring in skin manufacture process after the tanning stage [16].

The choice of the tanning method depends chiefly on the properties required in the finished leather, the cost of the alternative materials and the type of raw pelt. Mechanico-physical and merceological properties of finished leathers are evaluated with respect to requirements fixed by regulations.

Chrome tanning is the most important way to obtain light, inexpensive leather of high hydrothermal stability (Ts>100°C) and good merceological properties (feel, fullness, softness, dye affinity, flexibility). In the tanning step, trivalent chrome basic sulphate salts are employed at pH ranging between 2 and 3. The mechanism of chrome tanning involves the crosslinks formation (by coordinate covalent bonds) between bi- or polynuclear chrome ions and side-chain carboxyl sites of aspartic and glutamic acids of collagen fibrils [17 - 19]. However, the current methods of tanning allow exhaustion levels of only 40–70% of tanning agents to be achieved. Thus, chrome tanning process results in the presence of trivalent chrome in tannery wastewaters and sludges as well as in the by-production of solid residues. Such poor exhaustion levels would lead to a discharge of chrome equivalent to nearly 160,000 tons/year of tanning salts if not effectively managed [20]. In recent years, the environmental regulatory agencies in many countries have came out with stringent limitations for permissible chrome concentrations in wastewater [21, 22], while social awareness to the environmental effect of chrome discharge is growing. The reduction of chrome emissions can be obtained by (i) improvement of chrome fixation with higher float exhaustion, (ii) recovery and/or recycle of exhausted liquors [23], (iii) wet-white pretanning with Al, (iv) tanning with other mineral agents.

Aluminum tanning [24] is the most ancient process of mineral tanning producing good white leather. However, the leather shows poor stabilization (75-85°C) due to weak interaction aluminum-collagen and aluminum salts hydrolysis, low resistance to washing with loss of fullness and softness, and weakening of fibers during storage. Therefore aluminum is often used as pretanning agent, to obtain stabilized pelt materials that can be subjected to mechanical operations (splitting, shaving, and trimming) and then to any kind of tanning, eliminating chrome pollution from hide waste.

Another alternative to traditional process involves silicates as tanning agents. In sodium silicate tanning [25], silica monomer $Si(OH)_4$ polymerizes by condensation to form high molecular weight colloidal particles that stabilizes the skin through the formation of hydrogen bonds between the silanol groups, $-Si(OH)$, and the polar groups of collagen. In combined chrome tanning, polymeric silica interacts with chrome salts through coordination bonds, thus silicate pre-treatment, besides of stabilizing the skin, favours the chrome salts penetration and fixation in the collagen structure. In recent years, synthetic zeolite has been used in wet-white pretanning and pretanned leather [26] exhibiting better properties than other wet-white pretanning methods. Pretanned leather is color-neutral (white), durable and resistant, readily machine-processable (*shavable*), dimensionally stable substrate. Subsequent tanning processes makes it possible to manufacture different types of leather. The tanning properties of synthetic zeolites and their effectiveness in the exhaustion of chrome tanning float have been previously investigated: the combined use of zeolite and chrome results in both higher float exhaustion and higher shrinkage temperatures in shorter time than in conventional chrome tanning [27 - 29].

In this work we report the results obtained with different synthetic Na-zeolites (A, X, AX) in tanning processes, alone or in combination with chrome. Different tests were performed with sheepskins and calfskin pelts and the results compared to those obtained with conventional chrome tanning process. The optimization of the tanning process has been effected by evaluating chrome float exhaustion and hydrothermal pelt stabilisation as function of operating conditions. Finally, chrome-zeolite tanning process carried out at best operating conditions has been transferred from laboratory to pilot scale. Determination of mechanical properties of the finished leathers, such as tensile strength, elongation, tearing strength and ball bursting were also performed .

2. EXPERIMENTAL

2.1 Laboratory scale experiments

Sodium-zeolites, NaA, NaX, and mixed NaAX were supplied by Sasol Italia. Sodium carbonate and basic chrome sulphate 26/33 of high purity grade (supplied by Biokimica) were used. Pickled sheepskins and calfskins were processed in cylindrical laboratory drums (35 cm diameter, 20 cm length), using a bath containing chrome, alone or in combination with zeolite. Conventional chrome tannage (29.1g/l of Cr in the float, 8% of fleshed pelt weight), tannage with 3 wt % (referred to the fleshed weight) NaA zeolite and chrome tanning runs in presence of 3 wt% zeolite (NaA, NaX, and mixed NaAX) were performed. Pelt and chrome solution were loaded in the drum and after 15 min of rotation, zeolite was added and rotation maintained up to tanning times of 0.5; 2; 6; 24 hrs. After these times float and hide were recovered and analyzed. Using the same procedure, tanning runs at different operating conditions were performed. In each run chrome-zeolite NaA tanning tests were carried out at different values of the processing parameter examined. In the first run, tanning tests carried out at different initial bath chrome concentrations of 13.6, 18.0 and 27.1 g l^{-1} (3 wt% zeolite added and initial bath pH = 2). In the second run, tanning tests were carried out at initial zeolite amounts of 1, 2, and 3 wt% (27.1 g l^{-1} chrome concentration and initial pH bath = 2). In the final tanning run, tests were carried out at initial bath pH values of 2 and 3 (27.1 g l^{-1} chrome concentration and 3 wt% zeolite). Samples were processed at temperature of 25°C and bath volume/pelt mass ratio of 1. In each run, four hide pieces were processed at different tanning times (0.5; 2; 6; 24 h). Leather samples and exhausted chrome floats were characterised, as reported in [30], evaluating moisture and volatile substances content (I.U.C. 5), ashes content (I.U.C. 7), Al, Si, Cr contents in leather (UNI ISO 10887), shrinkage temperature (I.U.P. 16) Ts, Cr chemical analysis in the floats (I.U.C. 8). Differential scanning calorimetry (DSC) for determining denaturation temperature (Td) and thermogravimetric analysis (TG) were also used [27].

The content of Cr, Al, Si in tanning float as well as on leather was determined by ICP-AES and expressed on a dry basis as Cr_2O_3, Al_2O_3, SiO_2 amount with respect to pelt.

2.2 Pilot scale experiments

Adopting the optimal recipe obtained by laboratory tanning tests, chrome and chrome-zeolite tanning processes were performed on pilot scale in order to confirm the results obtained in laboratory scale. Sodium-zeolite A, sodium carbonate and basic chrome sulphate (33%

basicity) of high purity grade were used. Six pickled calfskins were processed in cylindrical pilot drums (3m diameter, 2m length), performing the same laboratory tanning procedure. The optimized operative initial tanning conditions were: zeolite concentration = 3 wt %, chrome concentration = 27.8 g l^{-1}, T = 25°C, pH = 2.8, bath volume/pelt mass ratio= 0.93 ml g^{-1}. Chrome-zeolite tanning tests were carried out employing sodium-zeolite A, fine-grained (NaAF sample) or coarse grained (NaAH sample). Industrial finishing was effected on tanned pelts with commercial products and deep brown dye. In order to determine mechanical properties of the final leathers, tensile strength, elongation, tearing strength and ball bursting were performed.

3. RESULTS AND DISCUSSION

3.1 Laboratory scale test
The main results of characterisation for chrome conventional tannage are reported in Table 1. The leather chrome content increased with tanning time, and the shrinkage temperature value required by regulations was obtained only after 6 hrs at about 4 wt% Cr_2O_3 in leather.

Table 1. Analyses of leathers and residual floats from the tests carried out in conventional chrome tannage. Operating conditions: Cr= 29.1g l^{-1}, R=1 mlg^{-1}, T= 25°C, pH= 2

t, h	Bath		Leather	
	C_{Cr}, g l^{-1}	pH	$Cr_2O_3^a$, wt%	Ts, °C
0.5	16.3	2.5	3.0	77
2	15.2	2.6	3.8	83
6	13.9	2.9	4.2	85
24	12.7	3.8	4.8	105

^{a} Dry basis

In Table 2 the results of chrome-zeolite tanning tests carried out with different zeolite are reported. The comparison with results of table 1 shows that chrome concentration in the residual floats was reduced from 12.7 g/l in conventional tannage to about 9 g/l for combined tannages at 24 hrs. In the presence of zeolite the amount of chrome in the samples was much higher (5.7wt%) than 4.8 wt% just after 0.5 h, this value being obtained in conventional system only after 24 h upon pH increase for bath basification. Moreover, chrome-zeolite tanned leather shows high values of shrinkage temperature in much shorter time with respect to conventional tanned leather (Tg >100 °C after 0.5 h in the presence of NaA or NaAX zeolite against 77 °C at the same time for chrome conventional tanning). In combined tannage, using NaAX and NaX the enrichment in Si parallels the slight increase of Cr in the leather with respect to NaA, without substantial effects on final shrinkage temperature. The float exhaustion, in agreement, is weakly higher. Chrome-NaA zeolite tanned leather has a chrome content higher than 6 wt% just after 2 h; a faster hydrothermal stabilisation, Ts = 103°C after 0.5 h, is also observed with respect to NaAX and NaX zeolites tanned samples. The analysed Si/Al ratio in the leather is about 1.1, similar to that of NaA zeolite (1.07).

Tanning with NaA alone increases the shrinkage temperature from 44°C (pickled samples) to 75°C, leading to quite full and thick white samples. High grain was also obtained. The amount of Al in the leather (1.8 %wt) is about the same with respect to hide substance.

Table 2. Analyses of leathers and residual floats coming from the tests carried out employing different synthetic Na-zeolites (NaA, NaX, NaAX). Operating conditions: $C_{Cr}= 29.1g$ l^{-1}, R = 1 ml g^{-1}, T = 25°C, pH = 2.

t, h	Bath $C_{Cr,}$ g l^{-1}	pH	Leather $Cr_2O_3{}^a$, wt %	Al^a, wt %	Si^a, wt %	Si/Al	Ts, °C
	3 wt% zeolite NaA						
0.5	20.0	3.9	5.7	0.7	0.7	1.0	103
2	12.5	3.6	6.1	0.7	0.8	1.1	103
6	11.3	3.6	6.1	0.6	0.7	1.1	103
24	9.0	3.7	6.4	0.5	0.5	1.0	104
	3 wt% zeolite NaAX						
0.5	14.0	3.6	5.1	0.7	0.8	1.1	102
2	12.7	3.4	5.3	0.7	0.7	1.0	101
6	12.7	3.2	5.7	0.7	0.9	1.3	104
24	8.2	3.2	6.3	0.7	1.0	1.4	104
	3 wt% zeolite NaX						
0.5	16.1	3.4	4.0	0.6	0.6	1,0	94
2	13.0	3.3	4.8	0.6	0.7	1,1	102
6	10.1	3.2	5.7	0.6	0.8	1,3	105
24	8.6	3.3	5.7	0.7	1.1	1,5	104

[a] Dry basis

In Table 3 the results of laboratory tanning tests at different operating conditions are reported. The data show that chrome exhaustion is about 70% after 24 hrs with 1 wt% zeolite, while with 2-3 wt% zeolite more than 70% chrome exhaustion is obtained after 6 hrs and the chrome oxide content in the leather is higher than 4 wt% just after 0.5 hrs. The leather tanned with 3 wt% zeolite shows faster hydrothermal stability, shrinkage temperature higher than 90°C only after 2 hrs, while the maximum Ts value already after 6 hrs is obtained. With 1 wt% zeolite, Ts values higher than 90°C are instead obtained after 24h.

The results of tanning tests at different chrome concentrations in combination with 3 %wt zeolite show that at 18.0 g l^{-1} chrome initial concentration, shrinkage temperature value is about 100°C after 2 hrs, while the same value is obtained in conventional tanning system only after 24 hrs and upon basification with NaOH. At 18.0 g/l chrome initial concentration, zeolite transfer is favoured and raises with tanning time, in particular after 6 hrs zeolite content in the leather is higher than 2 wt % due to a chrome bath exhaustion of about 80%. Data of tanning tests at different initial bath pH show that raising pH value from 2 to 3 higher shrinkage temperature (100°C) and chrome exhaustion (about 60%) values are obtained already after 0.5 hrs. Moreover, the chrome content in the leather is about doubled from 4.3% to 8.3%. Comparing chemical analyses difference between measured and calculated (by bath residual chrome concentration value) leather chrome content is lower than 0.2 %.

Table 3. Analyses of leathers and residual floats coming from the tests carried out at different operating conditions: R = 1 ml g^{-1}, T = 25°C, zeolite NaA

1 wt % zeolite					13.6 g l^{-1} Cr concentration			
t, h	C_{Cr}, g l^{-1}	pH	Cr$_2$O$_3$^{a}, wt%	Ts, °C	C_{Cr}, g l^{-1}	pH	Cr$_2$O$_3$^{a}, wt%	Ts, °C
0.5	19.4	2.9	3.3	68	9.2	3.6	1.9	82
2	15.0	2.9	5.3	82	8.3	3.5	2.3	90
6	13.6	2.9	5.6	87	2.9	3.5	4.9	94
24	8.3	3.1	8.8	95	1.9	3.6	5.4	105
2 wt % zeolite					18.0 g l^{-1} Cr concentration			
0.5	16.9	3.1	4.7	85	10.5	3.3	3.4	89
2	13.7	3.1	5.9	90	7.1	3.4	5.0	96
6	9.8	3.1	7.8	98	3.3	3.5	7.0	103
24	4.3	3.5	9.9	105	1.4	3.7	7.8	104
3 wt % zeolite, initial pH = 2					27.1 g l^{-1} Cr concentration, initial pH = 3			
0.5	17.7	3.3	4.3	78	8.5	3.4	8.3	97
2	14.5	3.3	5.7	92	5.2	3.4	10.0	104
6	7.1	3.4	9.2	102	3.4	3.5	10.9	106
24	4.4	3.5	10.6	105	2.1	3.5	12.0	108

^{a} Dry basis

It has been reported that in tanning with zeolite A [25] soluble aluminosilicates strongly react with pelts. At pH < 3.8 zeolitic fragments are rings, chains or reticular structures with 6-8 atoms of Si and Al, sizing from 2.6 to 15.8 Å. With an higher number of atoms, the reticular structures can result bigger in size, e.g. of 12.2 x 17.4 Å. The collagen interfibrillar space could accommodate these oligomers, considering that the distance between triple helical macromolecules ranges from 10 to 17 Å as found, respectively, in native collagen and in limed pelt. Moreover the smaller oligomers could approach the distance between complex reactive groups of collagen of 2-5 Å. The zeolite oligomers interaction with polypeptide chains is explicatedvia covalent, hydrogen or complex bonds, but polymerization, rearrangement to rings, chains and three-dimensional structure could also occur in the pelt.

However, the tanning with 3 wt% of zeolite A alone gives poor hydrothermal stabilisation, resulting in a final shrinkage temperature of 75°C [27,28]. So the geometrical factors should not be addressed as the most influent parameter in the tanning process. In the combined chrome-zeolite tanning, several hypothesis can be formulated. In the presence of zeolite, initial rate of chrome transfer is higher than in conventional chrome tanning, and also silicoaluminate is transferred into the pelt. Therefore, the enhanced chrome tanning rate obtained in this work could be due to the formation of chrome-zeolite oligomers complexes. However, pH control acted by zeolite should be also considered, because it improves chrome fixation into the pelt. Moreover, a shield effect of zeolite, favouring the chrome diffusion within collagen structure by limiting chrome-pelt interactions, cannot be excluded. Further studies are necessary to clarify the role of zeolite in the combined tanning.

3.2 Pilot scale test

In figure 1 the comparison of the bath chrome exhaustion and shrinkage temperature profiles as function of time in conventional and chrome–zeolite tanning process on pilot scale are reported. The addition of zeolite results just after 0.5 hrs in higher float exhaustion, about 75%, and high shrinkage temperature, about 110°C, confirming the good performance of chrome-zeolite tanning process. High chrome exhaustion (94%) is obtained after 24 hrs.

In conventional tanning test, chrome exhaustion values are lower than those obtained by chrome-zeolite process and high shrinkage temperature values are obtained only at longer times and higher pH values. The zeolite allows the control of bath pH, canceling the step of bath basification and obtaining a chrome content in residual float 55% less when compared to the conventional tanning system.

Pilot scale tanning tests confirmed results obtained in laboratory tanning tests. In figure 2 the comparison of the bath chrome exhaustion profiles and denaturation temperature profiles as function of time obtained by chrome-zeolite tanning tests on laboratory and pilot scale are reported. Transferring the chrome-zeolite tanning test from laboratory to pilot scale, chrome exhaustion is increased of about 10%. Moreover both tests show denaturation temperature values of about 112°C just after 0.5 h.

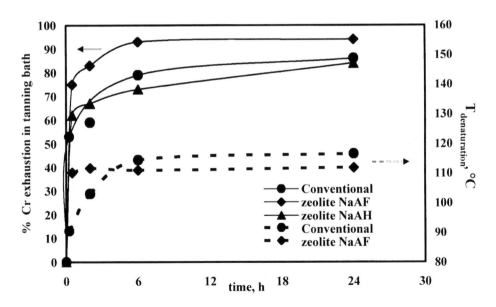

Figure 1. Chrome exhaustion and denaturation temperatures for scale up conventional and chrome-zeolite experiments as function of tanning time at better operating conditions examined (initial chrome concentration = 27.8 g l^{-1}, zeolite A initial concentration = 3 wt%, T= 25°C, bath volume-pelt mass ratio= 0.93 ml/g, initial pH =3).

The finishing of pelts was successfully performed: chrome-zeolite tanned leather has good merceological properties such as dyeability, grain stability, fullness, softness. In table 4 the results of leather mechanical tests are reported.

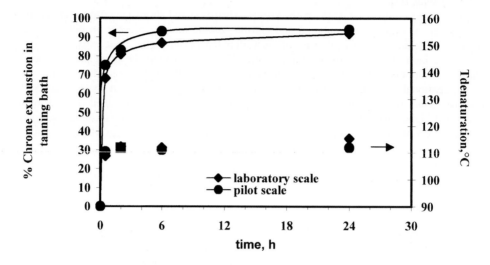

Figure 2. Comparison of chrome exhaustion and denaturation temperatures in laboratory and pilot chrome-zeolite experiments as function of tanning time

Table 4. Results of the mechanical tests

	Conventional	Chrome - Zeolite
Thickness (mm)	1.33	1.27
Tensile strength (N mm^{-2})	18.3	24.4
Elongation (%)	37.0	46.0
Tearing strength (N mm^{-1})	99.0	94.0
Ball Bursting Test		
Grain cracking (mm)	10.1	13.1
Grain resistance (Kg)	21.0	41.0
Grain distension (mm)	15.8	16.6
Grain strength (Kg)	54.0	71.0

Tensile strength values obtained are in the typical range 17.2-24.1 N mm^{-2}, in particular they should be never less than 13.8 N mm^{-2} (ASTMD 2209). Tensile strength (18.3 Nmm^{-2}) and tearing strength (99 Nmm^{-1}) referred to conventional tanned leather (table 4), confirmed the results obtained by mechanical tests performed on conventional tanned leathers with 2.8% Cr_2O_3 content (tensile strength of 18.4 Nmm^{-2} and tearing strength of 95 Nmm^{-1}) [31]. Chrome-zeolite tanned leather showed (table 4) higher tensile strength, elongation and bursting strength performances. Lower tearing strength with respect to finished chrome tanned leather was found. Morera et al. [31] have shown that tensile strength and tearing strength diminished almost linearly as the leather Cr_2O_3 content increased. Increasing the Cr_2O_3 leather content of 9.5% by using zeolite, resulted in increased elongation (from 37% to 46%) and decreased tearing strength (from 99 to 94 N mm^{-1}), confirming literature results. An opposite behaviour of the tensile strength was instead observed, as the presence of zeolite resulted in increased tensile strength (from 18.3 to 24.4 N mm^{-2}) in correspondence of increased Cr_2O_3 leather content. Moreover, the presence of aluminum and silicon containing species in the leather increased grain resistance of about 100% at similar grain cracking values and increased grain strength of about 30% at similar grain distension values.

4. CONCLUSIONS

While sheepskin tanning with zeolite alone leads to leather with lower hydrothermal stability and higher fullness with respect to chrome tanned leather, the combined use of zeolite and chrome sulphate results in both higher float exhaustion and higher shrinkage temperature (>100°C) just after a short time than in conventional chrome tannage. The performance of zeolites was weakly affected by their structure and chemical composition. Enhanced hydrothermal stability was observed, likely due to the greater chrome content in the leather.
The best results in laboratory scale tanning were found in the following operating conditions: zeolite A = 3 wt%, R = 1 ml g^{-1}, T = 25°C, Cr concentration = 27.1 g l^{-1}, bath pH = 3.
The promoting effect of zeolite A in chrome tanning has been confirmed on pilot scale: higher tanning rate as well as higher float exhaustion were found.
Finishing of leather was successfully performed, resulting in good merceological properties. Zeolite use allowed to obtain better mechanical characteristics of leather although Cr_2O_3 leather content was higher than that obtained by conventional tanning process.

Acknowledgements
MIUR for financial support to the project: "Studio e sviluppo di tecnologie innovative e pulite per il miglioramento e la razionalizzazione del ciclo produttivo conciario e di un sistema di recupero di sottoprodotti e dei rifiuti derivanti dalla pelle" , SASOL ITALIA, Milano, Italy for financial support and supplying zeolite samples, ARISTON TANNERY, Casoria (NA), Italy and RUSSO TANNERY, Casandrino (NA), Italy, for pilot scale tests, are gratefully acknowledged.

REFERENCES

1. K.J. Bienkiewicz, "Physical Chemistry of leather Making", R.E. Krieger Publishing Company, 1983, p.308
2. *ibidem*, p.26
3. D.J. Prockop, K.I. Kivirikko, Ann. Rev. Biochem., 64 (1995) 403
4. G.N. Ramachandran, J. Am. Leather Chem. Assoc., 63 (1968) 161
5. A.J. Bailey, J. Soc. Leather Technol. Chem., 75 (1992) 111

198

6. B. Brodsky, J.A.M. Ramshaw, Matrix Biol., 15 (1997) 545
7. E. Heidemann, "Collagen", M.E. Nimni (Ed.), Vol. 3, CRC Press, Boca Raton, FL, (1988) 40
8. J.W. Harlan, S.H. Feairheller, "Biochemical and Molecular Aspects", M Friedmann (Ed.), Vol.86A, Plenum Press, New York (1976) 425.
9. D.P. Speer, M. Chvapil, C.D. Eskelson, J. Ulreich, J. Biomed. Mater. Res., 14 (1980) 753
10. K.J. Bienkiewicz, "Physical Chemistry of leather Making", R.E. Krieger Publishing Company, 1983, p.247
11. G.S. Young, Stud Conserv., 43 (1998) 65
12. A.D. Covington, G.S. Lampard, R.A. Hancock, I.A. Ioannidis, J. Am. Leather Chemist Assoc., 93 (1998) 107
13. A.D. Covington, R.A. Hancock, I.A. Ioannidis, J. Soc Leather Technologists Chemists, 73 (1989) 1
14. A. Finch, D.A. Ledward, Biochim. Biophys. Acta, 278 (1972) 433
15. M. Komanowsky, J. Am. Leather Chemist Assoc., 86 (1991) 269
16. P.L. Kronick, P.R. Buechler, J. Am. Leather Chemist Assoc., 81 (1986) 213
17. K.H. Gustavson, J. Am. Chem. Soc., 74 (1952) 4608
18. K. Venkatachalapathi, M.S. Nair, D. Ramaswamy, M. Santappa, J. Chem. Soc. Dalton Trans., (1982) 291
19. S.G. Shuttleworth, J. Soc. Leather Technol. Chem., 34 (1950) 410
20. Sundar V. J., Raghava Rao J., Muralidharan C., J. Clean. Prod. 10 (2002) 69
21. Bureau of Indian Standards Bulletin, 2490, Part 3 (1995)
22. Tanneries & environment. A technical guide. UNIDO (1991)
23. Raghava Rao J., Leather Sci., 34 (1987)175
24. Manzo G., "Chimica e tecnologia del cuoio" Media Service edizioni, 1999, p.243.
25. A. D'Aquino, G. D'Elia, M. Seggiani, S. Vitolo, B. Naviglio, M. Tomaselli, JALCA, 99 (2004) 26
26. R. Zauns and P. Kuhm, JALCA, 90 (1995) 177
27. P. Ciambelli, B. Naviglio, L. Bianchi, D. Sannino, M.Tomaselli, Proc. XXV IULTCS, (1999) 153
28. P. Ciambelli, D. Sannino, A. Gallo, B. Naviglio, M. Tomaselli, R. Clerici, Prepr. AIZ-GIC Congress, Ravello, Italy, (2000) 407
29. P. Ciambelli, D. Sannino, A. Magliano, B. Naviglio, M. Geremicca, Prepr. AIZ Congress, Vietri sul mare, (2003) 106.
30. "Metodi Internazionali per l'analisi dei cuoi", Metodi I.U.C. Stazione Sperimentale per l'Industria delle Pelli e delle Materie Concianti
31. J. M. Morera, E. Bartoli, M. D. Borràs, S. Gongora, A. Marsal, JALCA 95 (2000) 29.

Studies in Surface Science and Catalysis 155
A. Gamba, C. Colella and S. Coluccia (Editors)

Structure and dynamics of hydrogen-bonded water helices in high pressure hydrated phase of natrolite studied by molecular dynamics simulations

P. Demontis[a], J. Gulín-González[b], G. Stara [a] and G. B. Suffritti [a],

[a] Università di Sassari, Consorzio Interuniversitario Nazionale per la Scienza e Tecnologia dei Materiali (INSTM), Unità di ricerca di Sassari, Via Vienna, 2, I-07100 Sassari, Italy

[b] Instituto Superior Politécnico José A. Echevarría (ISPJAE), Dpto. de Física, Marianao, La Habana, Cuba and Departamento de Matemática, Universidad de las Ciencias Informática (UCI), Carretera a San Antonio de los Baños, Torrens, Boyeros, La Habana, Cuba.

1. INTRODUCTION

Water in microporous materials and in particular in zeolites can be found in many different situations. In large - pore zeolites as LTA and FAU type zeolites at room temperature it can be liquid – like, as shown both by experimental values of diffusion coefficients and activation energy to diffusion, which are similar to those of bulk liquid water [1] and by computer simulations [2].

If the micropores are smaller, but no exchangeable cations are present, as in silicalite-1 [3, 4], water is still fluid, but at room temperature both experimental [5] and theoretical [6]. studies show that it is probably vapour-like, and computer simulations performed in this laboratory [7] confirm that finding and suggest that liquid-like behavior could be reached below about 280 K and freezing below 200 K.

In many medium and small pore zeolites water molecules are co-ordinated to exchangeable cations by electrostatic interactions stronger than water - water hydrogen bonds without assuming a definite structure [8]. An example is natrolite at ambient pressure [9-12].

On the contrary, in some cases stable hydrogen bonded water molecule structures are present. This happens, for instance, in Li containing zeolites as bikitaite [13] and Li-ABW [14, 15] and single walled carbon nanotubes (SWCN) [16], where linear water molecule chains are present.

The last chains in carbon nanotubes run along the axis of a hydrogen bonded water molecule nanotube, which, in turn is embedded in the SWCN [16].

Other water molecule hydrogen bonded structures are helices, which are present in aluminophosphate molecular sieves as VPI-5 [17], ALPO$_4$-5 [18], and, as we show in this paper, in high pressure superhydrated natrolite.

Natrolite (Na$_{16}$ Al$_{16}$ Si$_{24}$ O$_{80}$·16H$_2$O) is a natural zeolite belonging to the group of fibrous, small-pore zeolites [8-12]. It shows helical channels of elliptical section running along the c axis, containing, sodium cations and water molecules in 1:1 ratio. The water molecules form weak hydrogen bonds with the oxygens of the aluminosilicate framework *but not with each other*.

Recent diffraction studies of natrolite under hydrostatic conditions at high pressure (up to 5 GPa) [19, 20] found an abrupt volume expansion between 0.8 and 1.5 GPa without altering the framework topology. This anomalous swelling is due to the selective sorption of water molecules, which expands the channels along a and b unit cell axes and gives rise to a "superhydrated" ordered phase of natrolite, maintaining the Fdd2 symmetry, which contains two water molecules per sodium ion, with chemical formula $Na_{16} Al_{16} Si_{24} O_{80} \cdot 32H_2O$.

However, the experimental results do not include the position of hydrogens, or any dynamical information about this new phase of natrolite. Among the different approaches, the Molecular Dynamics (MD) simulation technique is very promising in order to complement the experimental information about structural and dynamical properties of zeolites [21] and we used this technique to attempt a prediction of the unknown properties of the recently discovered "superhydrated" phase of natrolite at different pressures (1.51, 3.58 and 5.01 GPa), after having ascertained that it performs satisfactorily for natrolite at ambient conditions.

2. MODEL AND CALCULATIONS

Natrolite structure in both phases belongs to the orthorhombic symmetry group Fdd2. For the simulations at ambient pressure we adopted the neutron diffraction structures, because this method yields also the positions of hydrogens. The MD simulation box corresponded to three crystallographic cells superimposed along the c axis in order to obtain an approximately cubic box of about $20 \times 20 \times 20$ Å^3, including 552 atoms for the ambient pressure natrolite, or 696 atoms for the high-pressure phase. As experimentally [10] the cell dimensions do not change appreciably with temperature, the cell parameters at room temperature were used also for higher temperature simulations.

The full hydration of the crystal corresponds to 48 molecules per simulation box for natrolite at ambient pressure and temperature and to 96 for the superhydrated natrolite, always lined in 8 parallel channels running along c. Superhydrated phases were simulated starting from the experimental structures and guessing the initial positions of hydrogens. Partially dehydrated natrolite was obtained by removing one ore two molecules per channel, while partially superhydrated natrolite was simulated by adding one or two molecules per channel, initially at positions corresponding to experimental positions in the superhydrated natrolite.

A sophisticated empirical model for simulating flexible water molecules [22] developed in this laboratory was adopted. To include lattice deformations and vibrations in the simulated system, a flexible zeolite framework model developed in this laboratory as well [7, 23] has been used. Moreover, new empirical potential functions have been elaborated for representing Na^+ - water interactions [22], as the ones previously proposed for simulating aqueous solutions containing sodium ions did not reproduce the structure of water in natrolite.

As in our previous papers [7, 22, 24], water was assumed to interact with Si and Al atoms *via* a Coulomb potential only, and the potential functions between the oxygens of the framework and the oxygens or hydrogens of water were derived from a simplified form of the corresponding O - O and O - H ones for water - water interactions.

The potentials for interactions between an oxygen atom of the zeolite framework and an oxygen or a hydrogen atom of the water molecule were the same adopted for the simulation of water in silicalite [7] and in bikitaite [24].

Finally, for the interactions between Na$^+$ and Si and Al atoms a simple Coulomb potential was considered, and the potential functions between Na$^+$ and the oxygens of the framework are reported in Ref. [22]. The evaluation of the Coulomb energy was performed using the efficient method recently proposed by Wolf *et al.* [25] and extended in our laboratory to complex systems [26]. The cut-off radius was R_c = 16.37 Å, equal to one half of the largest diagonal of the simulation box, and, correspondingly, the damping parameter was $\alpha = 2/R_c$ = 0.219 Å $^{-1}$.

We verified that runs lasting, after equilibration, 50 ps (using a time step of 0.5 fs) were sufficiently long to reproduce structural and vibrational properties, but they were carried out only at low temperature (about 50 K). At higher temperatures (in the range 250 - 800 K) the flip motion and the diffusion of the water molecules required much longer simulations to be observed and studied, namely 1 - 12 ns.

In order to compare our results with the experimental structures, the simulations at different pressures were carried out using the experimental cell dimensions in the NVE ensemble without any constraint on the co-ordinates. We used the same cell parameters in the explored temperature range, as the experimental ones do not change appreciably (see for instance Ref. [10], where for an X-ray diffraction study the same cell parameters were used in the range 298 - 471 K).

Besides structural properties (average co-ordinates, symmetry and radial distribution functions), the vibrational spectra, the time autocorrelation functions of water molecule rotations and the diffusion coefficients were evaluated using standard methods [7, 21, 24].

3. RESULTS AND DISCUSSION

3.1 Structures.

The simulated structure of natrolite results remarkably stable, like the experiment, showing thermal factors increasing with temperature, and with the correct order of magnitude. The symmetry of the crystal, which is not imposed, is conserved accurately, at least up to 500 K, as evidenced by narrow, symmetric and unimodal distribution functions of the co-ordinates of the asymmetric unit atoms [21, 24].

Average co-ordinates are close to the experimental ones. For the low temperature structure (20 K) at ambient pressure, as the average standard error of the computed co-ordinates with respect to the experimental ones is 0.15 Å. We verified that this error is due mainly to the orientation of (Al,Si)O4 tetrahedra and negligibly to an incertitude of mean interatomic distances, for which the error is an order of magnitude smaller.

At room temperature and at any pressure water molecules undergo a flip motion, which makes the computed distribution of the co-ordinates of the hydrogens bimodal, with two maxima corresponding to the co-ordinates of both hydrogens of the same molecule. Therefore, for high pressure structures the water molecules were "frozen" at 50 K to find the co-ordinates belonging to each hydrogen atom, in order to identify the corresponding maxima of the distribution functions at room temperature.

The experimental structures at 1.51, 3.58 and 5.01 GPa were reproduced with a reasonable accuracy. The average standard error of the computed co-ordinates with respect to the experimental ones is 0.19 Å for the three high pressure structures. Also in this case the above remark about the incertitude of the interatomic distances holds.

Indeed, the structural properties of water of the ambient pressure structures obtained by the simulations, which are compared with the experimental data in Table 1 (experimental O - H distances are those corrected for thermal vibrations), are well reproduced.

Table 1

Geometrical parameters of the water molecules adsorbed in natrolite at different pressures. Distances are in units Å, angles in degrees, pressures in GPa.

Pressures (GPa)	0.0001		0.0001		1.51	3.58	5.01
	21 K, Ref. [9]		298 K, Ref. [11]				
	cal.	exp.	cal.	exp.	cal.	cal.	cal.
OW1 - H11	0.977	0.995	0.980	0.98±0.02	0.978	0.978	0.980
OW1 - H21	0.977	0.991	0.980	0.94±0.03	0.978	0.978	0.979
H11 - OW1 - H12	108.2	107.9	107.6	108	106.2	106.3	106.8
OW2 - H21	--	--	--	--	0.979	0.980	0.978
OW2 - H22	--	--	--	--	0.979	0.980	0.979
H21 - OW2 - H22	--	--	--	--	106.6	106.1	105.6
Na - OW1	2.24	2.39	2.25	2.38	2.30	2.28	2.25
Na - OW2	--	--	--	--	2.80	2.70	2.67

In particular, as expected in view of the special potential model used, HOH angles are larger than in isolated water molecule and in liquid water and are close to the experimental results. Therefore, we expect that the structure of the two water molecules predicted for the superhydrated phase should be reliable.

It is interesting to remark that the HOH angles are smaller at high pressure than at ambient pressure, as the water molecules interact with each other (see below) and are less influenced by the sodium cations.

3.2 Hydrogen bond pattern and water molecule helices

Fig. 1 reports the average simulated positions of water molecule oxygens and hydrogens and of Na ions as seen from the c axis at 1.5 GPa. The results at the other higher pressures are visually indistinguishable.

The average positions of the oxygens obtained from the simulations seem to fit the hydrogen-bond geometry suggested by the experimentalists [19, 20] for the high pressure structures, but, by considering the positions of the hydrogens, the structure of "helical water molecule nanotubes", deserve some comments.

Ab initio computer simulations of ice nanotubes of pentagonal and hexagonal cross section were recently performed by Bai *et al.* [27]. In the simulated structures, for each water molecule two hydrogen bonds (HBs), one in donor and the other in acceptor configuration, lie in the cross section plane and two other HBs are parallel the nanotube axis, so that the nanotubes are kept rigid by a network of HBs.

More recently, as mentioned above, Kolensikov *et al.* [16] reported an experimental (Neutron Diffraction) and MD simulation study of nanotubes of water molecules encapsulated inside single-wall carbon nanotubes. The cross section of the water nanotube, with the same HB pattern as the one found by *ab initio* calculations, which Kolensikov *et al.* [16] call "square-ice" pattern, is octagonal.

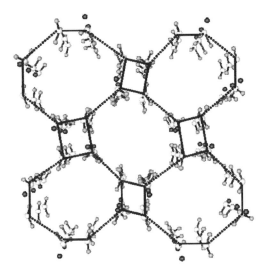

Fig. 1. Water molecules and sodium cations contained in the simulation box for the superhydrated phase of natrolite, as seen from *c* axis. Four adjacent helices of water molecules are visible near the center of the Figure. For sake of clarity only the foreground hydrogen bonds are shown. The solid lines represent internal hydrogen bonds of water molecule helices, dotted lines stand for weak hydrogen bonds between different helices. The dark spheres represent sodium ions.

On the contrary, in superhydrated natrolite each water molecule is hydrogen bonded to *only two* other molecule belonging to the same "nanotube" (O - O distance 2.80 Å) and the other hydrogen points toward a different "nanotube", forming a weak (O - O distance 3.40 Å) HB in donor configuration, as it appears in Figure 1 where four water "nanotubes" are clearly visible.

The fourth possible HB (in acceptor configuration) is clearly surrogated by the presence of Na ions, as it appears in Figs. 2 and 3, and cannot be formed with the oxygens of the framework either, because they bear a negative charge.

The views of one "nanotube" from *b* and *c* axes (Figs. 2 and 3, respectively) show that the HBs connect molecules which are shifted along *c* axis at a regular alternate pitches, about 2 and 1.25 Å, respectively, resulting in an *helicoidal* pattern coiled around the *c* axis.

This pattern is present in Figure 9 of Ref. [19], where the positions of oxygens only are considered, but there the HB pattern was not resolved, as the positions of hydrogens was not determined.

Fig. 2. A water molecule helix and sodium cations (dark spheres), as seen from *b* axis.

Fig. 3. A water molecule helix with sodium cations (dark spheres), as seen from *c* axis.

Therefore, we prefer to call these patterns "*water molecules helices*" instead of "helicoidal water molecule nanotubes" and we shall do so in the following. The difference may appear quibbling or even only semantic, but it corresponds to a HB pattern *different* from that supposed in Ref. [19], as the supposed weak HBs with O – O distance of 3.09 Å along the channels cannot be formed.

The above considerations are confirmed by the appearance of the radial distribution functions (not shown) and are consistent with the trend of interaction energies. In the ambient pressure structure, the computed water-water interaction energy per water molecule is -7.7 kJ/mol, while in the high pressure structures it decreases for increasing pressure, and its value is -15.5 kJ/mol at 1.51 GPa, -18.0 kJ/mol at 3.58 GPa and -18.4 kJ/mol at 5.01 GPa. Correspondingly, in the ambient pressure structure, the computed framework-water interaction energy (not including Na - water interactions) is -17.6 kJ/mol, while in the high pressure structures it increases with the pressure, and its value is -13.8 kJ/mol at 1.51 GPa, -5.9 kJ/mol at 3.58 and only GPa -1.3 kJ/mol at 5.01 GPa.

This trend can be easily explained by recalling that, as pressure is increased, the available space for the water molecules shrinks, so that the interactions of water with the framework become more repulsive and water-water interactions are favored.

As mentioned above, water helices are present in aluminophosphate molecular sieves containing straight channels as VPI-5 [17], with large hexagonal cross section, and ALPO$_4$-5

[18], again with a (smaller) hexagonal cross section ad low water loading. At higher loading, another helix of triangular cross section running along the channel axis is formed.

Another structure containing hydrogen bonded water molecule helices with square cross section, recently studied by MD simulations [28], is the helical peptide nanotube tryptophylglicine monohydrate, which contains water molecules in the core channels. At low temperature (40 K) the water molecules form helices very similar to those present in high pressure natrolite, but each water molecule is connected by a HB to a dipeptide molecule instead of to a water molecules of an adjacent helix.

3.3 Flip motion of water molecules

Whereas at low temperature the water molecules oscillate around their equilibrium positions, at room temperature flips of the molecules, exchanging the hydrogens, were observed in the simulations. This kind of motion had been detected by NMR experiments, [29, 30] and is a fairly common feature of zeolites, but experimental data for natrolite seem not clearly discussed. Nevertheless, the relaxation time of librational motion of water was evaluated, in order to find their trend against pressure. A series of MD computer simulations were performed for each pressure, at different temperatures in the range 250-500 K, in order to study the flip motion of the water molecules and to evaluate its activation energy.

The relaxation time for flip motion of water molecules in all structures follows an Arrhenius behavior only approximately, because as the temperature is increased above 400 K the equilibrium position of water and the framework - water interaction energy change slightly, so that the flip mechanisms are different at different temperatures. In particular, at high temperatures free rotation of the water molecules is observed. Therefore, a comparison among the different activation energies for the room temperature structures is meaningful only if evaluated in a small temperature interval around room temperature.

Table 2
Relaxation times τ_2 (ps) and approximate activation energies E_a (kJ/mol) for the flip motion of the water molecules adsorbed in natrolite at different pressures and temperatures

Pressures (GPa)	0.0001	1.51	3.58	5.01
τ_2 (300 K)	3.05	2.13	16.7	30.4
τ_2 (350 K)	1.33	1.00	6.0	10.8
E_a	12	18	23	25

The resulting values of τ_2 and of the activation energies are reported in Table 2. At 1.51 GPa the relaxation time is shorter than at ambient pressure, but the activation energy is slightly higher. This finding is not contradictory, because the environment of the water molecules in the two structures is different, so that the pre-exponential factor in the Arrhenius equation can be different. For high pressure structures the trends are more regular: both τ_2 and activation energy increase with pressure, as the motion is hindered by the narrowing of the channels and by the strengthening of intermolecular HBs, as above remarked.

3.4 Vibrational spectra

Fourier transforming the autocorrelation function of the total dipole moment of the system derived from MD calculations simulates the IR spectra. The comparison with experimental data [8, 31,32] for ambient pressure natrolite shows that the maximum shift of the vibrational bands with respect to the experiments does not exceeds a hundred of cm^{-1}.

The expected qualitative trend of vibrational spectra in passing from the ambient natrolite to the high pressure structures will be shown for the frequencies of the internal

206

vibrations of water molecules, which should be influenced by the formation the water molecule helices: the band around 1630 cm^{-1}, corresponding to bendings and the band in the range 3300 - 3600 cm^{-1} corresponding to stretchings.

In the frequency range of the bendings (Fig. 4), the formation of hydrogen bonds between water molecules and the larger Na - water distances in the superhydrated phase causes a red shift of the frequencies, which approach the value in the gas phase (1595 cm^{-1}, both experimentally and according to our model).

The O–H stretching band (Fig. 5) is slightly blue-shifted as the pressure increases, probably because of a weaker influence of the Na ions, but the most interesting features are the widening and the splitting of the high pressure bands, corresponding to the effect of different Na – water distances and the formation of HBs of different strength.

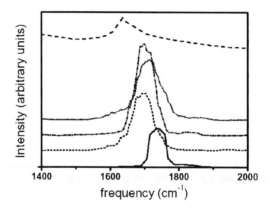

Fig. 4. IR spectra of water in Natrolite at different pressures in the bending frequency region. Solid line: ambient pressure structure; dashed line: 1.51 GPa; dashed-dotted line: 3.58 GPa; dotted line: 5.01 GPa. The long-dash line is the experimental IR spectrum at ambient conditions (from Ref. [8]).

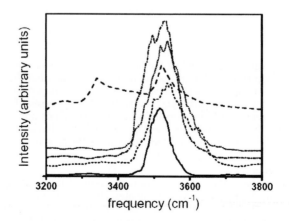

Fig. 5. IR spectra of water in Natrolite at different pressures in the stretching frequency region. Symbols as in Fig. 4.

3.5 Dehydration and superhydration mechanism

The water content of natrolite can be changed in two different ways: natrolite is dehydrated by heating at a temperature higher than 548 K [33] and it can be gradually superhydrated by applying a hydrostatic pressure [19, 20, 34].

MD simulation of these two phenomena could be accomplished in Grand Canonical ensemble, where temperature, pressure and chemical potential are kept constant, and the number of the particles as well as the volume of the system can change.

However, while Grand Canonical MD simulation technique itself is still subject to assessments [35], its application to the superhydration and dehydration of natrolite would require a remarkable amount of calculations, which would go beyond the scope of this work and possibly will be considered in future work.

Nevertheless, some suggestions about the initial stages of dehydration and superhydration mechanism can be derived from usual (microcanonical ensemble) MD simulations, and preliminary results will be shortly reported.

First, in order to enhance the mobility of the water molecules, simulations of ordinary natrolite, lasting more than 5 ns, were carried out at temperatures much higher than the dehydration phase transition temperature (548 K). Whereas at 720 K no clue of diffusion was detected after 5 ns, at higher temperatures (873 K) a diffusive process began to appear, with diffusion coefficient $1.2 \ 10^{-11} \ m^2 \ s^{-1}$, small (about two orders of magnitudes smaller that the diffusion coefficient of bulk water at room temperature), but still measurable.

Experimental diffusion coefficient of natrolite at room temperature is about $2 \ 10^{-16}$ $m^2 \ s^{-1}$ [34] and, assuming an Arrhenius dependence on temperature, activation energy should be of the order of 42 kJ/mol to derive the experimental diffusion coefficient at room temperature from the one computed at high temperature. The computed diffusion coefficient along the c axis is about three times larger than the ones along a and b axes, so that the diffusion occurs mainly along the c axis.

Moreover, some displacive disorder in the orientation of the framework tetrahedra is caused by the thermal motion, and the positions of the Na cations are changed accordingly, but their average mean square displacement does not exceeds $0.8 \ \text{Å}^2$ after 2.6 ns.

These results are in agreement with recent experimental NMR data [36, 37], showing that the diffusion process of water in natrolite involves at least three paths with different activation energies. The first one is parallel to the c axis, while the other two are in the plane normal to the c axis, with activation energies of 40 kJ/mol, 59 kJ/mol and 70 kJ/mol, respectively, indicating a preferential diffusion along the c axis. Moreover, diffusion of Na cations was not detected.

Experimental works suggest that the diffusion should be driven and enhanced by defects, both vacancies [30] (what is rather usual) and by excess interstitial molecules [34], what is less common and peculiar of fibrous zeolites as natrolite, where there is sufficient room to accommodate extra water molecules, as is proven by the pressure induced superhydration [19, 20].

Thus simulations of natrolite containing a deficit and an excess of water molecules were performed, always at high temperature (in the range 630 - 820 K). In particular, in a first series of simulations one or two molecules per channel were removed from positions chosen at random. The channels being eight in the simulation box, 40 or 32 water molecules instead of 48 remained.

In a second series, one or two molecules per channel were added in positions corresponding to those occupied in the superhydrated phase, at random along the channels, resulting in a simulation box containing 56 or 64 molecules. As remarked above, the position of water molecules in ordinary natrolite is approximately conserved in superhydrated natrolite, whereas the excess water molecules are accommodated in a new series of locations.

In agreement with experiment [34], in partially superhydrated natrolite the diffusion coefficients of water resulting from simulations were *larger* than in ordinary natrolite, whereas in partially dehydrated natrolite (for which experimental data are not available) were smaller. However, by considering their small value, to obtain a more accurate and reliable estimate of the diffusion coefficients much longer simulations would be required.

Unfortunately, at present is not possible to perform such long simulations, taking in account that a simulation run lasting one nanosecond required some hundreds of hours of computer time on a Compaq DS20 UNIX workstation. Therefore, the following discussion should be considered as preliminary and semi-quantitative.

In agreement with the experimental evidence for ordinary natrolite [36, 37] in all simulations of partially dehydrated and superhydrated natrolite the diffusivity along the c axis was larger than along the other axes.

A deeper insight into the diffusion mechanisms was gained by direct inspection of the trajectories of the sodium cations and water molecules. Indeed, it appeared that water molecule diffusion is induced by sudden jumps of the sodium cations involving also some disorder in the aluminosilicate guest framework.

These jumps, which are 2 - 4 Å long and occur in a very small time, of the order of some picoseconds, are mostly reversible and involve positions close to those occupied by the extra water molecules in superhydrated natrolite, which, in turn, are almost the same where the sodium cations migrate after dehydration. The cations drag with themselves the water molecules, which can transfer them from an ion to another, by means of the formation of HBs between water molecules. Indeed, Na - O radial distribution functions (RDFs) and consequently average Na - O distances remain practically unchanged with respect to ordinary natrolite.

Seemingly, this process is not detectable, as reported above, neither in experiments [30, 36, 37] nor in the simulation of ordinary natrolite, probably because of the symmetric environment, which is experienced by the cations. This situation should favour correlated collective oscillations of the cations allowing to water molecules to approach one another to form the HBs, which finally induce the diffusive process, without need of changing the average positions of the Na cations.

In partially dehydrated natrolite, where the average distance between the oxygens of water molecules is 4.2 Å, the motion of the sodium ions is necessary to make the formation of HBs possible. On the other hand, in the partially superhydrated material HBs between water molecules are always present, but the cations' jumps help the molecules to diffuse.

This mechanism is consistent with the changes of the O - O and O - H RDFs induced by partial dehydration or partial superhydration, as shown in Fig. 6.

We recall that, when a HB is formed between two water molecules, which involves an O - O average distance of 2.8 Å and two H - H distances of about 1.9 and 3.5 Å, in the RDFs peaks at these distances should be present.

In Fig. 6, one sees that in partially dehydrated natrolite O - O and O - H distances can become smaller than 3 and 2 Å respectively, what is necessary for HB formation, although with small probability and without a corresponding peak in the RDFs.

A peak at less than 3 Å begins to appear in the O - O RDF for ordinary natrolite and become more and more evident for partially superhydrated natrolite, for which also two peaks at less than 2 Å and at about 3.5 Å appear in O - H RDFs.

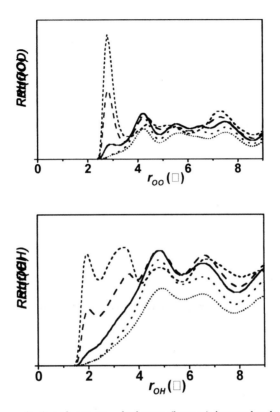

Fig. 6. Oxygen - oxygen (top) and oxygen - hydrogen (bottom) intermolecular radial distribution functions for water in natrolite at high temperature (750-870 K) for different water content (n water molecules per channel). Short dotted line: partially dehydrated natrolite ($n = 4$); long dotted line: partially dehydrated natrolite ($n = 5$); solid line: ambient pressure structure ($n = 6$);long dashed line: partially superhydrated natrolite ($n = 7$); short dashed line: partially superhydrated natrolite ($n = 8$).

4. CONCLUSIONS

Classical MD simulation technique was used to attempt a prediction of structural and dynamical properties of the recently discovered superhydrated phase of natrolite, which are still unknown. In particular, also in consideration of the severe experimental conditions, as the new phase is stable under hydrostatic pressure higher than 1.5 GPa, it was not possible to derive from X-ray diffraction experiments the positions of the water hydrogens, which are important in order to ascertain the hydrogen bond pattern. Moreover, NMR and vibrational spectroscopy data are still lacking, so that it is not known how much the dynamical properties of natrolite change due to the transition to the new phase.

Once it was verified that an optimized potential model for the interatomic interaction was able to reproduce reasonably the properties of natrolite at ambient pressure and the known part of the structure of the superhydrated phase of natrolite at high pressures, one could extend the calculations to some still unknown properties of the new phase.

It was predicted that hydrogen bonds are involved in building helicoidal chains of adsorbed water molecules coiled around Na cation rows, as well as in connecting adjacent helices lining along the channels. Moreover, relaxation times of the flip motion of water molecules and the corresponding activation energies are expected to grow with pressure.

The vibrational spectra should not be deeply influenced by the deformation of the zeolite framework caused by the phase transition, but more evident qualitative changes are predicted for the vibrational modes involving adsorbed water molecules and sodium cations.

Finally, a preliminary study of the possible dehydration and super hydration mechanisms was attempted, suggesting that both are defect driven, and helped by jumps of the sodium cations.

5. ACKNOWLEDGEMENT

This research is supported by the Italian Ministero dell'Istruzione, dell'Università, e della Ricerca (MIUR), by Regione Autonoma della Sardegna (Italy), by Università degli studi di Sassari and by Istituto Nazionale per la Scienza e Tecnologia dei Materiali (INSTM), which are acknowledged.

REFERENCES

[1]   J. Kärger and D. M. Ruthven, Diffusion in Zeolites and Other Microporous Solids, John Wiley & Sons, New York, 1992.
[2]   D. A. Faux, J. Phys. Chem. B, 103 (1999) 7803.
[3]   H. van Koningsveld, J. C. Jansen, H. van Bekkum, Zeolites, 10 (1990) 235 and references therein.
[4]   H. van Koningsveld, Acta Crystallogr. B, 46 (1990) 731.
[5]   A. Giaya and S. W. Thompson, J. Chem. Phys, 117 (2002) 3464
[6]   A. Giaya and S. W. Thompson, Micropor. Mesopor. Mater., 55 (2002) 265.
[7]   P. Demontis, G. Stara and G. B. Suffritti, J. Phys. Chem. B, 107 (2003) 4426.
[8]   G. Gottardi and E. Galli, Natural Zeolites, Springer-Verlag, Berlin, 1985.
[9]   G.; Artioli, J. V. Smith and Å. Kvick, Acta Crystallogr. C, 40 (1984) 1658.
[10] D. R. Peacor, Amer. Mineral., 58 (1973) 676.
[11] B. H. Torrie, I. D. Brown and H. E. Petch, Can. J. Phys., 74 (1981) 229.
[12] N. E. Ghermani, C. Lecomte and Y. Dusanoy, Phys. Rev. B, 53 (1996) 5231.
[13] K. Stahl, Å. Kvick and S. Ghose, Zeolites, 9 (1989) 303.
[14] E. Krogh Andersen and G. Plough-Sørensen, Zeit. Kristallogr., 176(1986) 67.
[15] P. Norby, A. Nørlund Chistensen and E. Krogh Andersen, Acta Chem. Scand. A, 40 (1986) 500.
[16] A. I. Kolensikov, J.-M. Zanotti, Ch.K. Loong, P. Thiyagarajan, A. Moravsky, R. O. Loutfy and C. J. Burnham, Phys. Rev. Lett., 93 (2004) 035503.
[17] E. Fois, A. Gamba and A. Tilocca, J. Phys. Chem. B, 106 (2002) 4806.
[18] N. Floquet, J. P. Coulomb, N. Dufau and G. Andre, J. Phys. Chem. B., 108 (2004), 13107.
[19] Y. Lee, T. Vogt, J. A. Hriljac, J. B. Parise and G. Artioli, J. Am. Chem. Soc., 124 (2002) 5466.
[20] Y. Lee, T. Vogt, J. A. Hriljac, J. B. Parise, J. C. Hanson and S. J. Kim, Nature (London), 420 (2002) 485.
[21] P. Demontis and G. B. Suffritti, Chem. Rev., 97 (1997) 2845.
[22] Cicu, P.; Demontis, P.; Spanu, S.; Suffritti, G. B.; Tilocca, A. J. Chem. Phys., 112 (2000) 8267.

[23] Demontis, P.; Suffritti, G. B.; Bordiga, S.; Buzzoni, R. J. Chem Soc. Faraday Trans., 91 (1995) 535.

[24] P. Demontis, G. Stara and G. B. Suffritti, J. Chem. Phys., 120 (2004) 9233.

[25] D. Wolf, P. Keblinki, S. R. Phillpot and J. Eggebrecht, J. Chem. Phys., 110 (1999) 8254.

[26] P. Demontis, P. Spanu and G. B. Suffritti, J. Chem. Phys., 112 (2001) 8267.

[27] J. Bai, C.-R. Su, R. D. Parra, X. C. Zeng, H. Tanaka, K. Koga and J.-M. Li, J. Chem. Phys., 118 (2003) 3913.

[28] Y. Pan, H. Birkedal, P. Pattison, D. Brown and G. Chapuis, J. Phys. Chem. B., 108 (2004) 6458.

[29] R. T. Thompson, R. R. Knispel and H. E. Petch, Can. J. Phys, 52 (1974) 2164.

[30] A. V. Sapiga and N. A. Sergeev, Cryst. Res. Technol., 36 (2001) 875.

[31] F. Pechar and D. Rykl, Can. Mineralog., 21 (1983) 689.

[32] C. M. B. Line and G. J. Kearley, Chem. Phys., 234 (1998) 20.

[33] W. Joswig and W. H. Baur, N. Jb. Miner. Mh., (1995) 26.

[34] N. K. Moroz, E. V. Kholopov, I. A. Belitsky and B. A. Fursenko, Micropor. Mesopor. Mater., 42 (2001) 113.

[35] S. Boinepalli and P. Attard, J. Chem. Phys., 119 (2003) 12769.

[36] A. V. Sapiga, Thesis on search of the scientific degree on a specialty solid state physics, Tavrida National University at the name of V. I. Vernadsky, Simferopol, Ukraine, 2003 (in Russian).

[37] A. V. Sapiga and N. A Sergeev, Abstracts of International Conference "Functional Materials" (ICFM 2003), Partenit (Crimea, Ukraine), 2003, p. 287.

Studies in Surface Science and Catalysis 155
A. Gamba, C. Colella and S. Coluccia (Editors)

Replication by molecular sieves: controlling the properties of porous carbon by the use of porous oxides.

Erica C. de Oliveira[a]**, Leonardo Marchese**[b]** and Heloise O. Pastore**[a]

[a]Instituto de Química – Universidade Estadual de Campinas – CP 6154 – CEP 13083-970 - Campinas – SP.

[b]Dipartimento di Scienze e Tecnologie Avanzate, Università del Piemonte Orientale "A. Avogadro", C. so Borsalino 54, I-15100, Alessandria, Italy.

Abstract: Carbon materials with controlled porosity can be prepared by the use of inorganic templates, such as molecular sieves or zeolites, to direct their syntheses and structures. These materials are interesting due to their magnetic and physical electronic properties as well as the numerous possibilities in the catalysis field. To master the properties of the nanostructured carbon materials it is paramount to learn to control the area and porosity of these solids. This knowledge will allow qualitative growth in the application of these exceptional solids in heterogeneous catalysis and the construction of devices, sensors and biosensors.

keywords: porous carbon, molecular sieves, zeolites, replication process.

1. Introduction

Until the end of the fifties, the petrochemical industry was based on the catalytic cracking of petroleum by acid amorphous silica-alumina. This material was characterized by a modest acidity and a large distribution of pores diameters. At the time zeolites were already known but were used in separation of linear and branched paraffins and their isomers [1].

In 1962, Mobil Oil began to use zeolite X as catalyst for the cracking of petroleum. In 1968, Grace described the first modification in zeolite Y to prepare the ultra stable zeolite Y. Few years later, in 1977, 22000 ton of zeolite Y were used in catalytic cracking. Nowadays, the fluid cracking catalysts composed of silica-rich zeolite Y represents more than 95 % of the total use of catalysts [2]. The use of zeolites in petroleum cracking is so important that if the oil was cracked only by a thermal process, the gasoline extracted would not be enough to move our vehicles [3].

This brief explanation puts in evidence the quantitative development in the petrochemical industry caused by the organization of the structure of the amorphous silica-alumina catalyst into a crystalline aluminosilicate zeolite with a narrow distribution of pores.

In fact, the expression molecular sieve describes materials such as zeolites, carbons, oxides and porous glasses. Some of them are crystalline with a narrow distribution of the pores and have a very well organized structure, as the zeolites. Others, as carbons and alumina, are amorphous and have a large distribution of pores diameter. The carbon molecular sieves, CMS, have an intermediate pore diameter between the zeolite A and zeolites X and Y, but a larger distribution of pores diameter, Figure 1.

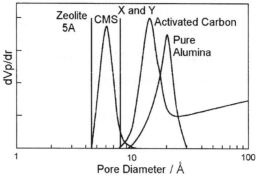

Fig. 1. Pore size distribution of some representative adsorbents. CMS = Carbon Molecular Sieves.

Zeolites and molecular sieves have been studied and applied in industrial catalytic processes due to their molecular sieve effects, large internal surface area and their high thermal stability. The molecular sieving properties also allow their application in adsorption processes for purification or separation of substances. Examples are the removal of CO_2 and H_2O from industrial gases and separation of linear and branched paraffins.

The zeolite structure determines to a large extend their possible industrial application, either in physical processes (separation and purification) or in chemical processes (catalysis or petrochemical) [4], because it is the core of the molecular sieve effect over reactants, products and transition states. This will depend on the size and the shape of the channels, cages of the structure and the relative dimensions of the reactants and products molecules. The majority of the industrial and technological important reactions would be not catalyzed by amorphous silica-alumina, without size sieving properties.

Nowadays, an interesting research area is that of development of carbonaceous materials, carbon nanotubes and nanofibers, due to their potential application as catalyst supports, modified electrodes and H_2 storage medium [5, 6, 7]. A more detailed discussion on applications will be made further in the paper.

For this kind of applications, the development of carbonaceous materials with controlled regular porosity, or a type of carbon molecular sieve is highly desirable. In order to obtain a regular porous carbon structure, zeolites and molecular sieves may be used as templates for the preparation of carbon materials [8, 9]. In the same way that happened in the passage from the amorphous silica-alumina to zeolites, another qualitative jump may be expected with the organization of the carbon pores.

2. The general approach to the preparation of carbon molecular sieves

The method for preparation of carbon materials with controlled porosity using inorganic molecular sieves as templates consists in the impregnation of the inorganic structure with a organic precursor (acrylonitrile [8, 10], furfuryl alcohol [11] and sucrose [9]), that will be polymerized in situ. The composite formed is carbonized and then the inorganic template is eliminated. The carbonaceous material synthesized is an inverse replica of the template used and its porosity, the replica of the one of the inorganic skeleton, Figure 2, thus, highly regular [11, 12]. It is easily envisaged that different carbon structures can be prepared depending on the kind of the inorganic structure used. The carbonization of the precursors in the cages and/or channels of the zeolite and molecular sieve structures permit the nanoscale control of the size and the shape of the carbon structure.

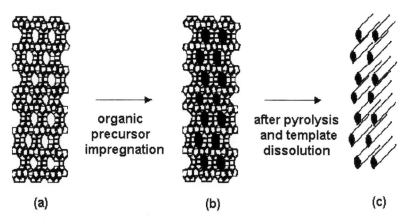

(a)   **(b)**   **(c)**

Fig. 2. Schematic syntheses procedure of carbon molecular sieve using an inorganic template: (a) a siliceous mesoporous molecular sieve, (b) after complete carbonization of organic precursor in the porous of the template, (c) carbon molecular sieve obtained by removal of the template.

3. Zeolites as templates

Enzel and Bein [10] have initiated the synthesis of carbonaceous materials from zeolites using the faujasite structure (zeolite Y) as template and acrylonitrile as the carbon source, with the purpose of obtaining porous carbon structures with a narrow distribution of pores. Recently, other zeolite structures were studied such as zeolite β, L, mordenite and ZSM-5 [13, 14, 15]. The results showed that although all structures yielded porous carbon, the best template was zeolite Y [15], because the carbon structure formed from it presents high porosity and a regular framework. The good results obtained from zeolite Y are due probably to the pore size of the α-cage (7.4 A [16]) and to the three-dimensional framework.

The zeolite Y-carbon structures present surface areas of the order of 3600 m^2g^{-1} and micropores volume of 1.5 cm^3g^{-1} [11]. The carbon material, however, presents only an intense and narrow peak in X ray diffraction, at about 14.73 Å, corresponding to {111} plane of the structure zeolite Y.

These materials presented ferromagnetic properties between 30 and 300 K. This feature is probably governed by the presence of weakly interactive or insulated carbon clusters as suggested by the increase in the ferromagnetism in lower temperature what indicates the existence of a percolative transition with the decrease in temperature. The high temperature ferromagnetism in this three-dimensional nanoarray probably originates from a topological disorder associated with graphene layers of the material [17].

The formation of dendritic lithium upon recharge of lithium batteries was a serious concern for the safety and for the duration of the charge/recharge cycles, what restricted their use as primary cells. The exchange of metallic lithium by a system that adsorbs/desorbs lithium reversibly on carbon at low electrochemical potential was the solution provided by Sony [18] to increase cycle duration and safety on this kind of batteries.

The carbon molecular sieves obtained from zeolites Y, β and ZSM-5 with acrylonitrile, furfuryl alcohol, pyrene and vinyl acetate also were characterized by cyclic voltametry [14] in the reaction of catechol oxidation to hydroquinone and quinone. The electrochemical response is dependent on the nature of the template and the carbon source used in the impregnation method. They were also evaluated as supports for lithium in the batteries but the results were not as good as the Li/C conventional systems [14].

4. Mesoporous molecular sieves as templates

M-41S family as template

The increase in carbon surface area and in pore sizes to the range of mesopores (2-50 nm) enlarge the range of potential applications of organized nanoporous carbons to double-layer supercapacitors [19] and and in fuel cells [12].

Ryoo and collaborators [9] were the first to report in the open literature a mesoporous carbon from MCM-48 as template and sucrose as the carbon source. This material was called CMK-1 (Carbon Molecular sieves from Kaist) and showed a BET area of 1380 m^2g^{-1} and pore diameter of 3 nm. Scanning electron microscopy revealed that the morphology of mesoporous carbon particles mimics the morphology of the template particles while transmission electron microscopy confirmed a regular pore arrangement.

The MCM-48 structure is highly organized [20]. The X ray diffractogram shows one intense signal at 2.40° 2θ (3.68 nm) and another less intense at 2.76° 2θ (3.20 nm) corresponding to the (211) and (220) planes. Moreover, there are a set of reduced signals at the range 3 to 6° 2θ (2.95 to 1.47 nm) that are assigned to (321), (400), (420), (322), (422) and (431) planes [21], Figure 3 – curve a.

The diffractograms of samples obtained after impregnation of carbon precursors and pyrolysis at 873 or 1173 K (Figure 3, curves b, c and d, respectively) still show essentially the diffraction pattern of the MCM-48 structure (curve a) although less intense and resolved and progressively displaced to smaller distances. After dissolution of the inorganic template another pattern is revealed for the sample pyrolysed at either temperature (curves e and f). This pattern is characteristic of carbon nanostructure obtained from MCM-48 and shows an intense signal at 1.76° 2θ (5.02 nm) and another less intense at 2.82° 2θ (3.13 nm), corresponding to (110) and (211) planes [9] respectively. The appearance of the (110) plane is due to the lower symmetry of the carbon structure when compared to that of the template [22, 23].

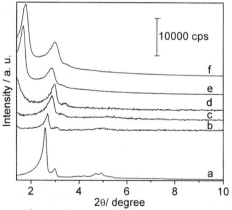

Fig. 3. X ray diffractograms after each stage of carbon molecular sieve formation: (a) MCM-48 template; (b) MCM-48 impregnated with sucrose according to the method of Ryoo [9]; (c) after pyrolysis at 873 K under vacuum; (d) after pyrolysis at 1173 K under vacuum; (e) after HF dissolution of the material pyrolysed at 873 K; (f) after HF dissolution of the material pyrolysed at 1173 K.

Continuous density function calculation was used to model the X ray diffraction pattern. The results agreed with that experimentally obtained for the CMK-1 material and allowed determination of geometric textural characteristics for this material that agree with TEM observations [23]. The comparison of the electron density distribution Fourier-map for the CMK-1 material and that of MCM-48 demonstrated displaced contiguous frameworks of CMK-1 that follows the shape of the MCM-48 mesopores.

In the Figure 4 the diffractograms of as-synthesized (curve a) and impregnated (curve b) MCM-41 are observed. They showed the crystallographic pattern of the MCM-41 structure with the (100), (110) and (200) planes that characterize a material with good textural uniformity that are present after the extraction and calcination processes.

The diffractograms of materials obtained after impregnation of carbon precursor (curve b) and pyrolysis at 873 K(curve c) still showed two diffraction signals that correspond to (100) and (110) planes of the MCM-41 structure (curve a). However, the sample pyrolysed at 1173 K (curve d) shows only one signal that is (100) plane, probably due to a structural disorder promoted by the high temperature pyrolysis.

After dissolution of the inorganic template, the materials lose the diffraction signals because the carbon nanotubes formed in the channels of the MCM-41 structure do not have a higher level of organization [24] as it was observed for MCM-48 (Figure 4, curve e) and that is due to the fact that the channels in the MCM-41 structure are not connected.

The images of scanning eletron microscopy of carbon structure obtained from MCM-48 after pyrolysis at 873 and 1173 K can be observed in the Figure 5. The carbon molecular sieves treated at 873 K (Figure 5(A) and 5(B)) shows prismatic crystals with dimensions in the range of 0.2 x 0.2 μm to 0.6 x 0.6 μm. The crystals aggregate to a large extent and show some intergrowth. The carbon material treated at 1173K (Figure 5(C) and 5(D)) shows crystals highly fused but with dimensions in the range of the ones treated at 873 K.

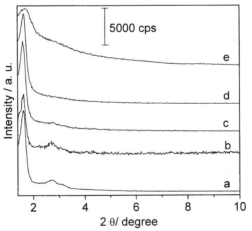

Fig. 4. X ray diffractograms after each stage of carbon molecular sieve formation: (a) MCM-41 template; (b) MCM-41 impregnated with sucrose according to the method of Ryoo [9]; (c) after pyrolysis at 873 K under vacuum; (d) after pyrolysis at 1173 K under vacuum; (e) after HF dissolution of the material pyrolysed at 1173 K.

218

(A)   (B)

(C)   (D)

Fig. 5. Scanning electron microscopy of CMK-1 obtained after pyrolysis at 873 K (A) and (B) and 1173 K (C) and (D) and HF dissolution.

Figure 6 shows images of scanning electron microscopy of carbon structures obtained from MCM-41 after pyrolysis at 873 and 1173 K. The carbon structure obtained from MCM-41 at 873 K (Figure 6(A)) shows undefined morphology while that obtained at 1173 K (Figure 6(B)) shows also cylindrical structures.

These images show that the carbon material with the most defined morphologies is that treated at 873 and 1173 K for the MCM-48 and MCM-41 templates, respectively.

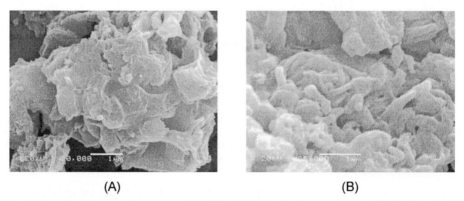

(A)   (B)

Fig. 6. Scanning electron microscopy of C-MCM-41 obtained after pyrolysis at 873 K (A) and 1173 K (B) and HF dissolution.

Barata-Rodrigues and collaborators showed the possibility of synthesizing carbon molecular sieves from furfuryl alcohol and bentonite clay, β zeolite, Al-MCM-48 as template [25]. The carbon molecular sieves obtained from Al-MCM-48 showed X-ray diffraction patterns different from that of the CMK-1 structure and with less ordered structure. The authors attributed these facts to the inorganic template structure itself. In the case of Al-MCM-48 it is formed by two gyroids systems interpenetrated but without connection, Figure 7, while pure silica MCM-48 does not present these systems. Naturally, the carbon replicas of these two different structures will present different X-ray diffraction patterns [25].

Fig. 7. Two possible gyroid representations of mesoporous channels in Al-MCM-48 template [23].

SBA-15 and MSU-H as templates

The use of SBA-15 as template and sucrose as carbon source [26] generated the CMK-3. This material showed a hexagonal mesoporous structure with BET area of 1520 m^2g^{-1} and pores of 4.5 nm.

The SBA-15 structure is very much alike the MCM-41 structure, it presents a regular hexagonal arrangement of mesopores. The difference in the two systems is that in the SBA-15 the mesopores are connected by micropores of 1-3 nm diameter, therefore its structure is formally three-dimensional [27]. This characteristic permits that the carbon formed, when SBA-15 structure is used as template, have a hexagonal mesoporous system where the tubes are still connected [28] on the contrary of the carbon structure obtained when MCM-41 is the template where their X ray diffractogram does not have any signal because their structure is formed by tubes without connection [24], Figure 8.

Studies of nitrogen adsorption in non-heat-treated CMK-1 and CMK-3 carbon molecular sieves show the presence of microporous smaller than 0.9 nm [29] even though these materials are essentially mesoporous. The CMK-3 material after thermal treatment at 1873 K showed very little microporosity.

(A)   (B)

Fig. 8. Pictorial representation of carbon structures obtained from MCM-41 (A) and SBA-15 (B).

Another mesoporous carbon material with hexagonal structure was also obtained from MSU-H structure, a siliceous template. The carbon structure showed 1230 m^{2}g^{-1} BET area and 3.9 nm of pore diameter [30]. The advantages of this procedure, as compared to the one starting from SBA-15, is that the production of MSU-H is made from sodium silicate while SBA-15 uses TEOS under highly acidic synthesis conditions. Since the silica nanostructure is dissolved during the preparation of the nanostructured carbon, the use of cost intensive silicon alkoxides is not highly recommended unless it brings particularly interesting properties.

If sucrose is exchange by furfuryl alcohol and the template is still SBA-15, CMK-5 is produced [31, 32.] In that case, the CMK-5 structure is formed by interconnected hollow cylinders of carbon atoms produced during the pyrolysis of furfuryl alcohol under vaccum, Figure 8(B). This structure have a higher BET surface area (2000 m^{2}g^{-1}), high pore diameter (3.3 nm) and an X-ray diffraction pattern similar to its template and different from the CMK-3 obtained from the same inorganic template.

Other works have focused attention on the carbon sources used. One of them was of a special interest because it used pitch as carbon precursor [33, 34].

Vix-Guterl and collaborators used the pitch to obtain mesoporous carbon from MCM-48 and SBA-15 templates [34]. These solids showed BET areas of 954 and 923 m^{2}g^{-1} for that obtained from MCM-48 and SBA-15, respectively. The materials presented less micropores formation and higher thermal stability (up to 1373 K) than that obtained by impregnation in solution and from other carbon sources.

Porous silica as template

Fuertes and Nevskaia synthesized carbon structures with different pore sizes distribution by the control of the degree of mesoporous silica impregnation with organic precursor [35]. When the silica pores are completely filled a carbon material with narrow pore distribution was obtained. However, in smaller filling degree carbon materials with bimodal pore sizes distribution in the range 3 to 18 nm were obtained. The smaller pores were derived directly from the silica walls while sizes of up to 18 nm were formed from the coalescence of unfilled silica pores when the silica walls were dissolved (Figure 9). The authors also observed that the nature of the organic precursors plays a fundamental role in the modulation of the pores sizes in the carbon nanostructured synthesized: furfuryl alcohol produces 3.0 nm pores sizes whereas phenolic resin produces pores of 2.6 nm diameter.

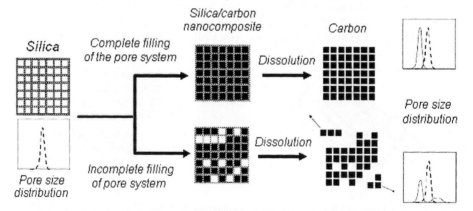

Fig. 9. The production of nanoporous carbon with monomodal and bimodal pore sizes distribution [35].

Hyeon and collaborators proposed the simultaneous polymerizations of the siliceous precursors and those of carbon as a mean to produce mesoporous [36] carbon-silica nanocomposites. The synthesis consists in adding HCl to the solution of carbon and inorganic precursors. After that, the material is pyrolysed and then the inorganic phase is dissolved. When the sucrose/SiO_2 molar ratio is 0.65 the mesoporous carbon obtained show BET area of 856 m^2g^{-1} and with narrow pore size distribution centered at 3 nm [36]. This new preparation method optimizes the synthesis time of the mesoporous carbon and therefore reduces the cost of the process.

Li and Jaroniec reported the synthesis of mesoporous carbon from silica gel and pitch [33]. The carbon material showed BET area of 680 m^2g^{-1} and pore sizes form 7 to 10 nm with almost no microporosity. These results are of interest for application of these materials in adsorption of large molecules and chromatographic separation. However, the carbon material produced showed no pores organization. The position of the pores produced from the silica-gel could be determined by the precise control of the synthesis temperature. Heating the reaction at temperatures higher than the softening point of the pitch caused the silica gel particles to enter the pitch particle causing bulk-imprinting (Figure 10) while lower temperatures caused the formation of pores in the surface, thus surface-imprinting, of the pitch aggregate [37].

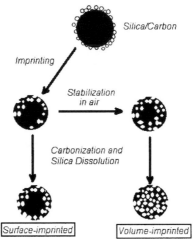

Fig. 10. Production of porous carbons from pitch and silica gel [37].

Yu and collaborators proposed a carbon polymerization procedure that used an inorganic template with acidic sites, thus trying to avoid large polymerization in the external surface of the inorganic template [38]. The process begins with previous impregnation of the silica nanospheres with aluminum to promote the formation of acidic sites in the external surface of the template that will be responsible for the carbon polymerization, Figure 11. The carbon phase obtained by this method can have its morphology controlled by the modulation of the acidity of the template better than when the acid agent is introduced in solution and promotes the polymerization in all empty spaces of the template.

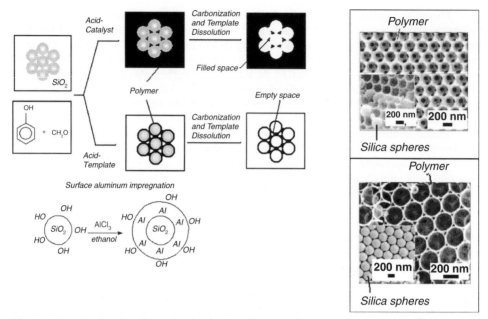

Fig. 11. Control of carbon formation by functionalization of template surface versus bulky reaction acidification. The carbon source was the phenol-formaldehyde resin and the template, silica nanospheres [38].

Figure 12 shows an overview of the different inorganic structures and organic precursors already used in the synthesis of carbon molecular sieves.

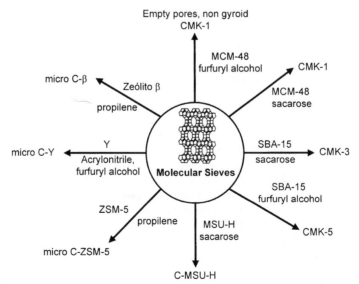

Fig. 12. Micro- and mesoporous carbons prepared by zeolites and molecular sieves templating method.

5. Applications

The material synthesized by Yu and collaborators, Figure 11, was impregnated with Ru and Pt precursors [38], and, after metallic reduction, the samples were used as catalysts in the reaction of methanol oxidation. The results showed a performance 15% better than that found for the couple Ru-Pt dispersed in commercial carbon (Vulcan XC-72) [38]. This was attributed to the higher surface area of the porous carbon for the catalyst dispersion, and also in part due to the three-dimensionally interconnected uniform mesopores, which favor efficient fuel and product diffusion in the porous carbon support.

The adsorption of cytochrome c has been widely studied due its large potential of application in biosensors and enzymatic catalysis [39]. Because of that, Hartmann and collaborators studied the adsorption of cytochrome c in different carbon molecular sieves all obtained from SBA-15 template [40] and that differed by pores diameter and surface areas.

The maximum adsorption cytochrome c was found for the carbon with a BET surface area of 1250 m^2g^{-1} and pore diameter of 4.3 nm [40]. The quantity adsorbed is also widely influenced by the pH in the solution, the higher adsorption value was found (18.5 $\mu mol.g^{-1}$) in pH of 9.6 near the isoeletric point of the cytochrome c (pI = 9.8) what indicates the point where liquid charge in the cytochrome is almost zero. In whatever situation, the maximum adsorption of cytochrome for carbon material was higher than that found for MCM-41 and SBA-15, 1.7 and 6.8 $\mu mol.g^{-1}$, respectively [41, 42]. These results clearly indicate the potential of application of mesoporous carbon in adsorption and separation process of large molecules such as proteins.

These works show the high level of development and characterization of carbon structures with regular and controlled porosity in the last years due to their high potential of application in diverse research areas. Although the application of these materials had focused in catalysis and biosensors another areas as hydrogen and methane (or natural gas) storage are still a challenge [43].

6. Thermal Stability of Nanoporous Carbons from MCM-48 and MCM-41

The effect of temperature on the inorganic template, MCM-48, and the carbon molecular sieves obtained from it was analyzed by X-ray diffraction under nitrogen flow, Figure 13.

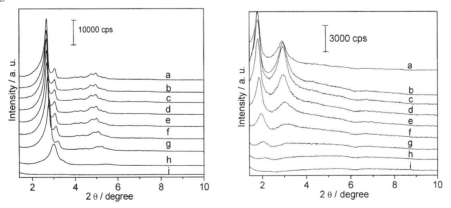

Fig. 13. X ray diffractograms of MCM-48 (A) and CMK-1 (B) at different temperatures (a. room temperature; b. 323 K; c. 573 K; d. 773 K; e. 873 K; f. 973 K; g. 1073 K; h. 1173 K; i. 1273 K) under nitrogen atmosphere.

MCM-48 template is thermally stable up to 1173 K, however, the increasing on the temperature causes a displacement of the characteristic diffraction signals to higher values of 2θ indicating a progressive contraction of the a_0 parameter. The cubic structure is completely lost at 1273 K (Figure 13A curve i).

The carbon structure obtained from MCM-48, Figure 13(B), is thermally stable up to 973 K. At 1073 K and 1173 K the characteristic diffraction signals are very weak and at 1273 K the structure is completely collapsed.

These results show that the carbon structure obtained from MCM-48 has high thermal stability and may be used in systems that reach or operate in temperatures up to 973 K.

The MCM-41 and MCM-48 templates keep their structural arrangements up to 1273 K, as seen by X-ray diffraction in high temperatures, however, the MCM-48 structure showed high contraction of the a_0 parameter. This explains the reason why MCM-48 makes more organized mesoporous carbon at 873 K. MCM-41 does note show any contraction of the framework and this produces more organized carbons at 1173 K.

7. Building New Structures: Organization of Inorganic Frameworks using Mesoporous Carbon Structures as Templates

A new approach in the synthesis of mesoporous molecular sieves has been reported recently. It uses the occlusion of carbon nanoparticles (12-18 nm) or tubes (1-20 nm diameter) in the gel during synthesis as a way to create cavities and pores. After crystallization, the carbon templates were removed by oxidation [44, 45].

In the same way, nanoporous three-dimensional carbons may be used as templates to recreate the initial molecular sieves or oxides. An interesting work reported a new structure from mesoporous carbon CMK-1 [46]. The nanocarbon material was impregnated with a silica precursor (tetraethoxyorthosylane – TEOS) and then exposed to HCl vapour. The template was finally removed by calcination.

The solid obtained was named HUM-1 (_H_annam _U_niversity _M_esostructure-1) [46] and shows a diffraction pattern similar to CMK-1, and without precedents in the molecular sieves family. This pattern indicates that its cubic channels arrangement is not interpenetrated as in the original MCM-48 structure. HUM-1 showed BET area of 472 m^2g^{-1} and pore sizes of 4.21 nm. This new synthesis route was preceded by the Fine-Cell and Viable Korea Co. patent that described the formation of carbon/metal oxides composites used as electrode material or double-layer capacitor [47].

Oxides other than silica, binary oxides, metal silicates, metal zirconates, phosphates, spinels and perovskites can also be prepared by nanocasting on carbon [48, 49]. The surface areas vary from 50 to approximately 450 m^2 g^{-1}. The crystallinity degree of the inorganic part can be controlled to some extent by careful control of ageing conditions after the infiltration of the carbon with the inorganic precursor solutions.

8. Conclusions

It is clear now that mesoporous silicas and aluminosilicates whose structure has been directed by surfactant supramolecular arrangements, may be used as templates themselves, to produce nanoporous carbons. These in turn, are useful as such for sensors and biosensors, double-layer capacitors and catalysts supports or, depending on their structure, be treated as to become crystalline. Apart from these applications, nanoporous carbons may be used to template other solids. Therefore, oxides, phosphates, spinels, perovskites, silicates and zirconates have already been produced from nanoporous carbons. This synthesis route allows the production of porous solids with compositions that were never produced before, by the conventional procedures. Properties such as narrow pore sizes distribution and high surface areas are now possible for a much larger range of materials; for some of them even crystalline and thick walls between the pores are also feasible.

What other creative and interesting procedures can come from this new way to prepare nanostructured materials, of several types of structures and compositions, is certainly only limited by our fantasy.

References

[1] Milton, R. M., in M. L. Ocelli e H. E. Robson, eds, Zeolite Synthesis, ACS Sympos. Ser. 398, American Chemical Society, Washington, DC, 1989, 1-10.

[2] Smart, M.; Esker, T.; Leder, A.; Sabota, K., Zeolites Marketing Research Report in Chemical Economics Handbook, SRI International, Ausgust 1999, 599, 1000.

[3] Neiva, J., Conheça o Petróleo e Outras Fontes de Energia, Ao Livro Técnico S/A, Rio de Janeiro, 1983, 115.

[4] Gianneto, G. P., Zeolitas – Caracteristicas, propriedades y aplicaciones industriales, EDIT, Caracas, 1989.

[5] Imai, J.; Suzuki, T.; Kaneko, K., Catal. Lett. 1993, 20, 133.

[6] Salman, F.; Park, C.; Baker, R. T. K., Catal. Today 1999, 53, 385.

[7] Dillon, A. C.; Jones, K. M.; Bekkedahl, T. A.; Kiang, C. H.; Bethune, D. S.; Hehen, M. J., Nature 1997, 386, 377.

[8] Kyotani, T.; Nagai, T.; Inoue, S.; Tomita, A., Chem. Mater. 1997, 9, 609.

[9] Ryoo, R.; Joo, S. H.; Jun, S., J. Phys. Chem. B 1999, 103, 7743.

[10] Enzel, B.; Bein, T., Chem. Mater. 1992, 4, 819.

[11] Ma, Z.; Kyotani, T.; Liu, Z.; Terasaki, O.; Tomita, A., Chem. Mater. 2001, 13, 4413.

[12] Joo, S. H.; Nature 2001, 412, 169.

[13] Johnson, S. A.; Brigham, E. S.; Ollivier, P. J.; Mallouk, T. E., Chem. Mater. 1997, 9, 2448.

[14] Meyers, C. J.; Shah, S. D.; Patel, S.C.; Sneeringer, R. M.; Bessel, C. A.; Dollahon, N. R.; Leising, R. A.; Takeuchi, E. S., J. Phys. Chem. B 2001, 105, 2143.

[15] Kyotani, T.; Ma, Z.; Tomita, A., Carbon 2003, 41, 1451.

[16] Breck, D. W., Zeolite Molecular Sieves, Wiley, Nova Iorque, 1974.

[17] Kopelevich, Y.; da Silva, R. R.; Torres, J. H. S.; Penicaud, A.; Kyotani, T., Phys. Review B 2003, 68, 092408.

[18] Nagawa, T.; Tosama, K., Prog. Batt. Solar Cells 1990, 9, 209.

[19] Yoon, A.; Lee, J.; Hyeon, T.; Oh, S. M., J. Eletrochem. Soc. 2000, 147, 2507.

[20] Pena, M. L.; Kan, Q.; Corma, A.; Rey, F., Micropor. Mesopor. Mater. 2001, 44-45, 9.

[21] Anderson, M. W., Zeolites 1992,19, 220.

[22] Kim, J. Y.; Yoon, S. B.; Yu, J.-S., Chem. Mater. 2003, 15, 1932.

[23] Solovyov, L. A.; Zaikovskii, V. I.; Shmakov, A. N.; Belousov, O. V.; Ryoo, R., J. Phys. Chem. B 2002, 106, 12198.

[24] Tian, B.; Che, S.; Liu, Z.; Liu, X.; Fan, W.; Tatsumi, T.; Terasaki, O.; Zhao, D., Chem Commun. 2003, 2726.

[25] Barata-Rodrigues, P. M.; Mays, T. J.; Moggridge, G. D., Carbons 2003, 41, 2231.

[26] Jun, S.; Joo, S. H.; Ryoo, R.; Kruk, M.; Jaroniec, M.; Liu, Z.; Ohsuna, T.; Terasaki, O., J. Am. Chem. Soc. 2000, 122, 10712.

[27] Kruk, M.; Jaroniec, M.; Ko, C. H.; Ryoo, R., Chem. Mater. 2000, 12, 1961.

[28] Solovyov, L. A.; Shmakov, A. N.; Zaikovskii, V. I.; Joo, S. H.; Ryoo, R., Carbon 2002, 40, 2477.

[29] Darmstadt, H.; Roy, C.; Kaliaguine, S.; Joo, S. H.; Ryoo, R., Micropor. Mesopor. Mater. 2003, 60, 139.

[30] Kim, S. S.; Pinnavaia, T. J., Chem. Commun. 2001, 2418.

[31] Joo, S. H.; Choi, S. J.; Oh, I.; Kwak, J.; Liu, Z.; Terasaki, O.; Ryoo, R., Nature 2001, 412, 169.

[32] Kruk, M.; Jaroniec, M.; Kim, T.-W.; Ryoo, R., Chem. Mater. 2003, 15, 2815.

226

[33]Li, Z.; Jaroniec, M., Carbon 2001, 39, 2080.

[34] Vix-Guterl, C.; Saadallah, S.; Vidal, L.; Reda, M.; Parnaentier, J.; Patarin, J., J. Mater. Chem. 2003, 13, 2535.

[35] Fuertes, A. B.; Nevskaia, D. M., Micropor. Mesopor. Mater. 2003, 62, 177.

[36] Han, S.; Kim, M.; Hyeon, T., Carbon 2003, 41, 1525.

[37] Jaroniec, M.; Li, Z., International Patent WO03/6372 (2002).

[38] Yu, J.-S.; Kang, S.; Yoon, S. B.; Chai, G., J. Am. Chem. Soc. 2002, 124, 9382.

[39] Dave, B. C.; Dunn, B.; Valentine, J. S.; Zink, J. L., Anal. Chem. 1994, 66, 1120A.

[40] Vinu, A.; Streb, C.; Murugesan, V.; Hartmann, M., J. Phys. Chem. 2003, 107, 8297.

[41] Deere, J.; Magner, E.; Wall, J. G.; Hodnelt, B. K., J. Phys. Chem. B 2002, 106, 7340.

[42] Deere, J.; Magner, E.; Wall, J. G.; Hodnelt, B. K., Catal.Lett. 2003, 85, 19.

[43] Bacsa, R. Laurent, C, Morishima, R. Suzuki, H., Le Lay, M. J. Phys. Chem. ASAP 06/08/04, DOI 101021/jp0312621.

[44] Jacobsen, C. J. H.; Madsen, C.; Houzvicka, J.; Schmidt, I.; Carlsson, A., J. Am. Chem. Soc. 2000, 122, 7116.

[45] Boisen, A.; Schmidt, I.; Carlsson, A.; Dahl, S.; Brorson, M.; Jacobsen, C. J. H., Chem. Commun. 2003, 958.

[46] Kim, J. Y.; Yoon, S. B.; Yu, J. S., Chem. Mater. 2003, 15, 1932.

[47] Fine-cell Co. Ltd. and Viable Korea Co., Ltd., International Patent, WO 01/89991 (2000).

[48] Dong, A. G., Ren, N., Tang, Y., Wang, Y. J., Zhang, Y. H., Hua, W. M., Gao, Z., J. Am. Chem. Soc. 2003, 125, 4976.

[49] Schwickardi, M., Johann, T., Schmidt, W., Schüth, F., Chem. Mater. 2002, 14, 3913

Studies in Surface Science and Catalysis 155
A. Gamba, C. Colella and S. Coluccia (Editors)
© 2005 Elsevier B.V. All rights reserved

Tailoring and Stabilization of Ultrafine Rhodium Nanoparticles on γ-Al₂O₃ by Troctylamine: Dependence of the Surface Properties on the Preparation Route

M.G. Faga[1], L. Bertinetti[1], C. Manfredotti[1], G. Martra[1a], Claudio Evangelisti[2], Paolo Pertici[2], Giovanni Vitulli[2].

[1]Dipartimento di Chimica I.F.M., Università di Torino, via P. Giuria 7, I-10125 Torino-Italy
[2]Istituto per la Chimica dei Composti OrganoMetallici, ICCOM-CNR, Dipartimento di Chimica e Chimica Industriale, Università di Pisa, via Risorgimento 35, I-56126 Pisa-Italy

ABSTRACT

Rhodium nanoparticles supported on γ-Al₂O₃ have been prepared by Metal Vapour Synthesis (MVS), using mesitylene solvated Rh atoms as starting material and working in the absence and in the presence of trioctylamine (TOA), as stabilizer. The size and surface features of the metal particles have been studied by Transmission Electron Microscopy (TEM) and by FT-IR spectroscopy of adsorbed carbon monoxide, and they have been compared with those of a commercial Rh/γ-Al₂O₃ sample. The results indicated that by MVS technique nanoparticles significantly smaller in size and with a narrower size distribution than those present on the commercial sample were produced. The addition of TOA during the MVS process resulted in a further decrease in the metal particle size. IR spectra of adsorbed CO indicated that metal particles produced *via* MVS expose at their surface a larger fraction of Rh atoms in low coordination. They can be extracted by CO, forming volatile Rh carbonyls which migrate onto the support, where the metal atoms are oxidised to Rh(I). For Rh particles produced *via* MVS in the presence of TOA, such low coordinated sites appeared stabilised by adsorbed amine molecules.

Keywords: Metal vapour synthesis; rhodium nanoparticles; TEM; IR spectroscopy of adsorbed CO.

1. INTRODUCTION

Metal particles of nanometric size are of interest in several fields, because of their optical and electronic properties and surface reactivity. The two first features attract growing attention towards these materials as components of innovative devices in microelectronics, information storage and sensing [1]. Because of the peculiar surface reactivity, nanosized model particles are used as supported metal catalysts. As such, they represent, together with heterogeneous catalysts based on nanosized oxides and sulfides, the oldest commercial application of nanotechnology [1, 2]. In this domain, the dispersion of the metal phase, i.e. the

[a] corresponding author: Gianmario Martra, phone : +39-011-670 7538; FAX: +39-011-670 7855; e-mail: gianmario.martra@unito.it

ratio between the number of atoms exposed at the surface of the particles and the total number of metal atoms, is a primary target. Furthermore, the size of the metal particles significantly affects also their surface structure, in terms of types of faces exposed and relative amount of atoms on faces and those in lower coordination on edges and corners [3]. Atoms on different faces and in different coordination exhibit different reactivity, with significant effects on the catalyst activity and selectivity.

Innovative preparation procedures have been developed to attain effective tailoring of the size of supported metal particles. Successful results have been obtained employing chemical vapour deposition [1] and nucleation and growth [4] techniques.

We recently reported that the clustering of rhodium atoms, from mesitylene solvated rhodium atoms, on oxide supports results in a gentle deposition of activated rhodium nanoparticles ranging 1.7-2.8 nm in diameter [5]. However, the mesitylene solvated rhodium atoms, obtained by reaction of rhodium vapour with vapour of mesitylene by the Metal Vapour Synthesis (MVS) technique [6-8], are stable only at low temperature (-40°C) and cannot be kept for a long time without further aggregation to rhodium metal. In the presence of surfactants, as trioctylamine (TOA), the above mesitylene solutions become thermally stable and they can be easily handled at room temperature.

We found that such solutions are suitable materials for the deposition of activated rhodium particles at all times, onto proper supports. In this work, we report the results of a characterisation study of $Rh/\gamma-Al_2O_3$ systems prepared via MVS, with and without stabilizer, compared with a commercial $Rh/\gamma-Al_2O_3$ sample. The investigation has been carried out by Transmission Electron Microscopy, in order to evaluate the size distribution of metal particles, and IR spectroscopy of adsorbed CO, used as molecular probe to test their surface features.

2. EXPERIMENTAL

2.1. Materials

Three systems were considered: i) commercially available $Rh/\gamma-Al_2O_3$ (5 wt % Rh); ii) $Rh/\gamma-Al_2O_3$ (1 wt %) prepared *via* MVS, hereafter referred to as Rh/Al_2O_3 (MVS); iii) $Rh/\gamma-Al_2O_3$ (1 wt% Rh) prepared via MVS in presence of trioctylamine (TOA), hereafter referred to as Rh/Al_2O_3 (MVS-TOA).

2.1.1. Metal Vapour Synthesis

All the operations concerning the Metal Vapour Synthesis (MVS) technique were performed under dry argon atmosphere. Solvents were purified by conventional method, distilled and stored under argon.

According to the MVS technique, rhodium atoms, generated by heating in a high vacuum apparatus cooled at –196°C, and mesitylene vapours, used in large excess, are cocondensed on the cold wall of the reactor. Warming up to –40°C produces melting of the solid matrix and a solution containing rhodium atoms stabilized by mesitylene is obtained (Solvated Metal Atoms, SMA).

2.1.2. Preparation of Rh/Al₂O₃ (MVS), (see scheme 1, route a)

The brown MVS Rh/mesitylene solution (5 ml, conc. 1 mg Rh/ml) was added to a suspension of $\gamma-Al_2O_3$ (0.5 g, Chimet product, surface area 110 m²/g) in mesitylene (10 ml). The mixture was stirred at room temperature for 24 hours. The colourless solution was removed and the light-brown solid was washed with pentane and dried under vacuum; the rhodium loading was 1 wt%.

2.1.3. Preparation of Rh/Al₂O₃ (MVS- TOA), (see scheme 1, route b)

Trioctylamine, TOA, (0.25 ml, 0.57 mmoles) was added to the brown MVS Rh/mesitylene solution (20 ml, conc. 1 mg Rh/ml, 0.19 mg-atoms) and the mixture was stirred for 30 minutes at –40°C. It was warmed to room temperature obtaining a brown Rh(TOA)/mesitylene solvated, stable at r.t. even for months. The Rh(TOA)/mesitylene solvated (10 ml, conc. 1 mg Rh/ml) was added to a suspension of γ-Al₂O₃ (10 g) in mesitylene (20 ml). The mixture was stirred at room temperature for 24 hours. The colourless solution was removed and the light-brown solid was washed with pentane and dried under vacuum; the Rh loading was 1wt%.

A scheme of the overall MVS process yielding to the Rh/Al₂O₃ (MVS) and Rh/Al₂O₃ (MVS-TOA) samples is reported below.

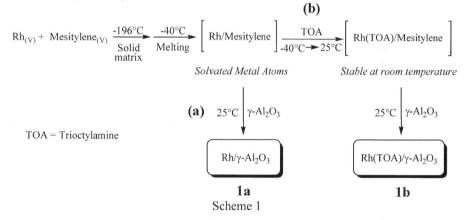

Scheme 1

2.2 Methods

Electron micrographs were obtained by a Jeol 2000EX microscope. The samples, in the form of powders, were ultrasonically dispersed in isopropyl alcohol and a drop of the suspension was deposited on a copper grid covered with a lacey carbon film. Histograms of the particle size distribution were obtained by counting at least 300 particles; the mean particle diameter (d_m) was calculated by using the formula $d_m = \Sigma d_i n_i / \Sigma n_i$ where n_i was the number of particles of diameter d_i.

For the IR measurements the samples were pressed into self-supporting pellet and put into an IR quartz cell with KBr windows, permanently attached to a vacuum line (residual pressure $< 10^{-5}$ Torr; 1 Torr = 133.33 Pa), outgassed at room temperature (r.t.) for 50 min, and reduced by contact with H_2 (100 Torr) for 12 hours at r.t. The samples were then outgassed at r.t. for 1h and put in contact with 100 Torr of CO, recording the IR spectra at increasing time until two hours of contact.

The spectra of adsorbed CO (Bruker IFS-28 spectrometer with MCT detector, 4 cm^{-1} resolution) are reported in absorbance, having subtracted the spectra of the samples before admission of CO as background.

High-purity grade H_2 and CO were used without further purification, except liquid nitrogen trapping.

3. RESULTS AND DISCUSSION

3.1. Metal particle size distribution (by TEM) of the Rh/γ-Al₂O₃ systems

In the case of the commercial Rh/Al₂O₃ (5 wt% Rh), metal particles with a size distribution spread over the 1.5 - 14 nm range were observed (Figure 1A), with a resulting average diameter of 7.0 ± 0.5 nm.

Conversely, metal particles in the Rh/Al₂O₃ (MVS) system exhibited a narrow size distribution, lying in the 1.0 - 4.0 nm range, with a mean diameter d_m = 2.5 ± 0.5 nm (Figure 1B). However, the actual mean size of Rh particles could be even smaller: as observed in previous studies [9], TEM suffers a lack of sensitivity in revealing particles smaller than 1 nm in size, and the occurrence of aggregation phenomena of sub-nanometric metal particles under the electron beam during TEM observation cannot be excluded [10].

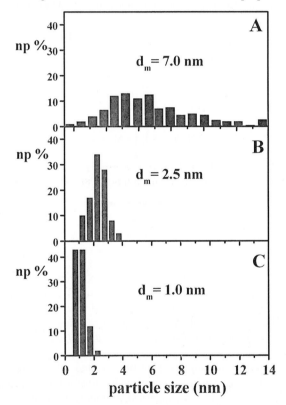

Fig. 1. Histograms of the particle size distribution of the samples: A) commercial Rh/Al₂O₃; B) Rh/Al₂O₃ (MVS); C) Rh/Al₂O₃ (MVS-TOA).

Noticeably, the same size distribution were obtained for a Rh/Al₂O₃ (MVS) 5% Rh (not reported), indicating that the differences observed with respect to the commercial Rh/Al₂O₃ sample are not related to the Rh loading.

Finally, in Figure 1C the metal particle size distribution obtained for Rh/Al₂O₃ (MVS-TOA) is shown. The distribution of the size of the metal particles appeared further shifted to lower size and highly asymmetric, with the maximum in correspondence of the smallest particles size detectable by TEM.

This feature suggests that metal particles even smaller should be present. Preliminary EXAFS analysis support this hypothesis. In conclusion, the data obtained indicate that the addition of a nucleophilic stabiliser such as TOA during the MVS synthesis yielded to the formation of smaller metal particles.

3.2. Monitoring of the surface properties of supported rhodium particles by IR spectra of adsorbed CO

The surface properties of supported metal particles can be conveniently investigated by IR spectroscopy of adsorbed CO, as the vibrational feature of the surface carbonylic adduct are highly sensitive to the properties of the adsorbing sites [11]. Moreover, in some cases CO behaves as a reactant towards metal sites. The possibility to monitor this kind of process *in situ* by IR spectroscopy results in an additional possibility to explore the surface properties of the metal particles [12-14].

In the following sections, the IR spectra of CO adsorbed on the three Rh/Al_2O_3 systems will be discussed.

Prior to the admission of CO, the samples were contacted with H_2 (100 Torr) and outgassed at room temperature. These mild conditions were adopted in order to limit possible effect of the reduction treatment on the original size and surface structure of the particles, which, though produced in a metallic zerovalent state by the MVS procedure, might have been oxidized by exposure of the samples to air before the insertion in the IR cell.

3.2.1. IR spectra of CO adsorbed on commercial Rh/Al_2O_3

In the case of Rh/Al_2O_3 commercial sample, a main peak at 2065 cm^{-1} with a shoulder at 2086 cm^{-1}, a partially resolved component at 2012 cm^{-1} and a band at 1870 cm^{-1} were observed (Figure 2, a) immediately after admission of CO. By increasing the contact time with CO (Figure 2, b-d) the intensity of the signals at 2086 and 2012 cm^{-1} progressively increased in a related way. Also the bands at 2065 cm^{-1} and at 1870 cm^{-1} exhibited some increase in intensity, and a weak broad component at ca. 1950 cm^{-1} became more evident. Such limited increase in intensity may be ascribed to the diffusion of CO in the inner layers of the pellettised sample [15].

On the basis of literature data [16-18], the peak at 2065 cm^{-1} can be assigned to the stretching mode of monocarbonyls linearly adsorbed on zerovalent sites on the faces of Rh particles. The slight asymmetry towards low frequencies may well be done to some heterogeneity among such centres.

Interestingly, also in this case it has been observed that the actual position of the peak is affected by the coupling among CO oscillators on the surface at high CO coverage, with a consequent shift towards higher wavenumber. In fact, at low coverage, the adsorbate-adsorbate interactions faded away and the band due to residual linearly adsorbed carbonyls, significantly decreased in intensity, appeared located at 2043 cm^{-1} (Fig. 2, inset, curve b).

Though extremely weak and broad, the band at ca. 1950 cm^{-1} is in a region were linear carbonyls on Rh^0 were observed to absorb [3]. The position at lower frequency indicates that the Rh^0 sites exhibit an enhanced electron backdonation ability towards adsorbed CO. Such electron feature should be ascribed to Rh atoms at the boundary between metal particles and the support, the oxygen atoms of which can enhance the electron density of the metal sites [19-20].

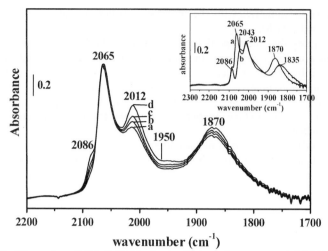

Fig. 2. IR spectra of CO (100 Torr) adsorbed on the commercial Rh/Al$_2$O$_3$ (5 wt% Rh): a) immediately after CO adsoprtion; b) after 5 min ; c) after 20 min, d) after 2 hours of contact with CO. Prior to CO adsorption, the sample was outgassed at room temperature (r.t.) for 50 min, reduced by contact with H$_2$ (100 Torr) for 12 hours at r.t. and finally outgassed at r.t. for 1h.

As for the band at 1870 cm^{-1}, it is due to carbonyls adsorbed in bridged form on pairs of rhodium atoms [21, 22]. This kind of interaction results in an even higher electron back donation to adsorbed CO molecule, which vibrational frequency shifts to lower wavenumbers. By decreasing the CO coverage this band decreased in intensity and exhibited a marked shift towards low frequency (Fig. 2, inset, curve b). More than to fading away of adsorbate-adsorbate interaction, such change in position can be ascribed to the conversion of carbonyls bridged on two Rh (0) atoms to carbonyls bridged on three Rh (0) sites [23].

As described above, the 2086 and 2012 cm^{-1} components appeared affected to a larger extent by the time of contact with CO with respect to the other components. By using a band fitting program they were extracted from the complex experimental profile, and their integrated intensity was calculated.

The growth in intesity at increasing time of contact with CO (I$_t$) with respect to the intensity exhibited immediately after CO adsorption (I$_0$) was found to be the same for both components (Fig. 3). This feature indicated that they are correlated and, consequently, they must be assigned to carbonyl species more complex than monocarbonyls.

Indeed, literature data on homogeneous rhodium carbonyl complex and on CO adsorbed on supported Rh particles, show that dicarbonylic adducts on Rh (I) [21-24] exhibit a couple of bands in this region.

However, the presence of oxidized rhodium sites must be understood. They should not result from the uncomplete reduction of the metal particles by treatment in H$_2$ at room temperature, as we observed them to be formed also by admission of CO on supported Rh particles which underwent a treatment in H$_2$ at high temperature, to guarantee full reduction of the metal phase.

An alternative and convincing explanation for the formation of Rh (I) (CO$_2$) species by interaction of CO with Rh(0) particles can be proposed in terms of a complex process, occurring through the following steps (scheme 2):

i)   adsorption of CO on a Rh (0) atoms in low coordination (on corners, edges, kinks, etc.) on the surface of rhodium particles (scheme 2, a);

ii)   extraction of such Rh (0) atoms from the surface by formation of a volatile rhodium carbonyl (the exact stoichiometry of which has not been clarified yet). (scheme 2, b);

iii)   migration of the volatile Rh(0) carbonyls to the support, where the metal atoms are oxidised to Rh (I) by reaction with the hydroxyls, with the consequent formation of H_2 (scheme 2, c).

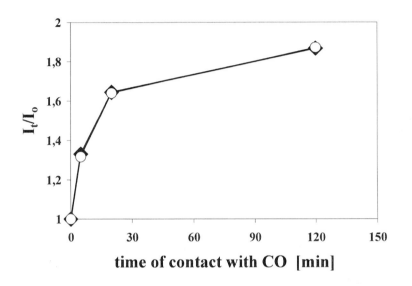

Fig. 3. Relative intensity, I_t/I_0, of the component at 2086 cm^{-1} (O) and of that at 2012 cm^{-1} (♦) at increasing contact time with CO, calculated as the ratio between the intensity at time t of contact with CO (I_t) and the intensity exhibited immediately after CO adsorption (I_0).

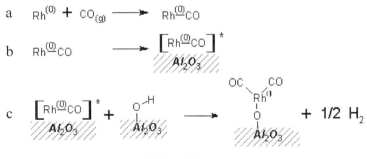

Scheme 2

3.2.2. IR spectra of CO adsorbed on Rh/Al₂O₃ (MVS)

The spectral pattern obtained by contacting CO with the Rh/Al₂O₃ (MVS) sample also exhibited components due to monocarbonyls (both linear and bridged) on metal particles and Rh (I) dicarbonyls resulting from the extraction of Rh atoms in low coordination, but with dramatic differences in relative intensities and position. As reported in Figure 4, the spectrum recorded immediately after admission of CO appeared dominated by the bands at 2086 and 2012 cm^{-1} due to Rh(I) (CO)₂ species, while the bands due to linear (at 2043 cm^{-1}) and bridged (broad and complex, in the 1900-1800 cm^{-1} range) carbonyls appear as minor components. These latter ehxibited a slight increase in intensity as the time of contact with CO increased, because of the diffusion of the probe in the inner layer of the pelletised sample, whereas the 2086 and 2012 cm^{-1} bands grew to a large extent (Fig. 4, b-d).

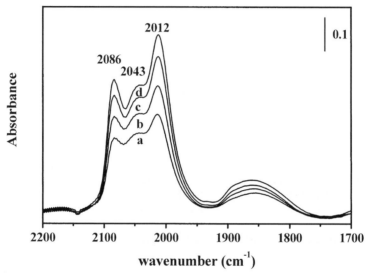

Fig. 4. IR spectra of CO (100 Torr) adsorbed on Rh/Al₂O₃ (MVS) (1 wt% Rh):   a) immediately after CO adsorption; b) after 5 min ; c) after 20 min, d) after 2 hours of contact with CO. Prior to CO adsorption, the sample was outgassed at room temperature (r.t.) for 50 min, reduced by contact with H₂ (100 Torr) for 12 hours at r.t. and finally outgassed at r.t. for 1h.

The predominance of the components due to Rh(I) (CO)₂ species indicated that in this MVS sample the metal particles exhibit a high number of surface metal atoms in low coordination, which, as commented on above, are those extracted by CO to form the volatile Rh(0) carbonyls, which are the precursors of the Rh (I) dicarbonyls. This feature is in good agreement with the small size of the metal particles of the Rh/Al₂O₃ (MVS) sample evidenced by TEM (see Fig. 1 B), as the number of surface atoms in low coordination on edges and corners is expected to increase as the size of metal particles decreas [3]. Furthermore, it must be noticed that, even at high CO coverage, the band due to linear monocarbonyls on Rh(0) atoms on the surface of metal particles appeared located in a position similar to that observed for the "singletone" species present at low coverage on the commercial Rh/Al₂O₃ sample (Fig. 2, inset, curve b). Also this feature can be related to the higher surface irregularity of very small metal particles, which prevented the dynamic coupling between the oscillators adsorbed on their surface. This effect is known to affect much less the monocarbonyls adsorbed in a bridged form [23] and, accordingly, the band related to these species was observed in a similar position for both the commercial Rh/Al_2O_3 (Fig 2) and Rh/Al_2O_3 (MVS) (Fig. 4).

For the sample prepared via MVS, the absorption due to bridged carbonyls appeared as the superposition of several components (Fig.4), each of them attributable to CO molecules bridged onto a pairs of Rh(0) atoms with a different local structure. This heterogeneity can be considered another evidence of the higher irregularity of the surface of Rh particles obtained via MVS. Finally, it must be noticed that the component at ca. 1950 cm^{-1}, related to CO adsorbed on Rh (0) atoms at the boundary between metal particles and Al_2O_3, with an enhanced electron density by interaction with the surface oxygens of the support, is either absent or extremely weak.

This fact might indicate that the interaction between Rh particles preformed via MVS and the support, where they were deposited in a second step is not strong enough to allow the occurrence of such electronic effect.

3.2.3. IR spectra of CO adsorbed on Rh/Al₂O₃ (MVS-TOA)

A further different spectral pattern, in terms of relative intensity and evolution with time of the bands due to the various Rh(0) and Rh(I) carbonyls was obtained by adsorbing CO on the Rh/Al_2O_3 (MVS-TOA) system (Fig. 5). The spectrum recorded immediately after admission of CO onto the sample (Fig. 5, a) exhibited as main components the band at 2043 cm^{-1} and the broad and complex absorption in the 1900 – 1800 cm^{-1} range due to linear and bridged monocarbonyls on Rh (0) at the surface of metal particles, while the signals at 2086 and 2012 cm^{-1}, due to Rh(I) $(CO)_2$ species resulting from the extraction of Rh(0) atoms in low coordination on the metal particles, appeared just as shoulders of the dominant peak at 2043 cm^{-1} (Fig. 5, a). Moreover, by increasing the time of contact with CO, these two components grew up only to a limited extent, as well as the bands due to carbonyls on the metal particles, which as commented for the previous system, are affected by the diffusion of CO into the inner layers of the pelletised sample.

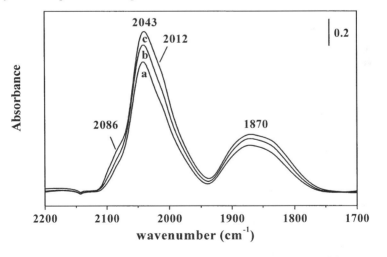

Fig. 5. IR spectra of CO (100 Torr) adsorbed on Rh/Al_2O_3 (MVS-TOA) (1 wt% Rh):   a) immediately after CO; b) after 5 min ; c) after 20 min, d) after 2 hours of contact with CO. Prior to CO adsorption, the sample was outgassed at room temperature (r.t.) for 50 min, reduced by contact with H_2 (100 Torr) for 12 hours at r.t. and finally outgassed at r.t. for 1h.

Such a pattern might appear quite surprising, by taking into account that TEM evidenced that in the Rh/Al$_2$O$_3$ (MVS-TOA) sample the metal particles are smaller than in the Rh/Al$_2$O$_3$ (MVS) one, and then they should expose at their surface an even higher number of Rh atoms in low coordination, easily attached by CO to finally form Rh(I)(CO)$_2$ species stabilised on the support. On the contrary, the inertness towards this process exhibited by the metal particles of the Rh/Al$_2$O$_3$ (MVS-TOA) sample can be readily understood as a consequence of the presence of trioctylamine molecules adsorbed on low coordinated surface Rh sites which, besides preventing sintering, also prevent the adsorption of CO on these centres and the consequent formation of volatile Rh carbonyls.

Conversely, TOA molecules should have been removed from rhodium atoms on the facelets of metal particles by washing and outgassing under vacuum, allowing the adsorption on such sites of CO, as monitored by the bands at 2043 and in the 1900-1800 cm^{-1} range.

Unfortunately, it was not possible to follow the evolution of IR bands due to TOA molecules in interaction with the metal particles because the vibrational modes of hydroxyls and carbonates groups on the surface of the support adsorbed almost completely the IR light in the 3500-300 and 1700-1300 cm^{-1} ranges, where distinctive bands of TOA molecules are expected to fall. Furthermore, as the sample was pelletised in a self supporting form, it was not transparent to the IR light at frequencies lower than 1000 cm^{-1}, because of the lattice absorption of Al$_2$O$_3$. EXAFS investigations, which may give evidence of TOA molecules adsorbed on the metal particles, by monitoring the presence of N and C atoms in the coordination sphere of surface Rh atoms, are in progress.

4. CONCLUSIONS

It has been shown that the deposition of Rh metal particles preformed via Metal Vapour Synthesis on an oxide support is an effective method to obtain a highly dispersed supported metal phase of nanometric particles with a narrow size distribution.

The size of such particles can be further decreased by adding a surfactant, in this case TOA, during the MVS procedure.

As a consequence of the very small size, the Rh metal particles prepared via MVS expose at their surface a relevant number of metal atoms in low coordination. These sites likely retain adsorbed TOA molecules when the Metal Vapour Synthesis is performed in the presence of such surfactant.

Interestingly, the high dispersion of the metal phase achieved by MVS, and likely the surface structure of the metal particles obtained by this way, seemed to significantly affect the catalytic performances of Rh/Al$_2$O$_3$ systems. In fact, preliminary experiments show that the Rh/γ-Al$_2$O$_3$ (MVS-TOA) system is largely more active than analogous Rh/γ-Al$_2$O$_3$ (MVS) system, as well as than the commercially available Rh/γ-Al$_2$O$_3$ sample, in the hydrogenation of a wide range of substrates, including substituted arenes and α,β-unsaturated carbonyl compounds, pointing out the relevance of additional long-chain amines in the preparation of a new class of very active supported metal catalysts.

ACKNOWLEDGMENTS

Financial support from CNR (PF-MSTA II and Agenzia 2000) is gratefully acknowledged.

M. G. Faga gratefully acknowledges the support by CNR- ISTEC.

REFERENCES

[1] Ultra-fine particles, Eds. C. Hayashi, R. Uyeda, A. Tasaki, Noyes Publication, Westwood, New Jersey, (1997).
[2] G. A. Somorjai, Y. G. Borolko, Catal. Lett. 76 (2001) 1.
[3] M. Che, C. O. Bennett, Adv. Catal. 36 (1989) 55.
[4] M. Che, Z. X. Zeng, C. Louis, J. Am. Chem. Soc. 117 (1995) 2008.
[5] G. Vitulli, E. Pitzalis, P. Pertici, P. Salvadori, S. Coluccia, G. Martra, L. Lampugnani, M. Mascherpa, Mat. Sci. Eng. C 15 (2001) 207.
[6] A. M. Caporusso, N. Panziera, P. Pertici, E. Pitzalis, P. Salvadori, G. Vitulli, G. Martra, J. Mol. Catal. A :Chemical 150 (1999) 275.
[7] G. Vitulli, E. Pitzalis, L. Aronica, P. Pertici, S. Bertozzi, A. M. Caporusso, P. Salvadori, S. Coluccia, G. Martra, in "Syntheses and Methodologies in Inorganic Chemistry, New Compounds and Materials" SAMIC 2000, vol. 9, p.43.
[8] G. Vitulli, M. Giampieri, P. Salvadori, J. Mol. Cat. 65 (1991) 21.
[9] A. Balerna, S. Coluccia, G. Deganello, A. Lungo, A. Martorana, G. Martra, C. Meneghini, P. Pertici, G. Pipitone, E. Pitzalis, A. M. Venezia, A. Verrazzani, G. Vitulli, Eur. Phys. J. D 7 (1999) 577.
[10] G. Martra, S. Coluccia, O. Monticelli, G. Vitulli, Catal. Lett. 29 (1994) 105.
[11] L. Marchese, G. Martra, S. Coluccia, in: C.R.A. Catlow, A. Cheetham (Eds.) New Trends in Materials Chemistry, Kluwer Academic Publishers, London, (1997), p. 79.
[12] C. Louis, L. Marchese, S. Coluccia, A. Zecchina, J. Chem. Soc., Faraday Trans. 1 85 (1989) 1655.
[13] A. Parmagliana, F. Arena, F. Frusteri, S. Coluccia, L. Marchese, G. Martra, A. L. Chuvilin, J. Catal. 141 (1993) 34.
[14] G. Martra, H. M. Swaan, C. Mirodatos, M. Kermarec, C. Louis, Stud. Surf. Sci. Catal. 111 (1997) 617.
[15] L. Marchese, R. Boccuti, S. Coluccia, S. Lavagnino, A. Zecchina, L. Bonneviot, M. Che, Stud. Surf. Sci. Catal. 48 (1989) 653.
[16] P. Basu, D. Panayotov, J. T. Yates Jr., J. Am. Chem. Soc. 110 (1988) 2074.
[17] E. C. Decanio, D. A. Storm, J. Catal. 132 (1991) 375.
[18] T. S. Zubkov, E. A. Wovchko, J. T. Yates Jr., J. Phys. Chem. B 103 (1999) 5300.
[19] C. T. Campbell, Surf. Sci. Rep. 27 (1997) 1.
[20] F. Boccuzzi, S. Coluccia, G. Martra, N. Ravasio, J. Catal. 184 (1999) 316.
[21] H. F. J. van't Blik, J. B. A. D. van Zon, T. Huizinga, J. C. Vis, D. C. Koningsberger, R. Prins, J. Am. Chem. Soc. 107 (1985) 3139.
[22] D. K. Paul, C. D. Marten, Langmuir 15 (1999) 4508.
[23] N. Sheppard, T. T. Nguyen, in: R. J. H. Clark, R. E. Hester (Eds.), Advanced Infrared Spectroscopy of Adsorbed Species, Heyden, London, 1978, Vol. 5, Ch. 2., p. 67.
[24] K. A. Almusaiteer, S. S. C. Chuang, J. Phys. Chem. B 104 (2000) 2265.
[25] G. Vitulli, C. Evangelisti, P. Pertici, A.M. Caporusso, P. Salvadori, M.G. Faga, C. Manfredotti, G. Martra, S. Coluccia, A. Balerna, S. Colonna, S. Mobilio, J. Organomet. Chem. 681 (2003) 37.

Studies in Surface Science and Catalysis 155
A. Gamba, C. Colella and S. Coluccia (Editors)

Spectroscopic characterization of Fe-BEA zeolite

P. Fejes[a], I. Kiricsi[a], K. Lázár[b], I. Marsi[c], R. Aiello[d], P. Frontera[d], L. Pasqua[d], F. Testa[d],
L. Korecz[e] and J. B.Nagy[f]

[a]Department of Applied and Environmental Chemistry, University of Szeged, Szeged,
Hungary

[b]Institute of Isotope and Surface Chemistry, CRC POB.77, H-1525 Budapest, Hungary

[c]Juhàsz Gyula Teacher's Training College, University of Szeged, Hungary

[d]Dipartimento di Ingegneria Chimica e dei Materiali, Università della Calabria, 87030 Rende
(CS), Italy
[e]Institute of Experimental Physics, Technical University, Budapest, Hungary

[f]Laboratoire de Résonance Magnétique Nucléaire, Facultés Universitaires Notre-Dame de la
Paix 61, rue de Bruxelles, B-5000 Namur, Belgium

Fe-BEA zeolite samples are characterized by Mössbauer and Electron Paramagnetic
Resonance (EPR) spectroscopies, ^{29}Si-, ^{27}Al- and ^{13}C-NMR spectroscopy and thermal
analysis. Particular attention is devoted to the determination of framework and extra
framework iron species.

1. INTRODUCTION

The isomorphous substitution of silicon for iron is very important in order to prepare highly
active zeolite based on the redox properties of Fe species [1-4]. In most of the cases, Fe
should be incorporated in the framework. Direct synthesis is never able to lead to 100%
incorporation. Substitution of silicon by iron up to 60-65% can be regarded as a real success.
Views concerning the catalytic component(s) of these interesting materials are ambivalent:
experimental results are accumulating which seem to prove that in Fe-containing zeolites not
the framework (FW), rather the extra-framework (EFW) Fe(III) is the active component.
Currently, Mössbauer and EPR spectroscopy, and temperature programmed reduction (TPR)
are the most promising techniques to get informed about the various Fe(III) species [5].
"Pristine" spectra as registered by various (optical-, Mössbauer-, EPR- etc.)
spectrometers are unsuited to obtain direct information about the matter studied unless the
spectra undergo first scrutinized evaluations. From among a large choice of computerized
spectra, deconvolution is a promising means, revealing how the individual sub-spectra are
superimposed to produce eventually the complex experimental spectrum. Even when this
method is widely used for the evaluation of the various optical and Mössbauer spectra since
several decades, apart from an idea by Lin [6] which has never been implemented in the
praxis, as far as is known, no attempt was ever made to apply spectrum deconvolution for the
analysis of Fe(III) EPR spectra. Fejes et al. published the first detailed study on this theme [5].

Nowadays various Fe-containing zeolites are in the focus of catalytic research, aiming at the selective oxidation of various organics, like benzene, methane etc., removal of NO_x from exhaust gases, just to mention the most interesting topics. The fate of the zeolite catalysts used in these reactions is very complex. In their as-synthesized state they contain Fe(III) incorporated in the zeolitic lattice, and Fe(III), as extra-framework species, in the state of single, magnetically not interacting ions; therefore, these zeolites are colorless [5]. Removal of the organic template by calcination brings about ejection of various amounts of Fe(III) from the framework and the heating in presence of water vapor causes agglomeration of the single ions into small Fe(III)-oxide clusters. Heat-treatment and use under catalytic conditions leads to further dislodgement of Fe(III), increase of the cluster size and deepening of the color (starting with buff and ending with deep brown).

"Pristine" spectra are excellent for various speculations, on the other hand apt to let draw from them dubious and even absolutely misleading conclusions. This is not surprising, because the end-effect of superposition of sub-spectra suggests an absolutely biased picture. This picture is the result of extremely complex effects, caused by the various "chemical" and "physical" forms of Fe(III) and the environments where they are emplaced.

The computerized spectrum convolution ("modeling") and deconvolution are fairly complex mathematical tasks. Companies selling Mössbauer spectrometers usually provide the respective softwares, as supplements, alleviating spectrum analysis. Unfortunately, no such programs are available for the optimized deconvolution of Fe(III) EPR spectra.

The sophisticated computer programs (so called "codes") used previously for parameter estimation in reaction kinetics and catalysis can be adapted to this task, which, in principle, is closely related to parameter estimation.

2. EXPERIMENTAL

The synthesis procedure was described in a previous paper [7]. In situ Mössbauer spectra were recorded on the calcined sample, obtained at 150°C in 8 hours from a gel having the following composition: 40 SiO_2:0.60 $Fe(NO_3)_3 9H_2O$:16.3 TEAOH:4 NaOH:676 H_2O, where TEAOH stands for tetra-ethylammonium hydroxide. Spectra were collected in constant acceleration mode, positional data are related to metallic alpha-iron. The EPR spectra were taken on a JEOL JES-FE3X spectrometer. For spectrum deconvolution a non-linear parameter estimation program was applied, making use of an objective function expressing the sum of squared errors. The EPR spectra were regarded in the constrained mode of operation (as composed of a defined number of individual Gaussian lines). The NMR spectra were recorded on a Bruker MSL 400 spectrometer. For ^{29}Si (79.4 MHz) a 4.0 μs ($\theta=\pi/6$) pulse was used with a repetition time of 6.0 s. For ^{13}C (100.6 MHz), a 7.0 μs ($\theta=\pi/2$) pulse, a single contact time and recycle time of 6.0 were used. For ^{27}Al (104.3 MHz), a 1.0 μs ($\theta=\pi/12$) pulse was used with a repetition time of 0.1 s.

3. RESULTS AND DISCUSSION

The ^{13}C-NMR spectra show clearly that TEA^+ cations are incorporated intact in the BEA zeolite channels: $\delta=53.3$ ppm for $-CH_2-$ groups and 8.4 ppm for CH_3-groups in Fe-BEA. The ^{27}Al-NMR spectra show that aluminum is incorporated mostly in framework tetrahedral sites ($\delta=52.7$ ppm), ca. 60-70 % and probably as framework octahedral species ($\delta=13.0$ ppm), ca. 30-40% in Fe,Al-BEA samples. In the pure Al-BEA the tetrahedral and octahedral species are 82% and 12%, respectively, the chemical shifts being equal.

The ^{29}Si-NMR spectra are most revealing. In all Fe-, Fe,Al- and Al-BEA samples three main NMR lines characterize the Beta zeolite structure. The −102 ppm line stems from both Si(1Al) configurations and some possible SiOM (M=H, Na or TEA) defect groups [8]. The other two NMR lines at −109 ppm and −111 ppm (shoulder) are due to Si(OAl) configurations of two crystallographically different sites [9]. Note also that some =Si(OM)$_2$ defect groups are also present at ca. −90 ppm essentially in the spectra of uncalcined material. If ferromagnetic oxides are present in the structure, no CP spectra could be taken due to the fast relaxation of both ^{1}H and ^{29}Si nuclei.

3.1. Nature of the iron species

3.1.1. Deconvolution of the EPR spectra of the zeolite samples Fe-BEA

For spectrum deconvolution non-linear parameter estimation program was applied, making use of an objective function expressing the sum of squared errors. The EPR spectra were regarded in the constrained mode of operation as composed of a defined number of individual (derivative) Gaussian lines.

The spectrometer sampled the (experimental) spectrum at 8000 discrete values of the magnetic field strength (measured in Gauss), the first 4000 of which (on the low-field side) represented the EPR spectrum. The objective function defined by the unweighted sum of the squares of deviations between the measured and computed intensities in these points, was a function of three parameters each (location of the peak maximum, H_0, the standard deviation, s, and the amplitude, C).

Taking into consideration the slight (sometimes even quite large) asymmetry of the (derivative) lines would have made the task more complicated or even intractable. In order to avoid local minima, before the actual optimization (the "local search"), a pattern search was carried out over a wide domain of the individual values of the parameters. In the local search a modified version of the Fletcher-Powell optimization procedure was applied [10]. The optimum so obtained was characterized by a relative error threshold $< 10^{-10}$.

The double integral of each sub-spectrum EPR (derivative) line, i.e. the intensity, I, was computed by the exact formula

$$I = ((2\pi e)^{1/2}/8) \times amp \times (width)^2 = 0.5166 \times amp \times (width)^2,$$

where amp is the peak to peak amplitude (in arbitrary units, a.u.) of the derivative line and $width$ is the peak to peak width measured in Gauss.

The deconvoluted Fe(III) EPR spectra are usually composed of one (or more) component(s) having g values near the free electron value ($g = 2.0023$), characteristic for framework (FW) Fe(III) in strictly "cubic" environment, and Fe(III) in extra-framework (EFW) position, but also in cubic ligand field. The two g values differ slightly, but only the computer is able to detect the difference. Slight distortion of the environment (caused by electric charge, solid surface tension, lattice defects, etc.) causes a shift of the peak maximum towards low field direction; first the symmetry becomes axial (the isotropic g becomes anisotropic), and, as the distortion increases, it turns to "fully rhombic". Axial and fully rhombic Fe(III) signals are always due to EFW iron-oxide clusters of various sizes. Fe(III) as "single ions" are magnetically non-interacting (and generate no color), but clustering under the influence of heat, water vapor and use under catalytic conditions leads to clustering and with that to development of magnetic interaction, first only within the cluster, later also between the clusters. This causes a widening of the EPR signal as the cluster size increases.

Above about 6-8 nm cluster size the clusters develop own magnetic domains with strong interaction between each other and in the EPR spectrum ferromagnetic resonance (FMR) appears which manifests itself as a sharp Dirac-delta type jump in the spectrum near to $g = 2.0$ [11].

One can raise the question, what is the Fe(III) spectrum deconvolution good for? First of all to get a general picture how the spectrum is composed from the sub-spectra: how much Fe(III) is sited in FW and EFW environments and what are the symmetries of these sites? By answering these questions Fe(III) EPR spectrum deconvolution may reach, or maybe even surpass the usefulness of temperature programmed reduction (using H_2, or CO) in some important aspects even that of Mössbauer spectroscopy.

Working on this theme for several years, and using always Fe-containing ZSM-5 zeolites, a great deal of experience has been accumulated [5, 11].

(i) Fe(III) incorporated into the silica lattice of the ZSM-5 structure (i.e. FW iron) produces an EPR signal which follows the Curie-law ($I \sim 1/T$) when the temperature changes. Deviation from this law is always indicative of extra-framework (EFW) iron. Increasing distortion of the respective site causes increasing deviation, at the extreme reaching a limiting value when the intensity will be proportional to the magnetic susceptibility of the sample (FMR).

(ii) The EPR transition probabilities for FW and EFW iron are different. At room temperature and under identical conditions, Fe(III) ions in an EFW environment give rise to an EPR signal which is about 13.75 times more intense than that originating from an FW site [5]. Unfortunately, this value is dependent on the temperature and on the (average) cluster size of the iron oxide (the previous 13.75 factor experiences a fast decrease as the particle size increases).

Fe containing ZSM-5 zeolites in their X-band EPR spectra produced an FW signal near $g = 2.0$ (we called this "F1-type"); an EFW signal (called "F2-type") also near, but slightly lower than $g = 2.0$, characteristic of small Fe_2O_3 clusters in weak (Heisenberg type) magnetic interaction, within the cluster; also an EFW signal at $g = 4.3$ (called "F3-type") due to "fully rhombic" environments (the amount of this component was always just a few tenth of a percent and in spite of the greater distortion of the field the signal was regarded and treated in the analysis as the F2-type sites).

At higher (> 2.5-3.0 wt %) concentrations the EPR spectra could not be approximated with 3 sub-spectra. Acceptable limiting value for the object function (residual value of the "sum of squares") could be attained only with the superposition of 4 sub-spectra. To this (so called "F4-type") signal a g value around 2.85 could be assigned and thus it might be one of the two axial components.

The validity of these observations and conclusions has been proven by the Mössbauer spectra registered on identical samples [5, 11]. The population data for the Fe(III) components located in FW and EFW positions have been found identical within the unavoidable experimental errors. This was at the same time a convincing proof that the Mössbauer "loudness" is nearly equal for the various Fe(III) species.

The determination of the FW/EFW ratio, or their individual values was a relatively easy task in the case of ZSM-5 zeolites. The stable F1 type framework EPR signal at about $g = 2.0$, being dependent only on the temperature, played here the role of the "Archimedean fix point" helping to string up the contributions from the various EFW iron components [5, 11]. In the case of Fe-beta this possibility does not exist: the F1 type signal is the resultant of two signals, that of the incorporated FW iron (F1/a) and that of magnetically not interacting EFW species

(F1/b) in the narrow zeolitic cavities (hexagonal prisms and sodalite cages) (this last type of Fe(III) might be in exchange position or simply a FeO(OH) entity).

In order to alleviate the estimation of the FW/EFW ratio, a constrained mode of deconvolution was carried out, limiting the number of sub-spectrum lines to three.

Fig. 1/a and 1/b show the deconvoluted spectra of the Fe-beta sample as-synthesized. (1/a) and in the calcined state (1/b). The double integrals, i.e. the respective intensities are summarized in Table 1.

Table 1
Constrained (F1, F2, F3) decomposition of the EPR spectra of as-synthesized and calcined Fe-BEA sample*

Sample	Intensities		
	F1	F2	F3
As-synthesized	420.22	4592.3	79.16
Calcined	485.98	3720.3	141.26

*F1: framework tetrahedral Fe; F2: extraframework octahedral Fe in Fe_2O_3 clusters; F3: extraframework rhombic Fe

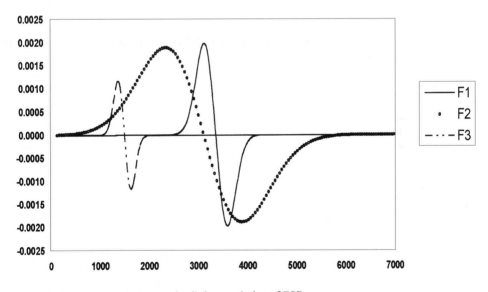

Fig. 1/a. Fe-BEA sample 'constrained' deconvolution of ESR spectrum

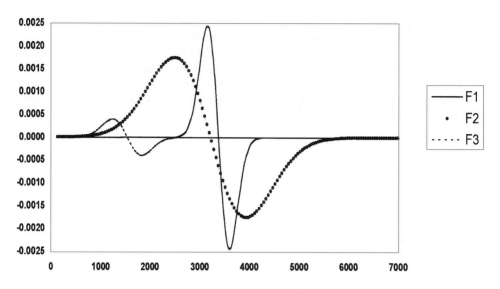

Fig. 1/b. Fe-BEA sample calcined – 'constrained' deconvolution of ESR spectrum

In the case of the as-synthesized. sample it was supposed, in first approximation, that the F1-type signal was due to the FW component alone and the FW and various EFW values were computed. The second step of the approximation was based on the assumption that in the as-synthesized state the FW/EFW ratio is the same over the whole zeolitic lattice. Thus, it was possible to calculate the contributions of the FW and EFW components to the F1-type signal once again.

The first approximation is based also on the premise that the F1 = F1/a + F1/b signal is caused exclusively by Fe(III) in the FW state:

$$420.22 + (4592.30 + 79.16)/13.75 = 760.0,$$

which is, thus, represented by its maximum value, being 420.22/760.0 = 0.553, i.e. 55.3%. Under such conditions the amount of the EFW component expressed as per cent is: 44.7% (note that F1/a and F1/b signals are not shown separately).

If this value is accepted, then of the 3.51 wt % iron 1.94 wt % is in the FW and 1.57 wt % in the EFW state.

We may look at this picture from another angle, too. An elementary cell of the zeolite beta in the as-synthesized state contains all together 7.38 FeO_2 (charged or uncharged) units on the average, of which 4.1 are located in FW, 3.24 in magnetically interacting, but not distorted EFW and finally 0.06 in fully rhombic EFW state. Note finally that the presence of Fe_2O_3 particles is also detected by ^{29}Si-NMR spectra, because no CP spectrum could be obtained for this sample.

In order to get the lower limiting value of Fe(III) in FW state, let it be supposed, as another premise, that the double F1 line in the as-synthesized state of the sample is generated by 55.3% FW and 44.7% EFW iron component (the amounts expressed in %, or in weights).

The resulting intensity, 420.22 a.u., is the sum of the contribution by the FW component: $0.553 \times I$, and by the EFW component: $0.447 \times I \times 13.75$:

$$0.553 \times I + 0.447 \times I \times 13.75 = 420.22,$$

where I is the intensity in a.u. produced by one quantity unit of the FW component.

From the equation above $I = 62.726$ a.u., and, thus the composite line is the result of the following two contributions:

$$34.69 + 385.53 = 420.22$$

Regarding now the whole deconvoluted spectrum (see Table 1) and expressing the respective intensities in "FW units":

$$34.69 + (385.53 + 4592.30 + 79.16)/13.75 = 402.49.$$

So, the lower limiting value of the FW component equals to $34.69/402.49 = 0.0862$, i.e. 8.62%; the amount of the magnetically non-interacting EFW component is $28.04/402.49 = 0.070$, i.e. 7.0%; that of the magnetically interacting EFW component: $334.0/402.49 = 0.830$, i.e. 83.0%; and finally the fully rhombic Fe(III) species is represented in $5.76/402.49 = 0.014$, i.e. 1.4%.

Also in the as synthesized state the Fe-beta zeolite sample contains 55.3%, or less incorporated Fe(III) component, and from the other approximation 8.62%, or more FW component. The real value seems to be closer to 55.3%.

3.1.2. Mössbauer spectra

In situ Mössbauer spectra recorded on the calcined $x = 0.6$, $y = 16.3$ sample are shown in Figure 2, the respective data are collected in Table 2.

Table 2
Mössbauer data extracted from spectra of Fig. 2*

Treatm/meas	Comp.	IS	QS	FWHM	RI
Calc	Fe^{3+}	0.21	-	1.32	82
As rec. / 300 K	Fe^{3+}	0.35	0.88	0.55	18
Evac.	Fe^{3+}	0.21	1.71	0.28	3
660 / 300 K	Fe^{3+}	0.24	1.02	1.05	84
	Fe^{2+}	0.92	2.23	1.11	13
H_2	Fe^{3+}	0.14	1.30	0.90	74
620 / 480 K	Fe^{3+}	0.12	0.57	0.45	18
	Fe^{2+}	1.02	1.76	0.67	9

*IS: Isomer shift, related to metallic alpha-iron, mm/s, QS: quadrupole splitting, mm/s, FWHM: line width, mm/s, RI: relative intensity of components in the spectra, %

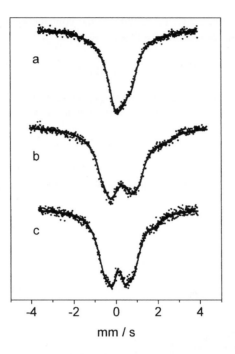

Fig. 2. In situ Mössbauer spectra recorded on the x=0.6, y=16.3 sample. (a) calcined – as received; (b) evacuated at 660 K, measured at 300 K; (c) in hydrogen, treated at 620 K, measured at 480 K

The as-received sample exhibits an asymmetric doublet of Fe^{3+} very similar to that reported on an NH_4-Fe-BEA [12]. The evacuation at 660 K removes the water chemisorbed in the vicinity of framework substituted iron – the quadrupole splitting of the dominant component increases to 1.02 mm/s. A small proportion (13 % in the spectra) can be assigned to extra-framework Fe^{2+} formed by autoreduction. The primary framework siting of iron is proven in the further step: this type of iron is not reduced even by 620 K treatment in hydrogen. The isomer shift values less than 0.3 mm/s also attest for the framework position: they are characteristic of tetrahedral sites. Brønsted acidic Si-O-Fe(OH)-O-Si groups are practically absent, the QS = 1.3 mm/s value show that charge compensation is provided by the voluminous sodium ions (and not by protons) [12].

However, this analysis was based on previously obtained in situ Mössbauer results on NH_4-Fe-BEA samples [12]. The present samples are mostly in Na^+ form, thus the reducibility and accessibility of Fe^{3+} components may differ. On the other hand it may be mentioned that in the Mössbauer technique each type of iron species contribute to the result spectrum, whereas in the ESR technique paramagnetic components with unpaired electrons are primarily seen. Thus, diamagnetic components, which may also be formed (e.g. symmetric Fe-O-Fe or (HO)-Fe-O-Fe(OH)) do not generate EPR signal [13]. The relative contribution of components also strongly depend on the coordination [11] - e.g. coordination of water to Fe^{3+} may increase the EPR signal by a factor of 1.5-2.0 [14].

Already in the analysis of the EPR spectra we have emphasized that the reducibility of the Fe^{3+} ion is strongly linked to their accessibility. On the other hand, the increasing size of the Fe-clusters could also lead to the reduction of the isomeric shift. More systematic work is necessary to be able to interpret more correctly both the EPR and the Mössbauer spectra.

4. CONCLUSIONS

The decomposition of the EPR spectra allowed us to propose the quantitative determination of the framework and extraframework Fe-content of the samples. The discrepancy between the EPR and the Mössbauer results stems probably from the fact that some of the Fe-clusters cannot be reduced by H_2, because they are not accessible.

ACKNOWLEDGMENTS
The authors are indebted to Mr. Guy Daelen for taking the NMR spectra. This work was partly supported by the Belgian Programme on Inter University Poles of Attraction initiated by the Belgian State, Prime Minister's Office for Scientific, Technical and Cultural Affairs (OSTC-PAI IVAP n° P5/01 on Quantum size effects in nanostructured materials). K.L. gratefully acknowledges the financial support provided by the National Scientific Research Grant (OTKA T 32249) enabling to perform the Mössbauer measurements.

REFERENCES
[1] A.S. Kharatinov, G.A. Sheveleva, G.A. Panov, V.I. Sobolev, P.A. Ye and V.N. Romannikov, Appl. Catal., A98 (1993) 33.
[2] C. Zhang, Z. Wu and Q. Kan, Cuihua Xuebao, 17 (1996) 34.
[3] S. Iwamoto, S. Shimizu and T. Inui, Stud. Surf. Sci. Catal., 125 (1994) 1523.
[4] G. Fierro, G. Ferraris, M. Inversi, M. Lo Jacono and G. Moretti, Stud. Surf. Sci. Catal., 135 (2001) 30-P-14.
[5] P. Fejes, I. Kiricsi, K. Lázár, I. Marsi, A. Rockenbauer, L. Korecz, J. B.Nagy, R. Aiello and F. Testa, Appl. Catal. A242 (2003) 247.
[6] D.H. Lin, G. Coudurier and J.C.Védrine, in: Zeolites: Facts, Figures, Future, R.A. van Santen et al. (eds.), Amsterdam, 1992, p. 1431.
[7] D. Aloi, F. Testa, L. Pasqua, R. Aiello and J. B.Nagy, Stud. Surf. Sci. Catal., 142 (2002) 469.
[8] R. Mostowicz, F. Testa, F. Crea, R. Aiello, A. Fonseca and J. B.Nagy, Zeolites, 18 (1997) 308.
[9] Z. Gabelica, N. Dewaele, L. Maistriau, J. B.Nagy and E.G. Derouane, in Zeolite Synthesis, M.L. Occelli and H.E. Robson (eds.), ACS Symposium Series, No. 398, American Chemical Society, Washington, DC, 1989, p. 518.
[10] R. Fletcher and M.J.D. Powell, Computer J., 6 (1963) 163.
[11] P. Fejes, K. Lázár, I. Marsi, A. Rockenbauer, L. Korecz, J. B.Nagy, S. Perathoner and G. Centi, Appl. Catal. A, submitted.
[12] A. Raj, S. Sivashanker and K. Lázár, J. Catal, 147 (1994) 207.
[13] A.V. Kucherov, C.N. Montreuil, T.N. Kucherova and M. Shelef, Catal. Lett, 56 (1998) 173.
[14] A.V. Kucherov and M. Shelef, J. Catal., 195 (2000) 106.

Studies in Surface Science and Catalysis 155
A. Gamba, C. Colella and S. Coluccia (Editors)

Thermally induced structural and microstructural evolution of barium exchanged zeolite A to celsian

C. Ferone, S. Esposito and M. Pansini

Laboratorio Materiali del Dipartimento di Meccanica, Strutture, Ambiente e Territorio, Facoltà di Ingegneria dell'Università, Via G. Di Biasio 43, 03043 Cassino (FR), Italy

A sample of Ba-exchanged zeolite A was thermally treated at temperatures up to 1400°C and its structural and microstructural evolution was investigated by room temperature X-ray diffraction, IR spectroscopy and SEM observations. The following are the main results: the zeolite structure thermally collapses at 200°C by a partial disruption of the D4R units. Thermal treatments in the temperature range 200-400°C result in a middle-range order which favors the crystallization of small crystallites of monoclinic celsian at 500°C. Thermal treatments in the temperature range 500-800°C do not give rise to the crystalline growth of monoclinic celsian, whereas crystallization of hexacelsian occurs above 800°C. At temperatures not lower than 1000°C hexacelsian begins to convert into monoclinic celsian. The conversion is completed by thermally treating the sample at 1400°C for 22 hours.

1. INTRODUCTION

Monoclinic celsian ($BaAl_2Si_2O_8$) has excellent physical properties: high melting point, low thermal expansion coefficient up to about 1000°C, and the absence of phase transitions up to 1590°C [1]. In view of these properties interest has grown in the use of celsian as a matrix reinforcement in high temperature structural composites such as in the hot sections of turbine engines [2-9].

The synthesis of monoclinic celsian gives rise to considerable problems [10]. On the one hand it requires high temperatures and/or long reaction times, which give rise to high process cost. On the other, the first polymorph to nucleate is hexagonal celsian (hexacelsian) even if it is stable at temperatures higher than 1590°C [11]. This behavior was ascribed to the simpler crystal structure of the high symmetry modification presenting a lower kinetic barrier to nucleation [12].

The early crystallization of hexacelsian appears a serious drawback on account of two different reasons: (1) hexacelsian cannot be used as a refractory because it, at 300°C, undergoes a reversible transformation into an orthorhombic form accompanied by a detrimental ($\geq 3\%$) volume change [11], (2) the transformation of hexacelsian to monoclinic celsian occurs after prolonged heating (more than 20 hours) at temperatures higher than 1500°C in the presence of mineralizers and monoclinic seeds particles [13].

One way round these difficulties is to use the technique proposed by Subramanian and co-workers [14-17] to obtain alkali-earth and alkaline feldspars, namely the thermal transformation of alkali-earth or alkaline cation exchanged zeolites. The main advantages arising from this technique are the following: (1) the desired starting composition of the system may be easily reproduced by selecting zeolites exhibiting the needed Si/Al ratio [18] and by properly adjusting the cation exchange operations; (2) the perfect homogeneity of composition at an atomic scale of the amorphous phase arising from the thermal collapse of the microporous zeolitic structure; (3) the low cost of many zeolitic terms which may act as starting materials.

Using this technique monoclinic celsian has been successfully prepared from a barium exchanged zeolite A precursors by thermal treatment at 1100°C [10, 19]. Significantly, only six hours of this treatment were sufficient to transform a sample of Ba-exchanged zeolite A, containing a Na residual amount of 0.60 meq^{-1}g^{-1}, into fully monoclinic celsian.

The very high reactivity of barium exchanged zeolite A precursors was partially ascribed to the mineralizing action of the residual Na [19, 20]. Actually an higher value of this quantity was found to result in lower temperatures and times of transformation. *ceteris paribus* [19].

In addition it was proposed that the amorphous phase arising from the thermal collapse of the microporous zeolitic structure contained, dispersed within it, small aggregates of monoclinic celsian [19].

Possible further explanations of the very high reactivity of Ba-exchanged zeolite A precursors could be found by characterizing the amorphous phase arising from the thermal collapse of the microporous zeolitic structure and the early stage of the crystallization process.

The aim of this work is to investigate the mechanisms involved in the thermal transformation of Ba-exchanged zeolite A into monoclinic celsian by the simultaneous use of XRD analysis, IR spectroscopy and SEM observations.

2. EXPERIMENTAL

Carlo Erba reagent-grade synthetic zeolite 4A ($Na_{12}Al_{12}Si_{12}O_{48} \cdot 27H_2O$) was used. This zeolite was subjected to exhaustive Ba-exchange. Exchange operations, which were terminated when their iteration did not give rise to a sensible decrease of the residual amount of the alkaline cation, were performed as follows.

Zeolite Na-A was contacted with a warm (60-70°C) 0.2 N Ba^{2+} solution with a weight solid/liquid ratio S/L = 1/25. The solution was prepared using bidistilled water and Carlo Erba reagent-grade Ba(NO$_3$)$_2$ (purity 99.5%). The solid was separated from the liquid through filtration and again contacted with the exchange solution for a total of four times. Then a fifth exchange was performed at a weight S/L = 1/50, all other conditions being equal to previous ones. The remaining exchange operations were performed contacting the zeolite with barium solutions prepared using extremely pure BaCl$_2$ (purity > 99.999%), provided by Aldrich. In particular a sixth exchange was performed at a weight S/L = 1/20 and a Ba^{2+} concentration equal to 0.5 g·dm^{-3}, a seventh at a weight S/L = 1/30 and a Ba^{2+} concentration equal to 7.5 g·dm^{-3}, and an eighth at a weight S/L = 1/40 and a Ba^{2+} concentration equal to 7.5 g·dm^{-3}, all other conditions being equal to the previous ones.

The resulting powder was washed with bidistilled water, dried for about one day at 80°C and stored for at least 3 days in an environment having about 50% relative humidity to allow water saturation of zeolite.

The residual Na content of zeolite A at the end of the exchange procedure was determined to be 0.20 meq·g^{-1} through the method hereafter described. The zeolite was chemically dissolved in a hydrofluoric and perchloric acid solution and its Na$^+$ concentration was determined by atomic absorption spectrophotometry (AAS), using a Perkin-Elmer AAnalyst 100 apparatus. The consistency of concentration measurements related to cation exchange operations was checked by determining Ba^{2+} concentration by titration with EDTA, using erio T as the indicator and a NH_3 and NH_4^+ solution at pH 10 as buffer [21].

The Ba-exchanged zeolite A sample was subjected to the following thermal treatments in a Lenton furnace, which ensures stable temperature to within $\pm 2°C$, using Al_2O_3 crucibles:

1) heated at 200°C and kept at this temperature for 2 hours;
2) heated at 500°C;
3) heated at 800°C;
4) heated at 1000°C;
5) heated at 1300°C;
6) heated at 1400°C and kept at this temperature for 22 hours.

At the end of the heat treatment the samples were quenched in air. The heating rate was 10°C/min. The polycrystalline sample of hexacelsian was obtained by thermal treatment No.5, whereas the polycrystalline sample of monoclinic celsian was obtained by thermal treatment No.6.

These products were characterized by XRD at room temperature using a Philips X'PERT diffractometer, Cu Kα radiation, collection of data between 20 and 40 ° 2θ, with a step width of 0.02 ° 2θ, and 1 s data collection per step.

The structure of the investigated samples was analyzed by room temperature Fourier transform infrared spectroscopy (FTIR). FTIR absorption spectra were recorded in the 4000-400 cm^{-1} range using a Nicolet system, Nexus model, equipped with a DTGS KBr (deuterated triglycine sulphate with potassium bromide windows) detector. A spectral resolution of 2 cm^{-1} was chosen. 2.0 mg of each test sample were mixed with 200 mg of KBr in an agate mortar, and then pressed into 200 mg pellets of 13 mm diameter. The spectrum of each sample represents an average of 32 scans, which were normalized to the spectrum of the blank KBr pellet.

Scanning electron microscopy (SEM, Philips XL30) was also used to observe crystal morphology in the powders.

3. RESULTS

3.1. X Ray Diffraction

In Fig. 1 the XRD pattern of zeolite Ba-A (Fig. 1a) is reported together with the XRD pattern of the sample thermally treated at 200°C for 2 hours (Fig. 1b), which appears completely amorphous.

In Fig. 2 the XRD patterns of the Ba-exchanged zeolite A heated at 500°C, 800°C or 1000°C and then quenched in air are reported (Figs. 2a-c, respectively). The same figure shows the XRD patterns of the polycrystalline samples of monoclinic celsian (obtained by thermally treating Ba-exchanged zeolite at 1400°C for 22 hours, Fig. 2d) and hexagonal celsian (obtained by heating the Ba-exchanged zeolite A at 1300°C and quenching it in air, Fig. 2e).

It must be noticed that the pattern of the hexacelsian sample reveals also the presence of a small amount of monoclinic celsian. Actually it is reported in ref. [19] that it was not possible to obtain a sample of hexacelsian absolutely free from monoclinic celsian through the thermal transformation of Ba-exchanged zeolite A.

In the XRD pattern of zeolite Ba-A heated at 500°C and then quenched in air (Fig. 2a) the presence of extremely broad diffraction peaks (denoted with M), which can be ascribed to monoclinic celsian, may be recorded. Raising the temperature of the thermal treatment to 800°C these peaks appear almost unaltered (Fig. 2b). Using the Scherrer formula, the

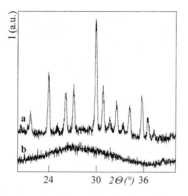

Figure 1. XRD patterns of the Ba-exchanged zeolite A (a) and Ba-exchanged zeolite A thermally treated at 200°C for 2 hours (b)

monoclinic celsian crystalline nuclei were estimated to be between 19.4 and 27.3 nm in size. In the XRD pattern of zeolite Ba-A heated at 1000°C and then quenched in air (Fig. 2c) the presence of diffraction peaks which can be ascribed to hexacelsian (denoted with H) is recorded together with the same extremely broad diffraction peaks of monoclinic celsian recorded in Figs. 2a and 2b. Thus, at this stage, very small crystals of both celsian polymorphs are dispersed in the amorphous matrix.

3.2. Infrared spectroscopy

Fig. 3 reports the room temperature IR spectra of Ba-exchanged zeolite A (Fig. 3a) and of the sample thermally treated at 200°C for 2 hours (Fig. 3b). The observed IR bands have been assigned on the basis of the structure of zeolite A and according to literature data [22-25]. The

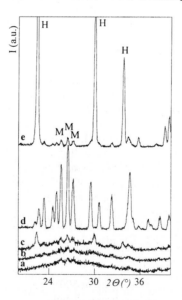

Figure 2. XRD patterns of the Ba-exchanged zeolite A heated at 500°C (a), 800°C (b), 1000°C (c) and then quenched in air. (d) = monoclinic celsian; (e) = hexacelsian.

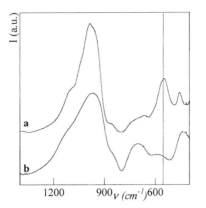

Figure 3. IR spectra of the Ba-exchanged zeolite A (a) and Ba-exchanged zeolite A thermally treated at 200°C for 2 hours (b).

structure of zeolite A is made up by tetrahedra SiO_4 and AlO_4 which have 4 bridging oxygen atoms. Such tetrahedra, which are the primary building units of framework aluminosilicates, are assembled in zeolite A so as to give rise to a structure characterized by a simple cubic cell in which eight truncated octahedral units (ß cages), present at the corner of a cube, are connected by twelve double-4-ring units (D4R), present in the middle of the sides, thus enclosing a great cavity having the shape of a truncated cuboctahedron (α cage) [18].

Moreover the electrostatic valence rule, as modified by Loewenstein, requires a rigorous alternation of the SiO_4 and AlO_4 tetrahedra, because the Si/Al ratio is equal to 1.

A careful review of recent literature data [24, 25] infers a reconsideration of some of the IR band assignments performed in a previous paper [23]. In the IR spectrum of zeolite Ba-A (Fig. 3a) the main absorption band centered at 983 cm^{-1} is assigned to the T-O (T = Si, Al) asymmetric stretching vibrations [24]. This band exhibits two shoulders at 1100 and 855 cm^{-1}. The corresponding internal bending modes of the Si-O-Al bonds give rise to the absorption band at 463 cm^{-1} [24]. Moreover two additional absorption bands were detected in the 500-800 cm^{-1} region: the first one at 668 cm^{-1} is broad and exhibits the lowest relative intensity and correspond to the symmetric stretching vibrations of tetrahedra [22], whereas the second one occurs at 550 cm^{-1} and can be likely assigned to the D4R units [24].

Figure 4 reports the IR spectra of the Ba-exchanged zeolite A heated at 500°C, 800°C or 1000°C and then quenched in air (Figs. 4a-c, respectively). The same figure shows the IR spectra of the polycrystalline samples of monoclinic (Fig. 4d) and hexagonal (Fig. 4e) celsian.

As one could expect, the line width of the absorption bands of the polycrystalline samples are narrower and have higher intensity than those of the thermally treated Ba-exchanged zeolite A samples on account of their amorphous or scarcely crystalline nature. In the examined wave numbers range (400-1400 cm^{-1}), the IR spectra of the samples of Ba-exchanged zeolite A, heated at 500, 800, and 1000°C and then quenched in air (Figs. 4a, 4b, and 4c, respectively) exhibit the higher absorption band in the 800-1200 cm^{-1} region. Its line width increases with the temperature of the thermal treatment. It is noteworthy that in the IR spectra of the sample heated at 500°C (Fig. 4a) the higher absorption band is split in two bands with two different peaks. A broad absorption band centered around 700 cm^{-1} is present in the 500-800 cm^{-1} region of the IR spectra of all thermally treated Ba-A samples (Fig. 3b, 4a, 4b, 4c), whereas the appearance of a band centered around 560 cm^{-1} in the spectra of the

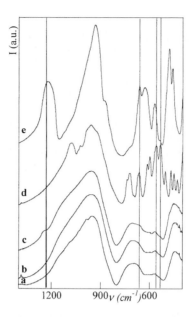

Figure 4. IR spectra of the Ba-exchanged zeolite A heated at 500°C (a), 800°C (b), 1000°C (c) and then quenched in air, monoclinic celsian (d) and hexacelsian (e).

samples treated at 500°C, 800°C and 1000°C (Fig 4a, 4b, 4c, respectively) is noticeable. The IR spectrum of the sample treated at 1000°C (Fig. 4c) shows three shoulders at 535 cm^{-1}, 660 cm^{-1} and 1220 cm^{-1}, respectively.

The bands present in the IR spectrum of hexacelsian (obtained by heating Ba-exchanged zeolite A at 1300°C and quenching it in air, Fig. 4e) are assigned on the basis of the structure of hexacelsian, which can be considered as double layer of mica-like sheets, held by the apexes of oxygen tetrahedra (Fig. 5) [20]. Barium atoms occupy the positions between such

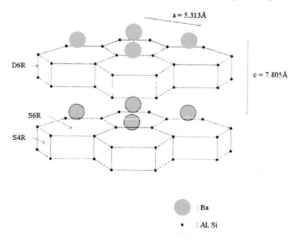

Figure 5. Hexagonal celsian structure (oxygen atoms were omitted)

double sheet and hold them together to form the three-dimensional framework of the crystals (Fig. 5). Therefore, according to the IR and Raman spectra of this phase reported by Scanu et al. [26], the absorption bands centered at 934 and 1223 cm^{-1} may be ascribed to the T-O stretching, whereas the absorption bands occurring in the 400-500 cm^{-1} region may be related to the bending modes of the Si-O-Al bonds [26], (Fig. 4e). Moreover three absorption bands were detected in the 500-800 cm^{-1} region which can be related to the characteristic Si-O and Al-O stretching of the rings forming the mica-like sheets.

The IR spectrum of monoclinic celsian (obtained by thermally treating Ba-exchanged zeolite A at 1400°C for 22 hours, Fig. 4d) differs from the one of the hexagonal polymorph (Fig. 4e). The former exhibits a higher number of absorption bands than the latter as a consequence of the reduction of the structure symmetry degree and is in complete agreement with the IR spectrum shown in ref. [26].

3.3. SEM observations

Fig. 6 reports the micrograph of Ba-exchanged zeolite A. Figure 7 reports the micrographs of the sample thermally treated at 200°C for 2 hours (Fig. 7a) and of the samples heated at 500°C (Fig. 7b), at 800°C (Fig. 7c) and then quenched in air. Figure 8 reports the micrographs of the sample heated at 1000°C (Fig. 8a), at 1300°C (Fig. 8b) and then quenched in air. The same figure shows the micrograph of the sample thermally treated at 1400°C for 22 hours (Fig. 8c).

Micrographs of the samples heated up to 800°C and then quenched in air show that the cubic morphology of the zeolite crystals (Fig. 7) is retained also in the amorphous samples (Figs. 7a-c). No evidence of the presence of small monoclinic celsian crystals can be detected. The shape of the crystals of the sample of Ba-exchanged zeolite A heated at 1000°C and then quenched in air (Fig. 8a) begins to be rounded, evidencing the upcoming crystallization of hexacelsian, which is complete at 1300°C (Fig. 8b), as confirmed by the relevant XRD patterns. Micrograph of the sample thermally treated at 1400°C for 22 hours (Fig. 8c) points out a microstructure composed of long and narrow crystals, belonging to monoclinic celsian.

Figure 6. Micrograph of Ba-exchanged zeolite A.

256

Figure 7. Micrographs of Ba-exchanged zeolite A thermally treated at 200°C for two hours (a), heated at 500°C (b) and at 800°C (c).

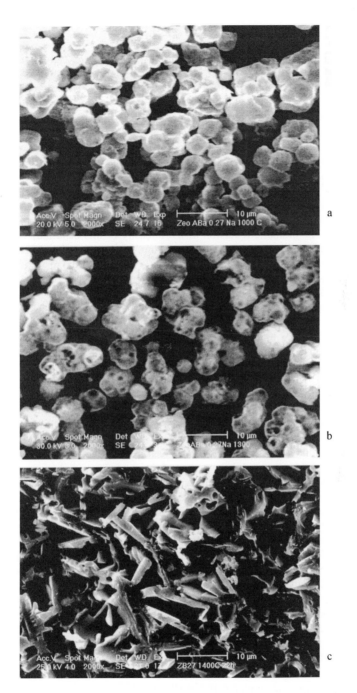

Figure 8. Micrographs of Ba-exchanged zeolite A heated at 1000°C (a), at 1300°C (b) and thermally treated at 1400°C for 22 hours (c).

4. DISCUSSION

A careful inspection of the XRD patterns reported in Figs. 1 and 2 gives rise to a series of considerations. First of all the zeolite structure has completely collapsed at a temperature as low as 200°C. This behavior was attributed [10, 19] to the zeolite framework distortion caused by the Na → Ba exchange and the subsequent loss of crystallinity of the Ba-exchanged zeolite. The thermal collapse of the zeolite structure is confirmed by the IR spectra reported in Fig. 3. Actually the fact that the bands, which were detected in the IR spectrum of Ba-exchanged zeolite A (Fig. 3a), appear far broader also in the IR spectrum of Ba-exchanged zeolite A thermally treated for 2 hours at 200°C (Fig. 3b) is consistent with the dramatic loss of crystallinity evident in Fig.1.

Then another finding which appears very noteworthy is the following: the intensity of the band at 550 cm^{-1} (Fig. 3a), which was assigned to the D4R units [24], decreases dramatically in the IR spectrum of the sample thermally treated at 200°C (fig. 3b), even if its presence may still be recorded. The dramatic reduction of the intensity of this band may be interpreted by ascribing the thermal collapse of the zeolite structure to the disruption of D4R units, which join the sodalite cages and act as pillars of the whole structure. This result is consistent with the most recent literature data [24, 27-29] and is somewhat different from the one proposed in a recent paper [23], in which a different mechanism for the collapse of the zeolite structure based on the disruption of sodalite cages was proposed.

The thermal evolution of the amorphous phase in the temperature range 200–800°C appears far more complex to interpret. In the IR spectrum of the sample heated at 500°C and then quenched in air (Fig. 4a) the main band centered around 950 cm^{-1} splits into two bands and its position is more similar to that of monoclinic celsian (Fig. 4d) than to zeolite (Fig. 3a) or hexagonal celsian (Fig. 4e). In the range 500–800 cm^{-1} the appearance of a band centered at 560 cm^{-1} (Fig. 4a) may be detected. This band remains almost unaltered in the IR spectra of samples thermally treated at 800°C and 1000°C. Its position corresponds to that of an intense band in the spectrum of monoclinic celsian (Fig. 4d), thus it might indicate the presence of a small amount of monoclinic celsian. This finding is supported by the XRD pattern of the sample heated at 500°C (fig. 2a) where broad peaks, which can be ascribed to monoclinic celsian, are clearly evident.

This interpretation appears consistent with the hypothesis that the presence of residual D4R units (band at 550 cm^{-1} in Fig. 3b), detected in the amorphous phase arising from the thermal collapse of the Ba-exchanged zeolite A structure, favors, at temperatures not lower than 500°C, the crystallization of small crystallites of monoclinic celsian [19].

Increasing the temperature of the thermal treatment from 500 to 800°C a marked evolution in the IR spectra can not be noticed, the exception being the disappearance of the doublet at 950 cm^{-1}. More evident differences are observed in the IR spectrum at 1000°C where three new shoulders at 535 cm^{-1}, 663 cm^{-1} and at 1220 cm^{-1}, respectively, are present. The first one corresponds to another intense band in the monoclinic spectrum (Fig. 4d), while the other two bands could be attributed to hexacelsian (compare spectra of Fig. 3c and 3e). The presence of hexacelsian is confirmed by the XRD pattern of the sample heated at 1000°C (fig. 2c) in which peaks of hexacelsian are clearly recognizable.

SEM observations can not confirm the presence of the monoclinic celsian nuclei in the amorphous phase arising from the thermal collapse of the zeolite structure owing to their extremely small dimensions ranging between 19.4 and 27.3 nm.

5. CONCLUSIONS

The simultaneous use of XRD analysis, IR spectroscopy and SEM microscopy allowed new insight into the thermally induced structural and microstructural transformation of Ba-exchanged zeolite A into monoclinic celsian. The interpretation of IR spectra allowed, first of all, to reasonably ascribe the thermal collapse of the microporous structure of Ba-exchanged zeolite A, clearly evidenced by XRD, to the disruption of D4R units which connect β cages. Secondly, the IR spectroscopy allowed to follow the evolution of the amorphous phase arising from the thermal collapse of the microporous zeolite structure. In particular it was found that a residual amount of D4R units was present even after the thermal collapse of the zeolite structure and created a middle-range order which favored the formation of crystalline nuclei of monoclinic celsian, clearly evidenced by XRD, in the temperature range 200-500°C. The very small dimensions of such nuclei (19.4-27.3 nm) did not allow to detect their presence by SEM microscopy. Thirdly, the IR spectroscopy gave strong indications that the further transformation of the amorphous phase arising from the thermal collapse of the microporous structure in the temperature range 500-800°C led to crystallization of hexacelsian rather than to the growth of the crystalline nuclei of monoclinic celsian formed at 500°C. Fourthly, the presence, in the IR spectrum of Ba-exchanged zeolite A heated at 1000°C and quenched in air, of bands that can be related to hexacelsian and bands that can be related to monoclinic celsian suggests that in the temperature range 1000-1400°C the presence of the crystalline nuclei of monoclinic celsian, formed at 500°C, favors the conversion of hexacelsian to monoclinic celsian, while the crystallization of hexacelsian from the amorphous phase arising from the thermal collapse of the zeolite structure keeps on occurring.

ACKNOWLEDGEMENTS
This investigation was carried out with the financial contribution of MIUR (Ministero dell'Istruzione, dell'Università e della Ricerca).

REFERENCES

[1] I.G. Talmy, D.A. Haught and E.J. Wuchina, in Proceedings of the 6th International SAMPE Electronic Conference, Society for the Advancement of Materials and Process Engineering, A.B. Goldberg and C.A. Harper, (eds.), Covina, CA, USA, (1992) 687.
[2] N.P. Bansal and C.H. Drummond III, J. Am. Ceram. Soc., 76 (1993) 1321.
[3] N.P. Bansal, J. Mater. Sci., 33 (1998) 4711.
[4] N.P. Bansal, J. Am. Ceram. Soc., 80 (1997) 2407.
[5] N.P. Bansal, US Patent No. 5,214,004, May 25, 1993.
[6] N.P. Bansal, US Patent No. 5,389,321, Feb 14, 1995.
[7] N.P. Bansal, NASA TM 106993, Aug 1995.
[8] N.P. Bansal, NASA TM 210216, Jun 2000.
[9] J.Z. Gyekenyesi and N.P. Bansal, NASA TM 210214, Jul 2000.
[10] G. Dell'Agli, C. Ferone, M.C. Mascolo and M. Pansini, Solid State Ionics, 127 (2000) 309.
[11] H.C. Lin, W. R. Foster, Am. Mineralogist, 53 (1962) 134.
[12] M. Chen, W.E. Lee and P.F. James, J. Non-Cryst. Solids, 130 (1991) 322.
[13] B. Hoghooghi, J. McKittrick, E. Helsel and O. Lopez, J. Am. Ceram. Soc., 81 (1998) 845.
[14] M.A. Subramanian, D.R. Corbin and R.D. Farlan, Mat. Res. Bull., 21 (1986) 1525.
[15] U.V. Chowdry, D.R. Corbin and M.A. Subramanian, US Patent 4,813,303, 21 March 1989.
[16] M.A. Subramanian, D.R. Corbin and U.V. Chowdry, Adv. Ceram., 26 (1989) 239.
[17] D.R. Corbin, J.B. Parise, U.V. Chowdry and M.A. Subramanian, Mater. Res. Symp. Proc., 233 (1991) 213.
[18] D.W. Breck, Zeolite Molecular Sieves: Structure, Chemistry and Use, Wiley, New York, 1974.

[19] C. Ferone, G. Dell'Agli, M.C. Mascolo and M. Pansini, Chem. Mater., 14 (2002) 797.

[20] K.T. Lee and P.B. Aswath, Materials Science and Engineering A, 352 (2003) 1.

[21] G. Schwarzenbach and H. Flaschka, Complexometric Titration, Methuen, London, 1969.

[22] E.M. Flanigen, H. Khatami and H.A. Szymski, Adv. Chem. Series, 101 (1971) 201.

[23] A. Aronne, S. Esposito, C. Ferone, M. Pansini and P. Pernice, J. Mater. Chem., 12 (2002) 3039.

[24] S. Markovic, V. Dondur and R. Dimitrijevic, J. Mol. Struct., 654 (2003) 223.

[25] Y. Huang and Z. Jiang, Microporous Mater., 12 (1997) 341.

[26] T. Scanu, J. Guglielmi and Ph. Colomban, Solid State Ionics, 70/71 (1994) 109.

[27] Y. Huang and E.A. Havenga, Chem. Phys. Lett. 345 (2001) 65.

[28] J. Djordjevic, V. Dondur, R. Dimitrijevic and A. Kremenovic, Phys. Chem. Chem. Phys., 3 (2001) 1560.

[29] N.J. Clayden, S. Esposito, C. Ferone and M. Pansini, J. Mater. Chem., 13 (2003) 1681.

Studies in Surface Science and Catalysis 155
A. Gamba, C. Colella and S. Coluccia (Editors)

Preliminary characterization of an unknown mineral from Mont Peylenc (Massif Central, France)

S. Ferrari[a], E. Galli[a] and A.F. Gualtieri[a]

[a]Earth Sciences Department, University of Modena and Reggio Emilia, L.go S.Eufemia 19, 41100, Modena, Italy

ABSTRACT

The discovery of a new natural zeolite phase is a rare event since only less than one hundred natural zeolites are known to date. We have been recently involved in a research project aimed at the investigation of a microscopic aggregate coming from the Massif Central (France) that appears to be a new zeolite species. The sample was collected more than 30 years ago by Professors Gottardi and Galli and since then, many unsuccessful attempts have been made to disclose the crystal chemistry of this new species. The quantity of available material is very limited, and single crystals are not available. For this reason, structure determination from powder diffraction data is attempted. This work presents the results of preliminary chemical, microscopy and mineralogical characterization of this apparently unknown mineral, mainly using X-ray powder diffraction. All data point to a new zeolite species and the results will be hopefully integrated in the future with the crystal structure solution by the powder methods.

1. INTRODUCTION

More than 30 years ago, during a field trip to the Massif Central area (France) that was organized to sample zeolite specimens, Professors Glauco Gottardi and Ermanno Galli collected a xenolite of about 20 cm diameter, with an extremely interesting appearance: the sample was in fact plenty of cavities filled with crystals which were almost certainly newly-formed minerals. Later, laboratory mineralogical characterization (above all, X-ray diffraction using a Gandolfi camera) confirmed this hypothesis.

Beside well known mineral species such as zeolites, carbonates and silicates, the diffraction pattern of small aggregates or a micro-crystalline phase with an apparently fibrous habit appeared to be different when compared to those reported in the literature databases. Since the amount of collected material was really small (only three small spherical aggregates of fibrous micro-crystals are available and no single crystals can be separated for determination of the structure), one of the authors (E.G.) organized in the past three more field trips in an attempt to find other specimens of this apparently new mineral species. Unfortunately, all the attempts were unsuccessful.

In the meantime, the techniques for the chemical and structural characterization of micro-volume materials have developed with the main stimulus given by the advent and fast development of synchrotron radiation (structure determination from powders is now more or less a routine analysis). Thus, a good opportunity is available to disclose the crystal chemistry of this species and reveal its crystal chemical nature. This prompted a research project aimed at the full crystal-chemical characterization of this new mineral species, which is believed to be a new natural zeolite.

It should be remarked that the discovery of a new species is a rare event as less than one hundred natural zeolites are known so far. On the one hand, the discovery of a new natural zeolite is very important mineralogically, on the other, it is very useful for the understanding of the chemical stability, formation conditions, and properties of the synthetic counterparts [1,2]. Moreover, besides the scientific interest, it should be said that zeolites have unique physical chemical properties (such as molecular sieving, cation exchange, dehydration/rehydration, etc) which can be used for a nearly infinite number of technological/industrial applications (catalysis, animal feeding, agriculture, molecular sieving for gas and liquid phase separation, and others). The huge interest regarding zeolites and their technological properties is unquestionably the rationale behind research to create zeolites in the laboratory. The synthesis of zeolites in the laboratory is now possible and, because the composition of synthetic zeolites may vary more widely (including Ga, Ge, Be, P in place of Si/Al in the framework, and alkali, alkaline-earth, rare earth elements and organic complexes as extraframework cations) the number of known synthetic zeolites nowadays is by far much larger than that of the natural species.

The aim of this work is to present the results of the preliminary characterization of this apparently unknown mineral, mainly using X-ray powder diffraction. Since no single crystals are available, these results will be hopefully integrated and completed by the crystal structure determination using the powder methods.

2. EXPERIMENTAL METHODS

One of the three small aggregates of fibrous micro-crystals available has been further divided into smaller fragments for the different analyses.

Preliminary chemical analysis using a small piece of the aggregate has been obtained by EMPA. The fragment of the specimen was embedded in an epoxy resin, polished to achieve a smooth surface and coated with a carbon film. Polishing has been performed with ¼ μm abrasive paste by a Struers DP-U2 instrumentation. Analyses were made using a WDS (wavelength dispersive) system ARL-SEMQ, with 15 kV and 10 nA and a 30 μm diameter beam.

Preliminary SEM morphology images and compositional maps have been collected with a Philips XL 40/604. Another small fragment was used for the analysis, mounted on an aluminum stub, coated with a 20 nm thick carbon film.

XRD analyses of the larger aggregate of micro-crystals were preliminary performed using a diffractometer Philips PW 1130/00 with a Gandolfi camera. Radiation used was Cu Kα, with Ni filter (1.5418 Å). In order to collect data, two different kinds of photographic film were used: 1- Kodak Industrex AA400, with 24 h pose; 2- Kodak Industrex MX, with 72 h pose; the results given show no important differences. All the procedures for working with the films used Kodak products.

High resolution XRD patterns have been collected with synchrotron light on the same aggregate used for the preliminary diffraction tests. Experiments were performed at the Italian beamline (GILDA, BM08) at the European Synchrotron Radiation Source (ESRF) in Grenoble (France). The patterns were collected using a Debye-Scherrer geometry and an Imaging Plate (IP) [3,4]. The experimental conditions were: distance sample-IP film = 280.9277 mm, tilt = 0.158°, 16 bunch beam, 90 mA current and λ = 0.688876 Å (calibration made with LaB_6 NIST standard). In order to obtain the random statistic of a powder, the aggregate sample has been mounted on a goniometric head and spun around the axis of the head. The raw data collected on the IP were corrected and transformed into digitalized data reporting the intensity and the peak position using the FIT2D software [5].

3. RESULTS AND DISCUSSION

The results of the XRD analyses revealed that the pattern was different from those present in literature databases although some similarities were found with the traces of some natural and synthetic zeolites. Similarities were found with the patterns of mordenite, mazzite, erionite and ECR-1 as shown in Table 1, which reports the *d-spacings* and intensities of the pattern of the unknown species compared with that of mordenite, mazzite, ECR-1 and erionite.

In Table 2, average values of the results obtained from the EPMA are shown. The results show a chemical analysis typical of a zeolite species. Assuming the sample to be a zeolite species, Table 3 reports an elemental analysis derived from the content determined with the EPMA. The value of the Si/Al ratio would be about 3.6, similar to that of many other zeolite phases [1]. The last column in Table 3 gives an important value: the balance error, which should be zero in a perfectly balanced analysis, and in any case should be generally lower than 10% [2]. Since the value is very large, a possible explanation could be the presence of light elements that could not be seen with EMPA. On the other hand, the large value of the balance error could be reasonably justified by taking into account the behaviour of two extra-framework cations, Na and K. Due to the presence of channels in the zeolitic structure, it is possible in principle to observe migration of the extra-framework cations along these channels, mostly Na and K [6]. The activation energy for the migration is given by the electron beam.

Table 1

Comparison between the pattern of the unknown phase and those of mordenite, mazzite, ECR-1 [7] and erionite. In the intensity column (I): **vs** = very strong, **s** = strong, m = medium, mw = medium weak, w = weak, vw = very weak

Unknown phase		Mordenite		Mazzite		ECR-1		Erionite	
d_{hkl}(Å)	I	d_{hkl}(Å)	I/I_o	d_{hkl}(Å)	I/I_o	d_{hkl}(Å)	I/I_o	d_{hkl}(Å)	I/I_o
14.9	s	-	-	15.92	98	14.84	13	-	-
13.1	vw	13.6	18	-	-	-	-	-	-
-	-	-	-	-	-	-	-	11.50	100
10.60	w	10.26	5	-	-	10.62	16	-	-
9.11	**s**	9.06	100	9.19	92	9.08	58	9.20	11
7.92	**s**	-	-	7.96	40	7.85	48	-	-
7.55	vw	-	-	-	-	7.45	13	7.52	7
6.75	m	-	-	6.89	25	6.76	35	-	-
6.65	w	6.59	14	-	-	6.55	13	6.62	73
6.32	vw	6.40	17	-	-	6.27	5	6.38	5
6.23	mw	6.07	4	6.02	45	-	-	-	-
5.90	m	5.80	18	-	-	5.89	11	5.74	16
5.48	vw	-	-	5.51	5	-	-	-	-
5.30	w	-	-	5.30	16	5.31	8	5.36	14
4.99	m	-	-	-	-	5.00	8	-	-
4.75	vw	-	-	4.73	50	-	-	-	-
-	-	4.53	31	-	-	-	-	4.55	12
4.43	w	4.46	2	4.41	9	4.43	11	-	-
4.36	w	-	-	-	-	4.34	5	4.32	67
4.25	m	-	-	-	-	4.23	20	-	-
4.16	mw	4.15	8	-	-	-	-	4.16	24
-	-	4.00	70	3.98	12	-	-	-	-
3.83	w	3.84	7	3.82	97	3.83	25	3.81	37
3.73	w	3.76	4	3.72	36	3.73	22	3.77	65
3.65	w	3.63	3	3.65	32	3.68	39	3.60	24
3.55	w	3.53	2	-	-	3.57	21-	-	-
3.50	**s**	3.48	43	3.53	100	3.49	100	-	-
3.42	w	3.42	11	3.44	9	-	-	-	-
-	-	3.39	33	-	-	-	-	3.37	4
-	-	-	-	-	-	-	-	3.30	29
3.27	m	3.29	3	-	-	3.26	51	3.28	25
-	-	-	-	-	-	-	-	3.27	25
-	-	3.22	40	-	-	-	-	-	-
3.18	**vs**	3.20	34	3.19	84	3.17	99	-	-
3.10	vw	3.10	48	3.10	35	-	-	3.11	12
3.07	vw	-	-	3.06	25	3.07	13	-	-
3.01	m	3.02	2	3.01	31	3.00	23	-	-
2.90	**s**	2.94	5	2.94	54	-	-	2.92	10

Table 2

The average result of the EMPA chemical analysis (percentage values)

SiO$_2$	Al$_2$O$_3$	Fe$_2$O$_3$	MgO	CaO	SrO	BaO	Na$_2$O	K$_2$O	H$_2$O
62.32	14.56	0.20	1.33	2.55	0.04	0.05	0.77	2.46	15.72

Table3

Elemental analysis derived from Table 2; values are per 100 O atoms

Si	Al	Fe^{3+}	Mg	Ca	Sr	Ba	Na	K	O	H$_2$O	E%
39.529	10.886	0.095	1.257	1.733	0.015	0.012	0.947	1.991	100	33.255	22.40

Fig. 1 (a,b) is a SEM image showing the morphology of the aggregate at two different scales. It is easy to see that the size of the crystals is very small and that the crystals form radially fibrous aggregate which compose more or less a sphere (here we see just a fragment) with a diameter of about 300 μm. Such radially fibrous aggregates have already been observed in the literature for natural zeolites such as mordenite and phillipsite. Unfortunately, it is not possible to see the terminations of the crystals, since the external surface is covered by a thin amorphous material. The single crystals have a length of about 100 μm, and a base smaller than 10 μm. Due to the size, it is impossible to separate a single crystal for diffraction studies. Since the data collected with the Gandolfi camera are of low resolution, synchrotron powder data were collected using an IP in a Debye-Scherrer geometry to get better peak intensity and better resolution. Fig. 2 shows one of the collected patterns (2θ vs. Intensity) from the small specimen aggregate. On the one hand, the peak position confirms the results previously obtained, on the other the peak intensity and better angular resolution allowed a tentative cell indexing and search for the possible space group.

The indexing of the unit cell was possible using TREOR [8]. The best indexing with zero unindexed peaks was obtained with an orthorhombic unit cell with cell parameters: $\underline{a}$ = 26.31(7) Å, $\underline{b}$ = 18.20(6) Å, $\underline{c}$ = 7.56(2) Å.

266

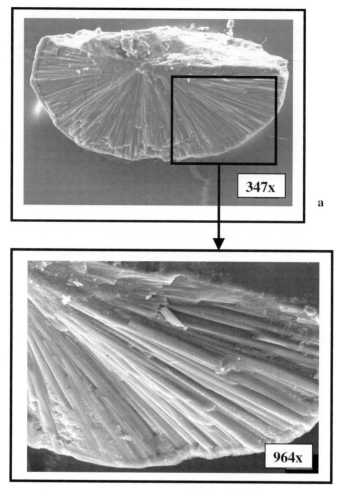

Fig. 1. Results of the SEM and EDS analyses: a) general aspect of the sample at 347x; b) higher resolution image, where it is possible to see the morphology of the crystals (964x)

A preliminary search for the best space group and cell refinement was attempted using the program Chekcell with the LMGP Suite [9]. The analysis lead to the conclusion that only a primitive orthorhombic cell was possible. Among the possible primitive space groups, Pmmm, Pmmn, Pmm2, P2$_1$2$_1$2, P222, and Pmma were the only ones compatible with the extinction conditions observed in the powder pattern. The unit cell was also refined with the Le Bail method [10] in GSAS [11] by imposing the different possible space groups. With this procedure, the best fit obtained is an indication also of the most likely space group. Among all, the best fit was obtained with the space groups Pmmm (R_{wp}=10.5%), Pmmn (R_{wp}=10.6%), Pmm2 (R_{wp}=10.5%). Fig. 3 shows the LeBail fit of the pattern using the space group Pmm2.

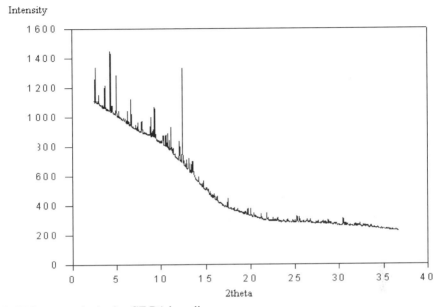

Fig. 2. XRD pattern obtained at GILDA beamline

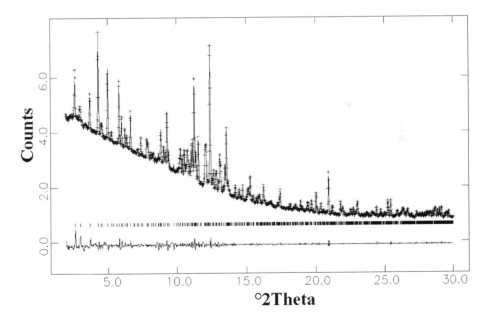

Fig. 3. The LeBail fit of the observed raw pattern using GSAS and the space group Pmm2. Crosses are the observed data points, the single line is the calculated pattern, with the peak position markers and the difference curve below

4. CONCLUSIONS

This paper reports a preliminary characterization of a specimen which could be a new natural zeolite species. This is a preliminary piece of work of a research project, still in progress, aimed at determining the crystal structure of this possible new zeolite by the powder methods. This is a particularly attractive task since this would be the first structure determination of a natural zeolite from powder diffraction data.

It should be remarked that the peak positions and the intensities are fully compatible with those of a synthetic zeolite: ECR-1 [7,12]. In fact, the cell parameters of ECR-1 are $\underline{a}$ = 7.31 Å, $\underline{b}$ = 18.144 Å, and $\underline{c}$ = 26.31 Å. Although the space group has not been unambiguously determined, the space group advanced for ECR-1 is one of the possible space groups that we found. The proposed structure of ECR-1 is composed by interposing layers of mordenite-like sheets between layers of mazzite-like cages. Determining the structure of the new natural zeolite would be also of great help to confirm the structure model of ECR-1 which has never been refined using diffraction data.

The next step in this project will be a new high resolution data collection with synchrotron radiation at ESRF (ID31) using a radiation of about 0.7-0.8 Å, and a narrow beam (less than half a mm width, if possible) on the sample. In fact, given the complexity of the structure (relatively low symmetry and large unit cell), the high resolution is extremely important for the extraction of intensities by the Le Bail method and structure determination. The cryostream cold-nitrogen-gas blower could be eventually used to dump the thermal motion and extra-framework molecule disorder of the zeolite phase to improve the chances to determine the structure correctly.

Acknowledgement

The authors would like to thank Francesco Di Renzo (ENSCM, Montpellier, France) for his important suggestions about the synthetic zeolite ECR-1.

REFERENCES

[1] D.S. Coombs, A. Alberti, T. Ambruster, G. Artioli, C. Colella, E. Galli, J.D. Grice, F. Liebau, J.A. Mandarino, H. Minato, E.H. Nickel, E. Passaglia, D.R. Peacor, S. Quartieri, S. Rinaldi, M. Ross, R.A. Sheppard, E. Tillmanns and G. Vezzalini, Am. Miner., 83 (1998) 935.

[2] G. Gottardi and E. Galli, Natural zeolites, Springer Verlag, Berlin, 1985.

[3] Y. Amemija, Y., Synchrotron Radiation News, 3 (1990) 21.

[4] C. Meneghini, G. Artioli, A. Balerna, A. F. Gualtieri, P. Norby and S. Mobilio, J. Synchrotron Rad., 8 (2001), 1162.

[5] Hammersley, A.P. (1998) FIT2D V10.3 Reference Manual V4.0 ESRF98HA01T.

[6] R. Rinaldi, La microsonda elettronica. In A. Armigliato, U. Valdrè, Microscopia elettronica a scansione e microanalisisi parte II microanalisi, Bologna, 1981.

[7] M. E. Leonowicz and D. E. W. Vaughan, Nature, 329 (1987), 819.

[8] P. E. Werner, L. Eriksson and M. Westdahl, J. Appl. Cryst., 18 (1985), 367.

[9] LMGP-Suite. Suite of Programs for the interpretation of X-ray Experiments, by Jean laugier and Bernard Bochu, ENSP/Laboratoire des Matériaux et du Génie Physique, BP 46. 38042 Saint Martin d'Hères, France.

[10] A. Le Bail, H. Duroy and J. L. Fourquet, Mat. Res. Bull., 23 (1988), 447.

[11] A. C. Larson and R. B. Von Dreele, General Structure Analysis System (GSAS), Los Alamos National Laboratory, document LAUR 86-748, 1994.

[12] D. E. W. Vaughan and K. G. Strohmaier, Proc. 7th Int. Zeolite Conf. (1986), 207.

Studies in Surface Science and Catalysis 155
A. Gamba, C. Colella and S. Coluccia (Editors)

High-pressure behaviour of yugawaralite at different water content: an ab initio study

E. Fois[a], A. Gamba[a], G. Tabacchi[a] , S. Quartieri[b], R. Arletti[c], and G. Vezzalini[c]

[a]Dipartimento di Scienze Chimiche e Ambientali, Università dell'Insubria at Como,
and INSTM, Via Lucini 3, I-22100 Como, Italy

[b]Dipartimento di Scienze della Terra, Università di Messina,
Salita Sperone 31, I-98166 Messina, Italy

[c]Dipartimento di Scienze della Terra, Università di Modena e Reggio Emilia,
Largo S. Eufemia 19, I-41100 Modena, Italy

ABSTRACT

The influence of the water content in a partially occupied water site in yugawaralite on the pressure-induced structural modifications of this zeolite has been studied by means of *ab initio* molecular dynamics simulations. The effect of the water site occupancy has been singled out by comparing results from simulations, performed with the cell parameters determined via XRPD at rP and 8.8 GPa, on two model systems the cell stoichiometry of which differs only by the presence of a water molecule in such a site. Our results show that the global P-induced changes of the framework structure upon compression are mainly attributable to the template effect of Ca extra-framework cations, while the presence or absence of water in a partially occupied site affects the framework structure only at short range distances from the site. However, the overall water content plays a fundamental role in stabilizing/destabilizing the zeolite structure as a function of pressure.

1. INTRODUCTION

The high pressure (HP) behaviour of zeolites has been the subject of growing interest in the recent years. X-ray diffraction studies with non penetrating pressure transmitting media have demonstrated that the compressibility of zeolites does not simply depend on the framework density; rather, the type, concentration and distribution of extra-framework species seem to play a fundamental role in governing the response of the system to an applied pressure [1,2]. However, we are still far from understanding the complex relationships between extra-framework content and P-induced behaviour in zeolites due to the lack of microscopic insight on the compression mechanism. So far little information is available on the atomic level details of the mechanism and the reversibility of the P-induced structural modifications in zeolites because of the difficulty of performing accurate HP X-ray diffraction experiments. In this respect, combined X-ray powder diffraction (XRPD) experiments and computer simulations may be of great relevance in providing information at atomic level resolution [3,4].

We present here an *ab initio* molecular dynamics study on the effect of the water content on the pressure behaviour of yugawaralite [1,5-7]. The cell parameters used as input for the simulations were obtained by accurate XRPD measurements at H*P* conditions. State of the art computational techniques were adopted to study at an atomic level the yugawaralite, thus providing detailed insight on the changes brought about by compression on the zeolite. In a previous study [8] we have shown that the *P*-induced structural modifications on yugawaralite were mainly determined by the Ca extra-framework cation: under increasing pressure, the Ca coordination polyhedron rearranges in such a way to reduce the free space in the channels [8]. Moreover the templating effect of Ca in driving the framework deformation has been highlighted.

The present study will be mainly focused on the understanding of the effect of a different number of water molecules per unit cell on the *P*-induced deformations in yugawaralite, and in particular, how the presence of a H_2O molecule in a partially occupied site may affect both the stability and the structural modifications of this zeolite under high pressures.

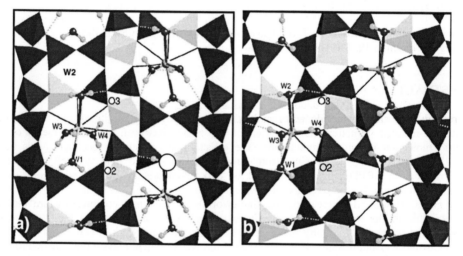

Figure 1. Projections on the *bc* plane of the average structures obtained from MD simulations of YU8W. a): r*P*; b): 8.8 GPa. Light grey tetrahedra: Al, dark grey tetrahedra: Si. Ca and H_2O are shown in ball and stick representation. The location of W5 site is represented by an empty circle.

2. YUGAWARALITE

Yugawaralite, $Ca_2[Al_4Si_{12}O_{32}]\cdot(H_2O)_9$, is a Ca-zeolite, the framework topology of which (framework type YUG) is characterized by channels parallel to [100] (Fig. 1a) and [001] with a 8-ring section [5]. The topological symmetry $C2/m$ is reduced to the real symmetry Pc by the (Si,Al) ordering, and the room pressure parameters of the unit cell are $a = 6.700(1)$, $b = 13.972(2)$, $c = 10.039(5)$ Å, $\beta = 111.07(2)°$. The extra-framework species, Ca and H_2O, are located in one Ca site and five water sites. The water sites W2 and W3 are 100% occupied, whereas W1 and W4 can be found in two partially occupied positions about 0.5 Å from each other. Such partially occupied sites, namely W1/W1A and W4/W4A, are not symmetry related. The water site W5 has an occupancy of 25%: H_2O molecules in this position are not coordinated with the cationic site, but form an hydrogen bond with the water molecule in W1. The most weakly bonded water molecules in W5 are thus the first to be lost upon dehydration [6,7]. The Ca cation is coordinated to four framework oxygen atoms and four water molecules.

The HP behaviour of yugawaralite was investigated by in situ synchrotron X-ray powder diffraction from room pressure to 11.4 GPa using a non-penetrating P-transmitting medium [1,8]. This zeolite shows a high stability upon compression, as no X-ray amorphization was detected in the studied P range. Moreover, the ambient pressure pattern is almost completely recovered upon decompression, thus indicating reversibility of the P-induced deformations. At the highest pressure, the a and c cell parameters contract by about 7%, while smaller variations were found for the b (2.4%) and β (1.3%) parameters. All the cell parameters show two discontinuities between 1.5 and 2 GPa and between 3.9 and 5.3 GPa. The much larger decrease of a and c parameters with respect to b has been rationalized by the presence of Ca-O$_{frame}$ bonds mainly oriented along b, hindering deformation along this direction, whereas the mobile water molecules may easily rearrange their hydrogen bonds network thus allowing a larger compressibility along a and c axes. *Ab initio* simulations on yugawaralite assuming a unit cell stoichiometry of $Ca_2[Al_4Si_{12}O_{32}]\cdot(H_2O)_9$, (YU9W from now on) and an occupancy of the W5 site of 50%, allowed to attribute the two discontinuities of the cell parameters vs. pressure curve to two P-induced changes of coordination of the extra-framework Ca cation along with rotations of quasi-rigid tetrahedral units [8]. Another interesting aspect resulted from the inspection of the Ca-water and Ca-framework distances behavior: at the lower pressures Ca-water distances are mostly affected (shortened), while at higher pressure the Ca-O$_{frame}$ are shortened. The goal of the present work is to study the effect of the weakly bonded water molecule in the partially occupied site W5 on the overall structural features of yugawaralite. To this aim, simulations on yugawaralite with 0% occupancy of this water site (unit cell stoichiometry $Ca_2[Al_4Si_{12}O_{32}]\cdot(H_2O)_8$, YU8W) are presented here.

3. SIMULATION MODEL AND COMPUTATIONAL APPROACH

Simulations on yugawaralite YU8W were performed with the Car-Parrinello molecular dynamics (MD) method [9], in the canonical ensemble with fixed number of atoms, volume and desired temperature (298 K). From the cell parameters *vs* pressure experimental curve, we have taken the values corresponding to a pressure of 8.8 GPa, and used them in the *ab initio* MD simulations. Reference simulations adopting the experimental cell parameters at room pressure (*rP*) were also performed.

The simulation cell adopted in all calculations corresponds to the unit cell of yugawaralite and contains two distinct sub-units (UNIT I and UNIT II from now on). In the present simulation system, YU8W, each sub-unit contains one Ca and four H_2O (W1,W2,W3,W4 in UNIT I and W1, W2, W3, W4A in UNIT II) whereas in YU9W five water molecules were present in UNIT I (W1, W2, W3, W4, W5) and four in UNIT II (W1, W2, W3, W4A) [8]. As the occupancy of W1A is about 25%, only the more populated W1 site (75%) has been considered in both sub-units. It is worth pointing out that both yugawaralite models do not reproduce exactly the experimentally found occupancy of W5, because the simulation of a system with 25% H_2O in W5 would have required a bigger simulation cell and much larger computational costs. On the other hand, comparison of the results on the YU8W and YU9W systems might help us to single out the effect of different water contents on the H*P* behavior of yugawaralite.

The density functional approximations and the simulation parameters (i.e. fictitious mass, integration time step, pseudopotentials) adopted in this study are the same reported in [8], and we refer to that paper for further technical details. With both sets of cell parameters (i.e. the r*P* ones and those at 8.8 GPa), simulations were performed for a total elapsed time of about 5 ps. Calculations of the minimum energy structures at 0 K were also performed by means of simulated annealing, adopting 5×10^{-4} a.u. as convergence criterion for the forces on atoms.

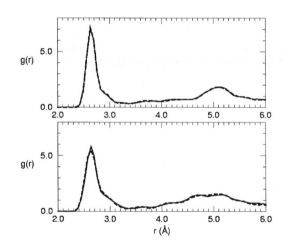

Figure 2. Radial distribution functions O_{frame}-O_{frame} calculated for the YU8W (solid line) and for the YU9W (dotted line) system. Top panel: r*P*, bottom panel: 8.8 GPa.

3. RESULTS AND DISCUSSION

3.1. *P*-induced modifications on the yugawaralite structure

Two simulations were performed for YU8W, adopting the r*P* and 8.8 GPa cell parameters, respectively. The room temperature structures of YU8W were obtained by *ab initio* MD trajectories performed using the above mentioned simulation cells. Structural equilibrium properties were obtained, for each simulation, by averaging (*e.g.* atomic coordinates) over the respective trajectory. In order to compare the YU8W results with those of YU9W [8], the calculated atomic coordinates were not averaged over the two sub-units because in YU9W such sub-units have different contents.

The average YU8W structures, which are very similar to those obtained from the geometry optimizations at 0 K, are represented in Fig. 1a-b, projected along the *a* axis. The figure shows that contraction of the cell volume due to pressure brings about large modifications on both the framework structure and the arrangement of the extra-framework species. In particular, the 8-rings channels' section is deformed into a lenticular-biconvex shape thus significantly decreasing the free space in the channels, and Ca is displaced to a position closer to the center of the channel. Comparison of the two structures also highlights that the extra-framework cation changes its coordination upon compression; namely, a new oxygen atom of the framework, O2, enters the Ca coordination polyhedron, while the O3 framework oxygen, which was coordinated to Ca at r*P*, leaves the Ca coordination environment. Geometry optimizations of the YU8W system with cell parameters corresponding to pressures in the range 2÷3.9 GPa evidenced that both framework oxygens (O2 and O3) belong to the coordination polyhedron of Ca, which is therefore coordinated, in this pressure range, to 5 framework oxygens and 4 water molecules. In other words, O2 enters the Ca coordination environment at low applied pressure, whereas coordination with O3 is lost only at pressures higher than that corresponding to the second discontinuity in the cell parameters *vs* pressure plot. These results are fully in line with those obtained for the YU9W system [8], thus confirming the hypothesis that the observed discontinuities found in the experimental *V vs P* curve are due to modifications of the coordination polyhedron of Ca. These changes are allowed by rotations of the quasi-rigid framework's tetrahedra [10], which, upon compression, may rearrange around Ca in order to minimize the free volume of the channels.

Comparison between the radial distribution functions may provide useful insight into the global structural changes, as the structural information is averaged over all atoms of the same species in the simulation cell. Some of the calculated radial distribution functions $g(r)$ for YU8W and YU9W at r*P* and 8.8 GPa are reported in Figs. 2-5. The $g(r)$'s of all framework species of the two yugawaralite model systems are nearly identical at each pressure. As an example, the O_{frame}-O_{frame} $g(r)$'s are shown in Fig. 2. It is evident that compression leads to a very modest shortening of the closest O_{frame}-O_{frame} average distances, indicating quasi-rigidity of tetrahedra. The long range behavior of such a $g(r)$ −on the other hand− is indicative of framework modifications, however the pressure induced change is not influenced by the different water content in the two investigated systems. Also the Ca coordination environment, in line with the previous discussion, seems to be only barely affected by the water content of the W5 site, as shown by the comparison of the Ca-O$_{frame}$ and Ca-O$_{water}$ $g(r)$'s in Fig. 3.

Therefore, these results further support the idea that the *P*-induced deformations of the yugawaralite structure are mainly determined by the template effect of Ca. By recalling that W5 is never coordinated to Ca, it should be evident that the occupancy of the W5 site may affect the yugawaralite pressure response only locally, i.e. close to the W5 site. To this respect, a detailed analysis of the *P*-induced deformations of UNIT I and II in YU9W has shown that only the T-O-T angles formed by the tetrahedra close to W5 undergo different *P*-induced modifications in UNIT I and II, because the presence of a water molecule in W5 in UNIT I (but not in UNIT II) leads to different hydrogen bond networks [8]. Moreover, by comparing the energies of the optimized YU9W geometry and of theYU8W system plus a water molecule in the same cell, we have found that the YU9W structure is energetically more stable than YU8W at r*P*, whereas the opposite occurs at 8.8 GPa [8]. It may be deduced that, even though the occupancy of W5 does not have detectable effects on the yugawaralite's global *P*-induced structural deformations, it may affect locally the framework geometry, and also influence significantly the overall stability of the system under different pressure conditions. One of the aims of the present work is indeed to rationalize the correlation between the water content and the stability *vs* pressure trend in yugawaralite.

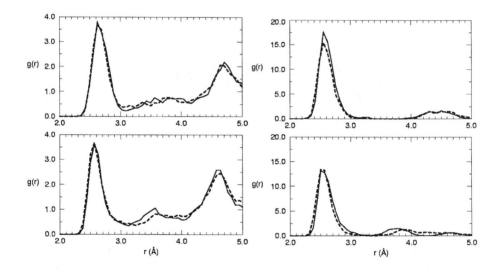

Figure 3. Left panels: Ca-O$_{frame}$ radial distribution functions calculated for the YU8W (solid line) and for the YU9W (dotted line) system. Top: r*P*, bottom: 8.8 GPa. Right panels: Ca-O$_{water}$ radial distribution functions calculated for the YU8W (solid line) and for the YU9W (dotted line) system. Top: r*P*, bottom: 8.8 GPa.

To this aim, insight may be provided by the H$_2$O's $g(r)$'s. As it should be expected, some significant differences are detected when inspecting the H$_2$O's $g(r)$'s calculated from the 8.8 GPa simulations of the YU8W and YU9W systems. The H-O_{water} and H-O_{frame} $g(r)$'s, reported in Fig. 4, indicate that in both systems compression increases the formation of hydrogen bonds between H_2O and framework oxygens. However, in YU8W the increased P-induced interaction with framework oxygens leads to a weakening of the water-water hydrogen bonds, whereas in YU9W a significant inter-water hydrogen bonding interaction is still present at 8.8 GPa, owing to the presence of a very strong hydrogen bond between W5 and W1. Let us now consider the O_{frame}-O_{water} and O_{water}-O_{water} $g(r)$'s in Fig. 5. At 8.8 GPa the first peak of the YU9W O_{frame}-O_{water} $g(r)$ is sharper and displaced toward shorter distances, indicating that, upon compression, the presence of H_2O in W5 forces the other water oxygens to occupy positions closer to framework's atoms. Moreover, differently from the YU8W O_{water}-O_{water} $g(r)$, which seems to be almost unaffected by pressure, the YU9W O_{water}-O_{water} $g(r)$ is significantly affected in passing from rP to 8.8 GPa. It might be deduced that, whereas the water molecules in YU8W are already, at rP, in an arrangement which can respond to applied pressure without undergoing, on average, energetically unfavourable readjustments, compression of the water-richer system YU9W leads to a more packed and significantly more perturbed arrangement of the water molecules. These results may thus explain why, with a cell volume corresponding to a pressure of 8.8 GPa, the YU8W system is energetically more stable than the YU9W one.

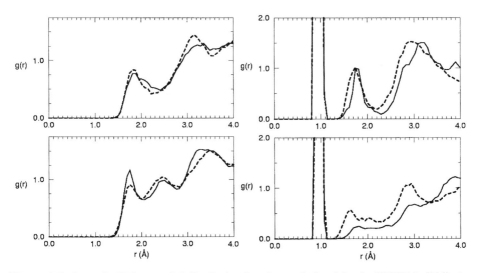

Figure 4. Left panels: H-O_{frame} radial distribution functions calculated for the YU8W (solid line) and for the YU9W (dotted line) system. Top: rP, bottom: 8.8 GPa. Right panels: H-O_{water} radial distribution functions calculated for the YU8W (solid line) and for the YU9W (dotted line) system. Top: rP, bottom: 8.8 GPa.

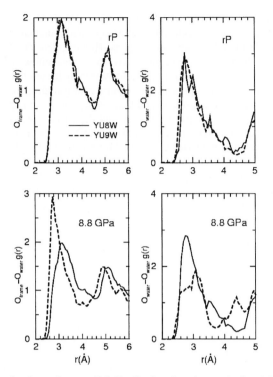

Figure 5. Left panels: O_{frame}-O_{water} radial distribution functions calculated for the YU8W (solid line) and for the YU9W (dotted line) system. Top: rP, bottom: 8.8 GPa. Right panels: O_{water}-O_{water} radial distribution functions calculated for the YU8W (solid line) and for the YU9W (dotted line) system. Top: rP, bottom: 8.8 GPa.

3.2. Detailed analysis of the water behavior under compression.

It should be pointed out that, in both YU9W and YU8W systems, compression does not affect significantly the molecular geometry of water (i.e. H_2O bond distances and angle). On the other hand, the response to pressure of the whole water system is rather complex, as pressure-induced changes in cell parameters lead to significant rearrangements in the details of the hydrogen bond network: in passing from rP to 8.8 GPa, all water molecules except W2 change their hydrogen bonding partner (cfr. Fig. 1a-b). Moreover, detailed inspection of the behavior of the water system along each simulation shows that the hydrogen bond network is not frozen, rather it fluctuates in time owing to rotations of H_2O molecules. Such equilibrium between different water arrangements indicates that there are moderate energy differences between the structures sampled by the MD simulations. Thus, the activation energies for the conversion between two different water arrangements should be of the order of kT.

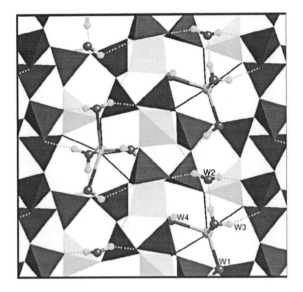

Figure 6. Projection on the *bc* plane of the alternative structural arrangement sampled during the 8.8 GPa MD simulations of YU8W in which a W2 molecule is hydrogen-bonded to W3. Light grey tetrahedra: Al, dark grey tetrahedra: Si. Ca and H_2O are shown in ball and stick representation.

As already discussed, at r*P* the average framework structures of YU8W and YU9W are very similar. However, the hydrogen bond network in YU8W is less stable than in YU9W: many changes of hydrogen bond partners were detected during the simulation, indicating that weak hydrogen bonds are broken and re-formed; the only stable hydrogen bond is the one between W2 and the framework oxygen O15. These results suggest that the water molecule in W5 plays a relevant role in stabilizing the yugawaralite structure at room conditions, thus rationalizing our finding that, with the r*P* cell parameters, the YU9W system is energetically more stable than the YU8W one.

As the occupancy of the W5 site influences the behavior of the whole water system, it follows that the details of the Ca coordination should be affected as well, even though, as discussed above, such modifications are too small to be detected from an inspection of the radial distribution functions. For instance, at 8.8 GPa the average Ca-OW2 distance for the YU8W system is 3.1 Å, significantly larger than the corresponding value obtained for the YU9W system in the same conditions (2.85 Å).

This finding can be rationalized by examining in detail the time evolution of the YU8W system in the simulation with cell parameters corresponding to 8.8 GPa. Indeed, two structural arrangements, characterized by qualitatively different distributions of the water molecules, were sampled along the simulation. In the former one, which is found in a larger portion of the total simulation time, both W2 molecules in the simulation cell are coordinated to Ca; in the latter, one W2 is bonded to Ca while the other forms a strong hydrogen bonds with W3, acting as a proton acceptor (see Fig. 6). As a consequence, in this alternative arrangement the average number of water molecules coordinated to Ca is lower than 4. The under-coordinated Ca compensates the loss of W2 from its coordination shell by occupying a position closer to the center of the 8-ring, thus allowing a larger number of framework oxygens to approach. Indeed, six framework oxygens are found at distances ≤ 3.3 Å from such Ca, indicating that the 2+ charge of the extra-framework cation is stabilized in this case by the weaker interactions with a larger number of framework oxygen atoms. These two different structural arrangements should have very similar energies; in other words, their energy difference should be of the order of kT.

4. CONCLUSIONS

In this work the effect of the water content on the pressure response of yugawaralite has been investigated by means of *ab initio* simulations. Such an effect has been singled out by comparing the behavior of two model systems, namely YU8W and YU9W, characterized by the presence or absence of water in the W5 site. Simulation results clearly demonstrate that water molecules in this partially occupied site have minor effects on the global P-induced structural deformations of the yugawaralite framework. On the other hand, the occupancy of W5 significantly modifies the energetic stability *vs* pressure trend, as well as, at a local level, the behavior upon compression of the framework atoms at a short distance (< 3 Å) from the W5 site. Moreover, the pressure behavior of the whole water subsystem, and in particular, the equilibrium between alternative arrangements characterized by small energy differences, is also affected by the water content.

REFERENCES

[1] R. Arletti, O.Ferro, S.Quartieri, A.Sani, G.Tabacchi and G. Vezzalini, Am. Mineral., 88 (2003) 1416.
[2] D.G. Gatta, P. Comodi, and P.F. Zanazzi, Microporous Mesoporous Mater., 61 (2003) 105.
[3]P. Ballone, S. Quartieri, A. Sani and G. Vezzalini, Am. Mineral, 87, (2002) 1194.
[4]O.Ferro, S.Quartieri, G.Vezzalini, E.Fois, A.Gamba and G.Tabacchi, Am.Mineral.,87 (2002) 1415.
[5]A.Kvick, G.Artioli and J.V.Smith, Zeitschr. Kristall. 174 (1986) 265.
[6]A. Alberti, S. Quartieri and G. Vezzalini, Studies in Surface Science and Catalysis, 84 (1994) 637.
[7]A. Alberti, S. Quartieri and G. Vezzalini, Eur. J. Mineral., 8 (1996) 1273.
[8] E.Fois, A.Gamba, G.Tabacchi, R.Arletti, S.Quartieri and G.Vezzalini, Am.Mineral.,90 (2005) 28.
[9]R. Car and M. Parrinello, Phys. Rev. Lett., 55 (1985) 2471.
[10]M.T. Dove, V. Heine and K.D. Hammonds, Mineral. Mag., 59 (1995) 629.

Studies in Surface Science and Catalysis 155
A. Gamba, C. Colella and S. Coluccia (Editors)
© 2005 Published by Elsevier B.V.

Synthesis of porous catalysts for Beckmann rearrangement of oximes

L. Forni[a], E. Patriarchi[a], G. Fornasari[a], F. Trifirò[a], A. Katovic[b], G. Giordano[b], J. B.Nagy[c]

[a] Dip. di Chimica Industriale e dei Materiali, Università di Bologna, viale Risorgimento 4, 40136 Bologna, Italy. Fax +39-51-6443680 e-mail: forni@ms.fci.unibo.it

[b] Dipartimento di Ingegneria Chimica e dei Materiali, Università della Calabria, Via P. Bucci,1-87030 Rende.

[c] Laboratoire RMN-Dept. Chimie- FUNDP-Rue de Bruxelles 51, B-5000 Namur.

It is generally known that vapor phase oximes rearrangement on solid catalysts is strongly dependent on the material acidic properties and porosity. Recent studies led to synthesize high performing catalysts with high silica content. In this paper we report the synthesis of several crystalline materials in order to understand the effect of the channels system and of the defective groups on the catalytic behaviour in Beckmann reaction. The MEL, MFI, MTW type zeolites were prepared and tested in the cyclohexanone-oxime conversion. Due to its unique channel system and to the presence of defective sites, the MFI and MEL type zeolites, in pure silica form (Sil-1 and Sil-2) and in boron form (B-MFI-(A), H_3BO_3/SiO_2=0.066 and B-MFI-(B), H_3BO_3/SiO_2=0.033) showed the best catalytic results. However silicalite-1 is still the most performing catalyst, the zig-zag channels structure has the same kind of acid sites (same desorption temperature) of three-dimensional straight channel silicalite-2, but an higher active sites distribution (Si-OH and Si-N groups). The introduction of boron atoms decreases the catalytic performances.

1. INTRODUCTION

Beckmann rearrangement of oximes is an acid catalysed reaction. Through the years, many researchers have studied the dependence of catalytic activity on the solid catalyst acidity in vapour phase Beckmann rearrangement of cyclohexanone oxime to ε-caprolactam. At first it was concluded that strong acid sites promoted the reaction more than weak acid sites, then it was found that Brönsted acid sites of intermediate strength were very important for the selectivity to caprolactam [1], however later Sato et al. found that very weak Brönsted acid sites in a MFI structure strongly affected the selectivity to caprolactam. Said catalyst is a microporous silicalite-1 which is characterized by a very low acidity and the presence of defects [2]. In addition it can carry out a shape-selective action on the formation of oligomers (tars) attributed to the size and morphology of its channels.

Some studies carried out on B-MFI [3] pointed out that the reaction takes place on the outer surface area, even though tars formation occurs inside catalysts pore. Recently it was proposed that cyclohexanone oxime rearrangement may occur in the inner surface area of silicalite-1 catalyst [4, 5] because caprolactam selectivity increases with increasing the crystal size and in addition, silicalite-1 microporous channels may act as molecular sieve, hindering tars formation. Last year Sumitomo announced the start-up of a new process for caprolactam production in vapour phase based on silicalite-1 catalyst with a Si/Al ratio of 147.000 [1].

In the work herein described we carry out a study in order to understand the effect of the channel systems and of the defect groups on the catalytic behaviour in Beckmann reaction of cyclohexanone oxime. For this reason we synthesised and characterized three different type of structures: (i) high-silica MTW which has non-intersecting one dimensional straight channels whose size is close to MFI and MEL structures, (ii) high-silica silicalite-2 (MEL structure) with a three-dimensional framework and straight channels of 5.4 Å x 5.6 Å, (iii) high-silica silicalite-1 and B-MFI with three-dimensional framework and zig-zag bidimensional channels (5.4 Å x 5.6 Å and 5.1 Å x 5.6 Å respectively).

2. EXPERIMENTAL

Sample	Molar Composition					Crystal Size (μm)
	Na_2O	Template	SiO_2	H_2O	H_3BO_3	
MTW	0.11	0.20 (MTBABr)	1	20	-	-
Silicalite-2	0.02	0.08 (TBABr)	1	20	-	3.9
Silicalite-1	0.08	0.08 (TPABr)	1	20	-	2.5
B-MFI-(A)	0.12	0.08 (TPABr)	1	20	0.066	10.1
B-MFI-(B)	0.12	0.08 (TPABr)	1	20	0.033	14.0

Table 1. Initial gel composition and crystal size of the catalysts prepared.
MTBABr: Methyl-triethyl-ammonium-bromide, TBABr: Tetra-butyl-ammonium-bromide, TPABr: Tetra-propyl-ammonium-bromide.

All catalysts are prepared according to the following procedure: a solution of NaOH is added to a solution of the structure directing agent and let under mixing at room temperature, then SiO_2 is added to the solution and finally (when required) H_3BO_3 is poured to the other reactants. SiO_2 source is a precipitated silica which is rich in Si-OH groups and should introduce a higher number of defective sites in the zeolitic structure. The mixture is let under stirring for 1 h, then put in an autoclave for hydrothermal treatment at temperatures varying according to the structure from 140 to 170°C [6] for about 48 hours. Then samples are filtered, washed and calcined at 450°C. The activation procedure is carried out by mean of an ammonia solution (pH≅ 10) and a following calcination at 450°C.

Beckmann rearrangement of cyclohexanone-oxime to caprolactam was carried out in a glass fixed-bed microreactor, packed with inert material (glass spheres). The reaction temperature was 350°C and weight hourly space velocity (WHSV) was 1.2 h^{-1}. A solution of cyclohexanone-oxime, methanol and toluene, in a flow of nitrogen, was utilized as the feed. The products were determined by gas chromatography.

IR spectra were obtained on a Nicolet Nexus FT-IR spectrometer equipped with a MCT (Hg/Cd/Tl) detector being the IR source made of CSi. Solid state Magic Angle Spinning

(MAS-NMR) spectroscopy was used to characterize the local Si environment in the samples. Spectra were recorded on a Bruker MSL 400 spectrometer working at 79.4 MHz (^{29}Si) a pulse length of 6 ms and a repetition time between acquisitions of 60 s, the reference at 0 ppm being taken as Si(CH$_3$)$_4$ (TMS). ^{11}B-MAS-NMR analysis, spectra were recorded at 128.3 MHz, the reference being BF$_3$.OEt$_2$ a pulse length of 5 μs and a repetition time between acquisitions of 5 s. X-ray diffractograms were recorded by means of a Philips PW 1050/81 being Cu K□ the radiation source. SEM analysis was carried out by means of a Philips XLS 30 with a ΔP of 25 KV.

3. RESULTS AND DISCUSSION

Zeolites composition was set in order to synthesise materials with the same reagent molar ratios, as reported in table 1. Samples crystal size are rather different and varie from a minimum of 2.5 μm to a maximum of 14.0 μm. In the case of B-MFI, the introduction of H$_3$BO$_3$ is balanced by the increase of Na$_2$O content, however B-MFI-(B) has a larger crystal size because the molar ratio Na$_2$O/H$_3$BO$_3$ is higher than B-MFI-(A) sample (see table 1). The use of the same silica source promotes the same crystals shape, hence the interface between the solid and the gas phase is dependent only on the crystals size.

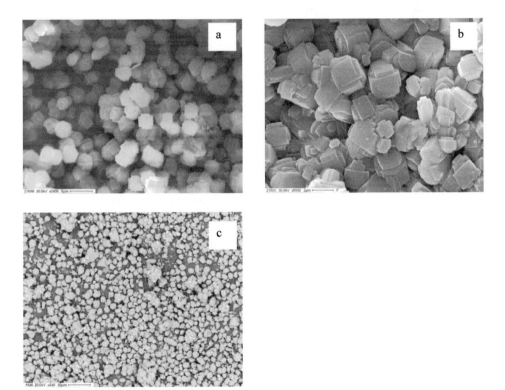

Figure 1. SEM images of zeolitic samples. a) Silicalite-2, b) B-MFI-(A), c) B-MFI-(B).

284

By means of X-ray powder diffraction it was possible to verify that all synthesised samples have the desired structure and even after activation treatment the crystallinity is unchanged. Silicalite-2 pattern shows the absence of the weak reflection at about 6.2° 2θ typical of the pure MEL phase. According to other authors [7, 8] when TBABr is used as the template the intergrowth with MFI-type structures occurs. In addition we also verified what Heitmann et al. [9] previously observed in silicalite-1 catalysts, the treatment with ammonia causes a change of the unit cell symmetry from monoclinic to orthorhombic. We suggest that the activation method and the thermal treatment at 450°C promotes a structural rearrangement towards an orthorhombic unit cell symmetry which is usually less thermally favoured, in silicalite-1 materials [10], than the monoclinic one which is observed for the same treatment at 550°C.

As a matter of facts by FT-IR spectroscopy under-vacuum (figure 2) we verify that all samples, after the activation treatment, show a shoulder at about 932 cm^{-1} attributed to Si-N stretching mode [11-13], other bands attributed to the Si-NH stretching modes are observed at about 3540 cm-1, while at 1550 cm-1 the weak band ascribed to the Si-O-NH$_2$ or NH$_3$ or Si-NH$_2$ deformation modes is also observed. Such a band is formed after treatment with ammonia solution, during Na$^+$ exchange with H$^+$, and calcination at 450°C. Said bonding gives the catalysts a thermally more stable structure and thus contributes to catalyst lifetime increase as reported in literature [4].

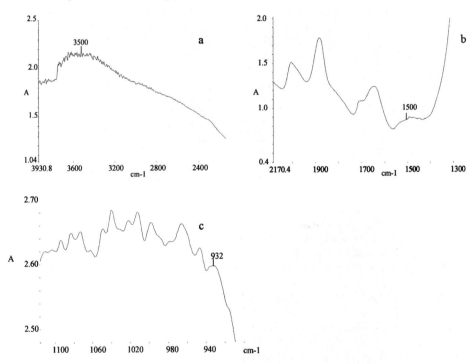

Figure 2. FT-IR spectrum of silicalite-1 sample, under vacuum at 200°C. The bands attributed to Si-N bindings are characterised by the band a) at 3500 cm^{-1}, b) the weak band at 1500 cm^{-1} c) and the shoulder at 932 cm^{-1}.

FT-IR spectra at room temperature of B-MFI samples show a double IR-band around this frequency with a signal at 950 cm^{-1} which is reasonably ascribed to Si-N bonding of the hydrated sample and another one at about 925 cm^{-1} which is attributed to BO$_4$ co-ordination [14]. We further investigated this bonding by ^{15}N-MAS-NMR spectroscopy and we observed a band at -72.2 ppm which cannot be assigned to the typical Si-N bonding (chemical shift at about -35 ppm). However such band cannot be attributed to NH$_3$ traces, for ammonia is characterized by a chemical shift at -400 ppm. Therefore this attribution needs to be further investigated.

Concerning the activity in vapour phase Beckmann rearrangement very important sites promoting a high caprolactam selectivity are defective silanols sites, namely nests of silanols and H-bonded silanols, which are very weak Brönsted acid sites. The silanol defective sites were determined both by ^{29}Si-MAS-NMR and ^{29}Si-CP-NMR spectroscopy, by the first one we were able to distinguish defective sites from the perfect tetrahedral SiO$_4$ (chemical shift – 113 ppm). All samples show the presence of defective sites with a chemical shift of –103 ppm, their abundance is reported in table 2. ^{29}Si-CP-NMR allows to distinguish between the different kind of defective sites (H-bonded or terminal silanols), however the results show that all catalysts are characterized by only one kind of defective site at –103 ppm which corresponds to the H-bonded silanols shift.

Sample	Defective sites per unit cell (%)	Acidity (mol NH$_3$ g^{-1}) x 10^{-5}	T$_{max}$ NH$_3$ desorption (°C)
MTW		4.33	190
Sil-2	6	4.32	175
Sil-1	7	7.15	175
B-MFI-(A)	5	51.2	160
B-MFI-(B)	5	18.6	170

Table 2. Percentage of defective sites and acidity in different samples.

In the case of B-MFI, ^{11}B-MAS-NMR on dried samples shows the presence of only tetra-coordinate boron (BO$_4$) this means that boron atoms are completely inserted in the zeolitic structure, however after calcination and activation two different kind of boron co-ordinations are found, in addition to BO$_4$ we identify extra framework BO$_3$ (~ 20-30%). Usually tetrahedral framework boron (BO$_4$) is balanced by a charge compensating ion such as Na$^+$ or NH$_4^+$ [15]. Extra lattice BO$_3$ is formed upon thermal treatment in an oxidizing atmosphere and consequently it may promote the formation of defective sites, it can also be washed away from the catalyst by means of NaOH solution or with distilled water [15, 16]. Nevertheless ^{29}Si-MAS-NMR shows only a negligible difference of defective sites amount between B-MFI and sil-1 samples. Hence, in agreement with Ruiter et al. [15], we suggest that deboronation of zeolites framework during calcination should be avoided, because silanol nests thus formed are supposed to be unstable due to condensation phenomena and to T atom reorganizations catalysed by water at high temperature. High-silica samples without heteroatoms are characterized by a very low acidity with weak acid sites, namely Brönsted acid sites. NH$_3$-TPD results are summarized in table 2. Only B-MFI samples show an higher acidity because of the introduction of boron atoms, for B(OH)Si bondings are stronger acids than SiOH groups [17].

Vapour phase Beckmann rearrangement on MTW sample shows low catalytic performances with low cyclohexanone oxime conversion values (31 %) and caprolactam yield (22 %). However the most relevant result is the very short lifetime, only one hour. The very quick deactivation rate may be ascribed to an inefficient diffusion in the non-intersecting mono-dimensional channels lattice consequently to tars formation which partially occlude the porosity. The by products identified on MTW as well as on the other catalysts are cyclohexanone, cyclohexenone, aniline and pentennitrile. However the low and weak acidity and the hydrophobic properties of some materials promote mainly the reaction towards caprolactam formation, while the by-products formation is limited. In scheme 1 we report the mechanism for some by-products formation according to Shouro et al. [18, 19].

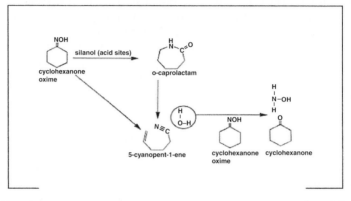

Scheme 1. Reactio   he main by-products (cyclohexanone and penten nitrile) according to Shouro et al and Aucejo et al [18, 19].

Different results have been obtained for the other structures. Despite of acidity values very close to the MTW material, silicalite-2 sample shows higher catalytic performances and a much lower deactivation rate. After 76 hours of reaction cyclohexanone oxime conversion is 66% and caprolactam selectivity is 97% (caprolactam yield is 64%) with an average caprolactam productivity of 2.80 kg lactam h^{-1} kg^{-1} catalyst. This behaviour may be mainly attributed to the difference of the channels arrangement. As a matter of facts the intersecting three dimensional channels system of Sil-2 catalyst improves the reactant and product molecules diffusion regardless of tars formation.

In addition Sil-2 catalyst has in the first hour of reaction an higher catalytic performance (see figure 3) than the MTW one, this is attributed to a different silanols distribution inside the zeolitic structure.

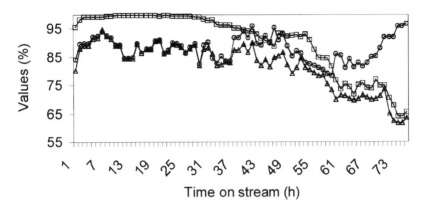

Figure 3. Catalytic performances of Sil-2 catalyst. (□) cyclohexanone oxime conversion, (Δ) caprolactam yield, (○) caprolactam selectivity.

It is therefore difficult to determine exactly the contribution of each phase (MEL and MFI) present in the Sil-2 structure in the catalytic performances. Probably the contribution of the MEL phase in the Sil-2 catalyst increases the deactivation rate, when compared to pure silicalite-1 sample (Sil-1), because of the lower defective silanols distribution inside the straight MEL channels. In fact the high-silica structure with zig-zag intersecting channels (silicalite-1) shows the best results. In figure 4 cyclohexanone oxime conversion for three MFI structures (silicalite-1, B-MFI-(A) and B-MFI-(B)) is reported. Silicalite-1 shows an almost complete conversion of the oxime for about 120 hours while for both B-MFI structures initial high conversion values are observed, but these decrease immediately and after about 10 hours conversions are around 20%.

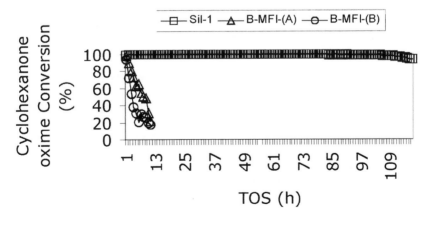

Figure 4. Oxime conversion as a function of time on stream for MFI samples.

The same behaviour is observed for caprolactam selectivity, figure 5. Silicalite-1 is an high performing catalyst with an average lactam selectivity of about 80% for over 120 hours on stream, while both B-MFI samples decrease rapidly to 70% and 40% respectively. This difference in catalytic activity can be explained by the presence of extra-framework boron species which promote the increase of by products formation, mainly tars. As a matter of fact tars amount (calculated at the 10th hour of reaction) is twice higher for B-MFI materials than for silicalite-1 (see table 3), this is also in agreement with penten-nitrile selectivity which is higher in B-MFI structures and may be the most important tars precursor [20]. The slight difference of caprolactam selectivity values between B-MFI-(A) and B-MFI-(B) can be explained by the different percentages of BO_4 atoms in the zeolitic structure. Thus the introduction of boron atoms in the H_3BO_3/SiO_2 molar ratio studied (0.033 – 0.066) increases the BO4 content and the MFI framework acidity. It has also been claimed that tetra-coordinated boron atoms have a rather good activity towards caprolactam selectivity. This may explain the higher B-MFI-(A) selectivity to the lactam than sample B-MFI-(B). During calcination boron partially leaves the MFI lattice, the extra-lattice boron has been evaluated to be about 20-30% of the boron total amount, however the quantity of defective sites is very close to Sil-1 sample. Probably new defective sites are formed after boron elimination during calcinations, however the calcinations temperature may promote the condensation of such sites, indeed. More drastic treatments are required in order to promote an higher amount of defective sites formation in B-MFI structures. The extra-framework boron is still mixed to the zeolitic phase while BO4 is part of the zeolitic structure.

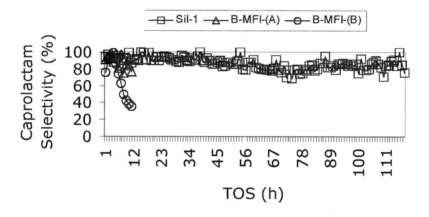

Figure 5. Caprolactam selectivity as a function of time on stream for MFI samples.

The high-silica structure with intersecting bi-dimensional zig-zag channels system (silicalite-1) seems to further improve molecules diffusion in comparison with silicalite-2, but most of all said catalyst has an higher number of active and selective sites (silanol nests and H-bonded silanols) because the acidity of silicalite-1 sample is higher than silicalite-2 and the TPD desorption maximum has the same temperature value (same kind of acid sites).
In table 3 the average productivities of the Sil-2, Sil-1 and B-MFI catalysts calculated on the basis of hours of reaction are reported. It summarizes the catalytic results discussed and they

confirm the high performances of Sil-1 and Sil-2 catalysts, the highest productivity value is obtained with silicalite-1 sample.

Sample	Caprolactam productivity (kg lactam h^{-1} kg^{-1} catalyst)	Time of reaction (hours)	Tars (%)
Sil-2	2.80	76	0.0027
Sil-1	4,12	120	0.0023
B-MFI-(A)	0.73	10	0.0054
B-MFI-(B)	0.41	10	0.0050

Table 3. Caprolactam productivity on the basis of catalytic tests duration.

4. CONCLUSIONS

Several crystalline materials with high silica content were synthesized in order to understand the effect of the channel systems and of the defective groups on the catalytic behaviour in Beckmann rerrangement.
The activation treatment with ammonia solution seems to promote in all the structure examined, the formation of Si-N groups which contribute to the higher stability of the structure. In addition all materials are characterised by a very low acidity, only B-MFI samples are more acidic because of the introduction of boron atoms. Said atoms are partially eliminated from the zeolitic lattice during calcination, but some are still mixed to the MFI phase and contribute to the poor catalytic performances of these materials. In addition the increase of defective sites after boron elimination is negligible, probably because during calcination the condensation of silanols of "new" defective sites occurs. However the presence of BO_4 atoms helps the rearrangement reaction towards the lactam molecule. In fact by comparing B-MFI-(A) and B-MFI-(B) we identify the higher caprolactam selectivity corresponding to the higher boron content. Among the catalysts prepared, high-silica silicalite-1 is still the best performing catalyst in Beckmann rearrangement of cyclohexanone oxime in vapor phase, the zig-zag channels structure has the same kind of acid sites (same desorption temperature) of three-dimensional straight channel silicalite-2 but an higher active sites distribution (Si-OH and Si-N groups). We suggest that the best performances of high-silica silicalite-1 may be ascribed to the kind of acid sites and to their distribution, while its low deactivation rate is promoted by the formation of Si-N bondings and by the shape selectivity of its bi-dimensional zig-zag channels.

REFERENCES

1. Curtin T., J.B. McMonagle, and B.K. Hodnett, Appl. Catal. A 93 (1992) 75.
2. Ichihashi H., Kitamura M., Catal. Today 73 (1-2) (2002) 23.
3. Albers P., K. Seibold , T. Haas, G. Prescher, and W.F. Hölderich 176 (1998) 561.
4. Forni L., Fornasari G., Giordano G., Lucarelli C., Katovic A., Trifirò F., Perri C., B.Nagy, J. Phys. Chem. Chem. Phys. 6 (2004) 1842.
5. Weitkamp J., Puppe L., Catalysis and zeolites: fundamentals and applications, (1999), 564, Springer, Berlin, (Germany).
6. Verified Synthesis of Zeolitic Materials, ed H. Robson, Elsevier, Amsterdam, (2001).
7. Perego G., Cesari M., Allegra G., J. Appl. Crystallogr., 17 (1984) 403.

8. Millini R., Berti D., Ghisletti D., Parker W.O. Jr., Carluccio L. C., Bellussi G., Sturdies in surface science and catalysis 142 A (2002) 61.
9. Heitmann G. P., Dahlhoff G., Hölderich W. F., J. Catal., 186 (1999) 12.
10. Ko Y. S., Ahn W. S., Microporous and Mesoporous Mater., 30 (1999) 283.
11. Morrow B. A., Cody I. A. and Lee L. S. M., J. Phys. Chem., 79 (1975) 2405.
12. Morrow B. A., Cody I., J. Phys. Chem.,79 (1975) 761.
13. Chen M., Zheng A., Lu H., and Zhou M., J. Phys. Chem. A 106 (2002) 3077.
14. Shibata M., Gabelica Z., Microporous Materials 11 (1997) 237
15. de Ruiter R., Kentgens A. P. M., Grootendorst J., Jansen J. C. and van Bekkum H, Zeolites, 13 (1993) 128.
16. Fild C., Shantz D. F.,. Lobo R.F and Koller H, Phys. Chem. Chem. Phys. 2 (2000) 3091.
17. Chu C. T-W, Chang C. D., J. Phys. Chem. 89 (1985) 1569.
18. Shouro D., Ohya Y., Mishima S., Nakajima T., Appl. Catal. A, 214 (2001) 59.
19. Aucejo A., Burguet M. C., Corma A. and Fornes V, Appl. Catal., 22 (1986) 187.
20. Albers P., Seibold K., Haas T., Prescher G., Hölderich W. F., J. Catal., 176 (1998) 561.
21. S. Sato, M. Kuroki, T. Sodesawa, F. Nozaki, G.E. Maciel, J. Mol. Catal. A: Chem. 104 (1995) 171.

Studies in Surface Science and Catalysis 155
A. Gamba, C. Colella and S. Coluccia (Editors)

Preparation and characterization of MoO_x-SnO_2 nano-sized materials for catalytic and gas sensing applications

G. Ghiotti [a], S. Morandi[a], A. Chiorino[a], F. Prinetto[a], A. Gervasini [b], M.C. Carotta[c]

[a]NIS-Dipartimento di Chimica I.F.M. Università di Torino, Via Pietro Giuria 7, I-10125 Torino, Italy

[b]Dipartimento di Chimica Fisica ed Elettrochimica, Università di Milano, Via Golgi 19, I-20133 Milano, Italy

[c]INFM-Dipartimento di Fisica, Università di Ferrara, Via Paradiso 12, I-4410 Ferrara, Italy

This work gives results about the preparation and characterization of a series of MoO_x/SnO_2 powders at increasing Mo content with the aim of studying their surface properties, their catalytic performances in the NO_x abatement by hydrocarbons and, eventually, their sensing abilities to NO_2/hydrocarbons mixtures. The goal of obtaining powders made by regularly shaped and nano-sized particles with a uniform spreading of MoO_x species at the surface of the SnO_2 particles is attained, as shown by XRD, SEM, HRTEM and XPS measurements. FT-IR spectroscopy gives information on the MoO_x species and on the way they are affected in the presence of NO/O_2 or NO_2 atmospheres. NO_x surface species formed were carefully investigated and individuated and some related surface reactions have been proposed. Catalytic measurements show that the activity seems slightly dependent on Mo concentration but that, for temperature up to 300 °C the N_2 production increases, while for temperature higher than 300 °C, the N_2 production decreases with the Mo loading. In any case the specific activity increases with the temperature, the maximum being at 350 °C. Electrical measurements show that all materials, in absence of hydrocarbons, are good NO_2 sensors at working temperatures in the range 150-250 °C and that the response to NO_2/hydrocarbon mixtures is still good and selective for the NO_2 gas at working temperature of 150 °C and 200 °C. At variance, at 250 °C the ability to sense NO_2 is drastically reduced.

1. INTRODUCTION

Tin oxide is the most extensively used n-type semiconducting oxide in gas sensing applications [1,2]. Its sensitivity towards CO, H_2, hydrocarbons, NO_2, has been tested and attempts have been made to elucidate the sensing mechanism by evidencing surface reactions responsible of conduction changes [2-7]. Tin oxide is also an interesting material for catalytic applications, even if its use in catalysis is much less diffuse. SnO_2 is known from long time as an effective low-temperature catalyst for the CO oxidation [8], in particular when supporting Pt [9]. Recently, pure SnO_2 turned out to be a good lean NO_x catalyst [10] and SnO_2 supported on γ-Al_2O_3 turned out to be one of the most active and selective alumina based lean catalysts [11-13].

Molybdenum oxides, supported on different oxides, have been from long time extensively used for catalytic applications, while they are much less diffused as sensor material. Molybdenum oxide species show important properties for various reactions like hydrodesulphurization, metathesis, hydrogenation and oxidation of olefins, NO_x abatement. This high flexibility as catalyst can be related to the different molecular structures that molybdenum oxide can simultaneously possess when supported and to the ability of Mo to assume various oxidation states depending on the pre-treatment and/or reaction atmosphere.

Recently MoO_x/SnO_2 systems have been proposed both in catalytic and chemical sensing applications (being n-type semi-conducting materials). MoO_x/SnO_2 systems have been shown to be very promising materials in the controlled catalytic oxidation of alcohols at low temperatures [14]. Not surprisingly similar systems have been studied as materials for alcohol vapour sensing [15]. Indeed, the capacity of semi-conducting solids to catalyze at low temperature the oxidation of combustible gases is, meanwhile, at the basis of its use in gas sensing elements because they offer high sensitivity at low-operating temperature. In the sensing applications MoO_x/SnO_2 systems have also shown promising properties as low-temperature NO_2 gas sensors. In particular, in the temperature range 150-350 °C, MoO_x/SnO_2 systems electrically respond to NO_2 decreasing their conductance, the maximum response being at 150-200 °C [7]. The same materials electrically respond, but increasing their conductance, to CO and to unsaturated hydrocarbons (like ethene or propene), but only at temperature higher than 250 °C [16]. This leaves hoping about the possibility to use these materials to develop chemical sensors selective to NO_2 in the urban atmospheres where, indeed, hydrocarbons and CO are simultaneously present with NO_x. The development of reliable and selective solid-state gas sensors for the urban air pollutants is one of the main desired results in the gas sensor application. For such a goal, a material should be selective for NO_2 in the presence of the above reducing gases At this point some questions arise: when NO_2 and hydrocarbons are simultaneously present in the atmosphere, will they interact independently with the surface or new reactions have to be taken into account? As previously remembered SnO_2 pure or supported on Al_2O_3 is known to be a good catalyst for the selective catalytic reduction (SCR) (reduction in the presence of O_2) of NO_x to N_2 by reducing agent like olefins. How the surface reactions leading to the NO_x abatement could influence the sensor response to this gas? The temperature of maximum activity in the SCR of the NO_x is generally higher than 300 °C, but a moderate activity is already present at lower temperatures. NO_x reduction by hydrocarbon in excess oxygen is a complex reaction that can be viewed as a competition between NO_x and O_2 in the hydrocarbon oxidation. What happens when Mo is added to tin oxide? The addition of MoO_x species on SnO_2 oxide should anticipates the reduction of NO_x to N_2 activating the hydrocarbons at lower temperatures than on pure SnO_2. If so, which should be the influence on the sensing capacity of the material towards NO_2?

These are the reasons that have driven us to focus our attention on the MoO_x/SnO_2 system. Therefore we prepared and characterized a series of MoO_x/SnO_2 powders at increasing Mo content with the aim of studying their surface properties, their catalytic performances in the NO_x abatement by hydrocarbons and, eventually, their sensing abilities to NO_2/hydrocarbons mixtures.

2. EXPERIMENTAL

2.1 Materials.

Pure SnO_2 was prepared following the method described by Harrison and Guest [17]: hydrolysis with NH_3 of $SnCl_4$ solution, followed by washing, drying first at RT, then at 200 °C for 12 h, and subsequent calcination at 550 °C. The powder as prepared (a.p.) was named

SN. The Mo containing powders were prepared by impregnation of SN with given amounts of aqueous solutions of ammonium heptamolybdate (Merck) to obtain three different Mo nominal loadings: 1, 3 and 5 mol %. The impregnated powders a.p., that are dried 3 h at 120 °C, are thereafter named Mo1SN, Mo3SN and Mo5SN. Powdered materials have been pelleted for FT-IR and catalytic measurements and used to prepare thick films for electric measurements. Subsequent oxidizing thermal treatments have been performed on powders, pellets and films at 550 °C or 650 °C. A figure reminding to the temperature of the thermal treatment will be added to the sample label. The same label has been used for powders, pellets or thick films (e.g.: SN-650, Mo1SN-650, etc.). Actually, a pre-treatment temperature of 550 °C was initially chosen for all the characterization techniques. Then we were forced to choose a pre-treatment temperature 100 °C higher in the case of the electric measurements, because the films pre-treated at 550 °C showed conductance values in air not reproducible. Therefore experiments by FT-IR and TEM were also repeated after pre-treatment at 650 °C.

2.2 Morphological and textural characterization.

The XRD patterns of the a.p. powders were collected on a CGR Theta 60 instrument using monochromatized CuK$_{\alpha 1}$ radiation (λ = 1.542 Å), (40 kV and 50 mA).

TEM and HRTEM images of the a.p. powders were obtained with a Jeol 2000 EX electron microscope equipped with a top entry stage.

SEM images of pellets and thick films activated at 550 °C and or 650 °C were done with a 360 Cambridge scanning electron microscope

B.E.T. surface areas, pore volumes and pore size distributions of the powders (a.p. and after thermal treatments) were determined by N$_2$ physisorption using a Micromeritics ASAP 2100 apparatus.

2.3 Spectroscopic characterization.

X-ray photoelectron spectra (XPS) of powders a.p., activated in dry flow of oxygen (36% O$_2$/He mixture at 550 °C), and discharged from the reactor (after use in the SCR of NO$_x$) were collected in an M-Probe apparatus (Surface Science Instruments). The X-ray source provided monochromatic Al K$_\alpha$ radiation (1486.60 eV). The residual pressure in the analysis chamber was typically 5 x 10^{-9} mbar. Charge effects were compensated by the use of a flood gun (3 eV). A spot size of 400 x 1000 μm and a pass energy of 150 eV were used for survey spectra, while for the single-region acquisitions a spot size of 200 x 750 μm and a pass energy of 25 eV with 0.74 eV resolution were used. The 1s level of hydrocarbon contaminant carbon (284.60 eV) was used as internal reference with an accuracy of ± 0.20 eV. Quantitative analysis was performed using the appropriate Scofield sensitivity factors [18]. Reference binding energies (BE) were Sn 3d$^{5/2}$ (486.7 eV, SnO$_2$), Mo^{VI} 3d$^{5/2}$ (232.6 eV, MoO$_3$) and Mo^{IV} 3d$^{3/2}$ (229.3 eV, MoO$_2$). The XPS peaks were deconvoluted using a peak-fitting routine. The lines used in the fitting of a peak envelope are defined according to their centred position, half-width, shape (Gaussian or Lorentzian distribution) and intensity. The best fit of the experimental curve by a tentative combination of bands was searched.

For the FT-IR measurements, the powders a.p. were compressed (pressure, 1x10^3 kg cm^{-2}) in self-supporting pellets of about 50 mg cm^{-2} and put in an IR cell that allowed thermal treatments in *vacuum* and in controlled atmospheres. Samples were submitted to activation consisting of alternate outgassing-oxidizing treatments in dry oxygen (20 mbar) at 550 °C (or at 650 °C), then cooled in oxygen at a chosen temperature. All IR spectra were run in situ at different temperatures (25÷450 °C) on a FT-IR spectrophotometer (Perkin-Elmer System 2000) with resolution of 2 cm^{-1}, firstly in O$_2$, then in a NO/O$_2$ 2:1 mixture (10 mbar) or in

NO$_2$ (5 mbar). High purity gases (from Praxair) were used: O$_2$ without further purification; NO was freshly distilled before use; NO$_2$ was prepared in laboratory by contacting NO (freshly distilled before use) with O$_2$, in excess of the 2:1 reaction stoichiometry, during 3 weeks at RT and then was purified from O$_2$ by freezing.

2.4 Catalytic experiments.

The catalytic activity in the NO$_x$ reduction with ethene was measured in a flow apparatus at atmospheric pressure. The fixed-bed quartz tubular downflow micro reactor was filled with ca. 0.5 g of catalyst and mounted in a tube furnace heated by a temperature controller (from Eurotherm) able to realize automatically programmed thermal sequences and isothermal steps.

The reaction was studied at temperatures between 150 and 450 °C; each examined temperature was maintained for 200 min, meanwhile four chromatographic analyses were performed. The reaction temperature was then increased (10 °C min^{-1}) to the successive temperature plateau, without intermediate activation. The overall test lasted about 35 h. Fresh portion of catalyst was activated at 550 °C in a flow of oxygen (30% O$_2$/He) for 3.5 h followed by He flow for 0.5 h. Reaction feed consisted of 0.5% of both NO and C$_2$H$_4$ and 5% of O$_2$ with He as balance. The mixture flowed at 5.5 NL/h, which corresponded to a space velocity of about 10000 h^{-1} (contact time: about 0.1 g h/L). The reactor outflow was analysed by gas-chromatography (C.P. 9000 from Chrompack) equipped with T.C.D. detector. A 60/80 Carboxen-1000 column (Supelchem) permitted the separation of N$_2$, O$_2$, NO, N$_2$O, CO, CO$_2$ and C$_2$H$_4$. NO in the fed gas mixture was partially converted to NO$_2$ in oxygen atmosphere, leading to a mixture of NO$_x$ as feed besides C$_2$H$_4$ and O$_2$. NO$_x$ reduction led to N$_2$ and N$_2$O, the total and partial reduction products, respectively. Conversion of C$_2$H$_4$ led to CO$_2$ as main product and to CO in very little amount. Selectivity to the main product, N$_2$, was evaluated by means of the competitiveness factor (c.f., %), which is defined as the ratio between the amount of C$_2$H$_4$ consumed to reduce NO to N$_2$ and the total amount of C$_2$H$_4$ consumed [18, 19].

2.5 Electrical measurements

The sensors were obtained starting from miniaturized laser pre-cut 96% alumina substrates (2.5 x 2.5 mm^2 and 0.25 mm thick for each device) provided with a heater element on the backside, a Pt-100 resistor for the control of the sensor operating temperature and a gold front interdigitated contacts. Three series of thick films were prepared starting from pastes obtained by adding to the above powders an organic vehicle together with a small percentage of glass fritt for improving the adhesion of the layers to the substrates. The films were then fired in air at 650 °C. All the conductance (G) measurements were performed in a sealed test chamber at a flow rate of 0.5 lt min^{-1}, at temperature ranging from 150 °C up to 400 °C in wet (40% R.H.) air, in wet air polluted with 10 ppm of a single gas (C$_2$H$_4$, C3H6 or NO$_2$) and in wet air polluted with the two mixed gases (C$_2$H$_4$: NO$_2$, 10 : 10 ppm; C$_3$H$_6$: NO$_2$, 10 : 10 ppm). The response is reported as G$_{gas}$/G$_{air}$ and G$_{air}$/G$_{gas}$ for reducing and oxidizing atmospheres, respectively; G$_{air}$ and G$_{gas}$ are the conductance in wet air or in polluted air, respectively.

3. RESULTS AND DISCUSSION

3.1. Morphology and texture

XRD patterns of all the a.p. powders showed the main reflexes reported in the literature (JCPDS 5-0467) for SnO$_2$ cassiterite and they appeared well crystalline. No phases distinct from cassiterite could be put in evidence.

TEM images of SN show aggregates of nano-sized particles ranging from 10 to 40 nm (medium size, 20 nm) with regular rounded borders. TEM images of MoSN samples show more densely packed aggregates of particles with more indented borders. For these two reasons the particle sizes are hardly measurable. However, they do not largely differ from 20 nm as medium value. The high resolution TEM images (see Figure 1a for a.p. Mo5SN) show plane fringe distances corresponding to two main reflexes reported in the literature for SnO2 cassiterite, i.e. those of (110) and (101) planes. No segregated particles distinct from SnO_2 could be put in evidence, according with XRD data.

On powders treated at 650 °C only scanning electron microscopies were performed. SEM images (see Fig.1b for SN-650 sample) show agglomerates of different thickness, made by particles with medium size between 30 and 50 nm.

The a.p. SN, Mo1SN and Mo5SN powders showed surface areas of 39, 44 and 41 $m^2 \cdot g^{-1}$, respectively, cumulative pore volumes of 9.8 x 10^{-2}, 1.6 x 10^{-1} and 1.3 x 10^{-1} cm^3 g^{-1}, respectively, and sharp pore size distributions centred on the average values of 104, 124 and 110 Å, respectively. The increase in the surface area after the impregnation with the Mo solution is in agreement with the more indented border of the particles observed for these materials by TEM.

SN-650 shows surface areas of 31 m^2 g^{-1}, pore volumes of 1.16 x 10^{-2} cm^3 g^{-1} and sharp pore size distributions centred on average value of 158 Å. As for MoSN-650 materials only the surface area of Mo1SN-650 has been measured, giving a value of 35 m^2 g^{-1}.

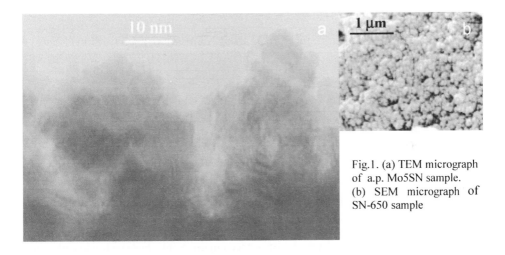

Fig.1. (a) TEM micrograph of a.p. Mo5SN sample. (b) SEM micrograph of SN-650 sample

3.2 Spectroscopic characterization.

3.2.1 XPS characterization.

Considering the way of Mo addition and the morphology of the a.p. powders, the molybdenum should be well dispersed on the surface as eptamolybdate-like anions, with NH_4^+ as counter ions. After treatments at 550°C the removal of surface OH⁻ and NH_4^+ as H_2O and NH_3 causes the formation of molybdate- and almost-bidimensional polymolybdate-like species or (if preferred) MoO_3 rafts grafted to the surface. This statement is in agreement with the examination of XPS results. All the Mo containing samples, either the a.p. or the high temperature activated (Mo1(3,5)SN-550) samples, showed an unique oxidation state for Mo,

Mo(VI). The BE values, Mo $3d^{5/2}$ = 232.7 ± 2 eV, are typical of MoO_3. On the discharged samples after reaction of NO_x reduction, besides the lines of Mo(VI) peaks at lower BE centred at 231,3 ± 2 eV could be observed. The values are shifted at higher BE with respect to the value typical of MoO_2 (Mo $3d^{3/2}$ = 229.3 eV). The shift could be indicative of the presence of MoO_x agglomerates containing Mo in different oxidation states. It can be put forward that some Mo(VI) atoms were reduced to Mo(IV) and the resulting electronic charge was spread to the surrounding atoms leading to MoO_x species with intermediate valence. The surface atomic concentration of the Mo species with BE at 231,3 eV with respect to the total surface Mo species was near 50% for the three samples independently of the Mo molar concentration. The dispersion of the Mo phase can be related with the ratio of the surface atomic concentrations of Mo and Sn evaluated by XPS as Mo/(Mo+Sn) atomic ratio. Figure 2 shows the plot of Mo/(Mo+Sn) versus the Mo molar loading on a.p., activated, and used samples.

A similar clear increasing trend was observed for the series of the a.p. and activated samples, each point of the activated samples lying over that of the a.p. ones. This behaviour points to a uniform spreading of the Mo species at the external surface of the SnO_2 particles even for high Mo loading (5 % of Mo molar concentration).

For the used samples, atomic ratios level off, lower values than those of the corresponding a.p. and activated samples could be calculated. This suggests either the segregation of Mo containing phase during reaction, which caused surface reconstruction or the preferential coverage of the Mo species by poisoning atoms, like C.

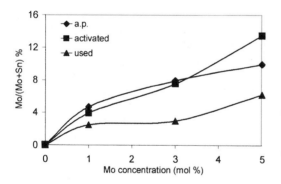

Fig. 2. Trend of the surface atomic concentration of Mo determined by XPS: Mo/(Mo+Sn)% vs. Mo loading is reported for the a.p. (♦), activated (■) and used (▲) samples.

3.2.2 FT-IR characterization.

Spectra of the materials in O_2. The FT-IR spectra of samples after activation will firstly discussed. The FT-IR spectrum of SN-550 and SN-650 samples cooled in O_2 at RT are similar (only a decreased intensity of the modes of surface impurities like hydroxyls and carbonates is observed, increasing the activation temperature) and very different from that expected for an ideal, stoichiometric bulk metal oxide. Looking at the expected shape of a metal oxide spectrum, the stoichiometric tin-oxide should be transparent in the medium and near infrared region; in particular no absorption should be expected at v > 800 cm^{-1}, whereas at v ≤ 800 cm^{-1} the absorptions of the longitudinal optical fundamental vibration fall. However, since we deal with particles of small size compared with the radiation wavelength, we will expect a Rayleigh scattering from them, with intensity increasing with wavenumbers [20]. Furthermore, the tin oxide under the examined conditions is not stoichiometric, mainly by oxygen vacancies (V_O). The first ionization and second ionization energy of the bulk V_O in the single-crystals of SnO_2 are 0.034 eV (274 cm^{-1}) and 0.145 eV (1170 cm^{-1}), respectively

[21]. The first ionization, $V_O \rightarrow VO^+ + e^-$ (c.b.), is thermally available at RT, therefore we will expect absorption by free electrons, of intensity decreasing with the radiation wave number [22]. The second ionization, $V_O^+ \rightarrow V_O^{++} + e^-$ (c.b.), is not thermally available at RT, therefore we will expect large absorption due to the photo-ionization of the mono-ionized oxygen vacancies, starting at about 1100 cm^{-1} and extending in all the IR region examined.

Indeed, the SN-550/650 sample spectrum show, besides the absorption edge of the fundamental optical modes, a low transparency in the medium IR (Figure 3a, thin solid curve shows the spectrum of the SN-650 sample) that we ascribe to the superimposition of: i) the absorption from free electrons; ii) the broad absorption due to photo-ionization of the V_O^+, and iii) the Rayleigh scattering from particles. However it is difficult to separate the three different contributions. In comparison with the spectra of SN-550/650 samples, those of samples containing Mo show increased transmittance at in the range 1000-2500 cm^{-1} and the presence, at $v \leq 1000$ cm^{-1}, of new peaks of increasing intensity increasing Mo content (Figure 3a, dotted and bold solid curves, show the spectra for Mo1SN-650 and Mo5SN-650 samples, respectively; Mo3SN-650 spectrum not reported to not overload the figure). The new peaks are more evident in Figure 3 b, where the spectra (in absorbance) of all samples activated at 650 °C are reported in the restricted 1100-800 cm^{-1} region. In particular, on Mo1sample (dotted curve) three peaks at 1006, 995 and 985 cm^{-1} are prominent, but very weak bands at lower wave numbers are also distinguishable. Passing from Mo1 to Mo3 sample (dot-dashed curve) a marked increase in intensity and a shift of the three peaks, now at 1015, 1007 and 980-70 cm^{-1} are observed accompanied by an upward shift of the absorption edge of the bulk fundamental vibrations. Passing from Mo3 to Mo5 sample (bold solid curve) six peaks at 1000, 980 (with two shoulders at 970 and 965), 940, 920, 890 and 875 cm^{-1} are prominent. In any case the FT-IR spectrum at the highest Mo loading does not show the typical features of the MoO$_3$ segregate particles (not reported). The precise assignment of all the peaks is not straightforward: in the 1020-900 cm^{-1} region the vibration stretching modes of O=Mo=O or Mo=O terminal groups with different double bond character are expected to fall; the 900-700 cm^{-1} is the region of the asymmetrical stretching modes of (Mo-O-Mo)$_n$ or (Mo-O-Sn)$_n$ bridges [24-25]. Furthermore, on the basis of the FT-IR technique only, it is hard to distinguish between the groups with Mo^{VI}, Mo^V and Mo^{IV}, but, on the basis of the XPS measurements, only Mo^{VI} should be present.

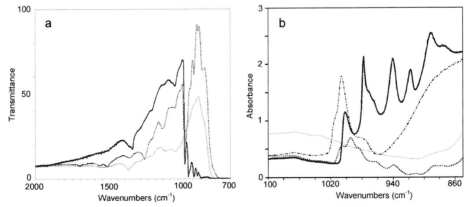

Fig. 3. FT-IR spectra of SN-650 (thin solids curves), Mo1SN-650 (dotted curves), Mo3SN-650 (dot-dashed curve) and Mo5SN-650 (bold solid curves); a) in transmittance, in the overall spectral region examined; b) in absorbance, in the 1100-800 cm^{-1} region.

However, we are interested to justify the observed transmittance increase in the 1000-2500 cm^{-1} region by Mo addition. Since the electrical measurements performed on the thick films show that molybdenum lowers (see *infra*) the conductance of the films in air, a reasonable explication of these findings, in spite of the XPS results, should be the following: during the activation processes a very small fraction of Mo^{VI} ions (with ionic radius of 0.42 Å) can migrate in the tin oxide surface and subsurface layers where can be reduced to Mo^{V} or Mo^{IV} (not detectable by XPS) by the free electrons or by electrons trapped in the oxygen vacancies. Mo^{V} and Mo^{IV} ions (atomic radii 0.63 and 0.65 Å, respectively) can easily occupy the Sn^{IV} (atomic radius 0.69 Å) lattice sites giving rise to a very dilute surface solid solution. As a consequence in the Mo-added materials the concentration of both the electrons trapped in oxygen vacancies and of free electrons should decrease, causing the observed transmittance increase and the conductance decrease (see *infra*). As support, we remember that recently Ivanovskaya et al. [15] have shown by EPR technique that, in MoO$_3$-SnO$_2$ solid solutions, increasing Mo^{VI} contents cause the decrease of the V$_O^+$ EPR signal in tin oxide and the parallel increase of the Mo^{V} signal. Actually, EPR technique is a much more sensible technique than the XPS one.

Concerning the spectra run at higher temperature, only minor differences in the sample transmittance were observed, some unit percent, starting from 350 °C.

Spectra in NO/O$_2$ mixture or in NO$_2$. We will examine now the spectroscopic results obtained after NO/O$_2$ or NO$_2$ interaction at RT. It is worthy to note that:

i) the type and the importance of the spectral features are dependent on the sample considered, while the pre-treatment temperature in O$_2$ does not significantly affect the results;
ii) for Mol(3,5)SN-550/650 samples the final spectroscopic results obtained using NO/O$_2$ mixture are comparable to those using NO$_2$ in the pressure range reported in the experimental and if the contact time was long enough (30 min). At variance, the results obtained for SN sample are slightly different in the two cases, as it will be underlined in the following.

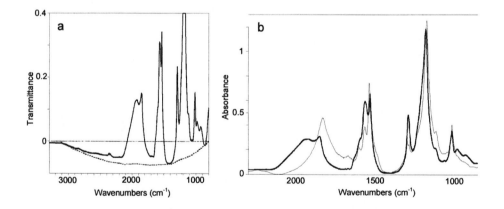

Fig. 4. a) FT-IR difference spectrum after NO$_2$ interaction (30 min, 5 mbar) with SN-650 sample in the overall spectral region. b) Comparison between difference spectra after NO$_2$ (30 min, 5 mbar, bold solid curve) and NO/O$_2$ (30 min, 10 mbar, thin solid curve) interaction with SN-650 sample in the spectral region of NO$_x$ surface species (2300-850 cm^{-1}). All spectra are reported as differences between the spectra run in NO$_2$ or NO/O$_2$ and the spectrum run in O$_2$.

The relevant changes induced by NO_2 or NO/O_2 in the FT-IR spectra of the SN material (Fig. 4) can be resumed in the following points: *a)* concerning the electronic absorptions, SN samples after interaction with NO/O_2 mixture or NO_2 showed a small increase in the sample transparency in the 1200-3000 cm^{-1} region, assigned to a decreased concentration of V_O^+, as revealed by the weak negative band with the shape expected for the broad absorption due to the electron photo-ionization from V_O^+ to the conduction band (c.b.) (see dotted trace in Fig. 4a for SN-650 sample); *b)* concerning the NO_x surface species formed, in Fig. 4b the comparison between the changes induced by NO/O_2 and NO_2 on the same SN-650 pellet are reported evidencing some differences. The large variety of peaks displayed and the fact that different NO_x species may show vibration modes in the same spectral region do not permit an easy assignment of each of them. However, the positions, relative intensities, and the different thermal stability of the main envelopes of peaks (together with data from literature [26-27]) are consistent with the correlations and assignments proposed in Table 1.

Table 1.

Band positions and related surface species formed upon NO_2 or NO/O_2 adsorption at RT

Samples				Band assignment [26-27]	
SN-550/650	Mo1-550/650	Mo3-550/650	Mo5-550/650		
Band positions (cm⁻¹)				Modes	Surface species
1174,1114 1320-1250[a]	1180, 1115 1320-1250[a]	1179, 1115 1320-1250[a]	- -	ν_{sym} (NO_2) ν_{asym} (NO_2)	Chelating nitrites
1835-1840	-	-	-	ν(NO)	NO^δ(ads)
1954 [b]	1890	1926-1926 [c]	1898, 2016 [d]	ν(NO) or ν_{asym}(NO_2)	$NO^{\delta+}$ or $NO_2^{\delta+}$(ads)
1670-1530 1299-1230 1014 -	1735-1535 1320-1222 1012- 930 893-826	1620-1530 [e] 1308- 1230 972,944 890	1660-1555 [e] 1299,1220 1010-962 -	ν(N=O) ν_{asym} (NO_2) ν_{sym} (NO_2) δ (NO_2)	Bridging and chelating nitrates
- -	- -	- -	1408, 1381 871	ν_{asym} (NO_3) δ (NO_3)	Ionic nitrates

(a) superimposed to ν_{asym} (NO_2) modes of nitrates, (b) not always present on all the SN pellets examined, (c) actually an envelope of three bands, (d) present in different amounts following the contact times, their amount decreases with the contact time, (e) with very low intensities.

The formation of NO^δ(ads) species requires the presence at the surface of sites able not only to coordinate NO through the electron pairs of its 5σ orbital but also to back-donate electrons. We cannot exclude the presence on tin oxide of Sn^{2+} ions or interstitial Sn atoms as stoichiometric defects, even if it is known that the V_O are the major responsible for the semiconducting properties of tin oxide. By NO_2 interaction with SN samples the peak at 1835-1840 cm^{-1} is still present although with low relative intensity while a very broad band (probably an envelope of overlapped peaks) is now well evident, showing an apparent maximum at 1954 cm^{-1}. In the case of interaction with NO_2 the formation of NO^δ(ads) species should require firstly the formation of NO following reactions like

$$NO_{2,gas} + e^- \rightarrow NO + O^-_s \quad (1)$$
$$2\,NO_2 + e^- \rightarrow NO + (NO_3)^-_s \quad (2)$$

Reactions (1) and (2) involve electrons (e^-) as reactant species. These electrons can come either by the conduction bands or by the V_O^+ vacancies. On the basis of their positions the overlapped peaks responsible for the broad absorption superimposed to the sharp 1835-40 cm^{-1} peak could be NO^δ^+ mono-nitrosyls. However they are never formed when samples are contacted with pure NO, but only by direct interaction with NO_2. So, we favoured their assignment to asymmetric stretching modes of $NO_2^\delta^+$(ads) species formed by coordination of NO_2 to strong Lewis acid sites [30-33]. On tin oxide strongly acid sites can be Sn^{4+} ions highly coordinatively unsaturated. Apart from NO^δ species related to the sharp peak at 1840-1845 cm^{-1} and $NO_2^\delta^+$(ads) species, nitrates and nitrite-like species are the other two main families of surface compounds formed by NO/O_2 or NO_2 interaction, but the relative amounts of the nitrate-like species are slightly different in the two cases.

The reactive oxygen surface species on SN samples after the thermal treatment at 550-650 °C are mainly surface lattice oxygen O^{2-}, very reactive because highly coordinatively unsaturated [1]. However, as the sample has been cooled in O_2 up to RT, O^- and O_2^- species may be present. A series of reaction giving nitrites and nitrates on SN-550/650 samples can be proposed:

$$2\,NO_{gas} + O_s^{2-} + 0.5\,O_{2,gas} \rightarrow 2\,(NO_2)^-_s \quad (3)$$
$$2\,NO_{gas} + (O_2)^-_s + e^- \rightarrow 2\,(NO_2)^-_s \quad (4)$$
$$NO_{,gas} + O_s^- \rightarrow (NO_2)^-_s \quad (5)$$
$$2\,(NO_2)^-_s + O_{2,gas} \rightarrow 2\,(NO_3)^-_s \quad (6)$$
$$2\,NO_{gas} + O_{2,gas} \rightarrow 2\,NO_{2,gas} \quad (7)$$
$$2\,NO_{2,gas} + O_s^{2-} \rightarrow (NO_3)^-_s + (NO_2)^-_s \quad (8)$$
$$NO_{2,gas} + e^- \rightarrow (NO_2)^-_s \quad (9)$$
$$NO_{2,gas} + O^- \rightarrow (NO_3)^-_s \quad (10)$$
$$2\,NO_{2,gas} + (O_2)^-_s + e^- \rightarrow 2\,(NO_3)^-_s \quad (11)$$

The formation of NO_2 in the gaseous phase according to reaction (7), although favoured by thermodynamics, is slow at RT so that the reactions (3)-(6) are suggested to occur mainly when a mixture NO/O_2 has used. Reactions (8) - (11) have to be taken into account when NO_2 is used. Reactions (4), (9) and (11) involve electrons (e^-) as reactant species and like reactions (1) and (2) can explain the transparence increase observed.

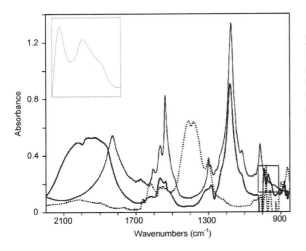

Fig. 5. Changes induced on the FT-IR spectrum of SN-650 (thin solid curve), Mo3SN-650 (bold solid curve) and Mo5SN-650 (dotted curve) samples by contact with 10 mbar of NO/O$_2$ for 30 min. In the window a zoom of the 900-1000 cm^{-1} region of spectrum relative to Mo3SN-650 sample is shown. All spectra are reported as differences between the spectra in NO/O$_2$ and the spectrum in O$_2$.

The relevant changes induced by NO/O$_2$ in the FT-IR spectra of the three materials containing Mo are practically identical to those induced by NO$_2$ and can be resumed in the following points: *a)* concerning the electronic absorptions, only the Mo5SN-550/650 samples after interaction with NO/O$_2$ or NO$_2$ showed a small increase in the sample transparency in the 1200-3000 cm^{-1} region, comparable to that observed by NO$_2$ interaction with SN samples, and assigned to a decreased concentration of V_O^+; *b)* concerning the NO$_x$ surface species formed on the materials supporting Mo, after NO/O$_2$ or NO$_2$ admission (see Figure 5), the 1835-40 cm^{-1} peak is no more present while in the 1850-2020 cm^{-1} region an envelope of heavily overlapped peaks, similar to that present on SN sample spectra after NO$_2$ admission, but with apparent maximum shifted towards higher frequency and more intense, appear. Further their intensity increases going from Mo1SN-550/650 to Mo3SN-550/650 samples. Again (see Table 1) we assign this band envelope to the asymmetric stretching modes of NO$_2^{\delta+}$(ads) species formed by coordination of NO$_2$ to Lewis sites of higher acidity than those present on SN samples. On Mo5SN-550/650 samples NO$_2^{\delta+}$(ads) are transient species: the related band envelope reaches its maximum intensity after 1 or 2 min contact (not reported), then the intensity decreases almost completely after 30 min of contact. Nitrite-like and nitrate-like species are the other families of surface compounds formed, but their variety (for nitrate-like species), their absolute and relative amounts depend on the Mo content. In particular, on Mo1SN-550/650 and Mo3SN-550/650 samples chelating nitrites are still present in consistent amounts while bridging or chelating nitrates are markedly decreased on increasing Mo loading, so that on Mo3SN samples the related peaks show intensity reduced of about one magnitude order. On Mo5SN-550/650 samples, nitrites are completely absent, and mainly nitrates of ionic type (absent on SN, Mo1SN and Mo3SN materials) are present at the equilibrium; *c)* concerning the stretching modes of Mo=O, O=Mo=O terminal groups (see Fig. 6), on Mo5SN-550/650 by interaction with NO/O$_2$ or NO$_2$ the total erosion of the 940 and 920 cm^{-1} peaks is observed, while the overlapped peaks in the 1020-970 cm^{-1} region present on all Mo containing samples are only partially eroded. In our opinion this erosion is not due to their reduction by NO or NO$_2$ to give nitrites or nitrates, but rather to their broadness and their shift towards lower frequency, in the 990-950 region cm^{-1}. This is the region where the ν_{sym}(NO$_2$) modes of

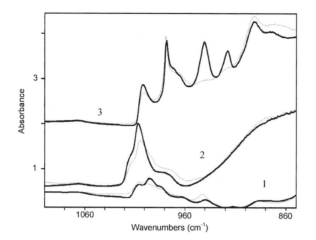

Fig. 6. Effects of 5 mbar of NO_2 (or 10 mbar NO/O_2) interaction for 30 min at RT on the different samples.
FT-IR spectra of Mo1SN-650 (curves 1), Mo3SN-650 (curves 2) and Mo5SN95-650 (curves 3) samples in the 1100-700 cm^{-1} region before (bold curves) and after (thin curves) interaction.

surfaces nitrates fall, so it is difficult to distinguish the two different contributions. However, this is particularly evident on Mo3SN sample spectra showing $v(N=O)$ and v_{asym} (NO_2) modes so weak that we could not expect to see the corresponding v_{sym} (NO_2) modes so intense as in inset of Figure 5. At the same time the envelope of band assigned to $NO_2^{\delta+}$(ads) species increases on passing from SN to Mo3SN samples. So we suggest the 990-940 cm^{-1} peaks for the major part are due to the stretching modes of molybdenyls strongly perturbed by the NO_2 adsorption as proposed in the following scheme:

$$
\begin{array}{ccc}
O & & O^{\delta-} \\
|| & & || \\
Mo & + NO_2 \Leftrightarrow & Mo \leftarrow NO_2^{\delta+} \\
/||\backslash & & /||\backslash
\end{array}
$$

We can conclude that the addition of Mo up to 3 mol % markedly decreases the amount of the surface oxygen species present on the tin oxide support able to give nitrates, while the surface molybdenyls are not easily reduced by NO or NO_2 at RT to give nitrates, even if some of them are strongly perturbed by the $NO_2^{\delta+}$(ads) species.

At variance, when the Mo content is increased to 5 mol % new molybdenyl peaks are present and two of them (at 940 and 920 cm^{-1}) are completely destroyed by NO/O_2 or NO_2 interaction and a new type of nitrates (of ionic type) are the most abundant species present at the surface. The formation of $NO_2^{\delta+}$(ads) species is transitory on Mo5SN-550/650 samples and they can be seen as intermediates in the ionic nitrates formation. We can conclude that on the Mo5SN samples the mechanism of nitrate formation is completely different from that observed on the other samples: on Mo5SN-550/650 samples nitrates are formed by reduction of particularly reactive molybdenyls, thus they can be seen as surface nitrates anchored to Mo ions. At variance on the samples at lower Mo loading nitrites and nitrates are mainly formed following the reactions proposed for SN-550/650 samples, that is involving the reactive oxygen species present on the tin oxide surface not covered by the MoO_x phase, and can be seen as nitrites or nitrates anchored to Sn ions.

The spectra recorded at increasing temperature up to 450 °C at the same gaseous pressures showed that the different NO_x surface species have different thermal stability: $NO^{\delta-}$(ads), $NO_2^{\delta+}$(ads), and chelating nitrites are stable only at RT, nitrates become to be decomposed at

150 °C. However, on SN and Mo1SN samples bridging or chelating nitrates are still present at 450 °C even if in very small amounts, while on Mo5SN sample very small amounts of ionic nitrates are still present at 250 °C, but completely absent by interaction at 300 °C.

3.3. Catalytic results

The series of pure and Mo-added SnO_2 catalysts were compared in the NO selective reduction by C_2H_4 at fixed contact time and feed concentration and variable temperature. The activity of pure SnO_2 in this reaction is high, as expected [9]. Curves representing N_2 formation as a function of reaction temperature are all volcano-shaped with temperature of maximum activity (T_{max}) around 350 °C, as shown in Fig. 7a. Conversion of C_2H_4 is regularly increasing up to quantitative formation of CO_2 starting from 400 °C, in any case. Among C-compounds, selectivity to CO_2 is always very high. Only very little amount of CO is detected (in the range 5-10%, at maximum) for temperatures lower than those corresponding to complete C_2H_4 conversion.

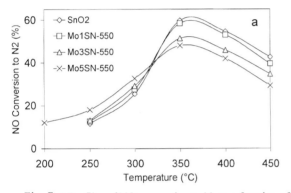
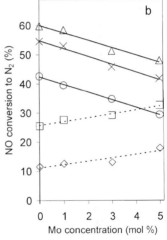

Fig. 7 a) Profiles of NO conversion to N_2 as a function of reaction temperature for pure SnO_2 and Mo-containing SnO_2 samples. b) Influence of Mo concentration on NO reduction to form N_2 at various reaction temperatures; (◊) 250°C; ( ) 300°C; (Δ) 350°C; (X) 400°C; and (O) 450°C.

Table 2.

NO conversion to N_2 in the selective reduction of NO by C_2H_4 (0.5% NO, 0.5% C_2H_4, 5% O_2).

Sample	NO conversion to N_2 (%)		Specific activity (μmol_{N2} g^{-1} s^{-1})		Selectivity, $c.f.$ (%)	
	300 °C	350 °C	300 °C	350 °C	300 °C	350 °C
SnO_2	25.5	59.8	0.358	0.327	12.0	9.4
Mo1SN-550	27.6	58.4	0.376	0.341	10.9	9.0
Mo3SN-550	29.2	51.3	0.359	0.321	9.0	9.6
Mo5SN-550	32.7	48.0	0.315	0.275	8.5	7.0

Table 2 collects the most significant results in terms of NO conversion to N_2, specific activity (expressed as mol of N_2 formed per unit mass of catalyst and per unit time) and selectivity (expressed as competitiveness factor, $c.f.$ %) at two temperatures, the one before (300 °C) and the other corresponding to T_{max} (350 °C). At 300 °C, the increasing loading of MoO_3 has a positive effect on the reduction of NO to N_2 whilst a negative effect is observed for higher temperatures (see Fig. 7a). Viewing the data in terms of specific activity, a more complicate trend appears. The activity seems slightly dependent on the Mo concentration. Surely not all the Mo deposited on SnO_2 is exposed at the surface and can be active in catalysis. Addition of low amount of Mo to SnO_2 (Mo1SN) leads to more active catalyst than those prepared at higher Mo loading. At high Mo concentration, the formation of large Mo-aggregates favours the reaction of C_2H_4 by O_2 (combustion) rather than that of NO by C_2H_4 (reduction). Due to the known ability of Mo to oxidize substrates, selectivity decreases as Mo concentration increases at any reaction temperature. Accordingly, selectivity results, expressed in terms of competitiveness factor, decrease with the Mo loading at temperatures both before T_{max} and at T_{max} (Table 2).

In Fig. 7b an interesting trend is reported in which the conversion of NO to N_2 has been plotted as a function of Mo concentration for various reaction temperatures. Lines with positive and negative slopes are found for low reaction temperatures (before T_{max}) and for temperatures equal to or higher than T_{max}, respectively. This behaviour shows the positive influence of addition of the MoO_3 phase on SnO_2 oxide, that is to anticipate the reduction of NO activating the hydrocarbon (C_2H_4) at lower temperatures than pure SnO_2. Unfortunately at high temperatures, Mo phase has better ability towards hydrocarbon combustion than SnO_2. Therefore, C_2H_4 is completely oxidized and it is no longer present in the feed for reducing NO.

3.4. Electrical characterization

We measured the conductance of the samples in air at various temperatures from 150° C up to 400° C. The conductance of the SN-650 sample ranges from 4.5×10^{-6} Ω^{-1} up to 2.4×10^{-5} Ω^{-1} on the above interval of temperature. The Mo addition decreases the conductance as already pointed out for SnO_2-MoO_3 ceramic sensors [34]. However, addition of Mo in amount of 1 mol % only moderately lowers the conductance, less than one order of magnitude, while we

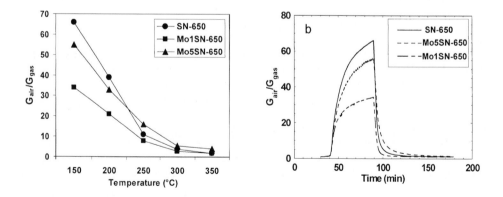

Fig. 8. a) Comparison between the response to NO_2 (10 ppm) in wet air (40% RH) of the three sensors at increasing temperatures. b) The electrical response of the three sensors at 150 °C.

measured a conductance decrease of two orders of magnitude when Mo is added in amount of 5 mol %. As anticipated, these results are in qualitative agreement with the increase of the sample transparency to radiation in the medium IR region.

The electrical responses (G_{air}/G_{gas}) to NO_2 are shown in Fig. 8. For the three films the response is very good and maximum at 150 °C, still good at 250 °C, while at 300-350 °C it is markedly lowered (see Fig. 8a). The measurements show that 1 mol % Mo markedly lowers the ability of the SnO_2 to sense NO_2, while Mo in amount of 5 mol % completely restores the ability to sense NO_2. Fig. 8 b shows as examples the very good responses (G_{air}/G_{gas} average values ranging between 66 and 34) of the three films at 150 °C.

In Fig. 9a the electrical responses (G_{gas}/G_{air}) to propene (C_3H_6) of the three films are reported, in the working temperature range 200-400 °C.

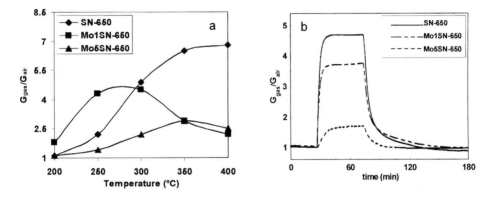

Fig. 9. a) Comparison between the response to (C_3H_6) (10 ppm) in wet air (40% RH) of the three sensors at increasing temperatures. b) The electrical response of the three sensors at 300 °C.

We notice that: i) the films only respond at working temperatures higher than 150 °C; ii) indeed, at 200 °C only the Mo1SN-650 film sensibly responds; iii) the response of SN-650 markedly increases on increasing temperature; iv) at variance that of Mo1SN-650 film first increases, reaching a maximum at 250-300 °C, then moderately decreases; v) up to 250 °C the highest response is given by Mo1SN-650 film, at 350-400 °C the SN-650 film has the highest response, vi) the response of Mo5SN-650 film slowly increases increasing the working temperature, at 350-400 °C it show a response very near to that of Mo1SN-650 film. The best properties as sensing materials for propene is thus shown by the pure tin oxide at 350 – 400 °C. Fig. 9b shows as examples the responses of the three films at 300 °C.

Concerning the response to ethene (C_2H_4), we want to underline that it is very low for all materials. The films only respond at working temperatures higher than 250 °C. The working temperature of the maximum response is 350-400 °C, at which actually the responses are very low: G_{gas}/G_{air} is 1.9 for Mo1SN-650 and 1.8 for Mo5SN-650, SN-650 films show even lower response.

The electrical responses to the C_3H_6/NO_2 and C_2H_4/NO_2 mixtures are now considered.

Table 3 gives the average responses to the C_3H_6/NO_2 mixture (reported as G_{air}/G_{gas} or G_{gas}/G_{air}) of the three films, in the working temperature range 150-400 °C. We notice that: i) at 150 and 200 °C they all respond to the mixture as to an oxidizing atmosphere; ii) as expected, at 150 °C the presence of propene does not affect the response, indeed all materials do not sense pure propene at 150 °C; iii) since the responses to pure propene of SN-650 and Mo5SN-650 materials are negligible at 200 °C we did not expect responses to NO_2 lowered by the presence of propene in the mixture.

Table 3.

The response (G_{air}/G_{gas} or G_{gas}/G_{air}) of the different films towards C_3H_6 (10 ppm) - NO_2 (10 ppm) in wet air (40% RH) at different operating temperatures.

Films							
	Working temperature / °C						
	150	200	250	300	300	350	400
	G_{air}/G_{gas}				G_{gas}/G_{air}		
SN-650	70 (66)*	27 (39)	3.3 (11)	0.21 (3.8)	4.7 [4.6]**	8.7 [6.5]	7.3 [6.8]
Mo1SN-650	31 (34)	2.5 (21)	0.22 (8)	0.27 (2.8)	3.8 [3.6]	2.5 [2.9]	2.2 [2.2]
Mo5SN-650	60 (55)	29 (33)	6.4 (16)	1.5 (5.4)	0.7 [1.8]	1.1 [2.9]	2.0 [2.5]

* All the values reported in round brackets are the response towards NO_2 (10 ppm).
** All the values reported in square brackets are the response towards C_3H_6 (10 ppm).

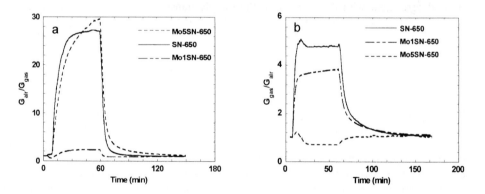

Fig. 10. The electrical response of the three sensors to (C_3H_6) (10 ppm)- NO_2 (10 ppm) mixture in wet air (40% RH) at: a) 200 °C (reported as G_{air}/G_{gas}) and b) at 300 °C (reported as G_{gas}/G_{air}).

Again, since the Mo1SN-650 sample show a measurable but very low response to pure propene we expected a response to NO_2 only moderately lowered by the presence of propene in the mixture. On the contrary for the three materials the responses are markedly lowered (in particular for Mo1SN-650 sample); iv) 250 °C is a borderline temperature, the SN-650 and Mo5SN-650 films still respond to NO_2 but their response is markedly decreased, while Mo1SN-650 already responses to propene; at 300 °C the unique films giving response to NO_2, but too low, is Mo5SN-650 sample; v) as expected, in the 350-400 °C range they do not respond or give response to the propene, the response being not too different from those to the pure propene, indeed all materials do not sense NO_2 at 400 °C. Fig. 10 shows the response of the three films to the mixture at 200 °C and 300 °C .

Concerning the electrical response to the C_2H_4 (10 ppm)-NO_2 (10 ppm) mixture in wet air (40% RH), the only two films examined in this case are SN-650 and Mo5SN-650. At the working temperature of 150 °C and 200 °C their responses are to an oxidizing atmosphere. In comparison with the responses to pure NO_2 (see Fig. 8), at these temperatures the responses to the mixture are only decreased of some percent. Furthermore, for the Mo5SN-650 film the response still remained very good doubling the ethene concentration. At 250 °C the responses of both SN-650 and Mo5SN-650 films are markedly reduced, but they still respond to an oxidizing atmosphere (G_{air}/G_{gas} being 11 and 13 respectively). At 300 °C Mo5SN-650 responds to an oxidizing atmosphere, but with a very low response (G_{air}/G_{gas} being 1.3), while SN-650 sensor does not show any response.

4. CONCLUDING REMARKS

The XRD, SEM and HRTEM measurements show that we have obtained powders and films made by regularly shaped and nano-sized particles, even after thermal treatments at 650 °C.

XPS measurements enlighten an uniform spreading of the Mo species at the external surface of the SnO_2 particles even for high Mo loading (5 % of Mo molar concentration).

Electrical and FT-IR spectroscopic measurements have been employed on films and powders respectively to obtain information on the electronic effect due to the Mo addition. The electrical data show that Mo markedly lowers (of about two orders of magnitude on passing from SN to Mo5SN sample) the conductance of the films in air. The conductance decrease is in good agreement with the increase of the sample transmittance attributed to a decreased concentration of free electrons and of electrons trapped in oxygen vacancies as a consequence of the Mo addition. This finding suggests the reduction of a fraction of Mo^{VI} to lower oxidation numbers, although XPS technique only revealed the presence of Mo^{VI}.

FT-IR spectroscopy was also employed to obtain information on the MoO_x surface species after treatments in O_2 and on the way they are affected in presence of the NO/O_2 or NO_2 atmospheres. NO_x surface species formed were individuated and some related surface reactions have been proposed. It is worth noting that we attained the demonstration that the surface species formed on the Mo5SN-550/650 sample are completely different from those occurring on samples at lower Mo loading and on SN ones. On Mo5SN-550/650 samples mainly nitrates of ionic type by reduction of surface molybdenyls are formed, while on the samples at lower Mo loading nitrites and nitrates similar to those present on pure tin oxide are formed, but in decreasing amounts with increasing Mo content.

Simultaneously electrical measurements show that all materials are very good NO_2 sensors at working temperatures of 150-250 °C. However, the response of Mo1SN-650 films is lower than the SN-650 ones, while Mo5SN-650 films show a response very near to that of SN-650 one.

We can conclude that at low molybdenum loading, the surface chemistry is that of the tin oxide with reduced activity caused by the Mo coverage. Therefore the decreased sensibility to NO_2 of the Mo1SN-650 sensor could be related to a decrease of those surface reaction centres of tin oxide able to interact with NO_2 giving NO_x surface species able localizing the free electrons. At variance when loading increase up to 5% Mo molar concentration, the surface chemistry is completely changed and the response to NO_2 is restored by the presence of new active reaction centres.

From electrical results it is evident that, at working temperature of 150 °C and 200 °C, the electric response to NO_2/hydrocarbon mixtures of the three materials is still good and selective for the NO_2 gas (at 200° in the case of ethene it is only slightly decreased with respect to the response in pure NO_2). At variance, at 250 °C the ability to sense NO_2 is drastically reduced. This happens even if at this temperature the response of the materials towards hydrocarbons is low, in particular at this temperature they do not sense ethene at all when alone. Looking at the catalytic results it is worthy to note that at 250 °C all the catalysts begin to be active in NO_x selective abatement to N_2.

Passing now to remark the catalytic performances of the materials, the specific activity increases with the temperature, the maximum being at 350 °C. Even if the activity seems slightly dependent on Mo concentration, for each working temperature up to 300 °C the N_2 production increases with the Mo loading, while for each working temperature higher than 300 °C there is an inversion in the influence of the Mo loading on the N_2 production. The presence of Mo thus increases the catalytic performance of the tin oxide under 300 °C. We want to underline that one of the goal in the NO_x abatement research is to obtain catalyst active at low temperatures.

ACKNOWLEDGMENTS

This work has been supported by Italian MURST PRIN: "Sviluppo di materiali nanostrutturati per sensori di gas selettivi ad altissima sensibilità per il monitoraggio di inquinanti atmosferici"

REFERENCES

[1] (a) J. F. McAleer, P.T. Moseley, J . O. W. Norris, D. E. Williams, J. Chem. Soc., Faraday Trans. 1, **83** (1987) 1323. (b) J. F. McAleer, P.T. Moseley, J . O. W. Norris, D. E. Williams, J. Chem. Soc., Faraday Trans. 1, **83** (1987) 3383.
[2] N. Barsan, M. Schweizer-Berberich, W. Göpel, Fresenius J. Anal. Chem., **365** (1999) 287.
[3] P.G. Harrison, M. J. Willet, Nature, **332**(1988) 337.
[4] G. Ghiotti, A. Chiorino, F. Boccuzzi, Sensors and Actuators, **19** (1989) 151.
[5] G. Ghiotti, A. Chiorino, W. Xiong Pan, Sensors and Actuators B, **15-16** (1993) 367.
[6] A. Chiorino, G. Ghiotti, F. Prinetto, M.C. Carotta, G. Martinelli, M. Merli, Sensors and Actuators B, **44** (1997) 474.
[7] A. Chiorino, G. Ghiotti, F. Prinetto, M.C. Carotta, M.Gallana, G. Martinelli, Sensors and Actuators B, **59** (1999) 203.
[8] M. Sheinthch, J. Schmidt, Y. Lechtman, and G. Yahav, Applied Catal., **49** (1989) 55.
[9] R. K. Herz, A. Badlani, D. R. Schryer and B. T. Upchurch, J. Catal., **141** (1993) 219.
[10] Y. Taraoka, T. Harada, T. Iwasaki, T. Ikeda, S. Kagawa, Chem. Lett., (1993) 773.
[11] P. W. Park, H. H. Kung, D.-W. King, M. C. Kung, J. Catal., **184** (1999) 440.
[12] A. Auroux, D. Spriceana, A. Gervasini, J. Catal, **195** (2000) 140.
[13] C. Guimon, A. Gervasini, A. Auroux, J. Phys. Chem., **105** (2001) 10316.
[14] N. G. Valente, L. A. Arrúa, L. E. Cadús, *Applied Catal. A*, **205** (2001) 201
[15] M. Ivanovskaya, P. Bogdanov, G. Faglia. P. Nelli, G. Serveglieri, A. Taroni, Sensors and Actuators B, **77** (2001) 268.
[16] A. Chiorino, G. Ghiotti, F. Prinetto, M.C. Carotta, G. Gnani, G. Martinelli,8th International Meeting on Chemical Sensors, Basel, Switzerland 2-5 July 2000, Abstract Book , p. 268.
[17] P. G. Harrison and A. G. Guest, J. Chem. Soc., Faraday Trans. 1, **83** (1987) 3383.

[18] J.H. Scofield, J. Elect. Spect. Relat. Phenom., **8** (1976) 129.

[19] K.A. Bethke, M.C. Kung, B. Yang, M. Shah, D. Alt, C. Li, H.H. Kung, Catal. Today, **26** (1995) 169.

[20] A. Gervasini, P. Carniti, V. Ragaini, Appl. Catal. B, **22** (1999) 201.

[21] G. Busca, Catal. Today, **27** (1996) 323.

[22] (a) Z. M. Jarzebski, J. P. Marton, J. Electrochem. Soc., **123** (1976) 229C
(b) Z. M. Jarzebski, J. P. Marton, *J. Electrochem. Soc.,* **123** (1976) 333C

[23] J. I. Pankove, "Optical Processes in Semiconductors", Dover, New York,1975, p.74.

[24] H. Knoezinger, Proc. of the 9th International Congress on Catalysis, M. J. Philips, M. Ternan Eds, Ottawa 1988, Vol. **5**, pp 20-53.

[25] J. Leyrer, B. Vielhaber, M. L. Zaki, Z. Shuxian, J. Weitkamp, H. Knoezinger, Mater. Chem. Phys., **13** (1985) 301.

[26] K. Nakamoto, *"Infrared Spectra of Inorganic and Coordination Compounds,* Wiley, New York, 1979.

[27] K.I. Hadjiivanov, *Catal. Rev.- Sci. Eng.,* **42** (2000) 71.

[28] A. Chiorino, F. Boccuzzi, G. Ghiotti, Sensors and Actuators B, **5** (1991) 189.

[29] G. Ghiotti, A. Chiorino, F. Prinetto, Sensors and Actuators B, **24-25** (1995) 520.

[30] E. Giamello, D. Murphy, G. Magnacca, C. Morterra, Y. Shioya, T. Nomura, and M. Anpo, J. Catal., **136** (199) 510.

[31] M. Iwamoto, H. Yahiro, N. Mizuno, W.-X.Zhang, Y. Mine, H. Furukawa, and S. Kagawa, J. Phys. Chem., **96** (1992) 9360.

[32] T.E. Hoost, K.A. Laframboise, and K. Otto, *Catal. Lett.,* **33** (1995) 105.

[33] W. Aylor, S.C. Larsen, J. A. Reimer, and A. T. Bell, *J. Catal.,* **157** (1995) 592.

[34] M. Ivanovskaya, E. Lutynskaya, P. Bogdanov, Proc. Eurosensors XI, Warsaw, Poland, September 21-24 (1997) 443.

[35] J. Laane and J.R. Ohlsen, Prog. Inorg.Chem., **27** (1980) 465.

Studies in Surface Science and Catalysis 155
A. Gamba, C. Colella and S. Coluccia (Editors)
© 2005 Elsevier B.V. All rights reserved

CD₃CN and NH₃ interaction with Ti(IV) catalytic centres grafted on mesoporous MCM-41

CD$_3$CN and NH$_3$ interaction with Ti(IV) catalytic centres grafted on mesoporous MCM-41

E. Gianotti[a], L. Marchese[b], M. Guidotti[c], N. Ravasio[c], R. Psaro[c] and S. Coluccia[a]

[a]Dipartimento di Chimica IFM, v. P. Giuria 7, 10125 Torino – Italy;

[b]Dipartimento di Scienza e Tecnologie Avanzate, Università del Piemonte Orientale, "A. Avogadro", v. Bellini 25/G, 15100, Alessandria – Italy.

[c]CNR – Istituto di Scienze e Tecnologie Molecolari and Centro di Eccellenza CIMAINA, via G. Venezian 21, Milano – Italy;

ABSTRACT

CD$_3$CN and NH$_3$ were used as molecular probes to monitor the nature of both Brønsted and Lewis acid sites as well as the coordination sphere of Ti(IV) centres in Ti(IV) grafted MCM-41 by means of FT-IR and Diffuse Reflectance UV-Vis spectroscopies. CD$_3$CN molecule was very effective to reveal the presence and the coordination changes of Ti(IV) Lewis acid centres, but was not helpful to evidence different type of OH acid centres in pure MCM-41 and Ti-MCM-41. On the contrary, ammonia adsorption on Ti-MCM-41 revealed the presence of Brønsted sites sufficiently acidic to protonate NH$_3$ molecules with the formation of NH$_4^+$ species.

1. Introduction

Ti-containing catalysts have shown high activity in various kind of selective oxidation reactions. In particular, Ti-functionalised silica based mesoporous materials are highly active in olefin epoxidation reactions [1]. Recently, titanium-based catalysts obtained by grafting titanium precursors onto the surface of siliceous materials have shown attractive performances in the epoxidation of unsaturated terpenes [2] and fatty acid methyl esters [3]. A detailed study of the surface acidity of Ti(IV)-based catalysts is therefore necessary to understand reaction mechanisms and determine strategies to enhance the selectivity towards epoxidised products, by minimizing the formation of acid-catalysed by-products. On the other hand, in some cases, the acidic features of these materials can be exploited, together with their oxidation ability, for bifunctional one-pot acid-redox syntheses [4].

Ti-grafted mesoporous MCM-41 was obtained by grafting titanocene dichloride (Cp_2TiCl_2) as precursor onto the walls of MCM-41. Whit respect to the one-pot synthesis, the grafting procedure leads to materials with a larger number of accessible Ti(IV) catalytic centres [5].

CD_3CN and NH_3 were used as molecular probes to monitor both the surface acidity and the Ti(IV) coordination in Ti(IV)-grafted mesoporous MCM-41 by means of FTIR and DR UV-Vis spectroscopies. Pure siliceous MCM-41 was also studied for comparison.

Ti-grafted MCM-41 displays different types of hydroxyl groups (silanols and titanols), acting as weak Brønsted acid sites, and the use of molecular probes with different basicity, such as acetonitrile and ammonia, allows to monitor their acidity [6].

Acetonitrile is widely used to characterize both Brønsted and Lewis acid sites. The interaction of the electron lone-pair located on nitrogen atom of CH_3CN with an acid site causes a high frequency shift of the C-N stretching mode with respect to the molecule in the liquid phase. The shift of the $v(CN)$ mode increases with the acidity of the centres involved. Deuterated acetonitrile (CD_3CN) is normally used instead of CH_3CN to avoid Fermi resonance effects [7,8]. Ammonia, a stronger base than acetonitrile, was also used to probe the acidity of the hydroxyl groups in Ti-MCM-41 [9].

2. Experimental

A purely siliceous mesoporous MCM-41 was synthesised according to literature method using cetyltrimethylammoniumbromide (CTMAB) as structure directing agent during the crystallization [10]. Before grafting titanium centres, the CTMAB surfactant was removed from the material at 550°C, first under inert gas flow and subsequently under oxygen. The purely siliceous MCM-41 was functionalised using titanocene dichloride [Cp_2TiCl_2] as precursor and following the procedure reported by Maschmeyer et al.[5]. After the grafting procedure, the material was calcined under O_2 at 550°C for 10h to eliminate the organic fraction of the Ti-cyclopentadienyl complexes grafted onto the MCM-41 silica walls. The titanium loading was 1.88 wt %, as obtained from AES-ICP analysis.

FTIR spectra of pellettised samples were recorded with a Bruker IFS88 spectrometer at a resolution of 4 cm^{-1}. Diffuse Reflectance UV-Vis-NIR spectra were collected by a Perkin Elmer (Lambda 900) spectrometer equipped with an integrating sphere attachment. CD_3CN and NH_3 adsorption were performed at room temperature by means of specially designed cells which were permanently connected to a vacuum line (ultimate pressure $<10^{-5}$ Torr) to perform in situ adsorption-desorption experiments. The samples were evacuated at 550°C for 1h prior to the adsorption of molecular probes.

In order to prevent any exchange of the deuterium atoms of CD₃CN with OH groups present at the surface of the materials, all the samples were submitted to several D₂O vapor adsorption-desorption cycles at room temperature and finally treated in D₂ at 400°C to deuterate all the surface OH groups.

3. Results and Discussion

The successful preparation of Ti-MCM-41 was monitored both by FTIR and DR UV-Vis spectra. It was, in fact observed that the IR absorption of silanols (3745 cm^{-1}), the OH groups located mainly at the surface of the inner walls of MCM-41 channels, decreased in intensity with respect to the one observed on purely siliceous MCM-41. This is a direct evidence that silanols are the anchoring site where titanium complexes are grafted. The presence in the IR spectra of a broad band at 935 cm^{-1}due to the asymmetric stretching mode of Ti-O-Si groups, and in the DR UV-Vis spectra of a strong and narrow band at 210 nm assigned to oxygen to tetrahedral Ti(IV) sites charge transfer [9, 11-13], fully confirms the success of the titanium grafting on the silica surface [9].

The accessibility of the Ti(IV) sites, and their adsorption properties, were monitored by CD₃CN probe molecule. In the DR UV-Vis spectra (Fig.1), the band at 210 nm broadened and shifted to 225 nm upon acetonitrile adsorption, clearly suggesting that the coordination sphere of Ti(IV) sites progressively expands from tetrahedral to octahedral [9] by insertion of two CD₃CN molecules (scheme 1).

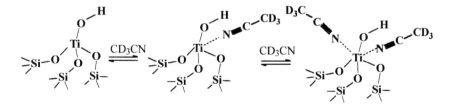

Scheme 1

314

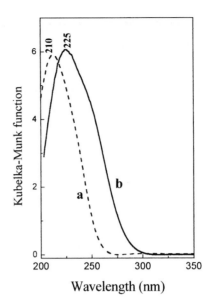

Fig. 1: DR UV-Vis spectra of CD₃CN adsorbed on Ti-MCM-41. Curve a: sample after calcination and evacuation at 550°C; curve b: after CD₃CN adsorption (vapour pressure ≈100 Torr)

FTIR spectra of CD_3CN adsorbed at room temperature on MCM-41 are reported in Fig.2. The narrow band at 2760 cm^{-1} in the spectrum of the sample in vacuum (curve 1) is due to the O-D stretching vibration of isolated deuteroxyl groups obtained after isotopic exchange with D_2O and D_2. Upon CD_3CN adsorption (vapour pressure ≈100 Torr, curve 2), the band at 2760 cm^{-1} completely disappears and new bands at 2525, 2275 and 2263 cm^{-1} are formed. The broad band centred at 2525 cm^{-1} is assigned to the O-D stretching mode of deuteroxyls D-bonded to CD_3CN molecules as shown in scheme 2. The band at 2275 cm^{-1} is due to the C-N stretching mode of CD_3CN interacting with OD groups and the band at 2263 cm^{-1}, particularly intense at high CD_3CN pressure, is associated to physically adsorbed CD_3CN. The small blue shift ($\Delta v_{CN} = 12$ cm^{-1}) of CN stretching mode produced by the interaction of CD_3CN with OD groups indicates that these OD groups have low acidity. In fact, materials with strong Brønsted acid sites, such as zeolite H-ZSM-5, show a shift of the v_{CN} of 30-35 cm^{-1} [7, 14, 15]. All the bands associated with $CD_3CN\cdots OD$ complexes and those associated with physically adsorbed species disappeared after outgassing the sample at room temperature and the band due to free OD groups (2760 cm^{-1}) was restored.

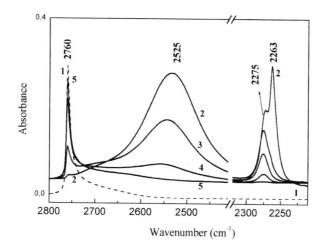

Fig.2: FTIR spectra of CD$_3$CN adsorbed on pure MCM-41. Curve 1: deuterated sample outgassed at 500°C in vacuum; curve 2: in contact with 100 Torr CD$_3$CN; curves 3,4: decreasing doses of CD$_3$CN; curve 5: after evacuation at room temperature.

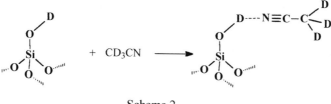

Scheme 2

Upon CD$_3$CN adsorption on Ti-MCM-41 (Fig.3), all the features present in the MCM-41 spectra, related to the formation of CD$_3$CN···OD complexes, are also formed and the shifts of the OD and CN stretching modes are similar to those observed on pure MCM-41. In addition, a new band at ca. 2300 cm^{-1}, particularly evident at high CD$_3$CN pressure, is formed. A band in a similar position is observed when CD$_3$CN is adsorbed on TiO$_2$ anatase due to the interaction of acetonitrile with Ti(IV) surface sites [7,8]. On this basis, the band at 2300 cm^{-1} can be assigned to the CN stretching mode of acetonitrile directly bound to Ti(IV) tetrahedral Lewis acid sites. All these species are completely reversible at room temperature.

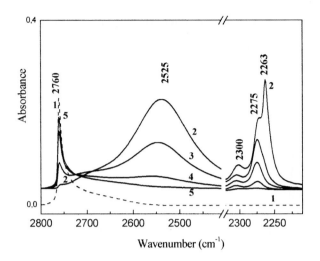

Fig.3: FTIR spectra of CD$_3$CN adsorbed on Ti-MCM-41. Curve 1: deuterated sample outgassed at 500°C in vacuum; curve 2: in contact with 100 Torr of CD$_3$CN; curves 3,4: decreasing doses of CD$_3$CN; curve 5: after evacuation at room temperature.

CD$_3$CN has not evidenced differences in the surface OH acidity on pure MCM-41 and Ti-MCM-41. However, the presence of Ti(IV) Lewis sites and their coordination changes were clearly displayed by CD$_3$CN adsorption.

As the introduction of Ti(IV) Lewis sites in mesoporous silica is expected to modify the nature of some surface hydroxyls and can produce new hydroxyl groups, e.g. Ti-OH groups, with different acidity with respect to Si-OH, we used ammonia, a stronger base than CD$_3$CN, to identify them.

The FTIR spectra of NH$_3$ adsorption on MCM-41 are reported in Fig.4. At high NH$_3$ pressure, (curve 2), bands at 3405, 3330 (very sharp), 3320 cm^{-1}, overlapped to a broader band centred at 3040 cm^{-1}, and at 1635 and 1625 cm^{-1} are formed. Simultaneously, the band at 3745 cm^{-1}, due to the O-H stretching mode of free silanols (curve 1), almost completely disappears (curve 2). The bands at 3330 and 1625 cm^{-1}, that disappear by decreasing the gas pressure (curves 2 to 4), are due to the asymmetric stretching and bending modes, respectively, of NH$_3$ molecules in the gas phase. The broad band at 3040 cm^{-1} is due to the OH stretching mode of silanol groups H-bonded to NH$_3$ molecules (scheme 3, structure A). The bands at 3405 and 3320 cm^{-1} are due, respectively, to the asymmetric and symmetric stretching N-H vibration of NH$_3$ bonded to silanols, whereas the band at 1635 cm^{-1} is assigned to the asymmetric bending mode of NH$_3$ adsorbed on silanols. All the bands related to adsorbed NH$_3$ completely disappear upon outgassing the sample at room temperature, and the band at 3745 cm⁻¹ due to free silanols is restored.

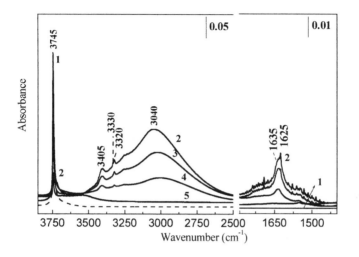

Fig. 4: FTIR spectra of NH$_3$ adsorbed on pure MCM-41. Curve 1: bare sample after calcination and evacuation at 550°C; curve 2: in the presence of 100 Torr NH$_3$; curves 3,4: decreasing NH$_3$ pressure; curve 5: after evacuation at room temperature

All the spectroscopic features of ammonia adsorption on MCM-41 are also present on Ti-MCM-41 (Fig. 5). Bands at 3405, 3320, 3040 and 1635 cm⁻¹, associated with NH$_3$ interacting with silanol groups, are present at high NH$_3$ doses (curves 2,3) and fade away as NH$_3$ is progressively pumped off (curves 4,5). Beside these bands, absorptions at 3390, 3290, 1605 and 1460 cm⁻¹ are present and become better defined when NH$_3$ pressure is decreased (curve 4). The bands at 3390, 3290 and 1605 cm⁻¹ are still present after outgassing the sample at room temperature (curve 5). The bands at 3390 and 3290 cm⁻¹ are assigned to the asymmetric and symmetric N-H stretching modes, respectively, while the band at 1605 cm⁻¹ is due to the asymmetric bending mode of NH$_3$ molecularly adsorbed on Ti(IV) Lewis acid sites (scheme 3, structure B). Ammonia adsorbed on Ti(IV) sites is not completely desorbed at room temperature, indicating that stronger surface complexes are formed in comparison to purely siliceous MCM-41. Moreover, the weak signal at 1460 cm⁻¹, not present in the MCM-41 spectra, can be assigned to the asymmetric bending mode of NH$_4^+$ species. The presence of this signal indicates that a fraction of hydroxyl groups in Ti-MCM-41 is acidic enough to

318

protonate ammonia molecules. Ti-OH groups or Si-OH groups close to Ti(IV) sites (scheme 4) may be responsible of this Brønsted acidity.

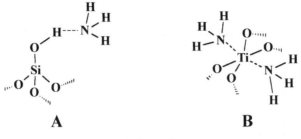

Scheme 3

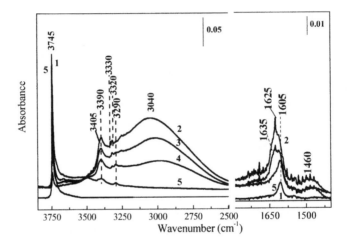

Fig. 5: FTIR spectra of NH₃ adsorbed on Ti-MCM-41. Curve 1: bare sample after calcination and evacuation at 550°C; curve 2: in the presence of 100 Torr NH₃; curves 3,4: decreasing doses of NH₃; curve 5: after evacuation at room temperature

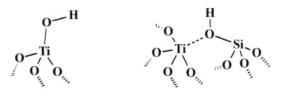

Scheme 4

CONCLUSIONS

CD_3CN and NH_3 were used as molecular probes to monitor the surface acidity of pure siliceous mesoporous MCM-41 and Ti-grafted MCM-41. CD_3CN was very effective to reveal the presence of Ti(IV) Lewis acid sites and their accessibility to probe molecules. By means of this methodology, it might be possible to evaluate the amount of exposed Ti(IV) sites accessible to peroxide reactants during the epoxidation reactions. However, acetonitrile was not useful to distinguish different type of hydroxyl acid sites in MCM-41 and Ti-MCM-41. Instead, ammonia, a stronger base, was capable to detect hydroxyl groups acidic enough to form NH_4^+ species. Because this proton transfer effect occurs only in Ti-grafted sample, it is proposed that these acidic hydroxyl groups are titanols (Ti-OH) or silanols (Si-OH) whose acidity is enhanced by the interaction with Ti(IV) Lewis acid sites.

Acknowledgements

Italian CNR and MIUR (FISR project "Materiali e processi per sistemi nanostrutturati") are acknowledged for financial support.

320

REFERENCES

[1] R.D. Oldroyd, J.M. Thomas, T. Maschmeyer, P.A. MacFaul, D.W. Snelgrove, K.U. Ingold, D.D.M. Wayner, Angew. Chem. Int. Ed. Engl., 35 (1996) 2787

[2] M. Guidotti, N. Ravasio, R. Psaro, G. Ferraris, G. Moretti, J. Catal., 214 (2003) 242

[3] M. Guidotti, N. Ravasio, R. Psaro, E. Gianotti, L. Marchese, S. Coluccia, Green Chem., 5 (2003), 421

[4] M. Guidotti, R. Psaro, N. Ravasio, Chem. Commun., (2000) 1789.

[5] T. Maschmeyer, F. Rey, G. Sankar, J.M. Thomas, Nature, 378 (1995) 159.

[6] L. Marchese, E. Gianotti, V. Dellarocca, T. Maschmeyer, F. Rey, S. Coluccia, J.M. Thomas, Phys. Chem. Chem. Phys., 1 (1999) 585

[7] F. Bonino, A. Damin, S. Bordiga, C. Lamberti, A. Zecchina, Langmuir, 19 (2003) 2155.

[8] P. Davit, G. Martra, S. Coluccia, V. Augugliaro, E. Garcia-Lopez, V. Loddo, G. Marcì, L. Palmisano, M. Schiavello, J. Molec. Catal. A: Chem., 204-205 (2003) 693

[9] E. Gianotti, V. Dellarocca, L. Marchese, G. Martra, S. Coluccia, T. Maschmeyer, Phys. Chem. Chem. Phys., 4 (2002) 6109

[10] C.T. Kresge, M.E. Leonowicz, W.J. Roth, J.C. Vartuli, J.S. Beck, Nature, 359 (1992) 710

[11] C.K. Jørgensen, Prog. Inorg. Chem., 12 (1970) 101

[12] J.A. Duffy, J. Chem. Soc., Dalton Trans., (1983) 1475

[13] L. Marchese, T. Maschmeyer, E. Gianotti, S. Coluccia, J. M. Thomas, J. Phys. Chem. B, 101 (1997) 4232

[14] A.G. Palmenschikov, R.A. van Santen, J. Jänchen, E. Meijer, J. Phys. Chem., 97 (1993) 11071

[15] C. Otero Areán, E. Escalona Platero, M. Peñarroya Mentruit, M. Rodriguez Delgado, F. Llabrés i Xamena, A. García-Raso, C. Morterra, Micropor. Mesopor. Mater., 34 (2000) 55

Studies in Surface Science and Catalysis 155
A. Gamba, C. Colella and S. Coluccia (Editors)
© 2005 Elsevier B.V. All rights reserved

Amorphous aluminosilicate precursors of mordenite: a case history in the formation of microporosity

F. Hamidi[a], M.F. Driole[b], B. Chiche[b], N. Tanchoux[b], A. Bengueddach[c], F. Quignard[b], F. Fajula[b] and F. Di Renzo[b]

[a]Département de Chimie, Faculté des Sciences, USTO, B.P. 1505 Elmenaouar, Oran, Algeria

[b]Laboratoire de Matériaux Catalytiques et Catalyse en Chimie Organique, ENSCM, 8 rue de l'Ecole Normale, 34296 Montpellier, France

[c]Laboratoire de Chimie des Matériaux, Université Oran Es-Senia, Oran, Algeria

1. INTRODUCTION

Zeolites are commonly formed through a process of self-organisation of silicate structural units in the presence of a template. The template, usually a hydrated alkali cation or an organic cation, fills a fraction of the structure and accounts for the micropore volume of the activated molecular sieve. At which point the formation of micropores is related to the ordering of the silicate framework?

In several instances, micropore-related properties of zeolites have been reported to appear at early stages of the hydrothermal synthesis, when no crystalline network can be observed [1-4]. In the case of tetrapropylammonium (TPA) template, gels formed in the first stages of the synthesis showed catalytic activity in the hydroconversion of n-decane similar to well-crystallized ZSM-5 [1]. The term "X-ray amorphous zeolites" has been used for these materials, which present an intense IR skeleton vibration at 550 cm^{-1} [1] and can probably be considered as nanocrystalline ZSM-5. However, other studies on calcined TPA-templated gels have indicated the presence of an important porosity in the size range 15-25 Å, at the borderline between micropores and mesopores [5, 6]. Such a porosity does not present the pore size expected for crystalline ZSM-5. This suggests that, at the earliest stages of the synthesis, an aperiodical local organization was formed by self-assembly of silicate units and organic cations. The disorder of these gels has been confirmed by ^{29}Si NMR [7].

In the case of tetraethylammonium (TEA) template, the formation of cation-containing amorphous gels in the earliest stages of the synthesis has been early reported [8]. These gels have been called TEA-permutite, from the name attributed to the amorphous aluminosilicate ion exchangers developed in the thirties, from which R.W. Barrer took inspiration for the earliest syntheses of zeolites [9]. The disorder of their structure is confirmed by the absence of IR skeleton bands, by the absence of any well-defined X-ray diffraction pattern, and by the ^{27}Al MAS-NMR band, broader than the corresponding band in the final crystallized zeolite, be it zeolite beta or mordenite [9]. TEA-permutites present, once calcined, a significant micropore volume, slightly lesser than the micropore volume of the zeolite obtained on a longer synthesis time. A broad band at 12 Å in the X-ray diffraction peak could be attributed to some local ordering of silicate units around TEA cations.

Instances of organisation of an amorphous solid around an organic template have been observed also beyond the field of microporous zeolites. In the case of micelle-templated mesoporous materials, the formation of a disordered porous network before the long-range ordering of the material has been reported in the system cetyltrimethylammonium-silica [10].

Less data are available in the case of zeolite precursors templated by hydrated inorganic cations. During the synthesis of zeolite Y, an X-ray amorphous gel with a composition similar to the final zeolite is formed in mild hydrothermal conditions and has been reported to present a significant activity in cumene cracking [11]. However, a [129]Xe NMR study indicated that this kind of gels, once calcined, present large mesopores and no micropores [5]. The hydrated sodium cations present in the gel are common templates for zeolite microporosity. Hence, the absence of microporosity of the gel is somehow puzzling.

This investigation was prompted by the hypothesis that micropores are indeed present in the structure of the aluminosilicate gels and cannot be observed due to modifications of the network during the dehydration process. The precursors of the synthesis of mordenite were studied, this zeolite being chosen as a representative example of zeolite with intermediate aluminium content templated by hydrated sodium cations. The method of preparation of the gel significantly affects the crystal size and morphology of mordenite [12].

2. EXPERIMENTAL

The reagents used for the syntheses are precipitated silica Zeosil 175MP (Rhône-Poulenc, BET surface area 175 $m^2 g^{-1}$, grain size 20-200 μm, H_2O 21%, Na 0.7%, Al 0.16 %), $NaAlO_2$ by Carlo Erba, NaOH by Prolabo and deionized water. The composition of the synthesis batches was 0.20 Na_2O / 0.031 Al_2O_3 / SiO_2 / 13 H_2O. Mordenite synthesis batches have been prepared in a thermosetted room at 298 K according to the reference [12]. The silica source was stirred 30 minutes in water (H_2O / SiO_2 11). The alkaline aluminate solution prepared with the remaining water was added afterwards. The gel was stirred 30 minutes before sealing the stainless steel autoclave and heating at 433 K without stirring. A primary gel sample was withdrawn before heating, a hydrothermal gel sample was withdrawn after 4 hours at 433 K and a crystalline sample was withdrawn after 40 hours at 433 K. After the allotted time of synthesis, the solid phase was separated by filtration and washed with an amount of H_2O twice the volume of the synthesis batch.

The washed samples have been dried in two alternative ways: drying in a ventilated oven at 80 °C or dehydration under supercritical CO_2 conditions. In this second method, the washed gels were dehydrated by immersion in a series of successive ethanol-water baths of increasing ethanol concentration (10, 30, 50, 70, 90, 100%) before drying under supercritical CO_2 conditions (73 bars, 31,5°C) in a Polaron 3100 apparatus. Dried samples have been kept in dry N_2 atmosphere and their texture has been characterised as early as possible.

N_2 adsorption isotherms at 77 K have been measured in a Micromeritics Asap 2010 apparatus after outgassing at 323 K until a stable $3 \cdot 10^{-3}$ Torr static vacuum was attained. External surface and micropore volume were measured by the α-S method by using a standard isotherm measured on Aerosil 200 fumed silica. Crystallised samples were characterised by powder X-ray diffraction (CGR Theta 60 diffractometer, Cu Kα radiation) and scanning electron microscopy (Cambridge Stereoscan 260). Sodium, aluminium and silicon content have been measured by EDX on the EDX coupled to the microscope. The water content has been determined by thermal gravimetry on a Netzsch thermal analyzer. ^{29}Si MAS-NMR spectras were measured on an ASX 400 Bruker apparatus at a 9.4 T field, π/6 4μs pulses with 60 s repeating time, 6 KHz spinning.

3. RESULTS

The composition of the conventionally-dried gels and final mordenite is reported in Table 1. The composition of the synthesis gel is also reported for sake of comparison. Both the primary and the hydrothermal gels present a Si/Al ratio intermediate between the initial batch and the final zeolite. The Na/Al ratio of the gels is about 3, indicating that nearly two sodium cations out of three compensate the charge of deprotonated silanols. The H_2O/Na ratio of the gels is about 3, lower than the hydration ratio 4.5 of the final mordenite.

Table 1
Composition (molar ratios)

	Na	Al	Si	H_2O
synthesis batch	0.400	0.062	1	13.000
primary gel	0.238	0.080	1	0.648
hydrothermal gel	0.200	0.070	1	0.701
mordenite	0.116	0.106	1	0.535

The thermogravimetry plots of the final mordenite and the conventionally-dried gels are reported in Fig. 1. The onset of the loss of water is delayed in the hydrothermal gel, suggesting a lower accessibility of the hydrated cations. The gels present a continuous loss of weight beyond 623 K, probably to be attributed to the dehydration of silanols. The loss of weight is more important for the hydrothermal gel. This suggests a higher silanol content which, in the case of silica, would correspond to a higher surface area of the material.

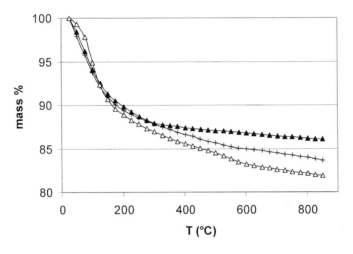

Fig. 1. Thermal gravimetry plots for mordenite (full triangles) and conventionally-dried primary gel (crosses) and hydrothermal gel (4 h at 433 K, open triangles).

The ^{29}Si MAS-NMR spectra of the conventionally dried gels and the final mordenite are reported in Fig. 2. The width at half height of the lines is nearly 15 ppm in the spectra of the gels, indicating a broad dispersion of the T-O-T angles, typical of disordered solids. In the case of the spectra of mordenite, the linewidth of about 6 ppm is justified by the crowding in each line of the contributions of four different crystallographic sites. X-ray diffraction (not shown) showed no diffraction line for the two gels and the typical pattern of mordenite for the final product.

The ^{29}Si MAS-NMR spectra of the primary gels, *viz.* the reagent mixture before hydrothermal heating, present three broad bands centered at -87, -100, and -111 ppm, to be attributed at silicate tetrahedra with, respectively, two, one or no OH groups (Fig. 2a). The $Si(OSi)_4$ band is much more intense than the $Si(OSi)_3OH$ band. The hydrothermal gel, formed after 4 h hydrothermal treatment at 433 K, presents a shoulder near -90 ppm and two largely superposed bands at -102 and -107 ppm (Fig. 2b). The $Si(OSi)_4$ and $Si(OSi)_3OH$ bands present the same intensity. The differerent distributions of silanols of the primary gel and the hydrothermal gel appears in agreement with the dehydroxylation behaviours observed by TG.

324

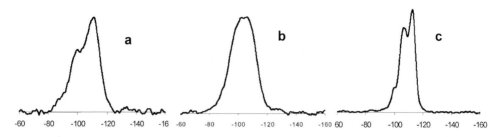

Fig. 2. ^{29}Si MAS-NMR spectra of primary aluminosilicate gel (a), hydrothermal gel (b, 4 h at 433 K), and crystallised mordenite (c).

The N$_2$ adsorption-desorption isotherms at 77 K for the source of silica, the final zeolite and the intermediate gels dried at 353 K are reported in Fig. 3. The textural data which can be calculated from the isotherms are reported in Table 2. The isotherm on Zeosil silica is at the threshold between type II and type IV, with large mesopores at the borderline of macroporosity. The texture of the solid can be interpreted as a loose assembly of silica spheres with 12 nm diameter. Comparison plot analysis by the α-S method, as well as an abnormally high value of the C$_{BET}$ parameter (not reported), indicate the presence of a small micropore volume. The isotherm of the primary gel is analogous to the isotherm of the source of silica. The preservation of the micropore volume suggests that the original grains of silica have not been substantially modified, notwithstanding some sintering among the grains, witnessed by the decrease in surface area.

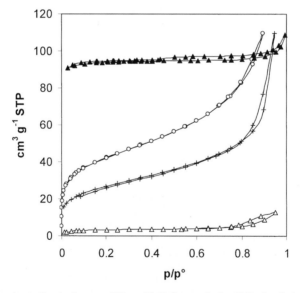

Fig. 3. Adsorption-desorption isotherms of N$_2$ at 77 K for mordenite (full triangles), Zeosil silica (empty circles), and conventionally-dried primary gel (crosses) and hydrothermal gel (4 h at 433 K, open triangles).

Table 2
Textural properties

	drying method	micropore volume cm³ g⁻¹	external area m² g⁻¹	particle diameter nm
Zeosil silica		0.007	132	12
primary gel	oven 353 K	0.006	80	19
	SC CO₂	0.008	118	13
hydrothermal gel	oven 353 K	0.002	6	270
	SC CO₂	0.109	1	1560
mordenite	oven 353 K	0.143	6	275

An additional hint on the preservation of the grains of the source of silica in the primary gel is provided by its handling behaviour. The dried primary gel retained the free-flowing behaviour of the Zeosil silica. On the contrary, the hydrothermal gel presented a hard glassy texture which was scantily altered also by calcination at 823 K. The isotherm of the dried hydrothermal gel is type IV, with no microporosity and extremely low surface area and mesopore volume. The formation of such an impervious solid in the hydrothermal conditions, which will lead some hours later to the formation of a microporous zeolite, is somehow puzzling, and prompted our investigation on alternative drying methods of the synthesis intermediates.

The N_2 adsorption-desorption isotherms of the gels dried by supercritical CO_2 are reported in Fig. 4. The isotherms of the source of silica and the final zeolite are also reported. Textural data are reported in Table 2. The isotherm on the supercritically-dried (SC-dried) primary gel is extremely close to the source of silica. The microporosity is still retained, but the decrease in surface area is negligible, suggesting that the sintering observed in the oven-dried sample is related to the presence of water during drying at 353 K or during the outgassing at 523 K in the adsorption apparatus.

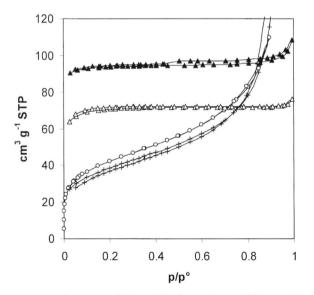

Fig. 4. Adsorption-desorption isotherms of N_2 at 77 K for mordenite (full triangles), Zeosil silica (empty circles), and supercritical CO_2-dried primary gel (crosses) and hydrothermal gel (4 h at 433 K, open triangles).

The SC-dried hydrothermal gel presents a well-defined type I isotherm, with a micropore volume 0.11 cm^3 g^{-1}, corresponding to more than 75 % of the micropore volume of the final zeolite. The striking difference from the isotherm of the equivalent oven-dried sample indicates that the presence of water in the drying process leads to complete collapse of the porosity of the hydrothermal gel.

The SC-dried gel presents a lower external surface and mesoporosity than the equivalent sample dried by water evaporation. This suggests that the surface evaluation by comparison plot is affected by the presence of a broad distribution of small mesopores, probably corresponding to microcracks openened during oven drying. The oven-dried hydrothermal gel and the final zeolite present the same external area. A value of surface area of 6 m^2 g^{-1} would corresponds to silica spheres of 275 nm diameter, a size which exceeds by an order of magnitude the external area expected for the 2 μm-thick crystals of the mordenite sample [12]. It seems likely that mechanisms of formation of mesoporosity during drying have been effective for both the hydrothermal gel and the crystalline zeolite.

4. DISCUSSION

The synthesis of mordenite in the experimental conditions of this communication seems to pass through three main steps. In the first one, the reagents interact at room temperature to form a primary gel. In the second one, a complete reorganisation of the gel takes place once the synthesis batch is heated in hydrothermal conditions. In the third step, a slower reorganisation in hydrothermal conditions leads to the nucleation and growth of crystalline mordenite.

The primary gel presents a texture extremely similar to the source of silica. It seems likely that virtually no dissolution of the source of silica takes place at room temperature, probably due to the preserving effect of a coating of sodium aluminosilicate. The presence of large silica particles with radial compositional heterogeneity can significantly affect the results of some characterisation techniques. The aluminium content evaluated by EDX probably represents a surface analysis and overestimates the global composition. The amount of silanols evaluated by ^{29}Si MAS-NMR is probably also overestimated, due to the very slow relaxation time of inaccessible ^{29}Si inside the grains of silica.

The formation of the hydrothermal gel corresponds to a homogenisation of the solid phase and probably occurs through a dissolution-precipitation mechanism in which the memory of the texture of the primary gel is completely lost. It can be assumed that the texture of the SC-dried aerogel represents the original texture of the hydrogel [13]. Under this assumption, the essential properties of the hydrothermal gel are
a) a zeolite-like microporosity templated by hydrated sodium cations;
b) an interrupted silicate framework, in which more than half tetrahedra bear silanol groups.

It seems clear that the organisation of aluminosilicate units around hydrated cations does not imply the formation of a periodical network. On the contrary, the haphazard aggregation of silicate units which do not form a fully connected network occurs more rapidly than the formation of an ordered zeolitic network. This is probably due to the higher probability of bonding in one of many disordered configurations instead of in a limited number of periodical arrays. It is questionable if the hydrothermal gel has to be considered as a true precursor of the zeolite structure. Surely it is a precursor in the sequence of events, but probably it is more a competitor than a logical precursor of the zeolite. The induction time in the nucleation of the zeolite depends on the presence of this early formed metastable phase.

The complete disappearance of the microporosity of the hydrothermal gel when dried by evaporation of water is a remarkable phenomenon, which takes place at the borderline of the common mechanisms of shrinkage of compliant gels [13]. In normal gel shrinkage, the so-called constant rate drying phase corresponds to a volume loss of the gel exactly corresponding to the amount of evaporated water. The tension of the meniscus is supported by the solid phase, which goes into compression. The network is contracted into the receding liquid and stiffens as the pore size decreases. When the network becomes solid enough to resist the tension, the meniscus enters the pores and the pressure increases no more. Only beyond this limit, defined as the critical drying pressure, the pore size is preserved [13]. The critical drying pressure is inversely proportional to the pore size.

The constant-rate drying mechanism can justify the fair preservation of the porosity of the primary gel. In large pores at the mesopore-macropore borderline, the capillary pressure is relatively low, while the presence of rigid silica spheres ensures a large modulus to the gel, which resists compression after a limited shrinkage. In the case of the hydrothermal gel, the micropores are two orders of magnitude smaller than the meso-macropores of the primary gel, and the presence of a large fraction of broken bonds weakens the thin silica walls. Very small pores and a compliant solid are the ingredients for the accumulation of an enormous capillary pressure and the total shrinkage of the gel. A crystalline zeolite would be subjected to the same capillary forces during drying. However, the completely connected aluminosilicate network of mordenite seems stiff enough to resist the bilateral compression from capillary forces.

This plain mechanism is probably only a qualitative approximation of the phenomenon. The main approximation in the application of the constant rate drying mechanism to zeolite-like pores is the assumption that water in the micropores presents the interface properties of liquid water. The formation of a liquid-gas interface in micropores has been early questioned by calculations suggesting that, in confined systems, the meniscus forces could exceed the tensile strength of the liquid [14], giving rise to an extraction in so-called supercritical capillary conditions [15, 16]. The capillary critical point was often identified with the threshold closure point of the hysteresis loop observed in adsorption-desorption isotherms [17]. In the case of water at 298 K, the closure point of the hysteresis loop was observed at p/p° 0.31 [18], suggesting that no meniscus can be formed in small mesopores nor in micropores. However, further studies indicated that the closure point of the hysteresis loop cannot be identified with the capillary critical point, which seems to lie inside the field of micropores [19-21].

The more puzzling questions about the trivial application of a capillary tension mechanism to the drying of microporous aluminosilicates are related to the strong interaction between water molecules and sodium cations. The degree of hydration of the sodium cations seems slightly lower in the hydrothermal gel than in crystalline mordenite, possibly due to the important fraction of sodium cations compensated by deprotonated silanols. However, it is clear that virtually all molecules of water in the hydrothermal gel, as well as in mordenite, are in direct interaction with sodium cations which are themselves in strong interaction with the aluminosilicate framework. It is clear that the liquid inside the micropores is an extremely concentrated brine. Very likely, significant osmotic phenomena take place during the evaporation of water and affect the interface behaviour.

The significance of the meniscus tension in the disappearing of the micropores of the hydrothermal gel is indirectly confirmed by the literature data about the microporous intermediates of zeolite synthesis in the presence of organic cations. A significant microporosity have been observed in hydrothermal gels templated by tetraethylammonium or tetrapropylammonium cations, albeit they have been dried and calcined in a conventional way [5, 6, 8]. The virtual absence of water in the tetraethylammonium-templated aluminosilicates [22] can account for the absence of capillary forces during the degradation and extraction of the organic moieties of these solids. It seems likely that microporous hydrothermal gels are a common occurrence in syntheses media of aluminosilicate zeolites templated by hydrated alkali cations, and are not easily observed due to their total shrinkage during evaporative drying.

5. CONCLUSIONS

The amorphous aluminosilicate formed at the beginning of the hydrothermal synthesis of mordenite presents a significant microporosity, corresponding to water loosely coordinated to sodium cations. The aluminosilicate network presents no local order and is poorly connected, with at least half tetrahedra featuring dangling bonds. The stability of this amorphous precursor is at the basis of the long induction time observed in the synthesis of mordenite: the rapid formation of the precursor depletes the species needed for the crystallisation of mordenite, which is later formed at the low supersaturation corresponding to the equilibrium concentration with the precursor.

The microporosity of the hydrothermal gel cannot be observed after evaporative drying but is only observed after supercritical CO_2 drying. This effect indicates that capillary forces are active in aluminosilicate micropores and suggests that the research of gas-solid interfaces in zeolite-like micropores is an acceptable field of investigation. Zeolites are probably subjected to the same forces during evaporative drying and the poor connectivity of the hydrothermal gel accounts for the collapse of its microporosity when water is evaporated from the solid.

ACKNOWLEDGMENTS
The authors appreciate the support of Didier Cot and Nathalie Masquelez for electron microscopy and Philippe Gonzalès for thermal analysis.

REFERENCES
[1] P.A. Jacobs, E.G. Derouane and J. Weitkamp, J. Chem. Soc. Chem. Commun., 1981, 591.
[2] G. Coudurier, C. Naccache and J.C. Védrine, J. Chem. Soc. Chem. Commun., 1982, 1413.
[3] Z. Gabelica, E.G. Derouane and N. Blom, ACS Symp. Ser., 248 (1984) 219.
[4] C.S. Cundy and P.A. Cox, Chem. Rev., 103 (2003) 663.
[5] T. Ito, J. Fraissard, J. B. Nagy, N. Dewaele, Z. Gabelica, A. Nastro and E.G. Derouane, Stud. Surface Science Catal., 49 (1989) 579.
[6] G. Bellussi, C. Perego, A. Carati, S. Peratello, E. Previde Massara and G. Perego, Stud. Surface Science Catal., 84 (1994) 85.
[7] C.D. Chang and A.T. Bell, Catal. Lett., 8 (1991) 305.
[8] M.A. Camblor and J. Perez-Pariente, Zeolites, 11 (1991) 202.
[9] M.A. Nicolle, F. Di Renzo, F. Fajula, P. Espiau and T. Des Courières, in Proc. 9th International Zeolite Conference, R. Von Ballmoos, J.B. Higgins and M.M.J. Treacy., Eds., Butterworth-Heinemann, Boston 1993, 313.
[10] A. Galarneau, F. Di Renzo, F. Fajula, L. Mollo, B. Fubini and M.F. Ottaviani, J. Colloid Interface Sci. 201 (1998) 105.
[11] B. Fahlke, P. Starke, V. Seefeld, W. Wieker and K.P. Wendlandt, Zeolites 7 (1987) 209.
[12] F. Hamidi, A. Bengueddach, F. Di Renzo and F. Fajula, Catal. Lett., 87 (2003) 149.
[13] C.J. Brinker and G.W. Scherer, Sol-Gel Science, Academic Press, Boston 1990.
[14] C.G.V. Burgess and D.H. Everett, J. Colloid Interface Sci., 33 (1970) 611.
[15] R. Evans, U. Marini Bettolo Marconi and P. Tarazona, J. Chem. Phys., 84 (1986) 2376.
[16] M. Thommes, R. Köhn and M. Fröba, Appl. Surf. Sci., 196 (2002) 239.
[17] K. Morishige, H. Fujii, M. Uga and D. Kinukawa, Langmuir, 13 (1997) 3494.
[18] K.S.W. Sing and J.D. Madeley, J. Appl. Chem., 4 (1954) 365.
[19] P.I. Ravikovitch, S.C. O' Domhnaill, A.V. Neimark, F. Schüth and K.K. Unger, Langmuir, 11 (1995) 4765.
[20] S. Gross, G.H. Findenegg, Ber. Bunsenges. Phys. Chem., 101 (1997) 1726.
[21] P. Trens, N. Tanchoux, D. Maldonado, A. Galarneau, F. Di Renzo, F. Fajula, New J. Chem., 28 (2004) 874.
[22] F. Vaudry, F. Di Renzo, P. Espiau, F. Fajula and P. Schulz, Zeolites, 19 (1997) 253.

Studies in Surface Science and Catalysis 155
A. Gamba, C. Colella and S. Coluccia (Editors)
© 2005 Elsevier B.V. All rights reserved

329

Iron species in FeO$_x$/ZrO$_2$ and FeO$_x$/sulphated-ZrO$_2$ catalysts

V. Indovina[a], M.C. Campa[a], F. Pepe[b], D. Pietrogiacomi[a] and S. Tuti[b]

[a]Sezione "Materiali Inorganici e Catalisi Eterogenea" dell'Istituto IMIP (CNR), c/o Dipartimento di Chimica, Università di Roma "La Sapienza", Piazzale Aldo Moro 5, 00185 Roma, Italy. Fax: +39-06-490324. E-mail: valerio.indovina@uniroma1.it

[b] Centro C.I.S.Di.C. c/o Dipartimento di Ingegneria Meccanica e Industriale, Università degli Studi "Roma Tre", Via della Vasca Navale 79, 00146 Roma, Italy.

Iron containing catalysts were prepared by (i) impregnation of ZrO$_2$ with aqueous solutions of Fe(NO$_3$)$_3$, Fe/Z$_i$, or aqueous solutions of NH$_4$Fe(SO$_4$)$_2$, FeS/Z; (ii) chemical vapour deposition of FeCl$_3$ on ZrO$_2$, Fe/Z$_{CVD}$; and (iii) impregnation of sulphated-ZrO$_2$ with toluene solutions of Fe(acetylacetonate)$_3$, Fe/SZ. After drying at 383 K and calcining at 823 K, samples were characterized by temperature programmed reduction with H$_2$ and Fourier transformed IR spectroscopy.

The FTIR characterization by CO adsorption evidenced the formation of various carbonyl-like species onto Fe^{n+} sites ($0 \leq n \leq 3$). The relative amount of the various iron carbonyl-like species depended not on the iron content, but on the catalyst preparation method.

The H$_2$-TPR experiments showed that iron species were much less reducible in FeO$_x$/sulphated-ZrO$_2$ than in FeO$_x$/ZrO$_2$. Their lower reducibility explains why FeO$_x$/sulphated-ZrO$_2$ samples achieve higher selectivity for the selective catalytic reduction of NO with hydrocarbons in the presence of O$_2$.

1. INTRODUCTION

The activity of MeO$_x$/ZrO$_2$ catalysts (MeO$_x$ = CuO$_x$, CoO$_x$, or FeO$_x$) strongly depends on MeO$_x$ loading, preparation method, and calcination temperature, because all these factors affect Me ion dispersion [1]. We previously found that low loading MeO$_x$/ZrO$_2$ samples with Me content up to about 2.5 Me atoms nm^{-2} contained a large fraction of isolated Me ions together with a small fraction of metal oxides (small particles, not detected by XRD). On more concentrated samples (> 2.5 Me atoms nm^{-2}) segregated metal oxides prevailed. Isolated metal ions (Cu^{II} [2,3], Co^{II} [4], or Fe^{III} [5]) in MeO$_x$/ZrO$_2$ were active and selective for NO reduction with propene in the presence of O$_2$. In particular, the turnover frequency number (NO molecules converted per second per site) on low loading CuO$_x$/ZrO$_2$ approached that on Cu-MFI [3], and the turnover frequency number on low loading CoO$_x$/ZrO$_2$ approached that on Co-MFI [4]. Metal ions on the surface of oxide particles were active for propene combustion, thus rendering the concentrated MeO$_x$/ZrO$_2$ non selective [3-5].

In previous papers, we showed that the presence of sulphates prevented metal oxide formation in concentrated samples. In particular, the presence of sulphates prevented CuO formation in high loading CuO$_x$/sulphated-ZrO$_2$ and CuSO$_4$/ZrO$_2$ [6,7]. Analogously, the presence of sulphates prevented Co_3O_4 formation in high loading CoO_x/sulphated-ZrO_2 and $CoSO_4/ZrO_2$ samples [8,9]. With propene as a reducing agent, sulphated catalysts were far more selective for NO abatement than the corresponding unsulphated CuO_x/ZrO_2 [6] and CoO_x/ZrO_2 [8]. The presence of sulphates (i) made Cu^{II} and Co^{II} less reducible in sulphated samples than in CuO_x/ZrO_2 and CoO_x/ZrO_2 and (ii) prevented the formation of CuO and Co_3O_4. Because both effects reduced the catalytic activity for the C_3H_6+O_2 reaction, sulphated samples were more selective than the corresponding unsulphated samples [6,8].

A possible analogous effect of sulphates in making Fe^{III} less reducible, and in preventing the formation of Fe_2O_3 in high loading FeO_x/ZrO_2 samples prompted us to investigate FeO_x/sulphated-ZrO_2 and $NH_4Fe(SO_4)_2/ZrO_2$ samples, and to compare their activity with that of FeO_x/ZrO_2 [5]. Because the reducibility and dispersion of the transition metal ion might depend not only on the presence of sulphates but also on the preparation method, in the present paper we compare samples prepared by various methods. Specifically, we compared and characterized (FTIR spectroscopy and TPR) samples prepared by (i) impregnation of ZrO_2 with aqueous solutions of $Fe(NO_3)_3$, Fe/Z_i, or aqueous solutions of $NH_4Fe(SO_4)_2$, FeS/Z; (ii) chemical vapour deposition of $FeCl_3$ on ZrO_2, Fe/Z_{CVD}; and (iii) impregnation of sulphated-ZrO_2 with toluene solutions of Fe(acetylacetonate)$_3$, Fe/SZ.

2. EXPERIMENTAL

2.1. Sample preparation

The zirconia support was prepared by hydrolysis of zirconium oxychloride with ammonia, as already described [2]. The precipitate hydrous zirconium was washed with water until the Cl^- test with $AgNO_3$ gave no visible opalescence. Before its use as a support, the material was dried at 383 K for 24 h and calcined at 823 K for 5 h. After calcination, the BET surface area of the ZrO_2 support (Z), measured by N_2 adsorption at 77 K, was 53 m^2 g^{-1}. XRD spectra showed that Z was in the monoclinic phase.

Sulphated zirconia was prepared by impregnating Z with an aqueous $(NH_4)_2SO_4$ solution and calcining at 823 K (SZ(b), with b = 2.4 SO_4 molecules nm^{-2}).

Iron containing catalysts were prepared by four different procedures. In the *first procedure*, FeO_x/ZrO_2 were obtained by impregnation of Z with aqueous solutions of $Fe(NO_3)_3$, and designated as Fe/Z_i(a), where a specifies the analytical iron content, 0.5 to 5.7 Fe atoms nm^{-2}. In the *second procedure*, FeO_x/ZrO_2 were obtained by chemical vapour deposition (CVD). A weighted amount of Z was maintained in the middle of a tubular quartz reactor by a porous frit. A weighted amount of anhydrous $FeCl_3$ was loaded in a tubular quartz reactor, which could be magnetically driven inside the tubular quartz reactor, without air contact. First, Z was heated at 773 or 553 K in a flow of He (100 cm^3 STP min^{-1}) for 2 h, maintaining the $FeCl_3$ powder in the RT region of the quartz reactor. Subsequently, Z was cooled to 553 or 573 K, and $FeCl_3$ was driven near Z and heated at the same temperature, 553 or 573 K. The sample was thereafter cooled to RT, and washed with water, until the Cl^- test with $AgNO_3$ gave no visible opalescence. Samples were designated as Fe/Z_{CVD}(a), where a specifies the analytical iron content, 0.2 to 2.5 Fe atoms nm^{-2}. In the *third procedure*, $NH_4Fe(SO_4)_2/ZrO_2$ were obtained by impregnation of Z with aqueous solutions of $NH_4Fe(SO_4)_2$, and designated as FeS/Z(a,b), where a specifies the analytical iron content and b the analytical sulphate content. In the *fourth procedure*, FeO_x/sulphated-ZrO_2 were obtained by impregnation of SZ with toluene solutions of Fe(acetylacetonate)$_3$, and designated as Fe/SZ(a,b), where a specifies the analytical iron content and b the analytical sulphate content.

After impregnation, Fe/Z$_i$, Fe/Z$_{CVD}$, FeS/Z, and Fe/SZ samples were dried at 383 K and calcined at 823 K. All analytical contents are expressed in molecules nm^{-2}.

Iron content was determined by atomic absorption (Varian SpectrAA-30). Sulphate content was determined by extraction of sulphates with NaOH 1 M, and ICP (Varian Vista-Mpx) of the resulting solution. Specific surface areas (SSA/m^2 g^{-1}) were measured by N$_2$ adsorption at 77 K.

Starting materials for catalyst preparation, catalyst name, analytical iron content, analytical sulphate content, and surface areas are reported in Table 1.

Table 1.
Catalysts and their main features

Starting materials	He at T/K [a]	FeCl$_3$ at T/K [b]	SSA/m^{2}g^{-1}	Catalysts [c]
ZrO$_2$			53	Z
ZrO$_2$ + (NH$_4$)$_2$SO$_4$			52	SZ(2.4)
ZrO$_2$ + Fe(NO$_3$)$_3$			53	Fe/Z$_i$(0.5)
			53	Fe/Z$_i$(1.9)
			53	Fe/Z$_i$(2.2)
			53	Fe/Z$_i$(2.8)
			53	Fe/Z$_i$(3.8)
			53	Fe/Z$_i$(5.7)
ZrO$_2$ + FeCl$_3$	773	553	50	Fe/Z$_{CVD}$(0.2) [d]
	773	573	51	Fe/Z$_{CVD}$(1.0)
	553	553	50	Fe/Z$_{CVD}$(1.4)
	773	553	48	Fe/Z$_{CVD}$(2.1)
	773	553	48	Fe/Z$_{CVD}$(2.5)
ZrO$_2$ + NH$_4$Fe(SO$_4$)$_2$			59	FeS/Z(0.3,0.6)
			59	FeS/Z(4.3,3.2)
SZ + Fe(acac)$_3$			52	Fe/SZ(2.3,2.3)

[a] Temperature at which Z was heated, before exposure to FeCl$_3$ (CVD).
[b] Temperature at which Z was exposed to FeCl$_3$ (CVD).
[c] For SZ, the figure specifies the analytical SO$_4$ content. For Fe/Z$_i$ and Fe/Z$_{CVD}$, the figure specifies the analytical Fe content. For FeS/Z and Fe/SZ samples, the two figures specify the analytical Fe and SO$_4$ content, *in that order*. All analytical contents are expressed as molecules nm^{-2}.
[d] For this sample the FeCl$_3$ amount was much lower than that used to prepare the other samples.

2.2. FT-IR measurements

FTIR spectra were recorded at RT on a Perkin Elmer 2000 spectrometer equipped with an MCT detector, collecting 100 scans at a resolution of 4 cm^{-1}. Powdered materials were pelleted (pressure 1.5×10^4 kg cm^{-2}) in self-supporting disks of ca. 50 mg cm^{-2} and 0.1-0.2 mm thickness. All samples were placed into an IR quartz cell allowing thermal treatments *in vacuo* or in a controlled atmosphere. Before experiments, samples were heated in O$_2$ from RT to 793 K, kept at this temperature for 1 h, and evacuated thereafter at the same temperature for 1 h. Absorbance spectra were obtained by subtracting the relevant background.

2.3. TPR measurements

H_2-TPR experiments were performed in a flow apparatus TPDRO 1100 ThermoFinnigan. The H_2 consumption was detected by a TCD detector and reduction products were trapped in a soda lime trap. A fresh portion of sample (0.1 g), placed into a quartz reactor, was pretreated in a flow of 5% O_2/He mixture (20 cm^3 STP min^{-1}) at 793 K for 30 min, and then cooled in He to RT. H_2-TPR measurements were performed in a flow of 5% H_2/Ar mixture (10 cm^3 STP min^{-1}) starting from RT to 793 or 1223 K, with a ramp of 10 K min^{-1}. Samples were maintained at the final temperature for 30 min. To calibrate the H_2 consumption, a known amount of pure Ar was introduced into the flow of 5% H_2/Ar mixture (10 cm^3 STP min^{-1}).

3. RESULTS AND DISCUSSION

3.1. Iron uptake in Fe/Z_{CVD}.

In samples prepared by chemical vapour deposition of $FeCl_3$, the iron uptake depended on the temperature at which ZrO_2 was dehydrated in the He flow, and on the temperature at which $FeCl_3$ was adsorbed. We obtained the maximum iron uptake (2.5 Fe atoms nm^{-2}) when ZrO_2 was dehydrated at 773 K, and $FeCl_3$ was adsorbed at 553 K. We obtained a lower iron uptake (1.4 Fe atoms nm^{-2}) when ZrO_2 was dehydrated at 553 K, and $FeCl_3$ was adsorbed at 553 K. We also obtained a lower iron uptake (1.0 Fe atoms nm^{-2}) when ZrO_2 was dehydrated at 773 K, and $FeCl_3$ was adsorbed at 573 K (Table 1). The iron uptake limit (2.5 Fe atoms nm^{-2}) approached that previously obtained with other transition metal ions on ZrO_2. With copper, in samples prepared by adsorption from aqueous solutions of Cu-acetate or toluene solutions of Cu-acetylacetonate, the uptake limit was 2.5 Cu atoms nm^{-2} [2]. With cobalt, in samples prepared by adsorption from aqueous solutions of Co-acetate, the uptake limit was 2.2 Co atoms nm^{-2} [4].

3.2. Surface sulphates and hydroxyls.

After evacuation at 773 K, the FTIR spectrum of pure ZrO_2 showed bands at 3777 and 3670 cm^{-1}, assigned to surface terminal and bridging OH of monoclinic ZrO_2 [2]. After evacuation at 773 K, the FTIR spectrum of sulphated-ZrO_2, SZ(2.4), showed no band of terminal-OH, and an extremely weak band of bridging-OH, occurring at a lower frequency (3640 cm^{-1}) than in pure ZrO_2 (3670 cm^{-1}). The disappearance of terminal-OH indicates that sulphates preferentially anchored to terminal rather than to bridged-OH of ZrO_2. The shift of the bridged-OH band to lower frequency indicates higher Brönsted acidity of these sites in SZ(2.4) than in pure ZrO_2. In addition to the band of bridged-OH, the SZ(2.4) spectrum consisted of several bands in the 1200-900 cm^{-1} region, typical of the ν_{S-O}, and a complex absorption in the 1420-1350 cm^{-1} region, typical of the $\nu_{S=O}$ of covalent organic-like sulphates, previously observed by other authors and assigned to tridentate sulphates and polynuclear sulphates [10].

FTIR spectra of all activated Fe/Z_i and Fe/Z_{CVD} samples showed bands of surface terminal and bridging OH, resembling those of the pure monoclinic ZrO_2. As the Fe-content in Fe/Z_i and Fe/Z_{CVD} increased, the intensity of both hydroxyl bands decreased, preferentially that of terminal-OH. We attributed the decreasing intensity of the hydroxyl-bands to the formation of Zr-O-Fe species in the iron anchoring process. The preferential decrease in intensity of terminal OH indicated that iron preferentially anchored to this hydroxyl type, as previously observed for Cu [2] and Co [4].

FTIR spectra of activated Fe/SZ(2.3,2.3) and FeS/Z(4.3,3.2) showed extremely weak bands of bridging-OH, and intense bands of covalent sulphates, similar to those detected on SZ(2.4). FTIR spectra of activated FeS/Z(0.3,0.6) showed weak bands of surface terminal and bridging OH, and weak bands of covalent sulphates.

3.3. Iron species in Fe/Z_i, Fe/Z_{CVD}, FeS/Z and Fe/SZ.

On ZrO_2 and sulphated-ZrO_2 samples, CO adsorption at RT yielded a composite band in the region 2230-2170 cm^{-1}, assigned to two Zr^{IV}-CO species. On pure ZrO_2, the two Zr^{IV}-CO bands formed at 2191 and 2185 cm^{-1} [11]. On sulphated-ZrO_2 these bands formed at higher wavenumber (2210 and 2200 cm^{-1}), indicating the involvement of Zr^{IV} endowed with higher Lewis acid strength.

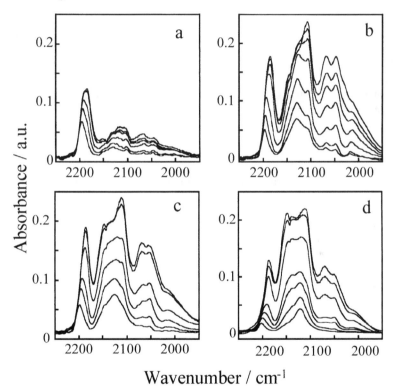

Figure 1. FTIR spectra after CO adsorption at RT on Fe/Z_i at increasing CO pressure from 0.5 to 80 Torr. Section a: Fe/Z_i(0.5); section b: Fe/Z_i(1.9); section c: Fe/Z_i(2.8); section d: Fe/Z_i(5.7).

On all Fe/Z_i (Fig. 1) and Fe/Z_{CVD} (Fig. 2), in addition to Zr^{IV}-CO species, CO adsorption at RT yielded (i) an envelope of bands at about 2160-2100 cm^{-1}, (ii) an envelope of bands at about 2100-1950 cm^{-1}, and (iii) bands in the region 1700-1200 cm^{-1} (spectral region not shown). The two envelopes at 2160-2100 cm^{-1} and at 2100-1950 cm^{-1} each consisted of at least three components. We assigned all components of these two envelopes to various carbonyl-like species formed onto Fe^{n+} sites ($0 \leq n \leq 3$), whose oxidation state n decreased with decreasing the wavenumber. Similar bands have been previously observed on FeO$_x$/ZrO$_2$

[5,12,13], and on pure and sulphated iron oxides [14]. We assigned the bands in the region 1700-1200 cm^{-1} to monodentate and bidentate carbonates arising from Fe^{III} reduction with CO at RT, in agreement with a previous report [5]. As the iron content in Fe/Z$_i$ (Fig. 1) and in Fe/Z$_{CVD}$ (Fig. 2) increased, the total amount of iron carbonyls increased. The relative amount of the various iron carbonyls (surface distribution of iron species) depended only on the sample preparation method, and not on the iron content (compare spectra in Fig. 1 with those in Fig. 2).

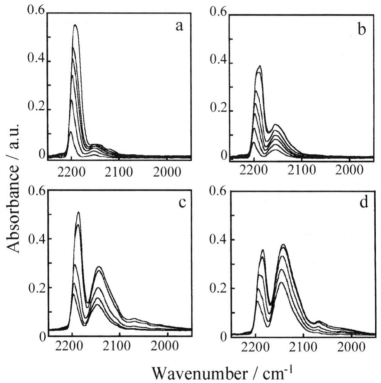

Figure 2. FTIR spectra after CO adsorption at RT on Fe/Z$_{CVD}$ at increasing CO pressure from 0.5 to 80 Torr. Section a: Fe/Z$_{CVD}$(0.2); section b: Fe/Z$_{CVD}$(1.0); section c: Fe/Z$_{CVD}$(2.1); section d: Fe/Z$_{CVD}$(2.5).

On sulphated samples (spectra not shown), in addition to Zr^{IV}-CO species, CO adsorption at RT yielded various carbonyl species, whose surface distribution depended on the iron and sulphate content. On FeS/Z(0.3,0.6), containing low amounts of both iron and sulphate, the same two envelopes of bands (i) at about 2160-2100 cm^{-1} and (ii) at about 2100-1950 cm^{-1} formed, as they did on unsulphated samples. These bands were weaker on FeS/Z(0.3,0.6) than on Fe/Z$_i$(0.5) and Fe/Z$_{CVD}$(0.2), despite the similar iron content. On Fe/SZ(2.3,2.3) and FeS/Z(4.3,3.2), containing high amounts of both iron and sulphate, only an envelope of weak bands at about 2160-2100 cm^{-1} formed, namely those corresponding to carbonyls with iron in a higher oxidation state.

To clarify better the dependence of the iron carbonyl amount and the iron carbonyl type on the preparation method, we compared samples containing nearly the same iron amount

(2.2-2.5 Fe atoms nm^{-2}) and prepared with different methods. The comparison shows that (i) a far larger amount of iron carbonyls formed on both Fe/Z$_i$ and Fe/Z$_{CVD}$ than on the sulphated sample; (ii) carbonyls of iron with a low n value (those in the 2100-1950 cm^{-1} region) formed in larger amounts on Fe/Z$_i$ than on Fe/Z$_{CVD}$, and did not form on the sulphated sample. In agreement with point (ii), carbonates formed in a larger amount on Fe/Z$_i$ than on Fe/Z$_{CVD}$, and none formed on the sulphated sample (Fig. 3).

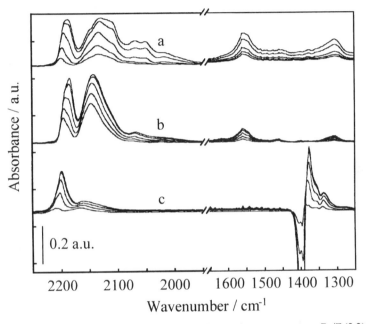

Figure 3. FTIR spectra of CO adsorbed at RT at increasing pressure on Fe/Z$_i$(2.2) (spectra a), Fe/Z$_{CVD}$(2.5) (spectra b), and Fe/SZ(2.3,2.3) (spectra c). In spectra c, because of side interactions of carbonyls with adjacent surface sulphates, the complex absorption in the 1420-1350 cm^{-1} region (covalent sulphates) shifts towards lower frequency. This yields a negative peak in the region 1420-1380 cm^{-1}, and an absorption at about 1380-1250 cm^{-1}.

3.4. The redox properties of iron and sulphates.

The H$_2$-TPR profile of all Fe/Z$_{CVD}$ samples (Fig. 4a) and that of Fe/Z$_i$(1.0) (Fig. 4b, spectrum 1) consisted of a broad peak, starting at 493 K and centred at 600 K, whose intensity increased with the iron content. Because the H$_2$ consumption corresponded to about e/Fe = 1, we assign this peak to the reduction of Fe^{III} to Fe^{II}. In addition to this broad peak, the H$_2$-TPR profile of Fe/Z$_i$ with iron content $\geq$ 2.2 Fe atoms nm^{-2} showed a shoulder (Fig. 4b, spectrum 2) or a peak (Fig. 4b, spectra 3 and 4) at 638 K and a weak peak at 763 K. Because a mechanical mixture Fe$_2$O$_3$ + ZrO$_2$ showed two reduction peaks at about 638 and 763 K (spectra not shown), we suggest the presence of Fe$_2$O$_3$ particles in Fe/Z$_i$ with iron content $\geq$ 2.2 Fe atoms nm^{-2}.

In the temperature region 373–1073 K, the H$_2$-TPR profile of SZ(2.4) consisted of an intense peak at 923 K. In the same temperature region, the H$_2$-TPR profile of FeS/Z(4.3,3.2) and Fe/SZ(2.3,2.3) consisted of an intense peak, starting at 648 K and centred at 728 K. On SZ(2.4), the H$_2$ consumption corresponded to e/SO$_4^=$ = 7.1. This value is consistent with the

336

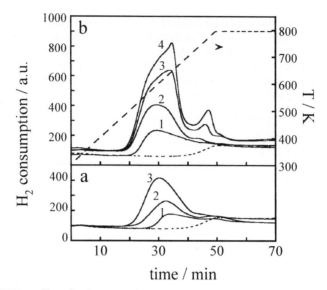

Figure 4. H_2-TPR profiles of Fe/Z_{CVD} (section a) and Fe/Z_i (section b). Section a: Fe/Z_{CVD}(0.2) (curve 1); Fe/Z_{CVD}(1.0) (curve 2); Fe/Z_{CVD}(2.1) (curve 3). Section b: Fe/Z_i(1.0) (curve 1); Fe/Z_i(2.2) (curve 2); Fe/Z_i(3.8) (curve 3); and Fe/Z_i(5.7) (curve 4). In both sections, we report the profile of blank experiments (dotted line).

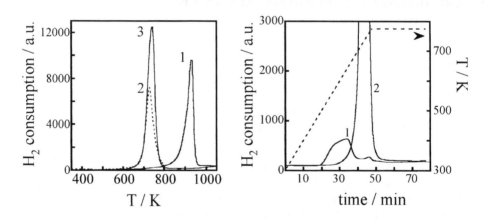

Figure 5. H_2-TPR profiles of SZ(2.4) (curve 1), Fe/SZ(2.3,2.3) (curve 2) and FeS/Z(4.3,3.2) (curve 3).

Figure 6. H_2-TPR profiles of Fe/Z_i(3.8) (curve 1) and FeS/Z(4.3,3.2) (curve 2).

reduction of $SO_4^=$ to SO_2 and $S^=$, as demonstrated by Xu and Sachtler [15]. The comparison of the H_2-TPR profile of SZ(2.4) with those of FeS/Z(4.3,3.2) and Fe/SZ(2.3,2.3) shows that the presence of iron shifted the temperature of the sulphate reduction peak from 923 to 728 K (Fig. 5). The copper in Cu/SZ caused an analogous shift [7]. The comparison of the H_2-TPR profile of FeS/Z(4.3,3.2) with that of Fe/Z_i(3.8) suggests that Fe^{III} reduced to Fe^{II} at a much higher temperature in the sulphated sample than in the unsulphated one (Fig. 6). In the sulphated sample no Fe^{III} reduced to Fe^{II} up to about 593 K, a far higher temperature than that needed for Fe^{III} to reduce to Fe^{II} in both Fe/Z_i and Fe/Z_{CVD}. We suggest that the intense sulphate reduction peak might obscure the Fe^{III} reduction peak.

4. CONCLUSIONS

The iron uptake limit in samples prepared by chemical vapour deposition (2.5 Fe atoms nm^{-2}) almost matched the copper uptake limit previously obtained in Cu/Z samples [2] and the cobalt uptake limit in Co/Z samples [4]. The finding that different transition metal ions yielded the same uptake limit suggests that the limit is an intrinsic property of the ZrO_2 surface. In Fe/Z_i with iron content higher than the iron uptake limit, Fe_2O_3 formed.

The relative amount of the various iron species in FeO_x/ZrO_2 and $FeO_x/sulphated-ZrO_2$ catalysts depends on the preparation method. In particular, the iron species more extensively reduced by CO at RT are present in larger amounts in Fe/Z_i than in Fe/Z_{CVD}, and are absent in sulphated samples. When samples prepared by the same method are compared, the surface distribution of iron species is independent of the iron content.

Iron in $FeO_x/sulphated-ZrO_2$ is much less reducible than iron in FeO_x/ZrO_2. The presence of sulphates stabilizes the iron oxidation state. The stabilizing effect of sulphates is analogous to that previously observed with copper [6] and cobalt [8]. In copper and cobalt samples, the presence of sulphates (i) made Cu^{II} and Co^{II} less reducible and (ii) prevented the formation of CuO and Co_3O_4. Because both effects reduced the catalytic activity for the $C_3H_6+O_2$ reaction, copper and cobalt sulphated samples were more selective than unsulphated ones for the $NO+C_3H_6+O_2$ reaction [6,8]. The lower reducibility of iron in sulphated-ZrO_2 accounts for its higher selectivity for the SCR reaction [16].

REFERENCES

[1]. K.A. Bethke, M.C. Kung, B. Yang, M. Shah, D. Alt, C. Li, H.H. Kung, Catal. Today, 26 (1995) 169.
[2]. V. Indovina, M. Occhiuzzi, D. Pietrogiacomi and S. Tuti, J. Phys. Chem B, 103 (1999) 9967.
[3]. D. Pietrogiacomi, D. Sannino, S. Tuti, P. Ciambelli, V. Indovina, M. Occhiuzzi and F. Pepe, Appl. Catal. B: Environ., 21 (1999) 141.
[4]. D. Pietrogiacomi, S. Tuti, M.C. Campa, V. Indovina, Appl. Catal. B: Environ., 28 (2000) 43.
[5]. S. Tuti, F. Pepe, D. Pietrogiacomi and V. Indovina, Catal. Today, 75 (2002) 373.
[6]. V. Indovina, D. Pietrogiacomi, M.C. Campa, Appl. Cat. B, 39 (2002) 115.
[7]. G. Delahay, E. Ensuque, B. Coq, and F. Figuéras, J. Catal., 175 (1998) 7.
[8]. D. Pietrogiacomi, M.C. Campa, S. Tuti, V. Indovina, Appl. Catal. B: Environ., 41 (2003) 301.
[9]. M. Kantcheva, A.S. Vakkasoglu, J. Catal., 223 (2004) 352, *ibid.* 223 (2004) 364.
[10]. C. Morterra, G. Cerrato, and V. Bolis, Catal. Today, 17 (1993) 505.
[11]. C. Morterra, L. Orio, C. Emanuel, J. Chem. Soc., Faraday Trans., 86 (1990) 3003.
[12]. E. Guglielminotti, J. Phys. Chem., 98 (1994) 4884.
[13]. E. Guglielminotti, J. Phys. Chem., 98 (1994) 9033.
[14]. G. Magnacca, G. Cerrato, C. Morterra, M. Signoretto, F. Somma, and F. Pinna, Chem. Mater., 15 (2003) 675.
[15]. B.-Q. Xu and W.M.H. Sachtler, J. Catal., 167 (1997) 224.
[16]. For the abatement of NO with propene in the presence of O_2, sulphated FeO_x/ZrO_2 catalysts were more selective than unsulphated FeO_x/ZrO_2, unpublished data from our laboratory.

Studies in Surface Science and Catalysis 155
A. Gamba, C. Colella and S. Coluccia (Editors)
© 2005 Elsevier B.V. All rights reserved

A thermodynamic model of chabazite selectivity for Pb^{2+}

F. Iucolano^a, D. Caputo^a, F. Pepe^b and C. Colella^a

^{a}Dipartimento di Ingegneria dei Materiali e della Produzione,
Università Federico II, Piazzale V. Tecchio 80, 80125 Napoli, Italy

^{b}Dipartimento di Ingegneria, Università del Sannio,
Piazza Roma 21, 82100 Benevento, Italy

Literature data relative to the cation exchange behavior of a Na–exchanged sedimentary chabazite for Pb^{2+} were analyzed by means of the recently proposed Double Selectivity Model (DSM). The DSM allowed to interpret the exchange isotherm by assuming that the exchangeable sites present in the zeolite could be divided in two groups: group "I", responsible for 15% of the overall cation exchange capacity, characterized by a very poor selectivity towards Pb^{2+} ($K_1 = 8.5 \cdot 10^{-3}$), and group "II", responsible for 85% of the cation exchange capacity, much more selective towards Pb^{2+} ($K_2 = 16.2$). The comparison between the results of the DSM and the experimental results provided very satisfactory and appeared to confirm the hypothesis that site "I" should be identified with the C1 site in the small cage of the chabazitic structure (6–6 pseudohexagonal prism), whereas site "II" should be identified with the C2–C4 cation sites in the large ellipsoidal cage (20–hedron chabazite cage).

1. INTRODUCTION

Unambiguous evaluation of cation exchange selectivity of a zeolite is a difficult task. Inspection of the profile of the exchange isotherm for a given cation pair ($z_B A^{z_A^+} \leftrightarrows z_A B^{z_B^+}$, where z_A and z_B are the A and B valences, respectively) provides only qualitative and often misleading information. In fact, the shape of an isotherm is affected by various factors, such as charge density of the anionic framework, i.e., framework composition [1], electrolyte concentration in the aqueous phase [2] and, in some instances, even nature and relative abundance of the extra–framework cations, as recently demonstrated for a natural clinoptilolite [3].
  A parameter which is often used to give a quantitative measure of the cation exchange selectivity is the selectivity quotient α [4]:

$$\alpha = \frac{z_A E_{A(z)} E_{B(s)}}{z_B E_{A(s)} E_{B(z)}},$$ (1)

where E_A and E_B are the equivalent cation fractions in zeolite (z) or solution (s). The selectivity quotient can be easily obtained from the isotherm data, but unfortunately it only gives a punctual estimation of selectivity.
  On the other hand, the thermodynamic equilibrium constant K:

$$K = \frac{a_{A(z)}^{zB} \, a_{B(s)}^{zA}}{a_{A(s)}^{zB} \, a_{B(z)}^{zA}}, \tag{2}$$

where the quantities denoted with a are the thermodynamic activities, is generally utilized as a measure of the cation exchange selectivity. However, the value of this parameter, while giving a comprehensive estimate of the preference that the exchanger exhibits for one cation compared with another, does not give unequivocal indications as regards selectivity over the whole composition range, because of possible selectivity reversals. Indeed, a change of selectivity is evidenced by a sigmoid–shaped isotherm and may be related to the presence in the zeolite framework of different cation sites [5].

Usually, equilibrium data do not enable to estimate specific cation site selectivities, because of the difficulty to correlate chemical behavior to structural features. However, a favorable case, in which the isotherm shape has unequivocally been related to specific framework features, occurred with the cation exchange behavior of a Na-exchanged sedimentary chabazite for Pb^{2+}, NH_4^+ and K^+ [6]. A good selectivity was observed for these cations and the selectivity sequence $NH_4^+ > K^+ > Pb^{2+} > Na^+$ was obtained. In the experimental conditions chosen, namely $T = 25°C$ and total solution normality $N = 100$ eq·m^{-3}, NH_4^+ and K^+ did not attain complete exchange. A conversion as high as 80–85% was obtained, despite several equilibrations of the exchanger with appropriate solutions of the ingoing cations. On the contrary, Pb^{2+} did attain complete exchange, but an inflection point at a Pb^{2+} equivalent fraction in chabazite roughly equal to 0.82, was evident in the relevant isotherm.

Such inflection point was interpreted by assuming that two types of cation sites were available in the chabazite framework: site (I), unselective for Pb^{2+}, being the C1 site in the small cage (6–6 pseudohexagonal prism), and site (II) highly selective for Pb^{2+}, coinciding with the C2–C4 cation sites in the large ellipsoidal cage (20–hedron chabazite cage) [7]. In the light of these results, it appeared worthwhile re-analyzing the above mentioned $2Na^+ \leftrightarrows Pb^{2+}$ exchange data using the Double Selectivity Model (DSM), which was recently proposed by Pepe et al. [8]. Indeed, the DSM is essentially based on the hypothesis that two different groups of sites exist in the zeolite, which undergo the exchange reactions independently of each other, and therefore it appears particularly suitable for the description of the available experimental data.

2. THE DOUBLE SELECTIVITY MODEL

The cation exchange between a solution containing $A^{z_A^+}$ cations and a zeolite containing $B^{z_B^+}$ cations can be described by means of the following reaction:

$$z_B A^{z_A^+} + z_A BL_{z_B} \leftrightarrows z_A B^{z_B^+} + z_B AL_{z_A}, \tag{3}$$

in which L is the portion of zeolite framework holding unit negative charge [8]. Therefore, if the Pb^{2+}/Na^+ exchange is considered, the reaction (3) reduces to the following:

$$Pb^{2+} + 2NaL \leftrightarrows 2Na^+ + PbL_2. \tag{4}$$

The main hypothesis of the DSM is that two different groups of sites are present in the zeolite, arbitrarily labeled as "type 1" and "type 2". However, since in general more than two groups of cation sites are actually available for exchange reactions, it is important to recognize that these two groups are actually to be intended as "pseudo–groups", lumping different "real" groups of sites. Sites "1" and "2" are assumed to represent two fractions, p_1 and p_2 of the overall cation exchange capacity (CEC) Q. Therefore, the exchange capacity attributable to site "1" is equal to $q_1 = p_1 Q$, and that attributable to site "2" is equal to $q_2 = p_2 Q$, with $p_1 + p_2 = 1$. Furthermore, it is assumed that each site undergoes the exchange reactions independently of the other and that the deviations from ideality in the zeolite-cation interactions are negligible, so that no activity coefficient is required to describe the behavior of the solid phase.

Following these hypotheses, for the exchange process Pb^{2+}/Na^+ mentioned above, the following two reactions are to be considered rather than Eq. (4):

$$Pb^{2+} + 2NaL_1 \leftrightarrows 2Na^+ + Pb(L_1)_2 \qquad K_1, \qquad (5a)$$

$$Pb^{2+} + 2NaL_2 \leftrightarrows 2Na^+ + Pb(L_2)_2 \qquad K_2, \qquad (5b)$$

where L_i ($i = 1, 2$) represents the portion of "type i" cation sites holding unit negative charge, and K_1 and K_2 are the equilibrium constants for the Pb^{2+}/Na^+ exchange reactions on cation sites "1" and "2", respectively. The equilibrium equations corresponding to reactions (5a, b) are the following:

$$K_i = \frac{[Na^+]^2 [Pb(L_i)_2]}{[Pb^{2+}][NaL_i]^2} \frac{\gamma_{Na^+}^2}{\gamma_{Pb^{2+}}} \qquad i = 1, 2, \qquad (6)$$

in which the quantities $[Pb(L_i)_2]$ and $[NaL_i]$ are the Pb^{2+} and Na^+ concentrations in zeolite and $\gamma_{Pb^{2+}}$ and γ_{Na^+} are the activity coefficients of the same cations in solution, respectively.

Since the total number of equivalents, both in solution and in the solid, is constant during the exchange process, it is useful to express the concentrations which appear in Eq. (6) as equivalent fractions, rather than moles. Therefore, the two molar concentrations $[Pb^{2+}]$ and $[Na^+]$ can be expressed as follows:

$$[Pb^{2+}] = \frac{NE_{Pb(s)}}{2}, \qquad (7a)$$

$$[Na^+] = NE_{Na(s)} = N(1 - E_{Pb(s)}), \qquad (7b)$$

where N is the total normality in solution, and $E_{Pb(s)}$ and $E_{Na(s)}$ are the equivalent fractions of Pb^{2+} and Na^+ cations in the solution phase. Similarly, in the solid phase, considering that two different groups of sites are taken into account, it is possible to write:

$$[Pb(L_i)_2] = \frac{q_i E_{Pb(z)i}}{2} \qquad i = 1, 2, \qquad (8a)$$

$$[NaL_i] = q_i E_{Na(z)i} = q_i(1 - E_{Pb(z)i}) \qquad i = 1, 2, \qquad (8b)$$

and, considering the total number of cation sites:

$$E_{Pb(z)} = p_1 E_{Pb(z)1} + p_2 E_{Pb(z)2} \qquad E_{Na(z)} = p_1 E_{Na(z)1} + p_2 E_{Na(z)2}.$$ (9)

In these two equations $E_{Pb(z)i}$ and $E_{Na(z)i}$ are the fractions of L_i cation sites exchanged with Pb^{2+} and Na^+ cations, respectively with $i = 1, 2$, whereas $E_{Pb(z)}$ and $E_{Na(z)}$ are the corresponding fractions evaluated with reference to the overall number of sites, and therefore to the total CEC.

Once Eqs. (7 a, b; 8 a, b) are substituted into the equilibrium Eq. (6), the following expressions are obtained, in which cation Pb^{2+} has arbitrarily been taken as a reference:

$$K_i = \Gamma \frac{N}{q_i} \frac{(1 - E_{Pb(s)})^2 E_{Pb(z)i}}{E_{Pb(s)}(1 - E_{Pb(z)i})^2} \qquad i = 1, 2.$$ (10)

Here Γ is the following combination of activity coefficients $\gamma_{Pb^{2+}}$ and γ_{Na^+}:

$$\Gamma = \frac{\gamma_{Na^+}^2}{\gamma_{Pb^{2+}}},$$ (11)

which, once N and $E_{Pb(s)}$ are assigned, can be computed by means of a "standard" thermodynamic model, such as the Pitzer model [9].

Eqs. (6–11) show that, once a set of experimental data relative to a binary exchange isotherm is available in the form of couples of equivalent fractions in solution and in the zeolite phase ($E_{Pb(s)}/E_{Pb(z)}$), the model parameters can be evaluated. Following the approach of Rodriguez and coworkers [10, 11], the parameters to be evaluated were considered to be the two thermodynamic constants K_1 and K_2, plus the fraction p_1 of "type 1" sites.

3. RESULTS AND DISCUSSION

The DSM Eqs. (6–11) were applied to the data obtained by Torracca et al. [6]. These authors studied the $2Na^+ \leftrightarrows Pb^{2+}$ exchange on a natural chabazite working at $T = 25°C$, with a total normality $N = 100$ eq·m^{-3} and using NO_3^- as counterion. They found a complete reversibility for the above reaction, and furthermore found that the natural chabazite considered had a CEC $Q = 2.75$ eq·kg^{-1}.

The estimation of the most likely values for the model parameters p_1, K_1 and K_2 was carried out by means of a non–linear regression. The sum of the quadratic deviations between calculated and experimental equivalent fractions in the zeolite was chosen as the objective function (U) for the regression [10]:

$$U = \sum_{i=1}^{m} [E_{1(z)_i} - E_{1(z)_i}^{exp}]^2,$$ (12)

in which m is the total number of experimental points.

In particular, a large number of values were attributed to p_1, and for each of these values the optimal couple (K_1, K_2), leading to the lowest value of U, was determined. Afterwards, the most likely triplet (p_1, K_1, K_2) was taken as the one which led to the lowest overall value of U. As shown in Fig. 1, the optimal value of p_1 turned out to be $p_1 = 0.15$. Correspondingly, the values of the equilibrium constants were $K_1 = 8.5 \cdot 10^{-3}$ and $K_2 = 16.2$.

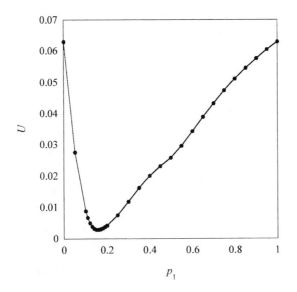

Fig. 1. Dependence of the objective function U (see Eq. 12) on the fraction of type "1" sites p_1.

The Pitzer model [9] was used to calculate the activity coefficients, and therefore Γ. Values of Γ ranging between 1.8–1.9 were obtained. This indicates that the solution has a remarkably non–ideal behavior and that the use of a thermodynamic model to properly take into account this non–ideality is fully justified.

The predicted isotherm, together with the experimental values obtained by Torracca et al [6], is reported in Fig. 2, in which the equivalent fraction of Pb^{2+} entering the zeolite phase is plotted versus the equivalent fractions of Pb^{2+} in the solution phase, when equilibrium has been reached.

Fig. 2 shows that the DSM is capable of describing the available experimental data with great accuracy. Furthermore, it appears possible to associate the two equilibrium constants obtained using the DSM, relative to *pseudo–sites* 1 and 2, to the two types of cation sites, I and II, individuated by Torracca et al. [7]. Indeed, site "I" ($p_1 = 0.15$, $K_1 = 8.5 \cdot 10^{-3}$) is a scarcely frequent site, characterized by a very poor selectivity towards Pb^{2+}. On the other hand, site "II" ($p_2 = 0.85$, $K_2 = 16.2$) is much more frequent in the zeolitic structure and is much more selective towards Pb^{2+}. The selectivity of Na–exchanged chabazite as a whole towards Pb^{2+} thus appears a sort of weighted mean of the selectivities of sites "I" and "II".

A comparison between the two equilibrium constants, K_1 and K_2, obtained by means of the DSM and those obtained by Torracca et al. [7] is reported in Table 1. There, together with K_1 and K_2, it is also reported the average equilibrium constant K_{ave}, defined as the weighted average between K_1 and K_2, i.e., $K_{ave} = (K_1)^{p1}(K_2)^{p2}$, which in this case turns out to be equal to

5.2. The good agreement between the previous data [7] and those of the present study is evident.

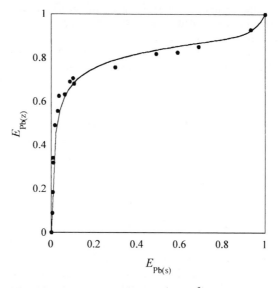

Fig. 2. Equilibrium data for the exchange reaction $2Na^+ \leftrightarrows Pb^{2+}$ in chabazite at 25°C. Symbols: experimental data; line: double selectivity model.

When analyzing the data reported in Table 1, it is important to keep in mind that, the analysis carried out by Torracca et al. [7] needs an *a priori* choice of the value of p_1. In fact, the value $p_1 = 0.16$ was derived from experimental evidences, i.e., the inflection point present in the isotherm of Pb^{2+} exchange for Na^+ and the value of the maximum exchange level of chabazite in the exchange isotherms of NH_4^+ and K^+ for Na^+ [6]. On the contrary, in the DSM, p_1 is one of the tree parameters of the model, and therefore no *a priori* information is required.

Table 1. Comparison of equilibrium constants for Pb^{2+}/Na^+ exchange in chabazite at 25 °C

Cation couple	Data from Ref. [7])		Present study	
	K	p_i	K	p_i
$2Na^+ \leftrightarrows Pb^{2+}$, whole exchanger	4.1^1	-	5.2^2	-
$2Na^+ \leftrightarrows Pb^{2+}$, site I	$8.5 \cdot 10^{-4}$	0.16	$8.5 \cdot 10^{-3}$	0.15
$2Na^+ \leftrightarrows Pb^{2+}$, site II	20.4	0.84	16.2	0.85

^{1}Experimental value; ^{2}Calculated value, $K_{ave} = (K_1)^{p_1}(K_2)^{p_2}$

This finding allows some interesting considerations to be made concerning the selectivity sequence $NH_4^+ > K^+ > Pb^{2+} > Na^+$ found in [6]. Actually it must be borne in mind that in $Na^+ \leftrightarrows NH_4^+$ and $Na^+ \leftrightarrows K^+$ exchanges only sites "II" participate to the cation exchange and, therefore, only 85% of the total CEC of chabazite is available. That is why the relative equilibrium constants, i.e., 8.2 and 11.7, respectively [6], were calculated, as it is usually made, from the "normalized" isotherms. Had the partial exchange occurred in $2Na^+ \leftrightarrows Pb^{2+}$ exchange too, considering the K value for site "II" in Table 1, the following selectivity sequence would have been observed: $Pb^{2+} > NH_4^+ > K^+ > Na^+$.

Therefore, almost paradoxically, the fact that site "I" too is available for $2Na^+ \leftrightarrows Pb^{2+}$ exchange, while allows the whole CEC of chabazite to be available for Pb^{2+}, results in a decrease of the selectivity of chabazite, as a whole, towards Pb^{2+}.

4. CONCLUSIONS

The exchange isotherm of Pb^{2+} for Na^+ on a sedimentary chabazite at $T = 25°C$ and $N = 100$ eq·m^{-3} was successfully analyzed in terms of the Double Selectivity Model (DSM). This model is essentially based on the hypothesis that two different groups of sites exist in the zeolite, which undergo the exchange reactions independently of each other. Its use allowed to describe the main features of the experimentally observed isotherm, and in particular: (a) that Na^+ is completely exchangeable for Pb^{2+} and (b) that an inflection point exists at a Pb^{2+} equivalent fraction in chabazite roughly equal to 0.82.

By using of the DSM, the exchange isotherm was interpreted by assuming that the exchangeable sites present in the zeolite could be divided in two groups: group "I", responsible for 15% of the total CEC, characterized by a very poor selectivity towards Pb^{2+} ($K_1 = 8.5 \cdot 10^{-3}$), and group "II", responsible for 85% of the total CEC, much more selective towards Pb^{2+} ($K_2 = 16.2$). The comparison between model and experimental results provided very satisfactory and appeared to confirm the hypothesis of Torracca et al. [7], according to which site "I" could be identified with the C1 site in the small cage of the chabazite structure (6–6 pseudohexagonal prism), while site "II" could be identified with the C2–C4 cation sites in the large ellipsoidal cage (20–hedron chabazite cage).

LIST OF SYMBOLS

a	thermodynamic activity
$E_{I(s)}$	equivalent fraction of cation I in solution
$E_{I(z)}$	equivalent fraction of cation I in the zeolite
K	thermodynamic equilibrium constant
m	number of experimental points
N	total normality in solution, eq·l^{-1}
p_i	fraction of sites of type i
Q, q	cation exchange capacity, eq·kg^{-1}
T	temperature, K
U	objective function
z_I	valence of cation I

Greek symbols

α	selectivity quotient defined in Eq. (1)
Γ	combination of activity coefficients defined in Eq. (11)
γ_I	activity coefficient of cation I

Subscripts and superscripts

1, 2	relative to sites of "type 1" and of "type 2"
ave	averaged value
exp	experimental

REFERENCES

[1] C. Colella, Mineral. Deposita 31 (1996) 554.
[2] R.M. Barrer and J. Klinowski, J. Chem. Soc., Faraday Trans. I, 70 (1974) 2080.
[3] A. Langella, M. Pansini, P. Cappelletti, B. de Gennaro, M. de' Gennaro and C. Colella, Microp., Mesop. Mater., 37 (2000) 337.
[4] R.M. Barrer, in Natural Zeolites. Occurrence, Properties and Use, L.B. Sand and F.A. Mumpton (eds.), Pergamon Press, Elmsford, N.Y., 1978, p. 385.
[5] R.M. Barrer and J. Klinowski, J. Chem. Soc., Farad. Trans. I, 68 (1972) 73.
[6] E. Torracca, P. Galli, M. Pansini and C. Colella, Microp. Mesop. Mater., 20 (1998) 119.
[7] E. Torracca, P. Galli, M. Pansini and C. Colella, in Proc. IV Convegno Nazionale di Scienza e Tecnologia delle Zeoliti, E. Fois and A. Gamba (eds.), Università dell'Insubria "A. Volta",Como, 1998, p. 220.
[8] F. Pepe, D. Caputo and C. Colella, Ind. Eng. Chem. Res., 42 (2003) 1093.
[9] K.S. Pitzer, in Activity Coefficients in Electrolyte Solutions, K.S. Pitzer (ed.), CRC Press, Boca Raton, Florida, 1991, p. 75.
[10] J.L. Valverde, A. De Lucas and J.F. Rodriguez, Ind. Eng. Chem. Res., 38 (1999) 251.
[11] A. De Luca, J.L. Valverde, M.C. Romero, J. Gomez and J.F. Rodriguez, Chem. Eng. Sci., 57 (2002) 1943.

Studies in Surface Science and Catalysis 155
A. Gamba, C. Colella and S. Coluccia (Editors)

Dehydrogenation of propane over Ga and Cr modified, "fresh" and steamed, MFI-type zeolites

T. K. Katranas[a], K. S. Triantafyllidis[b], A. G. Vlessidis[a], N. P. Evmiridis[a]

[a] Department of Chemistry, University of Ioannina, 451 10 Ioannina, Greece

[b] Department of Chemistry, Aristotle University of Thessaloniki and CERTH/CPERI, P.O. Box 116, University Campus, Thessaloniki 54124, Greece

In this work we synthesized MFI type zeolites with chromium and gallium. Hydrothermal synthesis was applied using, either aluminum with gallium or chromium sources in 1:1 mole ratios, or impregnation of H-ZSM-5 with the desired metal. The "as synthesized" samples were characterized by XRD, Chemical analysis, SEM-EDS, SSA, ammonia-TPD and subjected to severe steaming, in order to suppress their acidity and consequently decrease the oligomerization capability of the formed catalysts. The catalytic activity of the zeolite samples in propane dehydrogenation was investigated using a fixed-bed continuous-flow glass reactor. The catalytic test results revealed that propylene yield and selectivity over the steamed catalysts increased remarkably (e.g. propylene yield ~11% with selectivity of ~45%, for H-[Ga/Al]ZSM-5), as a consequence of the suppression of their acidity and therefore of the dramatic decrease of their oligomerization capability. The above results suggest that Cr- and Ga-modified steamed zeolites are promising catalysts for the dehydrogenation of propane to propylene.

1. INTRODUCTION

The increasing demand for the production of light olefins which are being used in several commercial processes, like the synthesis of high-octane gasoline additives, (e.g. methyltertiarybutyl ether, MTBE), induced the necessity of improving the up-today processes of catalytic dehydrogenation or oxidative dehydrogenation of alkanes towards alkenes [1-3]. Increase of olefin yield in the gaseous products of a Fluid Catalytic Cracking (FCC) unit is an alternative way for light alkenes production, especially attractive to refiners.

The large amount of light alkanes which are found in natural gas or may be obtained as by-products in the FCC process, renders the aromatization of C_3-C_4 alkanes an attractive alternative to the catalytic reforming of naphtha, which comprises the main route for BTX production, but is incapable of transforming light paraffins to aromatics.

The medium-pore zeolite ZSM-5 in it's proton-exchanged form or in the modified with metals form, such as gallium, platinum, zinc, vanadium, indium, chromium etc., is the most appropriate catalyst for the dehydrocyclization/aromatization of light alkanes [4-16]. With regard to commercial applications, the pioneering work of BP and UOP [17] resulted in the development of the Cyclar process which utilizes bifunctional H-[Al]ZSM-5 zeolites exchanged and/or impregnated with gallium ions, while in the M2 Forming process developed by Mobil [18], the catalyst is the acidic H-[Al]ZSM-5 zeolite.

Mechanistic investigations of the catalytic activity of metal-modified H-[Al]ZSM-5 catalysts have shown that the aromatization reaction proceeds through a bifunctional mechanism of combined function of metallic species with acid sites. Metals catalyze the dehydrogenation of light alkanes, oligomers, cyclic olefins and diolefins, whereas protonic sites are responsible for oligomerization, cyclization, and cracking [12, 19]. In addition it has been previously shown that under experimental conditions, that restricted the conversion of propane to primary products, the strongest Brønsted acid sites favoured cracking whereas the milder Brønsted acid sites favoured dehydrogenation of propane over H-Y and H-ZSM-5 zeolites, in accordance with the soft-soft and hard-hard acid-base interaction model [20].

Based on the above facts it can be suggested that, the suppression of the acidity of metal modified ZSM-5 zeolites, could lead to a decrease of the oligomerization capability of the catalysts, and therefore to an increase of the olefin yield and selectivity. As a result, hydrothermally treated (steamed) or deactivated ZSM-5 zeolites, may possibly be used alternatively as catalysts for alkane dehydrogenation. [21]

In the present work we synthesized a number of MFI type zeolites with gallium and chromium. Gallium-modified ZSM-5 zeolites are the most commonly used catalysts for the aromatization of light paraffins, whereas chromium although rarely used in this process [21, 22] is a strong dehydrogenative agent [1]. The synthesized zeolites were hydrothermally treated (steamed) and tested in the propane dehydrogenation reaction. The comparative catalytic results between the "fresh" and steamed samples reveal that propylene yield and selectivity over the steamed catalysts increased remarkably, suggesting that Cr and Ga modified steamed zeolites can be promising catalysts for the dehydrogenation of propane to propylene.

2. EXPERIMENTAL SECTION

The synthesis of the parent ZSM-5 zeolites of this work was based on a typical procedure [23, 24]. A solution containing partially dissolved silica [Colloidal silica HS-40 (Du Pont), 39.6% SiO_2 , 0.426 % Na_2O] in tetrapropylammonium (TPA) hydroxide reacted with a solution of sodium aluminate [(BDH), 54.2 % Al_2O_3, 38.5 % Na_2O] in water (Si/Al=40). The resulting gel was heated in an autoclave at 150°C for 8 days leading to crystallization of ZSM-5. The as-synthesized sample was calcined to combust the organic template and was further treated with dilute HCl solution (pH=3) to produce the H-form of the zeolite (labeled H-[Al]ZSM-5).

For the preparation of the chromium and gallium impregnated zeolite samples (labeled Cr_2O_3/ H-[Al]ZSM-5 and Ga_2O_3/H-[Al]ZSM-5 respectively), the H-[Al]ZSM-5 sample was impregnated with $Cr(NO_3)_3.9H_2O$ and $Ga(NO_3)_3.xH_2O$ aqueous solutions using vacuum rotary evaporation at 90°C (~0.5 mmoles of Cr and Ga per g zeolite). The samples were then dried in air at 120 °C for 12 h and calcined afterwards at 500°C for 4 h under the flow of dry air.

The samples labeled H-[Cr,Al]ZSM-5 and H-[Ga,Al]ZSM-5 were synthesized hydrothermally by using both Al^{3+} (sodium aluminate) and Cr^{3+} ($Cr(NO_3)_3.9H_2O$) or Ga ($Ga(NO_3)_3.xH_2O$) sources (mole ratios Cr/Al=Ga/Al=1 and Si/M=40, where M=Cr, Al = Ga, Al).

Parts of all the ZSM-5 samples were finally steamed at 790°C for 6 hrs, at a partial pressure of steam of 97.7 kPa. All the samples were stored over a saturated $MgCl_2$ solution to equilibrate with water vapor.

Chemical analysis of the zeolite samples was carried out by means of Atomic Absorption Spectroscopy (Shimadtzu 6800 AAS) and Electron Dispersive Spectroscopy (Oxford ISIS 300 EDS). Powder X-ray Diffraction (XRD) was carried out on a Siemens D-500 diffractometer with CuKa-radiation. Scanning Electron Microscopy (SEM) images were obtained with a JSM 840-A JEOL SEM. Specific Surface Area (SSA) was measured by nitrogen sorption isothermally at 77 K using a Sorptomatic 1900 instrument (multi-point BET). ^{27}Al-MAS-NMR measurements were carried out using a Varian Infinity plus AS400 spectometer. Temperature-Programmed Desorption (TPD) of ammonia tests were performed on a conventional apparatus which consisted of a cylindrical quartz micro-reactor, a vertical well-controlled high-temperature furnace and a gas chromatograph equipped with a thermal conductivity detector (TCD). The mass of the catalysts sample used was 0.2 g. Sorption of dry ammonia (Merck, H_2O-free) took place at 100°C, in a static system for 90 min at 1.5 bar ammonia pressure. Stripping was done afterwards for 40 min at 100°C under He flow. In this way, the weakly and physically adsorbed ammonia was minimized in the sample. Desorption of ammonia was done at a rate of 10°C/min from 100 up to 700°C under He flow (50 ml/min). The desorbed ammonia was detected on a Shimadzu GC-8A gas chromatograph (with TCD), and then it was trapped in an HCl aqueous standard solution (0.01N). The desorbed NH_3 was estimated by titrimetric determination of the excess HCl in solution, using a standard 0.01N NaOH solution.

The catalytic activity of the zeolite samples in propane dehydrogenation was investigated using a cylindrical fixed-bed continuous-flow glass reactor. The reaction was carried out between 400°C and 550°C. Temperature in the reactor was monitored by a thermocouple located at the center of the catalyst bed. The catalyst (mass 0.2 g) was placed into the glass reactor between two layers of glass-beads (Serva). Before reaction, the catalysts were outgassed at 500°C for 3 h under He flow. The reactant mixture had a molar composition He/C_3H_8 = 10/1 and space velocity of 2500-3000 h^{-1}. Products were analyzed using a Shimatzu GC-14b gas chromatograph equipped with a Supelco SP-1700 column and a Thermal Conductivity Detector (TCD).

3. RESULTS AND DISCUSSION

3.1. Compositional, structural and acidic characteristics of the ZSM-5 catalysts

The physicochemical characterization results of the "fresh" and steamed zeolites of this work are listed in Table 1.

Table 1

Compositional, Structural and Acidic characteristics of the zeolite catalysts

Sample	Cr	Ga	Al	SSA[1]	Total acidity
	(wt. %)			(m^2/g)	(mmoles NH_3/g)
H-[Al]ZSM-5 Fresh	-	-	1.51	367	1.05
H-[Al]ZSM-5 Steamed	-	-	1.51	367	0.09
H-[Cr,Al]ZSM-5 Fresh	0.95	-	0.23	372	0.29
H-[Cr,Al]ZSM-5 Steamed	0.90	-	0.18	372	0.11
Cr_2O_3/ H-[Al]ZSM-5 Fresh	2.91	-	1.53	353	1.11
Cr_2O_3/ H-[Al]ZSM-5 Steamed	2.90	-	1.51	353	0.09
H-[Ga,Al]ZSM-5 Fresh	-	0.58	0.72	370	0.44
H-[Ga,Al]ZSM-5 Steamed	-	0.44	0.59	370	0.12
Ga_2O_3/ H-[Al]ZSM-5 Fresh	-	3.87	1.52	345	1.17
Ga_2O_3/ H-[Al]ZSM-5 Steamed	-	3.87	1.51	345	0.13

[1] Specific surface area (multi-point BET)

From Table 1 it is revealed that the metal concentrations achieved by the hydrothermal synthesis method are about 0.6 wt %, for Ga, and 1 wt %, for Cr, while the amount of metals introduced by the impregnation method reached 3 (for Cr)-4 (for Ga) wt %. It is also clear that the introduction of aluminum into the zeolitic framework, in the presence of chromium, is difficult, thus the zeolites synthesized this way possess relatively low aluminum content.

The relatively severe steaming conditions of all the ZSM-5 samples reduced their framework Al (FAl) dramatically to less than 0.1 wt. %. The FAl content of the steamed samples was determined from the intensity of the peak at ~53 ppm in the ^{27}Al MAS NMR spectra, based on an appropriate calibration curve which correlated the total Al content of pure crystalline ZSM-5 samples, free from extra-framework Al (EFAl), with the intensity of the above ^{27}Al NMR peak [25].

The XRD patterns of the steamed zeolite samples (Fig. 1.) and SSA measurements (Table 1), revealed that they were highly crystalline and retained their structural integrity to a high degree. The various treatments (calcinations/combustion of the organic template, ion-exchange, impregnation, severe steaming) affected the crystallinity of the samples at a relatively low degree. The lower surface area observed for the impregnated samples is attributed to partial pore blockage by chromium and gallium oxides or oxy-hydroxides as a result of the preparation procedure.

SEM images of all the steamed modified samples (Fig. 2.) showed clear phases of zeolitic particles. Hydrothermally synthesized chromium and gallium samples consisted of

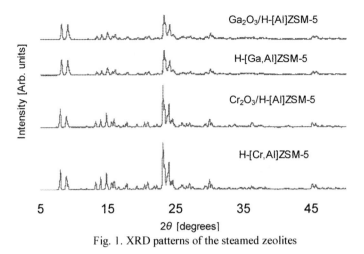

Fig. 1. XRD patterns of the steamed zeolites

large crystals or crystal aggregates (rectangular crystals or crystal aggregates for chromium, and typical for MFI type galloaluminosilicates [4] spherical crystals for gallium), while the impregnated samples consisted of characteristic small-sized ZSM-5 crystals along with some non-zeolitic particles of anomalous shape. These particles mainly consisted (according to EDS measurements) of metallic (chromium or gallium) species and are attributed to crystalline or amorphous Cr- or Ga- oxides/oxy-hydroxides

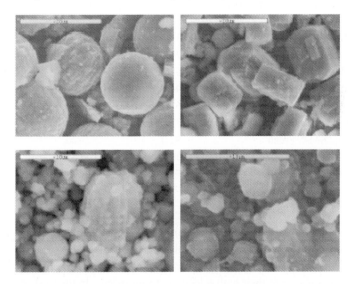

Fig. 2. SEM images of the steamed zeolites. Top left H-[Ga,Al]ZSM-5, top right H-[Cr,Al]ZSM-5, bottom left Ga_2O_3/H-[Al]ZSM-5, bottom right Cr_2O_3/H-[Al]ZSM-5. The climax of the top images is 20μ and of the bottom images is 10μ.

The strong Brønsted acidity of ZSM-5 zeolites depends mainly on the framework aluminum and the related framework hydroxyls. Consequently the acidity of the "fresh" samples of this work (see Table 1) is strongly related to the amount of the framework metals (Al, Cr and Al, Ga). The samples prepared by hydrothermal synthesis show lower total acidity in comparison to the impregnated ones (which had similar total acidity to their precursor H-[Al]ZSM-5), suggesting that their framework metal content is significantly lower.

On the other hand, the acidity of the steamed zeolites was significantly decreased upon steaming, due to framework dealumination. No significant variations in total acidity were detected between the steamed samples based on their different amounts and type of metallic species; they all had similar framework aluminum content.

3.2. Dehydrogenation activity of the ZSM-5 catalysts

The reaction of propane over MFI-type zeolites yields cracking products (e.g. methane, ethane, ethylene, coke), dehydrogenation products (e.g. propylene) and cyclo-oligomerization products (e.g. aromatics), according to reaction conditions and catalyst type. The maximum dehydrogenating activity values of the zeolitic catalysts of this work are shown in Table 2.

Table 2.
Maximum dehydrogenation activity of the tested catalysts

Sample	Propane conversion[a]	Selectivity to propylene	Propylene yield
		(%)	
H-[Cr,Al]ZSM-5 Fresh	39.1	11.2	4.37
H-[Cr,Al]ZSM-5 Steamed	29.0[b]	30.9	8.97
Cr_2O_3/H-[Al]ZSM-5 Fresh	58.5	3.90	2.30
Cr_2O_3/H-[Al]ZSM-5 Steamed	41.0	20.2	8.28
H-[Ga,Al]ZSM-5 Fresh	67.5	8.60	5.82
H-[Ga,Al]ZSM-5 Steamed	23.5[b]	46.0	10.80
Ga_2O_3/H-[Al]ZSM-5 Fresh	81.1	3.20	2.60
Ga_2O_3/H-[Al]ZSM-5 Steamed	30.0	27.1	8.13

[a] Reaction temperature 550°C unless otherwise stated, reactant mixture $He/C_3H_8=10$, space velocity 2500 h^{-1}
[b] Reaction temperature 500°C

From table 2 it is clear that the total conversion of all the "fresh" catalysts is higher in comparison to their steamed counterparts. This is easily explained by the very low acidity of the steamed samples. On the other hand this lack of acidity in the hydrothermally treated samples is the key-factor to their high selectivity to propylene. In the fresh samples the major products of the propane reaction are methane, ethylene and aromatic hydrocarbons. Propane reacts with the metallic species present in the catalysts creating dehydrogenation products which subsequently polymerize in the strong acid sites and lead to aromatic hydrocarbons.

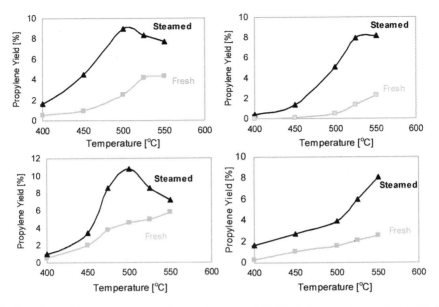

Fig. 3. Propylene yield versus reaction temperature. Top left H-[Cr,Al]ZSM-5, top right Cr_2O_3/H-[Al]ZSM-5, bottom left H-[Ga,Al]ZSM-5, bottom right Ga_2O_3/H-[Al]ZSM-5

In contrast the almost complete absence of strong acid sites in the steamed catalysts results to the avoidance of polymerization reactions and consequently the selective production of propylene.

The dependence of the propylene yield on the reaction temperature for all the tested catalysts is shown in Fig. 3. From Table 2 and Fig. 3. it is revealed that the best dehydrogenation results were obtained with H-[Ga,Al]ZSM-5 catalyst. This is natural as Gallium is a very strong dehydrogenative metal used in commercial aromatization catalysts introduced by BP and UOP. The results obtained with H-[Cr,Al]ZSM-5 were also very promising. The distinct higher dehydrogenative activity of the hydrothermally synthesized catalysts in comparison to the impregnated ones, regardless of their lower metal content, can be attributed to the fine dispersion and random distribution of the metallic species into the hydrothermally synthesized ZSM-5 framework. This allows metals to interact better with the remaining weak and medium acid sites on the catalysts framework (which favour dehydrogenation reactions [20]). In addition most of the metallic species in the impregnated samples are probably located in the external surface area, blocking the pores, and being unable to interact with the weak acid sites.

An interesting aspect of the activity of the hydrothermally synthesized catalysts is that the best results were obtained in lower temperature (500° C). This cannot be attributed to deactivation of the catalysts, at high temperatures, due to fouling (coke deposition) or framework collapse, since subsequent catalytic tests at lower temperatures (without previous reactivation of the catalyst under pure helium flow) resulted to identical activity as in the previous experiment.

4. CONCLUSIONS

The results of this work reveal that propylene yield and selectivity over Ga- and Cr-modified H-ZSM-5 zeolites increased remarkably with the steaming of the catalysts, as a consequence of the suppression of their acidity and therefore of the dramatic decrease of their oligomerization capability. The above results suggest that Ga- and Cr- modified steamed zeolites are promising catalysts for the dehydrogenation of propane to propylene.

ACKNOWLEDGEMENTS

The authors would like to thank Dr. V. A. Tsiatouras for his valuable assistance in the experimental section of the present work.

REFERENCES

[1] G. Ertl, H. Knözinger, J. Weitkamp (eds.), Handbook of Heterogeneous Catalysis, Vol. 5, VCH, Weinheim, (1997), 2140.
[2] F. Cavani, F. Trifiro, Catal. Today 24 (1995) 307.
[3] W. Schuster, J.P.M. W.F. Hoelderich, Appl. Catal. A 209 (2001) 131.
[4] R. Fricke, H. Kosslick, G. Lischke, M. Richter, Chem. Rev. 100 (2000) 2303.
[5] W. J. H. Dehertog, G.F. Fromen, Appl. Catal. A: General 189 (1999) 63.
[6] P. Meriaudeau, C. Naccache, J. Catal. 157 (1995) 283.
[7] B.S. Kwak, W.M.H. Sachtler and W.O. Haag, J. Catal. 149 (1994) 465.
[8] V.R. Choudhary, A. K. Kinage, C. Sivadinarayama, M. Guisnet, J. Catal. 158 (1996) 23.
[9] I. Nakamura, K. Fujimoto, Catal. Today, 31 (1996) 335.
[10] L. Brabec, M. Jeschke, R. Klik, J. Novakova, L Kubelkova, D. Freude, V. Bosacek, J Meusinger, Appl. Catal. A: General 167 (1998) 209.
[11] V. Choudhary, K. Mantri, C. Sivadinarayana, Micropor. Mesopor. Mater. 37 (2000) 1.
[12] A. Montes, G. Gianneto, Appl. Catal. A: General 197 (2000) 31.
[13] Z. Fu, D.Yin, Y.Yang, X. Guo, Appl. Catal. A: General 124 (1995) 59.
[14] H. Berndt, G. Lietz, B. Lücke and J. Völter, Appl. Catal. A: General 146 (1996) 351.
[15] G. Centi, F. Trifiro, Appl. Catal. A: General 143 (1996) 3.
[16] J. Halász, Z. Kónya, A. Fudala, A. Béres and I. Kiricsi, Catal. Today 31.
[17] E.E. Davis, A.J. Kolombos, GB Pat. 53012 (1976); AU. Pat., 509285 (1980).
[18] N.Y. Chen, T.Y. Yan, Ind. Eng. Chem. Process Res. Dev. 25 (1986) 151.
[19] M. Guisnet, N.S. Gnep, Catal. Today, 31 (1996) 275.
[20] J. Bandiera, M. Dufaux , Y. Ben Taarit, Appl. Catal. A 148 (1997) 283.
[21] V.A. Tsiatouras, T.K. Katranas, C.S. Triantafillidis, A.G. Vlessidis, N.P. Evmiridis, E.G. Pavlidou, Stud. Surf. Sci. Catal. 142A (2002) 839.
[22] E.S. Sphiro, R.W. Joyner, P. Johnston, G. J. Tuleuova, J. Catal 141 (1993) 266.
[23] Y.G.Li, W.H. Xie, S. Yong Applied Catalysis A: General 150, (1997), 231.
[24] R.G. Argauer and G. R. Landolt, U.S. Pat. 3702, 856, (1972).
[25] Z. Zhu, Z. Chang, L. Kevan, J. Phys. Chem. B, 103, (1999), 2680.

Studies in Surface Science and Catalysis 155
A. Gamba, C. Colella and S. Coluccia (Editors)

355

Templated and non-templated routes to mesoporous TiO$_2$

U. Lafont[a,b], P. Kooyman[b], A. Galarneau[a], F. Di Renzo[b]

[a]Laboratoire de Matériaux Catalytiques et Catalyse en Chimie Organique, ENSCM, 8 rue de l'Ecole Normale, 34296 Montpellier Cedex 5, France.
[b]Delft Chem Tech and National Centre for High resolution Electron Microscopy, Julianalaan 136, 2628 BL, Delft, The Netherlands.

1. INTRODUCTION

The photovoltaic properties of titanium dioxide are at the basis of its application as a deep oxidation catalyst for environmental protection. The obtention of high surface area titania has been the object of intensive investigation in the last years. The micelle-templating route, successfully applied to the obtention of mesoporous silicas, has been applied to the synthesis of ordered surfactant-titania mesophases [1, 2]. It has been shown that such mesophases are extremely unstable when the surfactant is removed, unless surface modifiers (*e.g.* phosphate groups) are used to prevent the growth of TiO$_2$ nanocrystals [3, 4]. A significant stabilisation of mesostructured titania film has been obtained by post-synthesis vapour treatment [5]. However, in the case of hexagonal mesophases, the accessibility of the mesopore system inside the film is limited by the orientation of the channels, parallel to the film surface.

The main obstacle in the adaptation of the micelle-templated synthesis method to TiO$_2$ is the faster condensation of Ti(IV) precursors compared to Si(IV) precursors. To avoid too fast condensation, the use of retarding agents is recommend. Chelating molecules (triethanolamine [6,7], 2,4 pentadione [3], propanediol [8], pentanediol [8], ethylene glycol [9] or hydrogen peroxide [10]) or highly acidic synthesis conditions are used [11-13]. The synthesis of ordered mesoporous TiO$_2$ presenting high surface area (250 m^2/g) and a tailored pore size distribution can be obtained using different structure directing agents: neutral surfactants like alkylamines [14-19], non-ionic surfactants like tri-block copolymers [13, 20-23] and cationic surfactants like quaternary ammoniums [6, 7, 9-11, 24]. Upon the years, many publications suggested that mesostructured titania presented a low structural stability in the absebce of surface modifiers and collapsed during the removal of the surfactant.

A possible drawback of the high dispersion of titanium dioxide is a large blue shift that can bring the useful electron transitions beyond the energy of solar radiation. From this point of view, the effectiveness of ordered titania mesostructures as photooxidation catalysts has yet to be compared with other forms of highly dispersed titania obtained by sol-gel method and, at some extent, already commercially available.

In this communication, three different synthesis procedures were applied in order to obtain highly dispersed TiO$_2$ [25, 26]. The first two methods imply the use of surfactant micelles as templates [6, 7, 27], while in the third method smaller organic molecules are used as porogene (pore-forming agent). The properties of the materials have been characterised before and after acivation by X-ray diffraction, transmission electron microscopy, thermal gravimetry and adsorption volumetry.

2. SYNTHESIS

2.1. The Atrane Route (Synthesis #1).
In this synthesis, cetyltrimethylammonium bromide (cationic surfactant, Aldrich, $C_{16}H_{33}N(CH_3)_3Br$, noted CTMA) is used as template. 1 mole of titanium isopropoxide (Aldrich, $Ti(OCH(CH_2)_2)_4$) is added to 3.5 moles of triethanolamine (Aldrich, $N(EtOH)_3$), which acts as a retarding agent of the hydrolysis/condensation reactions of the alkoxide precursor by forming an atrane complex [6,7]. The resulting complex is added to a $CTMA/NaOH/N(EtOH)_3/water$ solution with molar ratios 1/0.3/8/500 (

Scheme 1). The solution is stirred at room temperature during 48h. The as-synthesized material is recovered by filtration and washed with ethanol and deionised water. The material is activated by calcination at 350°C during 8h or by surfactant extraction by a warm acidic-alcoholic solution.

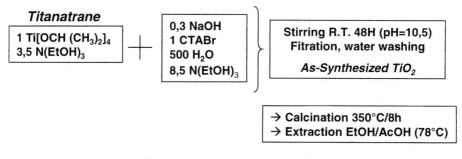

Scheme 1: Synthesis procedure #1, the Atrane Route. Molar ratios.

2.2. The P123 Route (Synthesis #2).
A non-ionic Pluronic® P123 triblock copolymer surfactant is used as template in this synthesis. The inorganic precursor is titanium isopropoxide (Aldrich, $Ti(OCH(CH_2)_2)_4$). A co-solvent (butylcellosolve, Aldrich) is added in order to control the reactivity of the inorganic precursor. After an aging step, which can be carried out at several temperature levels (Scheme 2), the as-synthesised white TiO_2 powder is recovered by filtration and washed with deionised water. The activation of the material is achieved by calcination at 350°C during 8h under air flow.

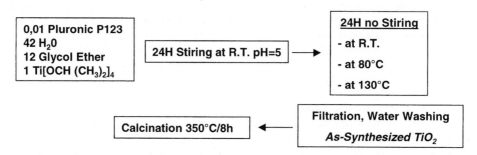

Scheme 2: Synthesis procedure #2, the P123 Route. Molar ratios.

2.3. The Aerogel and Xerogel Route (Synthesis #3).

In this synthesis, a classical acid-catalysed sol-gel method is applied. The inorganic precursor can be either titanium butoxide or ethoxide (Aldrich, $Ti[O(CH_2)_3CH_3]_4$ or $Ti[OC_2H_5]_4$). The precursor is diluted in an acidic solution of absolute ethanol (molar ratios $HCl/H_2O/EtOH/Ti$ 0.01/1.85/48/1). An urea/ethanol/water solution (molar ratios 0.8/5.3/2.7 per Ti) is added as a pore generator (porogene) [28-30]. The resulting pale yellow gel has been treated in two different ways (Scheme 3):

a) Dried at RT, then the porogene is removed by hot water extraction to form a xerogel.

b) Treated to form an aerogel by supercritical CO_2 drying (Polaron 3100 apparatus). This method allows to extract the porogene and dry the gel at the same time.

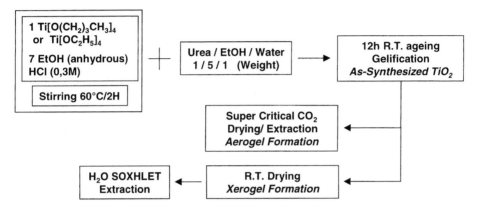

Scheme 3: Synthesis procedure #3, the Aerogel and Xerogel Route. Mass ratios.

3. CHARACTERIZATION METHODS

Adsorption/desorption isotherms of N_2 at 77K have been measured on a Micromeritics ASAP 2000 apparatus. Samples have been outgassed at 150 °C until a stable static vacuum of 0.3 Pa was reached. The surface area has been calculated by the BET method. The pore volume has been measured on the adsorption isotherm at the top of the capillary condensation step. The average pore size diameter has been calculated with de Broekhoff-de Boer method on the desorption branch of the isotherm. The X-Ray diffraction patterns have been registered on a $\theta/2\theta$ CGR Theta 60 diffractometer (Cu $K\alpha$, λ=1.54051Å). The thermogravimetric measurements were performed on a Netzsch TG209C apparatus in air flow from RT to 850°C with 5°C min^{-1} heating rate. The HRTEM images were registered using a Philips CM30T electron microscope with a LaB_6 filament. The source of electrons operated at 300 kV. The samples were mounted on Quantifoil® microgrid carbon polymer supported on a copper grid.

4. RESULTS

In the two first syntheses, the basic idea is the use of a surfactant mesophase in order to control the texture of the inorganic materials and increase the surface area. These self-assembly procedures are known from more than 10 years and give excellent results in the case of silica [25, 31].

4.1. The Atrane Route (Synthesis #1)

In this synthesis, a cationic surfactant with a 16-carbon hydrocarbon chain and a quaternary ammonium polar head is used. The same surfactant is used in the synthesis of the hexagonal mesoporous silica MCM-41 [31]. The thermogravimetric analysis of the as-synthesised titania material shows that some organic molecules are retained in the material until 500°C (Figure 1.). In the case of the silica analog, MCM-41, the CTMA is totally removed at 350°C. This indicates that the inorganic-organic interactions between the cationic surfactant and the oxide are much stronger in the case of titania than in the case of silica. These interactions play an important role during the activation process of the material. Stronger the interactions are, more difficult the removal of the surfactant is without damaging the mesostructure.

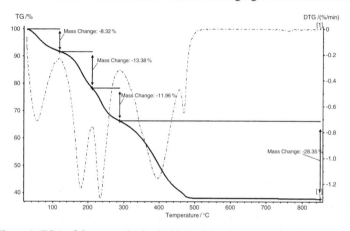

Figure 1: TGA of the as-synthetised TiO$_2$ by The Atrane Route (synthesis#1)

The low-angle X-ray diffraction pattern (Fig.2) of the as-synthesised titania material presents an ordered hexagonal structure. The position of the 100, 110, and 200 lines indicates a lattice parameter a° 4,5 nm. After the activation process (calcination or extraction) the hexagonal order clearly is no more visible.

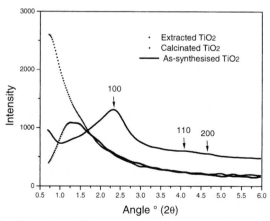

Figure 2: Low-angle XRD pattern of the as-synthesised, calcined and extracted Atrane Route TiO$_2$.

The high-angle X-Ray diffraction pattern indicates that no crystalline order is present for the as-synthesised nor for the extracted material. Amorphous TiO2 constitutes these materials. On Figure 3, the X-Ray diffraction pattern of the calcined material is reported. It corresponds to the anatase crystalline phase of TiO_2. The average particle size estimated using the Sherrer formula is 8.6 nm.

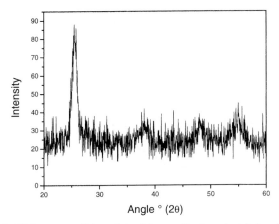

Figure 3: High-angle XRD pattern of the calcined Atrane Route material showing the anatase phase.

It is clear that an hexagonal mesophase is formed in the presence of cationic surfactant. However, this mesostructure disappears during the activation process. It seems that the nucleation of anatase crystallites destroys the mesostructure. A quantitative evaluation of the textural properties of the mesophase can shed some light on its instability. A geometric model allows to calculate the pore size and the wall thickness of the as-synthesised material from the low-angle XRD and the TGA data, In the case of a hexagonal array of pores, the average wall thickness t is obtained from the Eq.(1), the average pore diameter D from Eq.(2), the void fraction ε is estimated with the TGA data through the Eq.(3) and the a parameter is calculated from the d_{100} distance by Eq.(4). The results of the calculations are reported in Table 1.

$$t = a - a\sqrt{\varepsilon} \qquad (1)$$

$$D = 1.05a\sqrt{\varepsilon} \qquad (2)$$

$$\varepsilon = \frac{V_{Pores}}{V_{Tot}} = \frac{\dfrac{m_{CTMA}}{0.77} + m_{H_2O}}{\dfrac{m_{CTMA}}{0.77} + m_{H_2O} + \dfrac{1}{\rho_{TiO_2}}} \qquad (3)$$

$$a = \frac{2}{\sqrt{3}} d_{100} \qquad (4)$$

Table 1. Textural data on the as-synthetised TiO_2 from the Atrane Route.

Eq. (1)	t	0.3 nm
Eq. (2)	D	4.3 nm
Eq. (3)	ε	0.872
Eq. (4)	a	4.435 nm

The pore size of 4.3 nm is in good agreement with the size of the templating CTMA micelles [25]. The average wall thickness (0.3 nm) is extremely low. If the Ti-O distances (0.19 nm) are considered, the wall thickness of this material is close to a monolayer. This means that every titanium octahedron is in interaction with the polar heads of the surfactant. This interaction is a mixed blessing for the stability of the mesophase, because it is needed for the formation of an ordered mesophase and to avoid the precipitation of the bulk oxide but it prevents any extended cross-linking between titania octahedra. Such a highly-dispersed system is clearly unstable when the template is removed and this, together with the easy nucleation and growth of anatase crystals, fully justifies the collapse of the mesostructure.

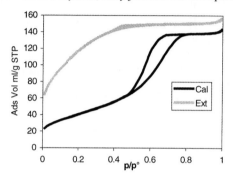

Figure 4: Isotherms of the calcined and extracted TiO_2 material from the Atrane Route.

From the nitrogen adsorption/desorption isotherms (Figure 4.) we can see that even if the mesostructure is not stable, the resulting calcined solid presents a type IV isotherm, typical of a mesoporous adsorbent. In the case of the calcined material, the pore size distribution is centred on 5.2 nm, the mesoporous volume is 0.21 ml g^{-1}, and the surface area 162 m^2 g^{-1}. The material from which the template has been extracted by ethanol washing presents a different isotherm, basically of type I, corresponding to a pore size distribution in between microporosity and mesoporosity.

In the HRTEM images (Figure 5), the anatase nanoparticles formed in the degradation of the hexagonal mesophase are clearly visible. The resulting anatase material presents a regular inter-particle mesoporosity and a size of nanocrystallite centered at 8-9 nm, in excellent agreement with the crystal size evaluation by the Sherrer formula.

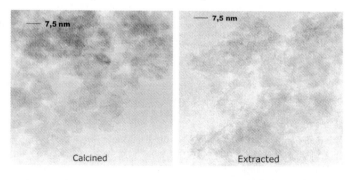

Figure 5: HRTEM images of activated TiO_2 from the Atrane Route.

4.2. The P123 Route (Synthesis #2)

The strong interaction between quaternary ammonium cations and titanium-bearing species prevents the titania to reach the condensation degree needed to stabilise the walls of the mesophase and represents a potential drawback of the atrane route. The use of non-ionic surfactants has been proposed to reduce the strength of the interaction between the inorganic and the organic species during the self-assembly process [13]. Soller-Illia *et al.* [20, 21] used different titanium alkoxides with triblock copolymers in acidic alcoholic conditions. They showed that between the polyethylene oxide groups of the surfactant and the titanium alkoxide precursor there are strong dative interactions in anhydrous conditions and weak hydrogen bonds in aqueous systems.

In the P123 synthesis route, the namesake non-ionic surfactant is used as template. This template, used in the synthesis of the highly ordered mesostructured SBA-15 silica [32,33], presents an interesting thermal behaviour. The micelle diameter increases with the temperature and it is possible to use this phenomenon to control the pore size of the resulting material, like is the case for SBA-15 silica [27]. In the case of titania, we adjusted the temperature of the aging step (RT or 80°C or 130°C) in the synthesis procedure in order to control the pore size. The small angle XRD patterns of the as-synthesised and activated materials (Figure 6) do not present any second order peak, suggesting that no ordered mesophase was formed.

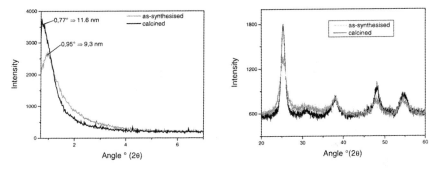

Figure 6: Small and wide angle X-ray diffraction patterns for the as-synthesised and calcined TiO_2 from the P123 Route.

A single broad peak was indeed observed in the XRD patterns and has to be considered an inter-particle correlation peak [34]. The wide angle X-Ray pattern clearly shows the presence of the TiO_2 anatase phase in the as-synthesized as in the calcined material. The presence of the anatase phase before the activation process is noticeable. The anatase phase is present foe every temperature of the aging step. The average crystal sizes estimated with the Scherrer formula are 4.5 and 7.1 nm for the as-synthesised and the calcined material, respectively. The growth of the anatase nanocrystals with calcination corresponds to an increase of the correlation distance, as evidenced by the shift of the XRD peak in Figure 6.

Despite the absence of observable mesoscopic order, the N_2 adsorption isotherms (Figure 7) indicate that the pore size depends on the temperature of the aging step, as it could be expected if micelles would play a template role. The average pore diameters are 4.6, 6.5 and 7.5 nm for the materials calcined aged at RT, 80 and 130°C, respectively. The type IVb isotherms with a H2 hysteresis loop suggest that the desorption is affected by percolation phenomena, as expected in the case of inter-particle porosity [35].

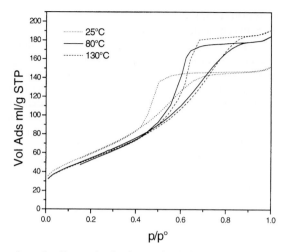

Figure 7: Nitrogen adsorption/desorption isotherms for P123 Route TiO$_2$ vs. the aging temperature.

HRTEM images confirm the absence of mesostructure (Figure 8.) and clearly show an inter-particular mesoporous system between 7 nm anatase nanocrystals.

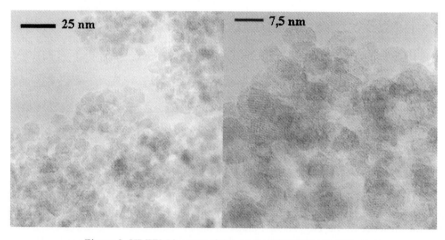

Figure 8: HRTEM images of the calcined P123 Route TiO$_2$.

4.3. The Aerogel and Xerogel Route (Synthesis #3)

The use of surfactant as structure directing agent in order to obtain ordered mesoporous material is not as effective for TiO$_2$ as it is in the case of silica-based materials. One wonder if materials with similar pore size and pore volume could not be obtained without any micelle-templating. This can be attempted by using urea as a pore forming agent (porogene) [28-30]. In this synthesis, HCl catalyses TEOS hydrolysis and silica condensation at 60°C during 2H. Urea is then added and the gel is aged at RT overnight. Two different activation procedures have been applied to the as-synthesized materials, leading to the formation of, respectively, xerogels or aerogels:

a) a RT drying during a week allows to form a TiO_2 xerogel. A water treatment is applied to the dried gel in order to extract all the organic molecules.

b) Super Critical CO_2 drying allows to obtain a TiO_2 aerogel.

The thermogravimetric results reported in Fig. 9 indicate that the mass loss on both aerogel and xerogel tales place at a temperature lower than 200°C. This suggests than no urea is retained in the activated materials.

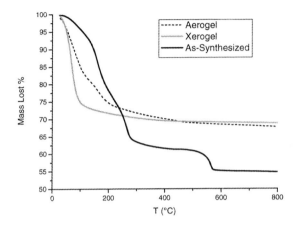

Figure 9: TGA mesurment of the As-synthesised, aerogel and xerogel TiO_2 material.

The wide-angle X-Ray diffraction patterns (not shown) indicate that no crystalline phase is present. TiO_2 is amorphous both in the aerogel and the xerogel. The small-angle X-Ray diffraction patterns (Figure 10) present one broad peak, very likely a correlation peak as commonly observed in the diffraction patterns of particulate gels [34].

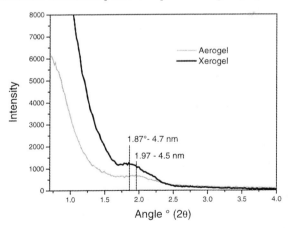

Figure 10: Small angles X-Ray diffraction patterns for the TiO_2 xerogel and aerogel.

The comparison of the adsorption isotherms of aerogels and xerogels, reported in Fig.11, indicates that, not surprisingly, the aerogels (left hand) present larger pore volume and pore diameter than the xerogels (right hand). The nature of the organic moieties in the titanium alkoxide also influences in some way the porosity. The use of titanium ethoxide instead of titanium butoxide in the aerogel formation gives a higher surface area (445 vs 337 m^2 g^{-1}), a higher mesopore volume (0.66 vs 0.65 ml g^{-1}) and a larger average pore diameter (14.0 vs 12.8 nm). In the case of xerogels, the use of titanium ethoxide gives a higher pore diameter (8.6 vs 7.4 nm) and pore volume (0.35 vs 0.31 ml.g^{-1}), but a lower surface area (216 vs 258 m^2.g^{-1}).

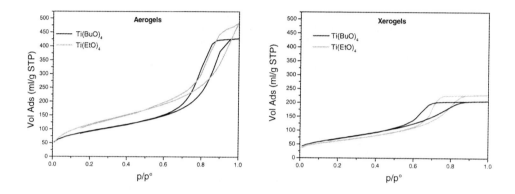

Figure 11: Nitrogen adsorption/desorption isotherms for TiO$_2$ aerogel and xerogel prepared from different titanium alkoxide precursor.

Both aerogels and xerogels represent low-density amorphous TiO$_2$ materials. The HRTEM images reported in Fig. 12 confirm the presence of mesoporosity through optical density variations.

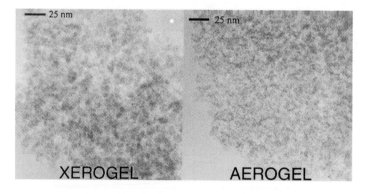

Figure 12: HRTEM images of TiO$_2$ xerogel and aerogel.

5. DISCUSSION

Both templated and non-templated routes to mesoporous TiO$_2$ allow to form mesopores of controlled size. For the synthesis #1, the degradation of the hexagonal mesophase produces aggregates of anatase nanoparticles. In the synthesis #2, the pore size can be adjusted by varying the aging temperature. The resulting TiO_2 crystallites form regular inter-particle mesoporosity. Surface areas as high as 200 $m^2 g^{-1}$ can be obtained. This result is very near to the dispersion of TiO_2 particles that can be obtained by sol-gel methods in the absence of surfactant. By such a route, surface areas of 250 and 450 $m^2 g^{-1}$ can be obtained, respectively by conventional drying or supercritical CO_2 drying. Textural data for samples obtained by these different methods are collected in Table 2.

Table 2
Textural data of templated and non-templated route to mesoporous TiO_2

	Synthesis Route	TiO₂ Structure	S_{BET} ($m^2.g^{-1}$)	V_p ($ml.g^{-1}$)	D_{BdB} (nm)
1	Atrane	Anatase	162	0.21	5.2
2	P123 RT	Anatase	203	0.22	4.6
	P123 80	Anatase	140	0.20	6.5
	P123 130	Anatase	145	0.27	7.5
3	Xerogel (But.)	Amorphous	258	0.31	7.4
	Aerogel (But.)	Amorphous	337	0.65	12.8
	Xerogel (Et.)	Amorphous	216	0.35	8.6
	Aerogel (Et.)	Amorphous	445	0.66	14.0

In Fig. 13, the pore volume of the TiO_2 materials presented in this study are compared with a commonly used commercial TiO_2: P25 from Degussa.

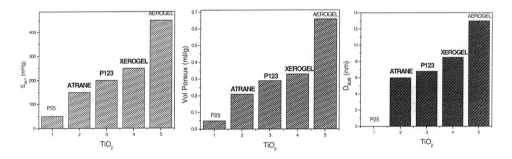

Figure 13: Comparative of TiO_2 material (Surface area, Porous Volume and Pore diameter).

6. CONCLUSION

The main conclusion is that the formation of a titania-surfactant mesophase is not enough to obtain an ordered mesoporous material. The activation process is at least as important as the synthesis step in the preparation of mesoporous TiO_2. Control of the synthesis and activation steps allows to change and adjust the structural and textural properties of the resulting material. Highly-dispersed forms of titanium dioxide are interesting for many applications in which the extent of the interface between TiO_2 and other species is an important parameter.

366

REFERENCES
[1]   Q. Huo, D.I. Margolese, U. Ciesla, T.E. Gier, P. Sieger, R. Leon, P.M. Petroff, F. Schüth, and G.D. Stucky, Nature, 368 (1994) 317
[2]   Q. Huo, F. Schüth, and G.D. Stucky, Chem.Mater., 6 (1994) 1176
[3]   D.M. Antonelli and J.Y. Ying, Angew.Chem.Int.Ed.Engl., 34 (1995) 2014
[4]   U. Ciesla, S. Schacht, G.D. Stucky, K. Unger, and F. Schüth, Angew. Chem.Int.Ed.Engl., 35 (1996) 541
[5]   D. Grosso, G.J.d.A.A. Soler-Illia, F. Crepaldi, F. Cagnol, A. Sinturel, A. Bourgeois, A. Brunet-Bruneau, H. Amenitsch, P. Albouy, and C. Sanchez, Chem.Mater, 15 (2003) 4562
[6]   S. Cabrera and al, Solid State Science, 2 (2000) 513
[7]   S. Cabrera, J.E. Haskouri, C. Guillem, J. Latorre, A. Beltran-Porter, D. Beltran-Porter, M.D. Marcos, and P. Amoros, Solid State Science, 2 (2000) 405
[8]   M. Fröba, O. Muth, and A. Reller, Solid State Ionics, 101-103 (1997) 249
[9]   D. Khushalani, G.A. Ozin, and A. Kuperman, Journal of Material Chemistry, 9 (1999) 1491
[10]   D.T. On, Langmuir, 15 (1999) 8561
[11]   G.J.d.A.A. Soler-Illia, A. Louis, and C. Sanchez, Chemistry of Materials, 14 (2002) 750
[12]   M. Thieme and F. Schüth, Microporous and Mesoporous Materials , 27 (1999) 193
[13]   P. Yang, D. Zhao, D. Margolese, B.F. Chmelka, and G.D. Stucky, Nature, 396 (1998) 152
[14]   D.M. Antonelli, Microporous and Mesoporous Materials, 30 (1999) 315
[15]   T.V. Anuradha and S. Ranganathan, NanoStructured Materials, 12 (1999) 1063
[16]   F. Leroux, P.J. Dewar, M. Intissar, G. Ouvrard, and L.F. nazar, Journal of Material Chemistry, 12 (2002) 3245
[17]   Y.D. Wang, C.L. Ma, X.D. Sun, and H.D. Li, Material Letters, 54 (2002) 359
[18]   Y. Wang, X. Tang, and L. Yin, Advanced Materials, 12 (2000) 1183
[19]   H. Yoshitake, T. Sugihara, and T. Tatsumi, Chemistry of Materials, 14 (2002) 1023
[20]   G.J.d.A.A. Soler-Illia, New Journal of Chemistry, 25 (2001) 156
[21]   G.J.d.A.A. Soler-Illia and C. Sanchez, New Journal of Chemistry, 24 (2000) 493
[22]   Y. Yue and Z. Gao, Chemical Communication (2000) 1755
[23]   W. Zhang and T.J. Pinnavaia, Chem.Commun. (1998) 1185
[24]   H. Hirashima, H. Imai, and V. Balek, Journal of Non-Crystalline Solids, 285 (2001) 96
[25]   F. Di Renzo, A.Galarneau, P.Trens, and F.Fajula, in Handbook of Porous Materials, F.Schüth, K.Sing, and J. Weitkamp, Eds., Wiley-VCH, 2002, 1311
[26]   F. Schüth, Chemical Materials, 13 (2001) 3184
[27]   A. Galarneau, H.Cambon, F.Di Renzo, and F.Fajula, Langmuir, 17 (2001) 8328
[28]   J.-Y. Zheng, K.-Y. Qiu, Q. Feng, J. Xu, and Y. Wei, Molecular Crystals and Liquid Crystals, 354 (2000) 771
[29]   J.-Y. Zheng, J.-B. Pang, K.-Y. Qiu, and Y. Wei, Microporous and Mesoporous Materials, 49 (2001) 189
[30]   J.-Y. Zheng, J.-B. Pang, K.-Y. Qiu, and Y. Wei, Journal of Material Chemistry, 11 (2001) 3367
[31]   J.S. Beck, J.C. Vartuli, W.J. Roth, M.E. Leonowicz, C.T. Kresge, K.D. Schmitt, C.T-W. Chu, D.H. Olson, E.W. Sheppard, S.B. McCullen, J.B. Higgins, and J.L. Schlenker, J. Am. Chem. Soc., 114 (1992) 10834
[32]   D. Zhao, Science, 279 (1998) 548
[33]   D. Zhao, J.Am.Chem.Soc., 120 (1998) 6024
[34]   T. Zemb and P. Linder, Neutron, X-Ray and Light: Scattering Methods Applied to Soft Condensed Matter, Elsevier, Amsterdam 2002
[35]   F. Rouquerol, J. Rouquerol and K. Sing, Adsorption by Powders and Porous Solids, Academic Press, San Diego 1999

Studies in Surface Science and Catalysis 155
A. Gamba, C. Colella and S. Coluccia (Editors)

Chromophore-bearing zeolite materials as precursors of colored ceramics

B. Liguori, A. Cassese and C. Colella

Dipartimento di Ingegneria dei Materiali e della Produzione, Università Federico II,
Piazzale V. Tecchio 80, 80125 Napoli, Italy

This work investigates the possibility to obtain colored ceramics by firing compacts of the synthetic zeolite A, pre-loaded with chromophore metal cations, and gives also a preliminary evaluation of their technical properties. Natural zeolite-bearing materials, i.e., clinoptilolite- and phillipsite-rich tuffs, which naturally contain chromofore agents, such as iron oxides and hydroxides, have also been considered as ceramic precursors. The results obtained appear very encouraging as regards either easiness of manufacture or technical properties of the resulting products. Obtaining ceramics with an uniform color distribution in a wide range of shades appears a major advantage of the this technique.

1. INTRODUCTION

One of the most promising applications of zeolites is their possible use, in the original cation form or after suitable pre-exchange, as precursors for the production of ceramics [1]. Firing of zeolite powders or pre-shaped manufacts results, in fact, in the breakdown of zeolite structure and, depending on its chemical composition, in its conversion into amorphous or crystalline ceramic materials. Advantages of this unusual route for ceramics manufacture, in alternative to the traditional methods based, e.g,. on firing of oxide mixtures, have thoroughly been discussed in earlier publications [2-7].

Introducing various cations into zeolite structure by ion exchange gives the possibility to produce different types of ceramics. Some cations, namely the transition metal cations, give rise to a distinctive coloration of zeolitic exchangers, which, although more or less deeply modified, is kept upon heating. It is, this way, possible to directly obtain colored ceramics or also powders, which may behave as ceramic pigments [8-9]. The homogeneous distribution of the extra-framework cations in zeolites, resulting in an uniform distribution of color centers within the zeolite structure with a consequent more uniform and safe coloring of ceramic manufacts after thermal treatment, is the most appealing aspect of this procedure.

This work is part of a larger multipurpose investigation aiming at exploring the feasibility of the chromophore-bearing zeolites to colored ceramic conversion through suitable firing treatments. Specific objectives of research are:

(a) evaluation of the ability of chromofore-cation pre-exchanged synthetic or natural zeolites to give rise to colored ceramic materials having equivalent of better features than those obtained by the conventional manufacturing methods;

(b) evaluation of the possibility to directly transform natural zeolitic materials, e.g., zeolitic tuffs, into colored ceramics, taking advantage of the presence of colored ancillary constituents, such as hydrated or anhydrous iron oxides;

(c) evaluation of the possibility to transform synthetic or natural zeolite powders, residual of toxic cations removal processes from wastewaters, into environmentally and ecologically safe ceramic materials.

Sodium zeolite A [LTA] [10], the most common and inexpensive synthetic zeolite on the market, has been selected as the white ceramic precursor, whereas Co^{2+} and Cr^{3+} have been chosen as chromophore agents, in consideration of their marked affinity for aluminous zeolites and also for their nature of cationic pollutants of specific wastewaters [11]. Considered natural zeolite-bearing materials have been those widespread in huge deposits all around the world, such as clinoptilolite- and phillipsite-chabazite-rich tuffs [12].

The environmental aspects related to the above point (c) have been treated in a previous publication [13]. This paper deals with the technical characterization of the fired compacts obtained either from the pre-exchanged Co^{2+}- or Cr^{3+}-rich zeolite A samples or, for reference and comparison, from naturally chromophore-bearing zeolitic tuffs.

2. MATERIALS AND METHODS

2.1. Loading zeolite A with Co^{2+} and Cr^{3+}

Zeolite A was a Carlo Erba Reagenti product, marketed as Molecular sieve 4Å powder. The reported grain size was 0.5-5 μm. Being a pure zeolite sample, its invariable chemical formula, $Na_2O \cdot Al_2O_3 \cdot 2SiO_2 \cdot 4.5 \ H_2O$, was assumed for true together with the calculated cation exchange capacity (CEC) equal to 5.48 meq·g^{-1}. Chromium and cobalt solutions were obtained from the relevant nitrates: $Cr(NO_3)_3 \cdot 9H_2O$ (a Baker Analyzed chemical) and $Co(NO_3)_2 \cdot 6H_2O$ (a Carlo Erba AnalytiCals chemical).

Co^{2+} and Cr^{3+} were loaded in zeolite samples by contacting under continuous stirring 50 g of zeolite with 1 l of variably concentrated solutions for 24 hours. Cation uptake by zeolite was estimated, by measuring residual Co^{2+} or Cr^{3+} concentrations in solution by atomic absorption spectrophotometry (AAS, AA2100 Perkin-Elmer apparatus). pH values (Radiometer mod. PHM220 apparatus) were also checked, to be sure of the absence of any precipitate (pH ≤ 6).

Three samples of each Co-rich and Cr-rich zeolite A were obtained, containing 2.0, 4.1, 7.3 w.-% Co^{2+} and 0.7, 1.4, 2.6 w.-% Cr^{3+}, respectively. Corresponding percent CEC coverages were 12.4, 25.4, 45.2 for Co^{2+} and 7.4, 14.7, 27.4 for Cr^{3+}.

Prepared samples were washed with distilled water, dried overnight at 80°C, and finally equilibrated and stored in an environment having near 50% relative humidity, realized with a saturated $Ca(NO_3)_2$ solution.

2.2. Natural zeolite-bearing samples

Two natural zeolite-bearing tuffs, representative of the most common materials in worldwide deposits, were selected: (a) a phillipsite-rich tuff from Marano (Naples, Italy), belonging to the huge formation of the so-called *Neapolitan yellow tuff* [14], referred to throughout the text as PHT; (b) a clinoptilolite-rich tuff from Eskişehir (Central Anatolia, Turkey) [15], indicated as CLT. The two main zeolite minerals constituting the tuffs belong to the framework types [PHI] and [HEU], respectively [10]. The variation ranges of their chemical compositions can be found in the recent Recommended Nomenclature for Zeolite Minerals [16]. Actually, phillipsite must be considered an intermediate silica zeolite, whereas clinoptilolite is decidedly a high silica zeolite.

The chemical composition of the two natural zeolite samples, determined with the usual methods of the quantitative chemical analysis, is reported in Table 1.

Table 1
Chemical composition of the tuff samples on dry basis (%)

Tuff sample	SiO_2	Al_2O_3	Fe_2O_3	TiO_2	CaO	MgO	Na_2O	K_2O
PHT	58.82	19.10	4.60	0.53	3.10	1.11	3.44	9.39
CLT	76.29	12.78	1.45	0.15	2.88	1.19	0.79	4.46

Inspection of Table 1 evidences that the reddish-brown Fe_2O_3 is the only chromophore agent of the two tuff samples. Actually, as Fe_2O_3 represents the total iron content of the samples and includes various Fe(II)-Fe(III) compounds, the colors of the two tuffs, depending on the different Fe contents too, turns out to be: brownish-yellow for PHT and pale greenish for CLT.

The mineral composition of the PHT sample was determined using the Reference Intensity Ratio (RIR) procedure [17], which is an improved version of the well known X-ray diffraction (XRD) technique based on the use of internal standards [18]. Instead, mineral contents of the CLT sample were measured using the Rietveld procedure (GSAS package [19]).

Results of the XRD quantitative analysis, limited to the exchanging phases, are as follows: PHT: phillipsite 46%, chabazite 5%, analcime 9%, smectite 10% (remaining 30% included feldspar, pyroxene, biotite and amorphous phases); CLT: clinoptilolite 79% (other minerals present were: silica phases, feldspar and biotite).

Tuff samples were ground to a fineness < 170 mesh, dried at 80°C overnight, and stored analogously as the zeolite A samples.

2.3. Preparation and firing of the compacts

Dry samples of either Co^{2+}- or Cr^{3+}-rich zeolite A samples and of PHT and CLT samples were dry-formed into cylindrical tablets ($\varnothing$ = 10 mm; h = 3 mm), using a suitable steel mould and a Carver Lab press. Compaction pressure was fixed at 120 MPa.

Firing of the compacts was carried out in an electric furnace. Green tablets were heated at 900°C, 1000°C or 1100°C (heating rate, 10°C/min), kept at the selected temperature for 1 hour, and finally slowly cooled down to room temperature in the furnace.

Side heating runs on the original or ion exchange modified powder samples in a wide range of temperatures were also performed in order to investigate the thermal evolution of the zeolite structures.

2.4. Characterization of the fired compacts

Changes in mineral composition of the various samples after thermal treatment were investigated by XRD (Philips PW 1730 apparatus, rad. CuKα1). Termogravimetry (TG) of the original and ion-exchanged zeolites was carried out using a thermal analyzer (Netzsch, model 409ST Luxx). The weight of the powdered samples was 25 mg; heating rate from 20 to 1200°C was equal to 10°C/min; the reference material was Al_2O_3.

Scanning electron microscopy (SEM, Oxford-Cambridge S440) was used to analyze the ceramic microstructures of the fired compacts.

Physical properties connected to porosity, i.e., open porosity and bulk density, were estimated using a mercury porosimeter (Micromeritics Autopore 9400). Linear shrinkage was determined, estimating with a gauge the variation of the tablets diameter after firing. Compressive strength measurements were performed using an Instrom 4204 instrument. All the physical and mechanical measurements were carried out in triplicate and the results averaged.

3. RESULTS AND DISCUSSION

Firing of the various green tablets gave rise to a more or less intense consolidation as a consequence of the different nature of the original material and the treatment temperature. Variation of chemistry, physical and mechanical properties, and color of the consolidated compacts were accordingly observed.

3.1. XRD characterization of fired samples

Results of the XRD characterization of the Co-loaded and Cr-loaded zeolite A compacts have been reported previously [11]. In summary, after a structure breakdown at temperatures $\geq 700°C$, a re-crystallization into nepheline (due to a residual sodium amount of the original zeolite A composition [20]) and a cobalt aluminate ($CoAl_2O_4$) for the Co-rich zeolite A sample and into nepheline and a sodium aluminum chromium oxide silicate ($Na_8Al_6Si_6O_{24}CrO_4$) or a sodium aluminum chromium oxide hydroxide ($NaAl_3(CrO_4)_2(OH)_6$) for the Cr-rich zeolite A sample.

Fig. 1 (a-b) summarizes the results of the XRD analysis of the heat-treated phillipsite- and clinoptilolite-rich tuff tablets. Mineral phases reported are limited to zeolitic phases or their transformation products by heating, being the non-exchanging ancillary phases present in the tuffs heat stable and/or inert.

In the PHT sample (Fig. 1a) phillipsite breakdown ends at about 600°C. Alkali feldspar crystallization is observed at about 900°C; some residual analcime and an amorphous phases are also present at this temperature. At 1100°C the sample appears almost amorphous.

The XRD patterns of clinoptilolite-rich tuff sample (Fig. 1b), heat-treated at different temperatures, evidences that a part of the heat-stable zeolitic phase remains up to about 800°C. The material is practically amorphous at 900°C. A moderate crystallization of high-temperature silica phases, e.g., cristobalite, starts from about 1000°C.

3.2. Color of the fired compacts

As regards the zeolite A compacts, color markedly depends either on the nature and concentration of the cation introduced into zeolite or on the treatment temperature (Fig. 2).

In the Co-bearing samples, which are pale pink at room temperature (pink is the color of the hydrated Co^{2+} ion), color changes are recorded towards dark blue as soon as zeolitic water is removed and the color of anhydrous Co^{2+} prevails. A partial oxidation to amorphous Cr_2O_3 (black) may have occurred. Dark blue gradation becomes more and more intense up to about 1000°C. A marked color change in a brilliant cobalt bleu is observed at about 1100°C.

Cr-bearing samples, which are pale bluish green at room temperature, turn towards light green and green, as the firing temperature increases.

Color of the natural zeolite tablets, starting from that of the original tuff samples, go towards dark brown in any case, which is more intense for PHT because of the higher iron oxide content. Different brown shades may be obtained, according to the reached temperature level.

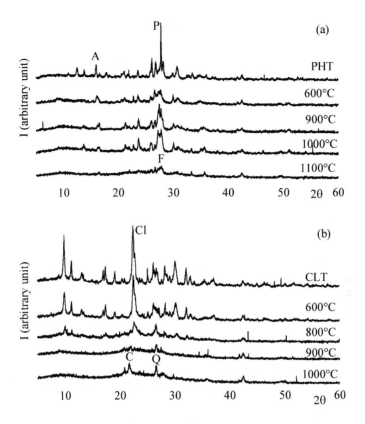

Fig. 1. XRD spectra showing the thermal evolution of phillipsite- (PHT, a) and clinoptilolite-rich samples (CLT, b), after treatment at different temperatures (the most intense peaks of A.=. analcime, C = cristobalite, Cl = clinoptilolite, F = feldspar, P = phillipsite, and Q = quartz, are indicated)

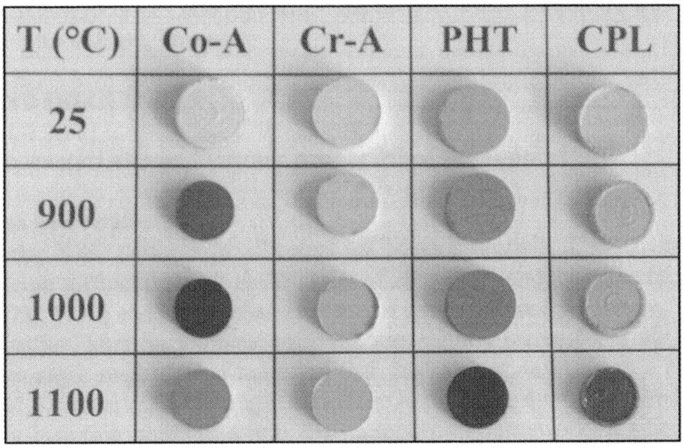

T (°C)	Co-A	Cr-A	PHT	CPL
25				
900				
1000				
1100				

Fig. 2. Colour changes in the fired zeolite compacts. Co-A (7.3% Co); Cr-A (2.6% Cr)

3.3. Physical properties of the heat-treated compacts

Table 2 summarizes the physical and mechanical properties of the heat-treated compacts. Data on the Co- and Cr-exchanged zeolite A samples are limited to those containing the highest metal concentrations (see Experimental). Data on zeolite A compacts are also reported for reference.

Table 2
Physical and mechanical properties of the heat-treated compacts

Sample	Firing temp. (°C)	Bulk density (g·cm^{-3})	Porosity (%)	Weight loss (%)	Linear shrinkage (%)	Compressive strength (MPa)
Zeolite A	900	1.42	34.55	20.9	14.5	66
	1000	1.36	33.23	20.9	14.7	78
	1100	1.49	31.92	20.9	15.6	81
Co-A[1]	900	1.73	25.15	24.2	19.9	92
	1000	1.84	24.30	24.3	21.2	95
	1100	1.90	17.44	24.3	23.3	60
Cr-A[2]	900	1.44	32.03	22.3	13.8	61
	1000	1.46	32.52	22.4	14.3	80
	1100	1.52	31.86	22.4	14.8	23
PHT	900	1.42	34.54	8.7	2.4	28
	1000	1.49	32.02	8.8	2.9	40
	1100	1.99	1.41	8.8	14.1	78
CLT	900	1.48	28.37	12.4	8.5	47
	1000	1.91	8.00	12.4	15.8	291
	1100	2.07	1.39	12.4	17.6	206

[1] Zeolite A containing 7.3 w.-% Co^{2+}. [2] Zeolite A containing 2.6 w.-% Cr^{3+}

Bulk density shows an increasing trend with firing temperature for all the samples. At 900°C, apart from the value of the Co loaded zeolite A sample (= 1.73 g·cm^{-3}), which is the highest one (a sign of an early sinterization [9]), the values relative to the synthetic zeolite samples and to the natural zeolite samples are very close to each other in the range of 1.42-1.48 g·cm^{-3}. At higher temperatures, whereas the data relative to the Co loaded zeolite A sample confirm a sharp increase (up to a value of 1.90 g·cm^{-3} at 1100°C), the bulk density of the Cr loaded zeolite A exhibits, as the reference sample, a slow increment. The natural zeolite bearing compacts, on their turn, show the highest increase of this parameter (up to values of 1.99 g·cm^{-3} and 2.07 g·cm^{-3} for PHT and CLT samples, respectively), due to an evident melting phenomenon (see the micrographs in Fig. 3 and the XRD spectra in Fig. 1).

Porosity (Table 2) varies accordingly. Note in particular the very low values relative to the PHT and CLT samples (1.41% and 1.39% at 1100°C, respectively), in agreement with the presence of a glassy, continuous phase.

Like bulk density, also linear shrinkage appears to increase with increasing temperature, as expected, although marked differences between the zeolite A samples and the natural zeolites samples may be detected. The diameter of the former compacts decreases, in fact, much more than the latter in the temperature range from room temperature up to 900°C, due essentially to the greater water loss of the more hydrated synthetic zeolite, compared to the natural zeolite-rich tuffs (see fifth column in Table 2).

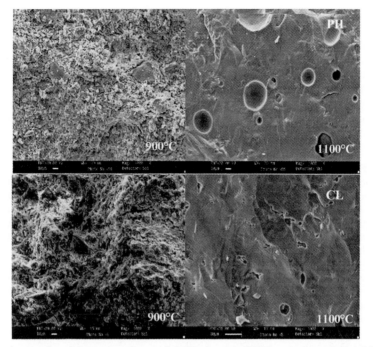

Fig. 3. SEM micrographs of fracture surfaces of phillipsite-rich (PHT) and clinoptilolite-rich (CLT) tuff compacts, heat-treated at 900°C and 1100°C.

A possible joint cause of the greater dimensional stability of the samples obtained from the natural products may be the presence in these of non zeolitic, dimensionally stable constituents, such as feldspar, mica, amorphous phases, etc., which may play the same role than the fillers play in the common clay ceramics. At temperatures higher than 900°C, a further gradual densification of the zeolite A samples results in a steady increase of linear shrinkage values. Unlike the synthetic zeolite samples, the mentioned melting phenomena of phillipsite- or clinoptilolite rich tuff samples in the temperature range 1000-1100°C give rise to very sharp shrinkage increases.

The last column of Table 2 reports the compressive strengths of the fired samples. Inspecting these data, which have however only an indicative value, because they have been measured in non standard conditions, allows an interesting observation to be made. Unlike the fired reference compacts of zeolite A, showing a constantly increasing trend in compressive strength as a function of temperature, the Co^{2+}-rich and Cr^{3+}-rich zeolite A compacts present a maximum for firing around 1000°C. This finding appears related to an incipient melting at temperatures higher than 1000°C. This phenomenon is particularly evident also in the CLT compacts, whereas it is absent in the PHT compacts, which gives strong indications of its dependence on several joint causes.

4. CONCLUSIONS

The results here presented show the great flexibility of zeolites as potential precursors of ceramic materials, in particular of colored ceramics. Variable color shades may be likely obtained, suitably selecting either the metal cation chromophore or, in dependence on the cation affinity, the proper zeolitic phase. The uniformity of the original framework cations and guest cations distribution in the zeolite framework, which results in an uniform distribution of color centers within the zeolite structure, gives rise to a perfectly uniform and safe coloring of the ceramic manufacts after firing. Interesting results have been obtained also starting from natural zeolite materials, although the chromophore agent is an external, but uniformly distributed, oxide. The physical and mechanical properties of the compacts are satisfactory, too, but further research on the control of shrinkage, due to water loss, is still needed.

REFERENCES

[1] J. Mckittrick, B. Hoghooghi and O.A. Lopez, J. Non-Cryst. Solids, 197 (1996) 170.
[2] B. Hoghooghi, J. Mckittrick, E. Helsen and O.A. Lopez, J. Am. Ceram. Soc., 81(4) (1998) 845.
[3] R.L. Bedard and E.M. Flanigen, Proc. 9th Int. Zeolite Conference, Vol. 2, R. Von Ballmoos, J.B. Higgins and M.M.J. Tracey (eds.), Butterworth-Heinemann, Stoneham, MA, 1993, p. 667.
[4] G. Dell'Agli, C. Ferone, M.C. Mascolo and M. Pansini, Solid State Ionics, 127 (2000) 309.
[5] S. Chandrasekhar and P.N. Pramada, Ceramics International, 27 (2001) 105.
[6] S. Chandrasekhar and P.N. Pramada, Ceramics International, 27 (2001) 351.
[7] C. Ferone, G. Dell'Agli, M.C. Mascolo and M. Pansini, Chem. Mat., 14 (2002) 797.
[8] S. Kowalak, M. Wrobel, N. Golebniak, A. Jankowska and B. Turkot, in Porous Materials in Environmentally Friendly Processes, I. Kiricsi, G. Pal-Borbely, J.B. Nagy, H.G. Karge (eds.), Studies in Surface Science and Catalysis, 125 (1999) 753.
[9] V.M. Pogrebenkov, M.B. Sedel'nikova and V.I. Vereshchangin, Glass and Ceramics, 55(1-2) (1998) 55.
[10] Ch. Baerlocher, W.M. Meier and D.H. Olson, Atlas of Zeolite Framework Types, Elsevier, Amsterdam, 2001, p. 96.
[11] B. Liguori, A. Cassese and C. Colella, J. Por. Mat., 2005, in press.
[12] C. Colella, in Handbook of Porous Solids, F. Schüth, K.S.W. Sing and J. Weitkamp (eds.), Wiley-VCH, Weiheim, Germany, 2002, Vol. 2, p. 1156.
[13] J.W. Patterson, Wastewaters Treatment Technology, Ann Arbor Science Publisher, Ann Arbor, Michigan, USA, 1975.
[14] M. de' Gennaro and A. Langella, Mineral. Deposita, 31(6) (1996) 452.
[15] M.N. Gündoğdu, H. Yalçin, A. Temel and N. Clauer, Mineral. Deposita, 31(6) (1996) 492.
[16] D.S. Coombs, A. Alberti, T. Armbruster, G. Artioli, C. Colella, E. Galli, J.D. Griece, F. Liebau, F. Mandarino, H. Minato, E.H. Nickel, E. Passaglia, D.R. Peacor, S. Quartieri, R. Rinaldi, M. Ross, R.A. Sheppard, E. Tillmanns and G. Vezzalini, Recommended Nomenclature for Zeolite Minerals: Report of the Subcommittee on Zeolites of the International Mineralogical Association, Commission on New Minerals and Mineral Names, Can. Mineral., 35 (1997) 1571-1606.
[17] S.J., Chipera and D.L., Bish, Powder Diffr., 10 (1995) 47.
[18] F.H., Chung, J. Appl. Cryst. 7 (1974) 519.
[19] A.C. Larson and R.B. Von Dreele, General Structure Analysis System (GSAS), Los Alamos National Laboratory, Report LAUR 86-748 (2000).
[20] C. Colella, in Characterization Techniques of Glasses and Ceramics, J.Ma. Rincon and M. Romero (eds.), Springer-Verlag, Berlin Heidelberg, Germany, 1998, p. 112.

Studies in Surface Science and Catalysis 155
A. Gamba, C. Colella and S. Coluccia (Editors)

Preparation and Spectroscopic Characterisation of Nitrogen Doped Titanium Dioxide

Stefano Livraghi, Annamaria Votta, Maria Cristina Paganini*, Elio Giamello

Dipartimento di Chimica IFM, Università di Torino and NIS, Nanostructured Interfaces and Surfaces Centre of Excellence, Via P. Giuria 7, I - 10125 Torino (Italy)

Photocatalytic oxidation constitutes one of the most promising methods for indoor and outdoor air purification. Solar energy contains only about 5% UV light and much of the rest is visible light. In order to utilize solar energy efficiently in photocatalytic reactions, it is necessary to develop a visible light reactive photocatalyst having smaller band gaps than TiO_2 rutile and anatase. In 1986 Sato reported that calcinations of NH_4Cl containing titanium hydroxide caused the photocatalytic sensitization of TiO_2 into the visible light region. The author proposed that the powder prepared according to the described method are actually NO_x-doped TiO_2 and that the sensitization of these materials is due to the presence of NO_x impurity. Several years after it has been reported that nitrogen-doped titania with a particularly high visible light photocatalytic activity had been prepared. The aim of the present work was to synthesize and characterize new materials based on N-doped TiO_2. The materials N/TiO_2 were prepared via sol-gel technique using solutions containing various kind of nitrogen compounds. UV-Vis diffuse reflectance spectroscopy and Electron Paramagnetic Resonance were the main experimental techniques used to characterized the materials. The latter technique was adopted to verify the presence of paramagnetic entities formed during the synthesis. In most cases the samples prepared via sol-gel reaction exhibit a pale yellow color.

1. INTRODUCTION

The search for a good titanium dioxide based photocatalytic system active with visible light has been the object of a great deal of research in recent years. Bare titanium oxide in fact is the most important photocatalyst for the decomposition of pollutants either in gas or in liquid phase because of its high activity and chemical stability. A limit of bare TiO_2 however consists in the amplitude of its band gap (3.2 eV for Anatase, the most catalytically active phase) which requires the use of UV electromagnetic radiation to produce the electron-hole separation. Among the materials based on TiO_2 a considerable interest was raised by those doped with non metal atoms. In particular, the most promising system seems to be that obtained doping the oxide with nitrogen. This reduces the amplitude of the gap between the valence and the conduction band [1].

In 1986 S. Sato [2] reported the preparation of a yellow material obtained by calcination of titanium hydroxide in the presence of ammonium chloride. The optical properties of this solid were attributed to rather undefined NO_x species supposed to produce impurity levels in the band gap of TiO_2. Later, in the nineties, several papers reported the preparation of N doped TiO_2 (N-TiO_2) according to various experimental procedures including mechanochemistry [3], DC magnetron sputtering [4], sol-gel synthesis [5], high temperature treatments in N_2 or NH_3 atmosphere [6] or oxidation of a precursors such as TiN [7]. However, despite the great deal of activity in the field, the typical features of N-TiO$_2$ causing its peculiar catalytic activity are not yet fully clarified. While Sato [2] points to the presence of rather undetermined nitrogen oxospecies (NO$_x$), Asahi et al. [1] suggested (based on theoretical calculations) the presence of nitrogen impurities in substitutional sites of the TiO$_2$ matrix as responsible of the band gap narrowing (caused by mixing of N and O 2p states) necessary for the onset of reactivity under visible light. More recently the presence of defective sites like oxygen vacancies and reduced centres [5,6] has been also suggested. In parallel with the debate on the reasons of the photocatalytic properties of N-TiO$_2$ an intense activity grew up aimed to characterize the typical features of this relatively new solid system. A controversial point of this debate concerns the presence and the role of a typical XPS feature at 396eV observed in the case of N-TiO$_2$ prepared by magnetron sputtering and assigned by the authors to N atoms present in the TiO$_2$ lattice in substitutional position. As a matter of fact such a feature was not observed in the case of other materials also active in visible light photocatalysis. Minor emphasis was devoted in the literature to the presence of paramagnetic defect species except for few poorly assigned signals reported by Sakatani et al. [8] and despite the fact that radical centres often are important defects of the solid state connected to important physical properties. The role of EPR in the characterisation of defects and impurities in the solid state is well known as, in many cases, the defective centres (dangling bonds, trapped electrons, atomic impurities) are paramagnetic [9].

The present paper describes the features of a series of N-TiO$_2$ materials prepared via sol-gel synthesis or alternatively via mechanochemical activation.

2. EXPERIMENTAL.

N-TiO$_2$ (100% anatase) samples were prepared via two different routes (sol-gel chemistry and mechanochemistry) using Ti compounds and various N containing compounds (NH$_4$Cl, N$_2$H$_4$) to obtain N doping. We will report hereafter procedure followed using NH$_4$Cl as nitrogen source. For both preparation routes the N/TiO$_2$ ratio in the starting mixture was the same (5 wt%). However the N content of the N/TiO$_2$ final samples was not determined. In both cases pale yellow samples were obtained with the typical optical absorption centred at about 450 nm. The solids show catalytic activity in visible light which was monitored according to the method proposed by Burda et al. [10] based on the methylene blue degradation under visible irradiation followed by U.V.-Vis. Spectrometer.

Sol-gel N-TiO$_2$ samples were prepared mixing a solution of titanium (IV) isopropoxide in isopropilic alcohol with a solution of water and keeping the mixture under constant stirring at room temperature until completed hydrolysis. The gel was left ageing for 15 hours at room temperature and subsequently dried at 70°C for 2 hours. The dried material was calcined in air at 770K from a minimum of one hour up to a maximum of 15 hours. After calcination the solid is pale yellow.

The mechanochemical preparation of N-TiO$_2$ was performed starting from an intimate mixture of powders of TiO$_2$ (synthesised via sol-gel method before described) and NH$_4$Cl which was put in a mechanical mill and ground using corundum balls for 1 hours. After 1 hour calcinations in air at 770K the activated material was put in the glass manifold for spectroscopic investigations.

For a better understanding of EPR spectra, isotopically label materials were also prepared using 70% enriched ^{15}NH$_4$Cl and following the same procedures described above. A reference sample of bare TiO_2 was also prepared via hydrolysis in pure water. XRD analysis of the sol-gel material after calcination indicates the exclusive formation of the anatase phase of TiO_2 with particle size of about 50 nm while in the case of the mechanically activated material a fraction of rutile wad detected by XRD whose formation has to be ascribed to the high energies developed during the mechanical milling.

The UV-Vis measurement were performed by means of a Varian Cary 5 spectrometer using a Cary win-U.V./scan software for data handling.

EPR spectra were run in the range between 77K and room temperature by a Bruker EMX spectrometer working in the X-band and equipped whit a cylindrical cavity. Computer simulation of the EPR spectra were obtained by the SIM32 program elaborated by Prof. Z. Sojka (Jagellonian University – Cracow).

3. RESULTS AND DISCUSSION.

The diffuse reflectance (DR) UV-Vis spectrum of a N-TiO_2 sample prepared via sol-gel is reported in Fig.1 where it is compared with that of bare TiO_2. Similar spectra (also in terms of adsorption intensity) were obtained using samples get by mechanochemical treatment. Fig. 1 shows that the N doped sample is characterised by an absorption in the visible region at about 450 nm and by a band gap narrowing of about 0.08 eV (derived from the plot of the square root of the Kubelka-Munk function $[F(R)E]^{1/2}$ against the photon energy E) [11].

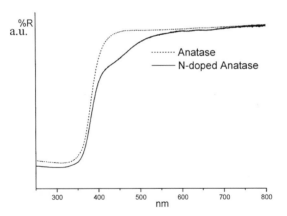

Figure 1. Diffuse Reflectance U.V.-Vis. spectra of pure and nitrogen doped TiO_2

All the samples exhibiting the described absorption in the visible also show a catalytic activity in the photodecomposition of methylene blue using visible light. The activity (details will be reported in a forthcoming paper) was measured by comparison with the activity of a sample of pure TiO_2 in the same experimental conditions.

A second important feature of N-TiO_2 is the presence of EPR signals not observed in the bare oxide. All the observed signals are due to N containing species and are thus related to the doping process. Two main EPR signals were observed after calcinations in air of the xerogel precursor (prepared using NH_4Cl as doping agent).

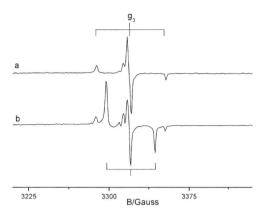

Figure 2. EPR spectra recorded at 298 K of species B present in the N-TiO$_2$ matrix. a: spectrum obtained using ^{14}N compounds; b: Spectrum obtained using ^{15}N enriched compounds.

The first one (species A) is observed at 77K and disappears rising the temperature at about 170K revealing the presence of a second signal (species B) which is observed also at room temperature. Both species were obtained using either ^{14}N or ^{15}N enriched (70%) compounds (Fig. 2 and 3).

The EPR spectrum of species B (Fig.2) is characterised by a g tensor with the three principal values extremely close one to the other (g$_1$= 2.003, g$_2$= 2.004, g$_3$= 2.005) and with a hyperfine structure dominated by a triplet of lines (the nuclear spin is I=1 for ^{14}N and a triplet of lines is expected for the hyperfine structure of a single N atom in the species) with 32 G separation. The whole tensor is as follows: A$_1$=32 G, A$_2$≅0 G, and A$_3$≅5 G. The second triplet at 5 G is not fully resolved in the experimental spectrum (Fig. 2a). These features are fully confirmed by the spectrum generated using ^{15}N enriched compounds (Fig. 2b, about 70% of enrichment). Since the nuclear spin of ^{15}N is ½ the main triplet is now a pair of hyperfine lines separated by 45 G as expected on the basis of the ratio of the two g nuclear factors of ^{14}N and ^{15}N which is 1.41. It has to be noticed that Fig.2b still shows the ^{14}N trace which contributes for some 30% to the whole spectrum. The illustrated features are typical of a species containing one N atom with half of the total spin density concentrated in the p orbital of this atom.

Species A signal shows up by cooling the system at 77K. As easily deduced from the spectra (Fig. 3) also A contains one nitrogen atom and is thus related to the doping process of the solid. Fig. 3 reports the EPR spectra of species A respectively obtained employing in the synthesis ^{14}NH$_4$Cl (3a) and ^{15}NH$_4$Cl (3b). The spectrum in Fig. 3a, whose main feature is a ^{14}N hyperfine triplet centered at about g = 2.0 is, differently from the case of species B, characterized by a rhombic g tensor (g$_1$ = 2.001, g$_2$ =1.998, g$_3$ = 1.927) with one value (g$_3$) pronouncedly lower than the free spin value (2.0023). Also the A tensor is anisotropic with A$_1$ ≅ 0 , A$_2$ = 32.12 G, A$_3$ ≅ 10 G. The latter value is not completely resolved in the spectrum and has been deduced by comparison of the two experimental spectra (3a and 3b). In the case of ^{15}N signal each hyperfine triplet is substituted by a doublet whose separation is again that expected on the basis of the ratio of the two μ nuclear factors.

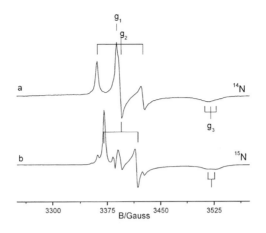

Figure 3. EPR spectra (recorded at 77K) of species A, trapped into the N-TiO$_2$ matrix. a: spectrum obtained with ^{14}N compounds. b: Spectrum obtained with ^{15}N enriched (70%) compounds.

The structure of species A signal is in agreement with what expected for a 11e$^-$ diatomic π radical and is assigned to nitric oxide (NO). The ground state configuration of this radical is: $1\sigma^2\,2\sigma^2\,3\sigma^2\,4\sigma^2\,1\pi^4\,5\sigma^2\,2\pi^1$ and its spectrum is not observed in the spectral region of the free electron unless the degeneracy between the two 2π antibonding orbitals is lift by the effect of some asymmetric electric field [12,13,14]. In this case the unpaired electron is confined, at first approximation, in the π orbital with lower energy. This leads to a rhombic g tensor whose elements are, at first order, as follows:

$$g_{xx} = g_1 = g_e$$

$$g_{yy} = g_2 = g_e - 2\,\lambda\,/E \qquad (1)$$

$$g_{zz} = g_3 = g_e - 2\lambda\,/\Delta$$

where λ is the the spin orbit coupling parameter associated to the investigated radical while Δ is the $2\pi_g^x - 2\pi_g^y$ separation and E is the $2\pi_g^x - 5\sigma_g$ separation. The g$_{zz}$ (g$_3$) component (z is now the direction of the internuclear axis) is a measure of the value of Δ and indirectly of the weak field perturbing the NO orbitals. The fact that both A and B species are not affected by outgassing up to about 470K and the absence of dipolar broadening when the EPR spectra are recorded in the presence of molecular oxygen indicates that the two N containing species here described are located into the bulk of N-TiO$_2$ microcrystals. The NO molecule is probably segregated in some morphological imperfection of the structure (microvoids) formed along the preparation of the solid.

The assignment of species B is not straightforward. The electronic structure and the g values of the species described above are compatible with the properties of an oxo-nitrogen radical ion like NO$_2^{2-}$, a 19-electrons bent species with the unpaired electron confined into a 2b$_1$ antibonding orbital mainly involving a p orbital of the central nitrogen atom and two parallel p orbitals of the oxygen atoms [15]. Formation of NO$_2^{2-}$ indicates a deep chemical interaction between the ammonium ions employed in the preparation and the oxide phase. However the features of species B are compatible also with other nitrogen species deeply interacting with the oxide lattice. In particular recent calculations [16] on the state of the N impurity in the TiO_2 anatase lattice point to the formation of an isolated paramagnetic state in the band gap close to the limit of the valence band edge. A similar isolated N center should carry, as indicated by preliminarily calculations, a spin density of about 0.5 with the remaining spin density distributed on the nearby oxygen and titanium ions. Such a results is in nice agreement with the features of species B. Further experiments (possibly EPR of single crystal) and elaborations are needed to reach a complete understanding of the nature and structure of this center. Whatever the exact nature of the B species the fact remains that the latter one is placed in the bulk of the system and is the result of a deep interaction with the matrix.

In conclusion we have shown that $N-TiO_2$, an intriguing photocatalytic system active in visible light, contains in its bulk two types of radical centers. These are formed during the synthesis of the material (performed by two different methods) by interaction of an oxidic phase with nitrogen containing compounds. The two centers are observed after calcination in oxygen up to 770K indicating that we are dealing with stable non transient species well stabilized into the oxide matrix. The first one is molecular NO permanently trapped into the crystal and the second one is a nitrogen based paramagnetic center deeply interacting with the oxide and characterized by a remarkable spin density in a p orbital of the nitrogen atom. The connection between one or both these species with the photocatalytic activity of $N-TiO_2$ is currently under investigation in our laboratory.

REFERENCES

[1] R. Asahi, T. Morikawa, T. Ohwaki, K. Aoki, Y. Taga, Science, 293 (2001) 269.
[2] S. Sato, Chem. Phys. Lett., 123 (1986) 126.
[3] S. Yin, Q. Zhang, F. Saito, T. Sato, Chem. Lett., 32 (2003) 358.
[4] T. Lindgren, J.M. Mwabora, E. Avendano, J. Jonsson, A. Hoel, C. Granqvist, S. Lindquist, J. Phys. Chem B, 107 (2003) 5709.
[5] a) S. Sakthivel, H. Kisch, ChemPhysChem ,4 (2003) 487. b) T. Ihara, M. Miyoshi, Y. Iriyama, O. Matsumoto, S. Sugihara, Applied Catal. B: Envir., 42 (2003) 403.
[6] H. Irie, Y. Watanabe, K. Hashimoto, J. Phys. Chem. B, 107 (2003) 5483.
[7] T. Morikawa. R. Asahi, T. Ohwaki, K. Aoki, Y. Taga, Japn. J. Appl. Phys., 40 (2001) L561.
[8] Y. Sakatani, J. Nunoshige, H. Ando, K. Okusako, H. Koike, T. Takata, J.N. Kondo, M. Hara, K. Domen, Chem. Lett., 32(12) (2003) 1156.
[9] F. Agullo Lopez, C.R.A. Catlow, P.D. Towsend, Point defects in materials, Academic Press pp 10-12. (1988).
[10] C. Burda, Y. Lou, X. Chen, A.C.S. Samia, J. Stout, J. L. Gole, Nanoletters, 3 (2003) 1049.
[11] D. G. Barton, M. Shtein, R. D. Wilson, S. L. Soled, E. Iglesia, J. Phys. Chem. B 103 (1999) 630.
[12] C. Di Valentin, G. Pacchioni, M. Chiesa, E. Giamello, S. Abbet, H. Heiz, J. Phys. Chem. B, 106 (2002) 1637.
[13] J. H. Lunsford, J. Phys. Chem., 72 (1968) 2141.
[14] M. Primet, M. Che, C. Naccache, M. V. Mathieu, B. Imelik, J. Chim. Phys., 67 (1970) 1629.
[15] P.W. Atkins, M. C. R. Symons, The Structure of Inorganic Radicals, Elsevier (Amsterdam) pp 148 (1967).
[16] C. Di Valentin, G. Pacchioni, A. Selloni, Phys. Rev. B, 70 (8) (2004) 85116.

Studies in Surface Science and Catalysis 155
A. Gamba, C. Colella and S. Coluccia (Editors)

Immobilization of Lipase on microporous and mesoporous materials: studies of the support surfaces

A. Macario[1], A. Katovic[1], G. Giordano[1], L. Forni[2], F. Carloni[2], A. Filippini[2] and L. Setti[2]

[1]Dipartimento di Ingegneria Chimica e dei Materiali, Università della Calabria - Via P. Bucci – 87036 Rende (CS) – Italy

[2]Dipartimento di Chimica Industriale e dei Materiali, Università degli Studi di Bologna – Viale del Risorgimento, 4 – 40136 Bologna - Italy

In this paper the performance of different supports respect to the lipase immobilization was investigated. The used enzyme was the lipase from *Rhizomucor miehei* (RML – commercial name: Palatase). The immobilization tests on mesoporous materials (MCM-41), delaminated zeolites (ITQ-2 and ITQ-6), Na-Silicalite-1, H-Silicalite-1 and FAU zeolites, were carried out by adsorption. The pore size, morphology, crystal dimension, acidity, hydrophobicity and chemical composition of the supports, strongly influence the amount of the enzyme adsorbed. The Na-silicalite-1 support shows the best lipase immobilization capacity, with an efficiency of c.a. 74% respect to the 41% of MCM-41 and the 27% of ITQ-2 type materials, while no enzyme is retained on the zeolite FAU (zeolite X). The preliminary activity tests using the lipase-support as a catalyst, were carried out for the reaction of hydrolysis of triglycerides. The activity observed for the lipase-Na-Silicalite-1 and the lipase-MCM-41 catalysts is 86% and 78%, respectively, respect to the activity of the free lipase enzyme in solution.

1. INTRODUCTION

Lipases catalyzing lipid modifications have attracted considerable attention over the past decade. The attractive aspects of this catalysis over chemical methods include: the high specificity of some lipases, the mild conditions under which the reactions take place thereby requiring minimal energy inputs, reduced levels of by-products generated during the reaction and more efficient conversion of labile substrates. A considerable number of processes use immobilized biocatalysts. The lipase enzyme catalyzes hydrolysis of long chain, insoluble triglycerides and other insoluble esters of fatty acids with varying chain length specificity. Apart from their biological significance, they play an important role in biotechnology, not only in food and oil processing but also in the preparation of chiral intermediates. In fact, about 30% of all bio-transformation reactions reported up to today have been performed with lipases. Immobilization of lipases has been done either by physical adsorption method or by chemical linkage using synthetic polymers or natural polymeric derivatives [1-10] as well as inorganic materials such as diatomaceous earth (Celite), controlled-pore glasses, silica, zeolites, phyllosilicates, mesostructured oxides, layered double hydroxides, ceramics, inorganic matrices based on sol–gel processes and microemulsion-based gels [11].
Immobilization by adsorption technique is a very attractive procedure because of its simplicity. In this context, the use of zeolite materials is of great interest. The zeolites and

related materials have potentially interesting properties, such as large surface area (between 200 and ~ 1000 m^2/g), hydrophobic or hydrophilic behavior, and electrostatic interactions, with the possibility of different ion-exchanged forms, mechanical and chemical resistance. Additional advantages such as their ease of water dispersion/recuperation, as well as the high water uptake capacity can also be mentioned. Therefore, the compositional and structural variances of molecular sieves offer a powerful tool for tuning the carrier properties. The main drawback of zeolites consists in their small pore sizes, i.e. the enzyme cannot enter the inner channels of the zeolite matrix. Later, this inconvenience was surpassed by the synthesis of mesoporous materials like MCM-41. As it can be seen, a few reports are exclusively dedicated to the lipase immobilization on zeolites and related materials [11]. In contrast to the majority of enzymes, which preferably adsorb on the materials having a polar surface, lipases are found to adsorb better on lipophilic carriers due to their peculiar physicochemical character [12, 13].

All lipases are members of the α/β hydrolase fold family [13]. The α/β hydrolase fold family consists of a central hydrophobic eight-stranded β-sheet packed between two layers of amphiphilic α-helices (coiled chain of amino acid units). Moreover, in most lipases, a mobile element (the 'lid') covers the catalytic site in the inactive form of the lipase. This 'lid' consists of one or two short α-helices linked to the body of the lipase by flexible structure elements. The configuration of the helix is maintained by hydrogen bonds. In the open, active form of the lipase, the lid moves away and makes the bonding site accessible to the substrate [14]. Therefore, in order to immobilize the lipase in/on an inorganic support (such as mesoporous or microporous materials), the value of the isoeletric point is not the only aspect to consider but, also, the hydrophobicity of the support is fundamental. In fact, preceding studies have shown that the pH value during the immobilization on the same kind of support, has no influence on the amount of the adsorbed lipase [4]. In this way lipases become adsorbed through a combination of hydrophobic, Van der Waals and electrostatic forces.

This paper presents our investigations on the synthesis of the micro- (such as Silicalite-1, delaminated zeolites ITQ, Faujasite X) and mesoporous materials (MCM-41) with different physical and chemical properties. The aim of this work, considering the mentioned lipase properties, was the selection of a suitable support for *Rhizomucor miehei* lipase immobilization. The selection of a suitable support must be preceded by characterization studies that yield fundamental information on the solid and surface properties. Therefore, various techniques of investigation (XRD, N_2 adsorption/desorption, FT-IR, NH_3-TPD, SEM) were applied in order to characterize the final support. The enzyme-supported material was tested as biocatalyst for the hydrolysis of triglycerides contained in olive oil.

2. EXPERIMENTAL

2.1 Materials

A commercial lipase from *Rhizomucor miehei* (RML), Palatase, from NovoNordisk, Denmark, was used. The synthesized supports are MCM-41 and Al-MCM-41, Silicalite-1 and FAU (zeolite X). The delaminated zeolites, ITQ-2 and ITQ-6, were kindly supplied by ITQ – Valencia. The MCM-41 materials were synthesized with different silica sources.

The chemicals used for the M41S type materials synthesis were: sodium silicate solution (water glass, Merck), Zèosil 175 (Rhodia), Tixosil (Rhodia) and Silica Fumed 200 (Aldrich) as silica sources, cetyl-trimethylammonium bromide (Aldrich) as template, tetraethylammonium hydroxide solution 40% (Fluka) and sodium hydroxide (Carlo Erba) as mineralizing agents and aluminum nitrate (Carlo Erba) as alumina source. The chemicals used for the Silicalite-1 synthesis were the silica gel precipitated (BDH) as silica source, sodium hydroxide (Carlo Erba) and tetrapropylammonium bromide (Fluka). For the synthesis of FAU zeolite, colloidal silica (Ludox AS-40, Aldrich), aluminium hydroxide (Aldrich) and sodium hydroxide (Carlo Erba) were used.

The pigments used for colorimetric tests were Eosin B (Sigma-Aldrich) as anionic pigment, and Methylene Blue (Sigma-Aldrich) as cationic pigment.

2.2 Mesoporous Materials and Zeolites synthesis procedure

The main synthesis conditions of the M41S and Silicalite-1 type materials are reported in the Table 1 and 2. The molar gel composition for preparation of the FAU zeolite was: 3 Na_2O-5 SiO_2 – 0.5 Al_2O_3 –100 H_2O. The crystallization time and temperature of the FAU type material were 24 h ed 80°C, respectively.

Table 1
Synthesis parameters of the mesoporous materials
(Molar gel composition: 1 SiO_2-0.12 CTABr- z X_2O- y Al_2O_3 - 30 H_2O)

Sample	Silica Source	Cation Type	X_2O (z)	Synthesis time (h)	Temperature (°C)	Final Phase
M1-Z	Zèosil	TEA$^+$	0.07	144	140	MCM-41
M2-T	Tixosl	Na$^+$	0.15	24	140	MCM-41
M3-W	Water Glass	Na$^+$	0.29	1	30	MCM-41
M4-F	Silica Fumed	Na$^+$	0.15	72	140	MCM-41
MA-W*	Water Glass	Na$^+$	0.29	24	30	Al-MCM-41

*with aluminum: y = 0.01 Al_2O_3 and 100

Table 2
Main synthesis condition of the Silicalite-1 type materials.
(Molar gel composition: 1 SiO_2 - x Na_2O – y Al_2O_3 - 0.08 TPABr – 20 H_2O)

Sample	Na_2O (x)	Al_2O_3 (y)	Ageing time (days)	Synthesis time (h)	Crystallization Temperature (°C)	Final Phase
NaS-1	0.12	-	5	24	170	Na-Silicalite-1
NaS-2	0.08	-	5	24	170	Na-Silicalite-1
NaS-3	0.04	-	6	24	170	Na-Silicalite-1
AlS-4	0.04	0.01	6	24	170	MFI

All the solid phases were recovered by filtration and washed with distilled water and dried at 110°C for 24 h. Samples were calcined in air at 550°C for 12 h in order to remove the template.

The H-Silicalite-1 was obtained by a two-fold ion exchange procedure at controlled basic pH. After separation, the solid sample was dried at 110°C for 24 h and calcined in air at 550°C for 12 h in order to obtain the proton form of the zeolite. The name of the exchanged sample is reported, for example, as HS-1.

2.4 Characterization of the supports

The final phases of all calcined samples were evaluated from the XRD measurements using a Phillips PW 1710 diffractometer with CuKα radiation. The specific surface area, the pore volume and the pore diameter of all samples, after calcination, were determined by N_2 adsorption/desorption analysis at 77K using an ASAP 2010 Micromeritrics instrument.

The final crystal morphology and dimension of the Silicalite-1 samples prepared with different sodium amount applying different ageing time, were studied by scanning electron microscopy (SEM).

The NH_3-TPD analysis were carried out in the conventional flow apparatus TPD/RO 1100 Thermofinningan, equipped with a thermal conductivity detector (TCD). The sample was pretreated at 450°C for 1 h under helium flow. After cooling, the sample was pretreated with a mixture of NH_3-He at isotherm conditions, 100°C, for 1 h in order to have only the chemi-adsorbed ammonia. Then the sample was heated at a rate of 10°C/min. The TPD curve of the desorbed ammonia was elaborated in order to obtain the area and the top temperature of the desorption curve.

For the FT-IR measurements, the calcined samples were pressed into thin, self-supporting wafers (surface density about 20 mg cm^{-1}) and the spectra were collected, at a resolution of 2 cm^{-1} in the 4000-450 cm^{-1} range, on a Perkin Elmer FTIR spectrophotometer equipped with MCT detector. Pre-treatments were carried out using a standard vacuum frame (residual pressure < 10^{-4} mbar) in an IR cell equipped with KBr windows. Wafers were outgassed at 500°C before each IR investigation.

2.5 Enzyme immobilization

The immobilized enzyme was prepared according to the following procedure: zeolite or mesoporous material powder (0.4 g) was added to 50 ml of a 0.5 M phosphate buffer (pH 7) containing 1 g of RML. The mixture was stirred for 24 h at 0°C and, then, filtered. The support with immobilized lipase was washed two times with de-ionized water and dried at 25°C overnight. The amount of immobilized enzyme was calculated by the UV Adsorption Methods [15]: protein characteristically exhibit an ultraviolet absorption maximum at 280 nm due primarily to the presence of tyrosine and tryptophan. This method for determination of total protein should meet most general as well as specialized applications. The advantages of all the UV absorption techniques for total protein determination is that the sample is retrieved intact. However, the presence of non-protein substances that absorb in the UV region usually invalidate the procedure. In this paper the possible presence of other molecules that may absorb at 280, as phenols and/or poly-phenols, is evaluated by appropriated analysis published by Setti *et al.* [16] and Folin & Ciocaulteau [17]. Also the amount of the glucose presents in the lipase solution was determined by the spectrophotometric method proposed from Bailey et al. [18]. The presence of the glucose in a enzyme solution is generally necessary in order to preserve the enzyme stability.

The enzyme concentration of the initial solution and filtrate was calculated from the calibration curve determined by Perkin-Elmer UV Spectrophotometer and using the BSA as a standard protein. A linear regression equation is applicable to the obtained data: *Concentration = 2,2423 * Absorbance* with a correlation coefficient of 0,999. The amount of enzyme adsorbed on the support (W_{IL}, [mg]) was determined from the following lipase mass balance: $W_{IL} = C_0 V_0 - C_f V_f$, where C_0 is the initial enzyme concentration (mg ml^{-1}); V_0 is the initial volume of lipase solution (ml); C_f is the enzyme concentration of the filtrate (mg ml^{-1}); V_f is the volume of the filtrate (ml).

2.6 Phenols determination methods by Setti *et al.* [16]

This qualitative method involves the oxidative connecting reaction between the 3-methyl-2-benzotiazolinone hydrazine (MBTH – Carlo Erba) and the ortho-methoxyphenol (Guaiacol - Merck) in presence of hydrogen peroxide and peroxidase enzyme. Hydrogen peroxide and peroxidase are in excess respect to the substrate. The reaction is carried out at room temperature for a few minutes. The absorbance peak of the obtained red complex is read at 505 nm. Since this method concerns the evaluation of the peroxidase activity, if no phenols are present in the used substrate (in our case: commercial lipase) no red complex is formed.

2.7 Total Poly-phenols determination by Folin-Ciocalteau method [17]

This method utilizes a reagent (Folin-Ciocalteau's phenols reagent) that in presence of phenols or sugar forms a blue compound. In order to determine the total phenols in the lipase solution, the interferences due to the sugar molecules have been eliminated using a column C-18 end-capped Phenomenex-Strata, 1 g/ 6 ml, preliminary washed with 2 ml of methanol and 5 ml of sulfuric acid 0.01 N. After column activation, 1 ml of the sample is adsorbed on the column with 2 ml of sulfuric acid in order to eliminate the present sugar. The sample adsorbed into the column, without the sugar molecules, is washed with 2 ml of methanol and 5 ml of distilled water and put into a graduated cylinder. The mixture is diluted to a volume of 20 ml. 1 ml of Folin reagent is added to the mixture and after five minutes 4 ml of 10% w/v Na_2CO_3 are introduced. The reaction time is 1 hour and 30 minutes. The blank is obtained with 20 ml of a solution containing 2 ml of methanol, 1 ml of Folin reagent, 4 ml of 10% w/v Na_2CO_3 and distilled water. The absorption peak of the final complex is at 700 nm and the phenols' concentration is calculated from the calibration curve obtained with catechin (Sigma) as a standard phenol. The obtained calibration equation, with a correlation coefficient of 0.9945, is *Catechin Concentration [mg/l] =212.8 ABS_{700}*.

2.8 The spectrophotometric determination of reducing sugars by Bailey *et al.* [18]

This method, known also as DNS (3,5-dinitrosalicylic acid) assay, consists in a spectrophotometric determination of the reduced acid (red-brown color) obtained from the oxidation-reduction reaction between glucose and DNS acid leading to gluconic acid and 3-amino-5-nitrosalicylic acid (reduced acid) formation. DNS solution is prepared solving 0.25 g of DNS and 75 g of NA/K tartrate in 50 ml of NaOH 2 M adjusting the volume at 250 ml with distilled water. The 3-amino-5-nitrosalicylic acid to sugars ratio is a stoichiometric one. 600 μl of DNS solution are added to 400 μl of glucose solution at a concentration between 0,2 and 0,8 mg/mL. The reaction is carried out at 100°C for 7 minutes. The absorbance is read at the λ value of 550 nm. The amount of the present sugar is obtained by the calibration curve made with solutions having a known concentration of glucose. The obtained calibration equation, with a correlation coefficient of 0.9934, is *Glucose Concentration [mg/ml] = 1,0998 ABS_{550} – 0,0363*.

2.9 Pigment Adsorption

The amount of pigment adsorbed depends on the temperature, pH value and concentration of the solution. For all tests, the pH of the solution was the same as the pH used for the immobilization test. 200 mg of adsorbent (only the support or the support-lipase system) were added to 40 ml of an aqueous solution of the pigment (0.01 mmol ml^{-1}). The mixture was stirred at room temperature for 18 h. The choice of this temperature depends on the thermal stability of the enzyme. In order to determine the pigment amount adsorbed, the weight loss of the filtered and washed samples of adsorbents containing pigment was evaluated between 120 to 800°C in air flow by thermal analysis.

2.10 Activity of the immobilized enzyme

By the Tietz and Fiereck method [19], the activity of the immobilized lipase, compared to the activity of the lipase in its native form, was tested. This method involves hydrolysis of triglycerides contained in olive oil into fatty acid, diglycerides and to some small extent into monoglycerides and glycerol. This activity assay procedure was the same for both, soluble and immobilized, lipases.

3. RESULTS AND DISCUSSION

3.1 Structural characteristics and surface properties of the support

3.1.1 Mesoporous Materials

Figure 1 shows the XRD patterns of siliceous MCM-41 materials. The calcined materials exhibit a very intense diffraction peak in the range of $2\theta = 1 \div 3$ and the reflections which can be indexed as [100], [110] and [200], characteristic of the hexagonal symmetry. The intensity of the [100] diffraction, after the template removal, was slightly increased, while the peak was shifted to the low-angle region. This indicates that the calcination leads to a better-defined structure of MCM-41 materials and a slight shrinkage in pore size due to the condensation of -SiOH groups. By inserting aluminum into the mesoporous framework, a silicon-aluminum MCM-41 is obtained. The pore size, specific surface area and pore volume of the calcined mesoporous materials were reported in Table 3. These results show that typical MCM-41 materials with a d_{100} value between 25 – 35 Å were formed. The wall-thickness of calcined samples was calculated considering the difference between the unit cell and pore diameter calculated from data obtained by N_2 adsorption/desorption technique. For all samples the large pore volume indicates that no structural collapse occurs during calcination. The main difference between the syntheses parameters of the mesoporous materials is the type of the utilized silica source.

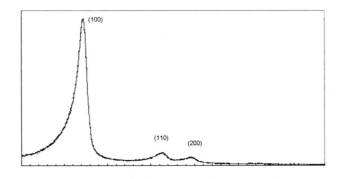

Fig. 1. XRD pattern of Si-MCM-41 calcined sample (M4-F)

Table 3
Pore structure parameters of the MCM-41 supports

Sample	Specific Surface Area (BET) [m² g⁻¹]	Mesoporous Volume [cm³ g⁻¹]	d_{100} [Å]	a_0 [Å]	Pore diameter [Å]	Wall-Thickness [Å]
M1-Z	1028	1,08	45	52	35	17
M2-T	1289	0.89	36	42	25	17
M3-W	1333	1,11	35	40	30	10
M4-F	1306	0,86	33	38	26	12
MA-W	1199	1,21	38	44	30	14

Table 4
NH_3-TPD results of the Si-MCM-41 samples

Sample	Silica Source	mmol NH3 DES/g	NH_3-TPD Desorption Peak Area (%)	Peak Temperature [°C]
M1-Z	Zèosil	0,100	12	183
M2-T	Tixosl	0,190	16	178
M3-W	Water Glass	0,012	27	263
M4-F	Silica Fumed	0,057	6	185

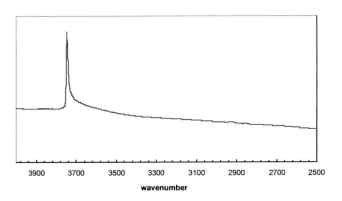

wavenumber

Fig. 2. FT-IR spectra of the hydroxyl region for Si-MCM-41 support (M1-Z sample) outgassed at 500°C for 2 h

The top temperature of the desorption peak of NH_3 on MCM-41 synthesized with Zèosil as silica source (sample M1-Z), is about the same as that of the samples M4-F and M2-T, synthesized with silica fumed and Tixosil as silica sources. But the peak area increases when the precipitated silica (Zèosil or Tixosil) are used, that is to say the number of weak acid sites increases. Between the two precipitated silicas, the amount of NH_3 adsorbed is higher for the sample synthesized with Tixosil than the sample synthesized with Zèosil. This aspect indicates that the acid sites of the sample M2-T are stronger than those of the M1-Z sample. The top temperature of the desorption peak (263°C) and the amount of the adsorbed NH_3

(0,012 mmol/g) on the M3-W sample (synthesized with Water Glass) indicate that this support has few acid sites but they are stronger than those of the other silica mesoporous materials.

3.1.2 Zeolites
The first aspect to observe in the synthesis procedure of the Silicalite-1 type supports is the effect of the sodium amount in the initial gel. The amount of sodium strongly influences the crystal dimension of the final products. The crystal size decreases by increasing the sodium amount as it can be observed from SEM micrographs (see Figures 3, 4 and 5). Moreover, a lower amount of sodium allows obtaining a more regular crystal morphology. Another interesting aspect regarding the structure of the microporous supports is the presence of silanol group types. From the results obtained by FT-IR, a higher quantity of free -SiOH groups (3740 cm^{-1} band) in the case of ITQ-2 delaminated zeolites (having high hydroxilated external surface [21]) respect to the Na-Silicalite-1 type support (Figure 6) can be observed. Between the Na-Silicalite-1 type supports, the amount of these terminal silanols decreases with the increase of the sodium amount (a weak band at 3740 cm^{-1} is observed for NaS-2 sample), but from the NH$_3$-TPD results it emerges that the acid strength of all Na-Silicalite-1 samples is quite the same (Table 5).

After ion exchange, the NaS-3 sample lost several terminal silanols while the quantity of the bridged silanols increased as it can be clearly observed from the FT-IR results (Figure 7). Finally, when the aluminum is present in the zeolite framework, the hydrophilic character and acidity of the support increase.

Table 5
NH$_3$-TPD results of the Na-Silicalite-1 samples

Sample	Na$_{uc}$ (Atomic Adsorption Results)	mmol NH3 DES/g	NH$_3$-TPD Desorption Peak Area (%)	Peak Temperature [°C]
Na-S-1	2,18	0,19	73	163
Na-S-2	1,87	0,13	74	166
Na-S-3	1,17	0,11	64	160

Fig. 3. (a) SEM image of the silicalite-1 (NaS-1 sample) – (b) magnification of the same sample

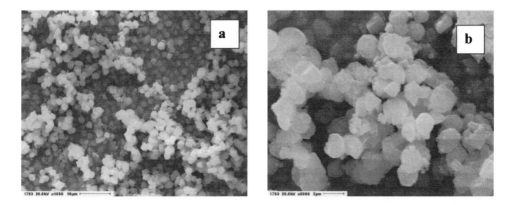

Fig. 4. (a) SEM image of the silicalite-1 (NaS-2 sample) – (b) magnification of the same sample

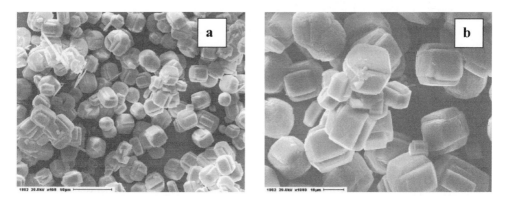

Fig. 5. (a) SEM image of the silicalite-1 (NaS-3 sample) – (b) magnification of the same sample

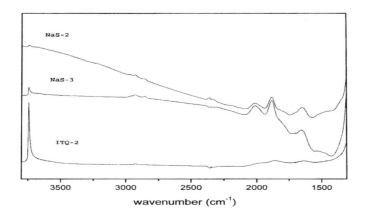

Fig. 6. FT-IR spectra of microporous type supports outgassed at 500°C for 2 h (confront between Na-Silicalite-1 and ITQ type zeolites)

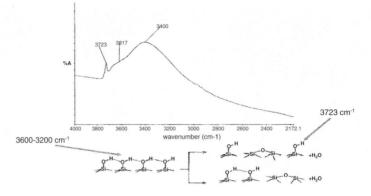

Fig. 7. FT-IR spectra of HS-3 sample outgassed at 500°C for 2 h

3.2 Adsorption of the enzyme on the supports

The immobilization of enzymes on/into the surface area of supports by physical adsorption is one of the simplest methods of immobilization and involves low energy binding forces, e.g. Van der Waals interactions, hydrophobic interactions, hydrogen bonds, ionic bonds. Thus, the enzymes can be simply retained on the surface of the support without forming covalent bonds between the enzyme and the support. It is of interest to note that lipase display a statistically significant enhanced occurrence of non-polar residues close to the surface, clustering around the active site [11]. Consequently, in contrast the majority of enzymes, which preferably adsorb on materials which have a polar surface, lipases are better adsorbed on hydrophobic carriers due to their peculiar physicochemical character [4]. When zeolites are used as support, the enzyme can be adsorbed only on the external surface. In the case of the mesoporous materials, if they have pores enough large to accommodate the molecules of the enzyme, the adsoption can occur inside their pores. The estimated spherical molecular diameter of the used enzyme (Palatase) is 41 Å .The pore size of the synthesized MCM-41 is always lower than the enzyme dimension (25 ÷ 35 Å). Therefore, no internal adsorption can take place. These mesoporous materials were synthesized without taking into account the possibility to adsorb the enzyme on their internal surface because, from our point of view, a so-prepared catalyst could be involved considerably in the mass transfer resistances during the reaction.

The adsorption results are reported in the Table 6. Regarding the immobilization of lipase on microporous mterials, the amount of the enzyme adsorbed on the Na-Silicalite-1 ($\approx$ 400 mg/g or 0,4 mg/mg) is higher than those reported for the other zeolite, such as MCM-22 (20 mg/g), MCM-36 (4 mg/g) [11]   and zeolite Y (8,2 mg/g) [8].

Among the Na-Silicalite-1 type supports, the amount of the adsorbed enzyme increases with the increase of the amount of the -SiOH groups. For the exchanged NaS-3 sample (HS-3), the amount of the adsorbed enzyme decreased due to the decrease of the terminal silanol groups.

The aluminum present in the zeolite framework generates Brönsted acid sites (sample AlS-4). Their presence in the support that has the same morphology properties as the NaS-3 sample caused a lower adsorption amount of enzyme. Moreover, the presence of aluminum induced a higher hydrophillicity respect to the pure silica forms and thus a lower affinity towards high

Table 6
Lists the adsorbed amount of RML for the various tested supports

Sample	Immobilized Lipase [mg]	Imm. Lipase/Supp. [mg/g]	% of Immobilized Lipase
M1-Z	68,2	170,5	41
M2-T	0	0	0
M3-W	18,9	47,2	14
M4-F	0	0	0
MA-W	14,0	35,0	7
NaS-1	79,6	199,0	60
NaS-2	120,1	300,2	69
NaS-3	159,1	397,8	74
HS-3	94,0	235,0	66
Al-S-4	76,1	190,2	61
ITQ-2	53,0	132,5	27
ITQ-6	33,4	83,5	17
FAU	0	0	0

hydrophobic molecules such as the lipase enzyme. For the same reason, no enzyme was retained on the FAU type support.

The delaminated zeolites presented a high amount of adsorbed enzyme due to the presence of a high amount of hydroxyl groups. Probably, their lower performance with respect to the Silicalite-1 type supports was due to their amorphous surface therefore to the different silanols distribution that seemed to be optimal in the case of the sample NaS-3. In fact, the presence of silanols is necessary for the lipase attachment, but their distribution has to be appropriated for the physical conformation of the enzyme. Moreover, the hydroxylated cage surface is not accessible to the enzyme due to the cages having a smaller dimension than the enzyme one [21]. The morphologic aspect might explain the general adsorption difference between the microporous and mesoporous materials when the adsorption occurs only on the external adsorbent surface. Moreover, for M41S type materials, the different acidity among the samples obtained utilizing different silica sources can clarify the different adsorption results.

The evaluation of the FT-IR and NH_3-TPD results indicate that there is an abundance of weak acidic hydroxyl groups (weak acid sites) on the MCM-41 type supports, especially in the case of samples M1-Z and M2-T. These acid sites of the M2-T sample are two times stronger than the acid sites of the M1-Z support. Probably, for this reason no enzyme was attached on the M2-T support. The M4-F sample, synthesized with silica fumed, presents a lower amount of acid sites than sample M1-Z, however these sites are stronger and then no enzyme was retained. Finally, the M3-W sample, synthesized with water glass, possess a low concentration of weak acid sites that are capable of adsorbing a considerable amount of enzyme.

From the results of the DNS assay [18], the evaluated amount of the glucose in the commercial lipase solution is 6.12 mg/ml. After the immobilization tests no glucose is retained on the support, because the concentration of glucose in the solution, before and after the contact with the support, remains unaltered. During the application of the MBTH assay [16] on the lipase enzyme solution no peroxidase activity was observed. Therefore, no phenols are present in the used commercial lipase solution. By using the Folin & Ciocalteau method [17], the calculated amount of poly-phenols in the lipase solution results to be 0.124 mg/ml. This value is not competitive with the total protein present in the solution during the adsorption, because the results from the UV Adsorption Methods [15] at 280 nm, using the BSA as a standard protein, indicate that the amount of total protein in the commercial lipase solution is equal to 306,4 mg/ml. Since the method chosen to determine the amount of enzyme retained on the support was the UV Adsorption Method at 280 nm, the assay results confirm the elimination of the interference due to the presence of non-protein substances.

3.3 Results of the pigment adsorption

The adsorbed amounts of the two pigments were very different when the lipase was immobilized on the support (Table 7). This aspect was well observed when Methylene Blue (a cation molecule having a positive charge) was the adsorbed pigment. In the case of the anionic pigment (Eosin B) no differences in the adsorbed amount were observed. These results suggest that the lid of the enzyme is negatively charged and then the carboxyl acid group of the lipase are exposed. The slight difference between the adsorbed amounts of the two pigments by the same support, confirmed that the used support had a non-polar surface, while, in the case of the Silicalite-1 sample (NaS-3), the small differences indicate the presence of the free Na^+ cations in the framework   Indeed, the expected adsorption mechanism was different, due to the fact that the interactions between the enzymatic carboxyl groups and the positive charge of the NaS-3 framework were the main cause of the high differences in the amount of adsorbed lipase with respect to the other support types. Probably, the presence of Na^+ on the surface support allows to the lipase enzyme to assume the right open conformation for the attachment.

Table 7
Adsorbed amounts of pigments on the supports

Support	Adsorbed Methylene Blue (mmol/g)	Adsorbed Eosina B (mmol/g)
M1-Z (Si-MCM-41)	0,187	0.166
M1-Z+lipase	0,505	0.095
NaS-3 (Na-Silicalite-1)	0,004	0,079
NaS-3+lipase	0,512	0.105

3.4 Activity Results

Although the lipase lost about 26% of its original activity after immobilization, these results (Table 8) are satisfactory if compared with other studies in which the same enzyme was used on other type of supports [2, 4, 22].
When the attachment is linked to the hydrogen bonding between the -SiOH terminal groups of the support and terminal amino acid groups of the enzyme lid, the lost of the activity results lower than in the case in which the attachment between lipase and support involves the hydrophobic centers of the enzyme. Probably, the connection between lipase and Silicalite-1 support is very intense and this type of attachment can hinder the substrate access to the active site of the enzyme. In fact, the higher activity of the RML-Silicalite-1 system respect to other supports such as M41S and delaminated zeolites, and compared with the amount of immobilized enzyme, may be explained by the type of the attachment. For the mesoporous materials and delaminated zeolites, the lower amount of the adsorbed enzyme is attached at the support surface and, probably, the active site of the enzyme is not involved. The Silicalite-1 is the support that showed the highest affinity towards the adsorption of the RML enzyme

and its final activity is anyway good. The stability of this biocatalyst should be evaluated in further studies in order to determine the leaching level occurring during the reaction.

Table 8
Activity of RML immobilised on various supports

Lipase-Support System	% Immobilized Lipase	Activity [U/L]	Relative Activity [U/L]
M1-Z	41	12040	78%
ITQ-2	27	8120	74%
ITQ-6	17	7280	81%
NaS-2	59	15680	86%
NaS-3	74	19040	85%

4. CONCLUSIONS

The amounts of enzyme immobilized on the external surface of the best supports studied in this work are higher than those adsorbed on other support types reported in previous studies. This result is uncontaminated because the absence of competitive molecules during the enzyme adsorption process has been proven.

The high-hydrophobic character of the support surface promotes the immobilization of the lipase. On the contrary, the enzyme attachment is disabled by high acidity and hydrophillicity of the support.

The -SiOH groups of the support are the main attachment sites for the lipase enzyme, while their amount, distribution and acidity strongly affect the extent of the enzyme immobilization.

The high-ordered morphology and large crystal dimensions allow adsorption of a greater amount of enzyme respect to the amorphous support. From the activity tests, it emerges that when a high amount of enzyme is adsorbed on the most suitable support (Silicalite-1 type support), the type of attachment may over-involve the active sites of the enzyme.

Acknowledgements: The authors are grateful to Prof. Avelino Corma (from Instituto de Tecnología Química (UPV-CSIC) Universidad Politécnica de Valencia, Spain) for the supply of ITQ-2 and ITQ-6.

REFERENCES

[1]   H. Takahashi, B. Li, T. Sasaki, C. Miyazaki, T. Kajino, S. Inagaki, Chem. Mater., 12 (2000) 3301

[2]   I.E. de Fuentes, C.A. Viseras, D. Ubiali, M. Terreni, A.R. Alcántara, J. Mol. Catal. B: Enzymatic, 11 (2001) 657

[3]   A.P.V. Gonçalves, J.M. Lopes, F. Lemos, F. Ramôa Ribeiro, D.M.F. Prazeres, J.M.S. Cabral, M.R. Aires-Barros, J. Mol. Catal. B: Enzymatic, 1 (1996) 53

[4]   A.C. Oliveira, M.F. Rosa, J.M.S. Cabral, M.R. Aires-Barros, Bioprocess Engineering, 16 (1997) 349

[5]   H.P. Yiu, P.A. Wright, N.P. Botting, Micropor. Mesopor. Mater., 44-45 (2001) 763

394

[6]   D.T. Nguyen, M. Smit, B. Dunn, J. I. Zink, Chem. Mater., 14 (2002) 4300
[7]   J.F. Díaz, K.J. Balkus Jr., J. Mol. Catal. B: Enzymatic, 2 (1996) 115
[8]   Z. Knezevic, L. Mojovic, B. Adnadjevic, Enzyme Microb. Technol., 22 (1998) 275
[9]   J. Deere, E. Magner, J.G. Wall, B.K. Hodnett, Chem. Commun., (2001) 465
[10]   A. Corma, V. Fornés, J.L. Jordá, F. Rey, R. Fernandez-Lafuente, J.M. Guisan, C. Mateo, Chem. Commun., (2001) 419
[11]   E. Dumitriu , F. Secundo, J. Patarin, I. Fechete, J. Mol. Catal. B: Enzymatic, 22 (2003) 119
[12]   K. Faber, Biotransformations in Organic Chemistry, Springer, Berlin, 1997, p.88
[13]   Z. S. Derewenda, U. Derewenda, J. Mol. Biol., 227 (1992) 818
[14]   R. Fernandez-Lafuente, P. Armisén, P. Sabuquillo, G. Fernandez-Lorente, J.M. Guisan, Chem. Phys. Lipids, 93 (1998) 185
[15]   Determination of total protein – Methods in enzymology, Academic Press Inc., Londra, Vol. 91, 1987, pp. 95-119
[16]   L. Setti, S. Scali, I. Degli Angeli, P.G. Pifferi, Enzyme Microb. Technol., 22 (1998) 656
[17]   Analysis of total phenols and other oxidation substrates and antioxidants by means of Folin-Ciocalteau – Methods in Enzymology, Academic Press Inc., Londra, Vol. 299 (1999), pp. 152-178
[18]   M.J. Bailey, P. Biely, K. Poutanen, J. Biotech., 23 (1992) 257
[19]   N.W. Tietz, E.A. Fiereck, A specific method for serum Lipase determination, Clin. Chim. Acta 13 (1996) 352
[20]   X.S. Zhao, G.Q. Lu, A.K. Whittaker, G.J. Millar, H.Y. Zhu, J. Phys. Chem., 101 (1997) 6525
[21]   A. Corma, V. Fornes, F. Rey, Adv. Mater., 14 (2002) 71
[22]   C.V. Suresh Babu, K.R. Kiran, S. Divakar, World J. Microb. Biotechnol , 17 (2001) 659

Studies in Surface Science and Catalysis 155
A. Gamba, C. Colella and S. Coluccia (Editors)

Adsorption of carbon dioxide and nitrogen on silicalite-1 and M41S type materials

A. Macario[a]**, A. Katovic**[a]**, G. Giordano**[a]**, F. Iucolano**[b] **and D. Caputo**[b]

[a]Dipartimento di Ingegneria Chimica e Materiali, Università della Calabria,
Via P. Bucci, 87036 Rende (CS), Italy

[b]Dipartimento di Ingegneria dei Materiali e della Produzione, Università Federico II,
Piazzale V. Tecchio 80, 80125 Napoli, Italy

The ability of silicalite-1 and M41S type materials to adsorb pure CO_2 and N_2 has been experimentally studied by determining the adsorption isotherms at 20°C. The experimental data have been analyzed using the Langmuir and Freundlich adsorption isotherm models. In both adsorbents N_2 adsorption is strongly reduced in connection to the sodium amount in the synthesis gel, while no significant effects have been observed in the CO_2 adsorption performance. This way, a CO_2/N_2 selectivity increase is obtained. The silicalite-1 type adsorbents are characterized by good CO_2/N_2 selectivity (5.9-10.2) and high working capacity for CO_2 (4.0-4.8 w-%). In comparison, a very high CO_2/N_2 selectivity (> 7.4), but a significantly lower working capacity for CO_2 (1.4-1.9) has been recorded for M41S type materials.

1. INTRODUCTION

Fossil fuels still remain the primary energy source all over the world, but, as it is well known, their use is cause of serious and worrying global warming, because of the exhaust gases deriving from combustion [1]. Carbon dioxide (CO_2) is the main pollutant responsible for this "greenhouse" effect [1-2]. Therefore its capture and sequestration are receiving growing attention in order to mitigate the warming danger [3-5].

Among the various methods proposed to separate CO_2 from gas mixtures, selective adsorption seem to be one of the most promising [6]. The critical point of the adsorption process is the choice of the adsorbent. Since in the exhaust gases from combustion process carbon dioxide is typically accompanied by N_2 and H_2O, adsorbents having hydrophobic character should be preferred.

Silicalite-1 is the aluminum-free member of the MFI-type materials [7]. Respect to the isotype ZSM-5, this crystalline microporous silica exhibits a similar shape-selective behavior, but has a higher hydrophobic character. These features indicate silicalite-1 to be a potential adsorbent for selective gas separations in the presence of mixtures containing water.

The M41S type mesoporous materials, discovered by the Mobil researchers at the beginning of 1990s [8, 9], due to their high porosity, high specific surface area, controllable and narrowly distributed pore sizes, are considered very promising adsorbents for selective gas separations [10-12]. The adsorption capability of MCM-41(a member of the M41S family) for polar molecules (e.g., H_2O, CO_2, N_2) is related to the density of the surface silanol groups [13]. Since MCM-41 surface is quite hydrophobic, this material should be regarded as a possible candidate for the separation of gas molecules in the presence of water [14].

Adsorption features of zeolites and mesoporous materials in gas separations depend on the structure and composition of the framework, as well as on the nature of extra-framework cations [15, 16]. As an example, the replacement of sodium for divalent cations in the FAU-type zeolite adsorbents increases the N_2/O_2 selectivity [17-19]. Moreover, the extra-framework cation that replaces sodium also influences the adsorbent regeneration step in the adsorption process [20].

Most syntheses of zeolite or mesoporous materials are performed under basic conditions, using OH⁻ as a mineralizing agent. pH and solubility of reactants in the synthesis mixture are governed by the amount of the OH⁻ ions [21], coming, e.g., from NaOH. Actually, the crystallization rate is influenced by the alkalinity of the synthesis mixture and by the presence of sodium [22]. In particular, sodium amount in the synthesis mixture of the mesoporous materials influences the final quality of the product (specific surface area and pore volume), whereas in silicalite-1 synthesis it influences the crystal growth and the number of defect groups [22, 23].

Referring to these last remarks, the aim of this work is to compare the performances of silicalite-1 and M41S type materials as adsorbent for CO_2 and N_2 and to investigate the influence of sodium present in the synthesis gels on their adsorption properties. To fulfill this goal, the main adsorption parameters, such as CO_2 and N_2 adsorption capabilities, CO_2/N_2 selectivity and working capacity for CO_2 (an useful index of the easiness of adsorbent regeneration) have been evaluated and discussed.

2. EXPERIMENTAL

2.1. Material synthesis

The chemicals used for the M41S synthesis were: Tixosil (Rhodia) as silica source; DTABr (dodecyltrimethylammonium bromide, from Aldrich) as structure directing agent, and NaOH (Carlo Erba) as mineralizing agent.

The silica mesoporous materials were synthesized from gels having the following molar composition: 1 SiO_2 : 0.12 DTABr : y Na_2O : 30 H_2O, where y ranged from 0.08 to 0.30. The mixture was aged for 5 days at room temperature and then heated at 140°C in a Teflon-steel autoclave. The crystallization time was 1 day.

The following chemicals were used for the synthesis of silicalite-1: silica gel precipitated (BDH) as silica source, TPABr (tetrapropylammonium bromide from Fluka) as structure directing agent, and NaOH (Carlo Erba) as mineralizing agent.

Na-silicalite-1 samples were synthesized starting from gels having the following molar composition: 1 SiO_2 : z Na_2O : 0.08 TPABr : 20 H_2O, where z ranged from 0.04 to 0.12. The gel was aged for 5 days at room temperature. The hydrothermal treatment was carried out at 170°C for 1 day.

The solid phases obtained were recovered by filtration, washed with distilled water and dried at 110°C for 24 h. Afterwards samples were calcined in air at 550°C for 12 h in order to remove the organic compounds.

H-silicalite-1 was obtained by a two-fold ion exchange procedure at controlled basic pH. 1 g of calcined Na-silicalite-1 sample ($Na_2O = 0.04$) was contacted at room temperature with 500 ml of the following solution: 10 g of NH_4Cl in 200 ml of distilled water and 300 ml of 30 vol.-% ammonia. After drying at 110°C for 24 h, the exchanged samples were calcined in air at 550°C for 12 h in order to obtain the H-form of the zeolite.

2.2. Characterization of the adsorbents

The final phases of all calcined samples were evaluated by XRD technique using a Philips PW 1710 diffractometer, CuKα radiation. Samples of the mesoporous materials were scanned from 1 to 10° (2θ) in steps of 0.02° with a count time of 2 seconds at each point, whereas the microporous materials were scanned from 5 to 40° (2θ) in steps of 0.02° with a count time of 1 second at each point. The specific surface area, pore volume and pore diameter of all samples after calcination were determined from N_2 adsorption/desorption isotherms obtained at 77 K using an ASAP 2020 Micromeritics instrument.

The adsorption isotherms of N_2 and CO_2 of calcined samples at 20°C, were obtained using a gravimetric technique based on a McBain–type balance [24]. The working capacity was calculated as difference between the amounts of gas adsorbed at 1 bar and 0.2 bar [20].

Na concentration in Na-silicalite-1 was estimated by atomic absorption spectroscopy (GBC 932 AA instrument).

3. DATA MODELING

Adsorption isotherms were modeled using the Langmuir and Freundlich equations.

The Langmuir model is based on the hypothesis that all sites of the solid surface are energetically equivalent and that the adsorbed molecules are distributed on the adsorbate surface as a monomolecular layer ("mono-layer"). If x is the amount of the adsorbed gas at a given pressure P and x_{max} is the maximum amount that the surface can take up, the parameter θ, expressed as follows:

$$x/x_{max} = \theta, \tag{1}$$

is defined as the fractional surface coverage. The Langmuir model is represented by the following equation [25]:

$$\theta = \frac{KP}{1 + KP}, \tag{2}$$

where K is the adsorption coefficient, which reflects the intensity of adsorption. To obtain x_{max} and K, Eq. (1) may be arranged into the form:

$$\frac{P}{x} = \frac{1}{Kx_{max}} + \frac{P}{x_{max}}. \tag{3}$$

From the linear plot of P/x vs. P, the slope $1/x_{max}$ and intercept $1/Kx_{max}$ can be readily calculated. Not all adsorptions data obey the Langmuir isotherm expression. There are several reasons for this, but the most important reason is that the assumption concerning the energetic equivalence of all adsorption sites is rarely found to be true in practice. More often, the enthalpy of adsorption decreases with increasing the surface coverage. This is possibly due to the fact that in general surfaces are not atomically smooth, therefore there are adsorbing molecules that will react first and most energetically with atoms or ions having low coordination numbers and other molecules, that arriving later are unable to give strong adsorption bonds.

Other model isotherms were devised to eliminate the assumption of energetic equivalence, one of this is the Freundlich isotherm [25], which assumes a logarithmic decrease of adsorption capacity and no energetic equivalence of adsorption sites. Its common forms are:

$$C = K_f P^{1/n},\qquad(4)$$

$$lnC = ln K_f + \frac{1}{n} ln P,\qquad(5)$$

where C is the adsorbed gas amount and K_f and n are constants (the latter being greater than unity).

4. RESULTS AND DISCUSSION

4.1. Main physical properties of the adsorbents

Fig. 1 shows the XRD patterns of the calcined mesoporous materials, synthesized according to the recipe reported in the Experimental, by varying the (Na_2O/SiO_2) in the original gel (see the second column in Table 1).

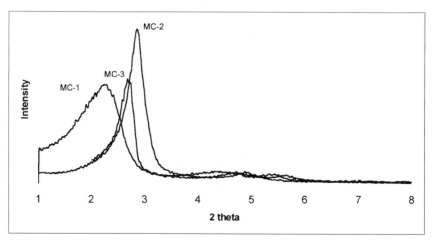

Fig. 1. XRD pattern of the synthesized mesoporous materials after calcination

Table 1
Main physical properties of calcined mesoporous samples

Sample	$(Na_2O/SiO_2)_{gel}$	BET area $m^2 \cdot g^{-1}$	Pore volume (BJH method) $cm^3 \cdot g^{-1}$	Pore diameter (BJH method) Å
MC1	0.08	888	0.88	32
MC2	0.15	1321	0.80	29
MC3	0.30	1166	0.93	29

Sample MC2, obtained with the middle amount of Na in the initial gel (0.15 as Na_2O/SiO_2 molar ratio), is characterized by a sharp profile of its main diffraction peak, which suggests a higher order in the mesoporous structure. When the amount of Na in the initial gel is lower (sample MC1 obtained with a Na_2O/SiO_2 molar ratio equal to 0.08) or higher (sample MC3 obtained with a Na_2O/SiO_2 molar ratio equal to 0.30), the intensity of the maximum peaks decreases and shifts towards lower angle values, as a consequence of a decrease of the order in the mesoporous structure and a likely increase of the pore size, respectively.

Table 1 reports also specific surface area, pore volume and pore diameter of the calcined mesoporous materials prepared.

Fig. 2. SEM images of the mesoporous materials synthesized with different amount of sodium: (a, b) sample MC1 with $Na_2O/SiO_2 = 0.08$; (c) sample MC2 with $Na_2O/SiO_2 = 0.15$; (d) sample MC3 with $Na_2O/SiO_2 = 0.30$

Inspecting Table 1 reveals that, among the mesoporous materials synthesized, the molar ratio Na_2O/SiO_2 equal to 0.15 (sample MC2) is optimum to obtain the highest ordered mesoporous structure having the highest specific surface area (1321 $m^2 \cdot g^{-1}$). Na amount in the initial gel may also affect the crystal morphology as observed from the SEM micrographs. When Na amount increases, the morphology becomes more defined (see Fig. 2).

Results of chemical and physical characterization of the silicalite-1 are reported in the Table 2 As expected form the literature [22], the presence of sodium in the initial gel strongly influences the crystal size and morphology. Moreover sodium cations remain in the silicalite-1, likely as extra-framework counter-cations of surface defects [23].

Table 2
Main chemical and physical properties of calcined silicalite-1 type samples

Sample	Type of adsorbent	$(Na_2O/ SiO_2)_{gel}$	$(Na)_{uc}$	Crystal dimension μm	BET area $m^2 \cdot g^{-1}$
SZ1	Na-silicalite-1	0.04	1.17	20 - 30	402
SZ2	Na-silicalite-1	0.08	1.87	2 - 3	402
SZ3	Na-silicalite-1	0.12	2.18	~1	402
HSZ1	H-silicalite-1	0.04	-	20 - 30	384

4.2. Adsorption isotherms

The adsorption isotherms at 20 °C for CO_2 and N_2 on M41S and silicalite-1 type adsorbents are presented in Figs. 3 and 4, respectively.

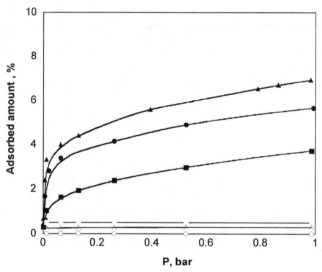

Fig. 3. Adsorption isotherm at 20°C of CO_2 (filled symbols) and of N_2 (open symbols) on M41S type samples: MC1 (square); MC2 (triangle); MC3 (circle); model (line).

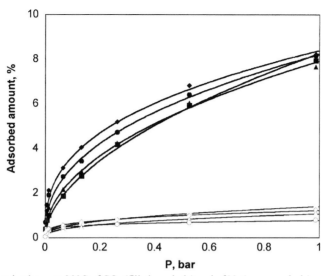

Fig. 4. Adsorption isotherm at 20°C of CO_2 (filled symbols) and of N_2 (open symbols) on silicalite-1 type samples: SZ1 (square); SZ2 (triangle); SZ3 (circle); HSZ1 (rhomb); model (line).

Experimental data were fitted using the models described in the Section 3. The Freundlich equation, succeeded to fit well experimental points in all the investigated adsorbent-adsorbate pairs, except for the adsorption of N_2 on MC1 and MC2 mesoporous samples. Here the amount of N_2 adsorbed sharply reached a saturation value which remained roughly constant over the whole pressure range. The resulting "step-shape" adsorption isotherms were well fitted by Langmuir equation. No adsorption of N_2 was exhibited by the sample MC3. Correlation coefficients and estimated value of the parameters of the best fitting models, are reported in Tables 3 and 4.

The results of the regression reported in Tables 3 and 4 were used also to calculate the main parameters of the adsorption tests, which are reported in Table 5, such as the CO_2 and N_2 adsorption capacities, the CO_2/N_2 selectivity and the working capacity of CO_2. Adsorption capacities were calculated at 1 bar, whereas the working capacity was evaluated, as reported in the Experimental, as a difference between adsorption capacities calculated at a pressure of 1 bar and 0.2 bar.

As general remark, silicalite-1 type adsorbents exhibit higher adsorption capacities, respect to M41S type ones. This result can be explained, considering the close values of pore diameter of silicalite-1 (~6 Å [26]) and the dimensions of the adsorbate molecule (kinetic diameters of CO_2 and N_2 are 3.3 Å and 3.64 Å, respectively [26]), in comparison to the pore sizes of M41S type samples reported in Table 1 (~30 Å). In the case of the small pores of the zeolite structures, an enhancement of the adsorbate-adsorbent interaction energy, due to the overlap of the adsorption fields from neighboring walls, can be expected [27].

402

Table 3
Correlation coefficients (r^2) and estimated values of the Freundlich model parameters (K_f and n) for the adsorption of CO_2 and N_2 on silicalite-1 and CO_2 on M41S samples at 20°C

Sample	Adsorbed Gas	r^2	K_f bar^{-1}	n
MC1	CO_2	0.9953	3.67	3.28
MC2	CO_2	0.9926	6.79	4.98
MC3	CO_2	0.9895	5.51	5.63
HSZ1	CO_2	0.9985	8.40	2.78
	N_2	0,9886	1.22	3,58
SZ1	CO_2	0.9973	8.27	1.85
	N_2	0.9982	1.40	2.58
SZ2	CO_2	0.9979	7.93	2.10
	N_2	0.9860	1.07	2.65
SZ3	CO_2	0.9971	8.43	2.43
	N_2	0.9907	0.77	5.35

Table 4
Correlation coefficients (r^2) and the estimated value of the Langmuir model parameters (K and x_{max}) for the adsorption of N_2 on silicalite-1 samples at 20°C

Sample	Adsorbed gas	r^2	K, bar^{-1}	x_{max}, w-%
MC1	N_2	0.9999	$4.6 \cdot 10^2$	0.5
MC2	N_2	0.9999	$2.9 \cdot 10^2$	0.3

Table 5
CO_2 and N_2 adsorption capacities at 1 bar, CO_2/N_2 selectivity and CO_2 working capacity of the synthesized adsorbents, calculated from the Freundlich model

Sample	CO_2 ads. capacity w-%	N_2 ads. capacity w-%	CO_2/N_2 selectivity	Working capacity of CO_2 w-%
MC1	3.7	0.5[a]	7.4	1.4
MC2	6.8	0.3[a]	22.7	1.9
MC3	5.5	-	∞	1.4
SZ1	8.3	1.4	5.9	4.8
SZ2	7.9	1.1	7.2	4.2
SZ3	8.2	0.8	10.2	4.0
HSZ1	8.4	1.2	7.0	3.7

[a] Calculated by Langmuir model

Among the studied mesoporous materials, the different performances in the CO_2 adsorption experiments is mostly due to their different specific surface areas. The main result of the adsorption runs on these adsorbents is the very low amount of the adsorbed N_2 (0-0.5 w-%), leading to a very high CO_2/N_2 selectivity (> 7.4). The inability of the MC3 sample to adsorb N_2, as well as the low ability observed for the other two samples, are likely due to a weak interaction between N_2 and adsorbent, whereas the large pore size of the mesoporous channel does not permit adsorption and retention of N_2 molecules on the adsorbent surface. In the case of CO_2, the interaction adsorbate/adsorbent is stronger, due to its quadrupole moment (-14.3·10^{-40} C·m^2 [28]) higher than N_2 molecule (-4.99·10^{-40} C·m^2 [29]), as absolute magnitude.

It can be observed that when, in Na-silicalite-1 samples, Na per unit cell increases, the amount of adsorbed N_2 decreases (from 1.4 to 0.8 w-%). Instead, for the same adsorbents, the Na amount does not affect neither the CO_2 adsorption performance, nor the adsorbed amount and the working capacity. As a consequence, the selectivity value increases from 5.9 to 10.2.

It must be noted that, when Na^+ is completely removed from the silicalite-1 sample SZ1, obtaining the sample HSZ1, the adsorption capacity for CO_2 and N_2 remains substantially unvaried. Even if Na^+ is present in the zeolite structure, likely as extra-framework cation counter-balancing surface defects (see Table 2), the latter experimental evidence seems to suggest for it a role essentially as mineralizing agent in the synthesis gel, similar to its action in the mesoporous materials synthesis, with only a minor contribution to the adsorbate-adsorbent interaction.

For an ideal PSA adsorbent, high adsorption capacity, selectivity and working capacity are required [30]. While searching for the "best" adsorbent, it is often necessary to make compromises with practical aspects, so an adsorbent allowing easy regeneration (high working capacity) could be preferable to one with higher adsorption capacity and selectivity. From this point of view, the Na-silicalite-1 samples, showing the highest working capacity, seem to be better adsorbents for CO_2 separation, even if, the mesoporous materials synthesized in this work should give good performances mainly for their much higher CO_2/N_2 selectivity.

5. CONCLUSIONS

Na amount in the synthesis gel of the silicalite-1 type materials reduces the N_2 adsorption capacity without effect on the CO_2 adsorption performances. For a hypothetic application in a PSA process, the Na-silicalite-1 type adsorbents here synthesized, allow to achieve an easy regeneration (high working capacity), even if the CO_2/N_2 selectivity is not very high. At the same time, the mesoporous material appears to be a very promising adsorbent for the CO_2 separation, because of its very high CO_2/N_2 selectivity.

Further studies will deal with the porosity control of the mesoporous channels, in order to improve the pore diffusion and the adsorption/desorption kinetics, two important aspects that lead to an easier adsorbent regeneration.

REFERENCES

[1] M.M. Maroto-Valer, C. Song and Y. Soong, Environmental Challenger and Greenhouse Gas Control for Fossil Fuel Utilization in the 21st Century, Kluver-Academic/Plenum Publishers, New York, 2002, p. 447.
[2] C. Azar, H. Rodhe, Science, 276 (1997) 1818.
[3] C. Song, A.M. Gaffney, K. Fujimoto, Am. Chem. Soc. Symp. Ser., 809 (2002) 448.

404

[4]   S.C. Shen, S. Kawi, J. Catal. 213 (2003) 241.
[5]   S.C. Shen, S. Kawi, Appl. Catal. B: Environ., 45 (2003) 63.
[6]   C. Song, Am. Chem. Soc. Symp. Ser., 809 (2002) 2.
[7]   E. M. Flanigen, J.M. Bennet, R.W. Grose, J.P. Cohen, R.L. Patton, R.M. Kirchner and J.V. Smith, Nature, 271 (1978) 517.
[8]   C. T. Kresge, M.E. Leonowicz, W.J. Roth, J.C. Vartuli, J.S. Beck, Nature, 359 (1992) 710
[9]   J. S. Beck, J. C. Vartuli, W. J. Roth, M. E. Leonowicz, C. T. Kresge, K. D. Schmitt, C. T.W. Chu, D. H. Olson, E. W. Sheppard, S. B. McCullen, J. B. Higgins, J. L. Schlenker, J. Am. Chem. Soc., 114 (1992) 10834.
[10]   P.J. Branton, P.G. Hall, K.S.W. Sing, J. Chem. Soc., Chem. Commun., (1993) 1257.
[11]   P.J. Branton, P.G. Hall, K.S.W. J. Sing, H. Reichert, F. Schüth, K.K. Unger, J. Chem. Soc., Faraday Trans., 90 (1994) 2965.
[12]   P.J. Branton, P.G. Hall, M. Treguer, K.S.W. Sing, J. Chem. Soc., Faraday Trans., 91 (1995) 2041.
[13]   X.S. Zhao, G.Q. Lu, A.K. Whittaker, G.J. Millar, H.Y. Zhu, J. Phys. Chem. B, 101 (1997) 6525.
[14]   A. Corma, Chem. Rev. 97 (1997) 2373.
[15]   S.U. Rege, R.T. Yang, M.A. Buzanowski, Chem. Eng. Sci., 55 (2000) 4827.
[16]   M.W. Ackley, S.U. Rege, H. Safena, Micr. Mes. Mat., 61 (2003) 25
[17]   C.C. Chao, J.D. Sherman, J.T. Mullhaupt, C.M. Bolinger, U.S. Patent 5, 413, 625 (1995).
[18]   C.C. Chao, J.D. Sherman, J.T. Mullhaupt, C.M. Bolinger, U.S. Patent 5, 174, 979 (1992).
[19]   C.G. Coe, J.F. Kirner, R. Pierantozzi, T.R. White, U.S. Patent 5, 152, 813 (1992).
[20]   T. R. Gaffney, Current Opinion in Solid State & Materials Science, 1 (1996) 69
[21]   J. C. Jansen, in Introduction to Zeolite Science and Pratice, Elsevier, Amsterdam, 2001, vol. 137, p. 175.
[22]   D. T. Hayhurst, A. Nastro, R. Aiello, F. Crea, G. Giordano, Zeolites, 8 (1988) 416.
[23]   S. Bordiga, P. Ugliengo, A. Damin, C. Lamberti, G. Spoto, A. Zecchina, G. Spanò, R. Buzzoni, L. Dalloro, F. Rivetti, Topics in Catalysis, 15 (2001) 43.
[24]   D. Caputo, B. de'Gennaro, M. Pansini, C. Colella, in NATO Science Series, Series E: Applied Sciences, Vol. 362, P. Misaelides, F. Macasek, T. Pinnavaia and C. Colella (Eds.), Kluwer Academic Publishers, Dordrecht (The Netherlands) 1999, p. 225.
[25]   G.C. Bond. In: Heterogeneous Catalysis: Principles and Applications, Oxford University Press, Oxford, 1987.
[26]   D.W. Breck, Zeolite Molecular Sieves, John Wiley & Sons, New York, 1974.
[27]   D.M. Ruthven. Principle of Adsorption and Adsorption Processes, John Wiley & Sons, New York, 1984.
[28]   J.N. Watson, I.E. Craven. G.L.D. Ritchie, Chem. Phys. Let., 274 (1997) 1.
[29]   A. Halkier, S. Coriani. P. Jørgensen, Chem. Phys. Let., 294 (1998) 292.
[30]   G.V. Baron, in Gas Separation Technology, E.F. Vansant and R. Dewolfs (Eds.), Elsevier, Amsterdam, 1990, p.137.

Studies in Surface Science and Catalysis 155
A. Gamba, C. Colella and S. Coluccia (Editors)

Nanostructure and catalytic properties of gold supported on group IV oxides

M. Manzoli, A. Chiorino and F. Boccuzzi

Department of Chemistry I.F.M., University of Turin, Via P. Giuria 7, Torino, 10125, Italy

INCA, Interuniversity Consortium Chemistry for the Environment, Viale della Libertà 5/12 Marghera (Venice) Italy

The effect of the support on the HRTEM features and on the FTIR spectra of adsorbed CO at 90 and 300 K of Au/TiO_2, Au/ZrO_2 and Au/CeO_2, both oxidised and reduced, has been examined. All catalysts have been prepared by the deposition-precipitation method. The aim of this work is to obtain information on the surface structure of the three catalysts in order to explain the differences observed in the catalytic activity in the CO oxidation reaction. The size and the size distribution of the metal particles produced on Au/TiO_2 and on Au/ZrO_2 are almost the same, $d \approx 4$ nm. On the contrary, the ceria supported sample exhibits Au particles that are significantly smaller, $d \approx 1$ nm, that are difficultly detected by HRTEM. FTIR data of CO adsorbed at 90 K on the three samples, either oxidized or reduced, and the spectroscopic evolution by heating up to RT, are presented and correlated with the different catalytic performances. In addition, data concerning the $CO-O_2$ interaction at different temperatures will be discussed. The size of gold nanoparticles together with the size and the chemical nature of the oxidic supports are shown to be relevant in determining firstly their durability and then the chemisorption and catalytic properties.

1. INTRODUCTION

The increased interest and the rapidly growing number of investigations on supported nano-gold catalysts is due to their potential applicability to many reactions of both environmental and industrial importance. The most widely studied are the CO oxidation at temperature as low as 200 K [1, 2], the PROX reaction [3], the WGS reaction [4] and the propene epoxidation [5].

The synthesis of nanosized gold catalysts is highly sensitive towards the preparation technique which determines a great difference in both size and shape of gold particles. Moreover, the preparation method also influences their interaction with the support. Indeed, Haruta and co-workers have shown that the incipient wetness impregnation is unsuitable to produce highly dispersed gold catalysts. On the contrary, active catalysts are produced, mainly on n-type oxidic semiconductors, when gold catalysts are prepared by co-precipitation (CP) or by deposition-precipitation (DP) [5, 6]. It appears quite well established that the presence of low coordinated Au atoms is a key factor for the catalytic activity of gold when the CO activation is required. In addition, the nature of the support on which gold nanoparticles are dispersed plays a crucial role in determining the catalytic activity and appears in some extent involved in the oxygen activation. Transition and post-transition metal oxides and, in general, reducible metal oxides have shown good performances, as previously evidenced [7]. Nevertheless, the role of the support is not fully clarified as for the chemical features and the dispersion up to now. Here we investigated, by HRTEM and by FTIR spectroscopy of adsorbed CO at 90 and 300 K, the effect of the group IV oxides as gold supports on three samples, Au/TiO_2, Au/ZrO_2 and Au/CeO_2. The three oxides have different electronic properties (TiO_2 and CeO_2 are n-type semiconductors, ZrO_2 is an insulator) and different morphologies. However, gold has been deposited on all the samples by the same method. The aim of this work is to obtain information on the surface structure and composition of the three catalysts in order to explain the differences observed in the CO oxidation activity.

2. EXPERIMENTAL

2.1. Materials
The samples were prepared by the deposition-precipitation method [8]. The experiments were performed on the samples preliminarily heated up to 673 K in dry oxygen and cooled down in the same atmosphere (oxidised samples) and/or reduced in hydrogen at 523 K. Au/ZrO_2 and Au/TiO_2 were then outgassed at the same temperature down to RT, while Au/CeO_2 was cooled down in hydrogen and finally outgassed at RT (reduced samples). The main characterization data of the samples are summarized in Table 1.

Table 1
Morphological properties of the samples

Supporting oxide	Surface area (m^2/g)	Size of support (nm)	Au loading (wt%)	Au size (nm)
TiO_2	86	10-30	3	4.1±0.5
ZrO_2	26	20-200	3	4.1±0.5
CeO_2	118	3-6	3	1.±0.5

2.2. Methods
HRTEM analysis was performed using a Jeol JEM 2010 (200kV) microscope equipped with an EDS analytical system Oxford Link. The powered samples were ultrasonically dispersed in isopropyl alcohol and the obtained suspensions were deposited on a copper grid, coated with a porous carbon film.

The FTIR spectra have been collected on a Perkin-Elmer 1760 spectrometer equipped with a MCT detector. The samples, in self-supporting pellets, were introduced in a cell allowing thermal treatments in controlled atmospheres and spectrum scanning at controlled temperatures (from 90 up to 300 K). The spectra have been normalized in respect to the pellet weights.

3. RESULTS AND DISCUSSION

3.1 HRTEM and EDS analysis data
Figure 1 show some typical HRTEM micrographs, the metal particle size distribution and EDS data relative to Au/TiO_2, Au/ZrO_2 and Au/CeO_2. The images were taken at an original magnification of 500 000 for Au/TiO_2 and Au/ZrO_2 and of 800 000 in the case of Au/CeO_2.

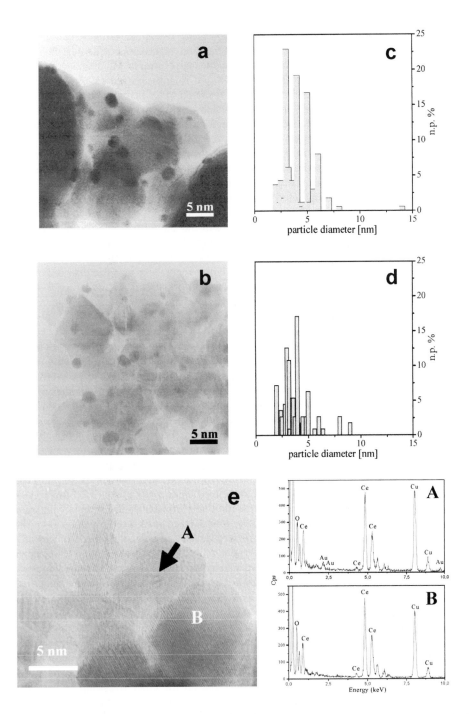

Figure 1. HRTEM images and size distributions of Au particles of Au/TiO_2 (sections a and c) and of Au/ZrO_2 (section b and d). HRTEM image and EDS spectra on two different regions of Au/CeO_2.

Gold particles are seen as dark contrasts on the two first samples, being their size distribution, shown in Figs. 1b and 1d is almost the same, 4.1±0.5 nm. The morphology of the two oxides is quite different: microcrystals with sharp borders for TiO_2, (Fig 1a) larger particles, with a much broader size distribution and roundly shaped for ZrO_2 (Fig. 1c) (as summarised also in Table 1). As for the Au/CeO_2 sample (Fig. 1e), no metallic particles larger than 1 nm can be detected, being only very thin and small clusters put in evidence with some difficulty. No dark dots are evidenced in any region of the sample. However, the intensity of gold signal is higher in the region where very small gold clusters can be evidenced in the transmission micrographs by EDS analysis, as illustrated in Fig.1e. The support appears highly homogenously dispersed and nanocrystalline (size: 3-6 nm). A significant effect of the calcination temperature and of the ageing of the catalysts on the growth of gold particles has been found on the sample supported on titania [2, 9]. On the contrary, the size of gold and ceria nanoparticles remains unchanged even after use [10,11], suggesting that the gold cluster coalescence is inhibited, in this case, by the size of the support particles. Similar results and conclusions were presented recently by Carrettin et al. in a paper [12] on which they also report that the specific rate for CO oxidation on this system is almost twice respect to the one observed for the reference Au/TiO_2. However, both average size and size distribution of their gold particles, as shown in Fig. 2 of the Supporting Information [12], are larger, 3-4 nm, and broader than in our case.

3.2. FTIR characterization by spectra of adsorbed CO

The FTIR spectra of CO adsorbed at 90 K on Au/TiO_2 and Au/ZrO_2, both calcined in O_2 at 673 K, are shown in Figs. 2a and 2b. They are very similar to those detected on the same samples reduced after previous use (not shown for sake of brevity). At full CO coverage, an absorption band at ≈ 2100 cm^{-1}, assigned to CO chemisorbed on Au^0 step sites of metallic particles, is evident. Moreover, bands at 2169 cm^{-1}(with a shoulder at 2181 cm^{-1}) on Au/ZrO_2, and at 2163 and 2177 cm^{-1} on Au/TiO_2, related to differently coordinated cationic support sites, are observed [13]. All the IR bands present on the two samples, both calcined and reduced after use, behave quite similarly by decreasing the CO pressure or by increasing the temperature up to RT. The main difference between the two samples is the relative intensity of the peaks of the cationic carbonyls in respect to the one of 2100 cm^{-1} absorption band of CO adsorbed on gold sites, clearly related to their different surface areas (see Table 1). As for the shape, quite surprisingly, the peaks are definitively broader on the sample with the lower surface area, Au/ZrO_2. In previous studies, low surface area ZrO_2 samples [14] showed narrow bands, probably as a consequence of well developed and defect free surfaces of ZrO_2, as shown by HRTEM images. The broadness, in our case, is coherent with a lower degree of surface order of our sample, put in evidence also by the HRTEM measurements.

CO adsorption at 90 K on the oxidised gold/ceria catalyst (Fig. 2c, bold curve) produces a very weak band at 2100 cm^{-1} together with two bands at 2151 and 2170 cm^{-1}, ascribed to CO on Ce^{4+} cations with different co-coordinative unsaturation [15]. The same experiment performed on the Au/CeO_2 sample reduced at 523 K produces a broad and strong absorption extending from 2100 to 1950 cm^{-1}, with a maximum at 2055 cm^{-1} (Fig. 2c, fine curve).

The big difference between the FTIR spectra of the sample either reduced or oxidized (Fig. 2c) can be understood taking in account that on ceria, as shown by HRTEM, the gold particles are very small, 1 nm. At the same time, on the other two samples, we detected particles whose dimensions are larger than 3 nm. Oxygen is quite strongly bonded on very small gold particles as evidenced by Bondzie et al. [16]. Therefore, a mild reduction is needed in order to obtain the exposure of the clean metallic gold .

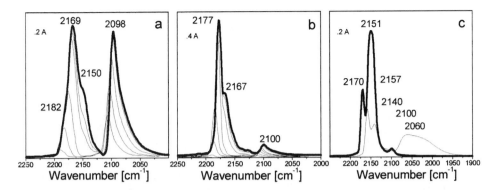

Figure 2. FTIR absorption spectra of CO adsorbed at 90 K on oxidised Au/TiO$_2$ (section a) and on oxidised Au/ZrO$_2$ (section b) at full coverage (bold curves) and at decreasing pressures (thin curves); FTIR absorption spectra of CO adsorbed at 90 K on oxidized (bold curve) and on reduced (thin curve) Au/CeO$_2$.

Figure 3a shows the evolution of the bands of CO adsorbed on reduced Au/CeO$_2$ catalyst by raising the temperature from 90 up to 300 K. A similar behaviour is also observed by decreasing the CO pressure (not shown for sake of brevity). Bands in the 2050 - 1950 cm^{-1} range are reported on gold electrodes during the electrooxidation of CO at negative potentials [17]. On this basis, the absorption observed on reduced Au/CeO$_2$ can be assigned to CO adsorbed on very small gold clusters, negatively charged as a consequence of an electron transfer from the reduced supports to the small clusters. Moreover, the bands at 2157 and 2140 cm^{-1} gradually reduce their intensities. The intensity of the band at 2157 cm^{-1} decreases slower than the intensity of the band at 2140 cm^{-1} because of stronger bond between CO and Ce^{3+} sites appeared after reduction. At the same time, the maximum of the adsorption assigned to CO adsorbed on Au$^\delta$ sites blue-shifts up to 2070 cm^{-1} and slightly reduces its intensity. The erosion from the low-frequency side could be an indication that a component at 2040 cm^{-1} is depleted going from 90 up to 300 K. This behaviour is opposite to that observed on the freshly reduced Au/TiO$_2$ sample. As shown in Fig. 3b, the interaction with CO at RT on this sample produces a band at 2110 cm^{-1}, fully reversible to the outgassing, and some more strongly bonded ones at 2055 and at 1990 cm^{-1}. It has been proposed that the 2055 cm^{-1} band is related to CO adsorbed on top on gold small clusters, negatively charged as a consequence of an electron transfer from the reduced support, while the 1990 cm^{-1} one has been assigned to CO bridge-bonded on the same small clusters. We observed quite similar features also on Au/Fe$_2$O$_3$ [9]. Therefore, a decrease of CO adsorbed linearly on top on the small clusters likely occurs (band at 2055 cm^{-1}) and CO bridge-bonded grows up (band at 1990 cm^{-1}) by decreasing the CO coverage.

The electrons transferred from the support to these small clusters allow a back donation from gold to the carbon monoxide and a bridge-bonded configuration is stabilized at low coverages. CO adsorbed on these sites is more strongly bonded than CO adsorbed on larger gold particles.

Negatively charged small gold clusters have been observed at RT after reduction only on fresh Au/TiO$_2$ and Au/Fe$_2$O$_3$ catalysts and never on the used ones, where probably they are not present anymore as a consequence of the gold sintering on the supports.

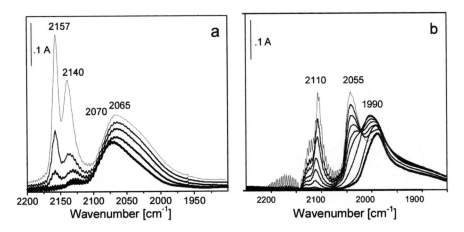

Figure 3. Evolution of the FTIR absorption spectra from 90 K (thin upper curve) up to 300 K (bold curve) of CO adsorbed on reduced Au/CeO$_2$ (section a). FTIR absorption spectra of CO adsorbed at RT on freshly reduced Au/TiO$_2$ (section b) at decreasing CO equilibrium pressures from 20 mbar (thin upper curve) to ≤ 1 mbar (bold curve).

On the contrary, these features are still present on Au/CeO$_2$ after many oxidation-reduction treatments and after re-reduction following WGS reaction. The different stability of the small negatively charged clusters on the two n-type semiconductors, TiO$_2$ and CeO$_2$, is probably related to the long-term stability of the Au/CeO$_2$ catalysts shown in a number of papers by different groups [11, 12,18].

The small size of the ceria supports favours the long-term stability, preventing cluster coalescence, as suggested by the already cited ref [12]. As commented above, our gold deposited on CeO$_2$ is formed by very thin gold clusters of 1 nm, that are smaller than the 3–4 nm particles present on the sample of Carrettin et al. [12]. These clusters are covered by adsorbed oxygen at the end of the calcination pre-treatment. For this reason, the authors detected the usual band at 2100 cm^{-1} in the FTIR spectra of adsorbed CO, while on our sample this band is very weak. By reduction in H$_2$, both oxygen on gold clusters and oxygen species on ceria react with hydrogen, giving rise to the formation oxygen vacancies and/or Ce^{3+} defects on ceria. Oxygen covering the small gold clusters is removed after a very mild reductive treatment and CO can interact with gold sites, more or less negatively charged as a consequence of an electron transfer from the reduced support (absorption bands 2060 –1950 cm^{-1}).

3.3. Temperature and support effect on the CO oxidation rate on different gold catalysts

There is no evidence of carbon dioxide production (not shown for sake of brevity) in the FTIR spectra of CO – ^{18}O$_2$ interaction at 90 K on Au/ZrO$_2$. On the contrary, a quite strong band at 2323 cm^{-1}, with a weaker component at 2332 cm^{-1} have been observed on Au/TiO$_2$ (see Fig 4a).

The band at 2323 cm^{-1} is related to C^{16}O^{18}O, while the 2332 cm^{-1} one is assigned to C^{16}O$_2$.

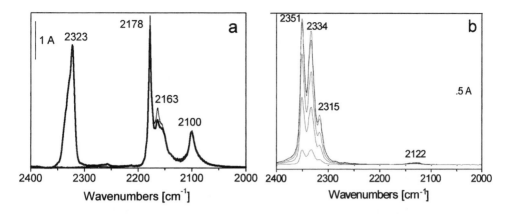

Figure 4. FTIR absorption spectra of CO – $^{18}O_2$ interaction on Au/TiO_2. Section a: inlet of $^{18}O_2$ at 90K (fine curve) on CO preadsorbed at the same temperature (bold curve) after 20'; section b: spectra recorded at increasing contact times after the inlet of CO at RT on $^{18}O_2$ preadsorbed at the same temperature.

At 90 K, mainly the atoms of the $^{18}O_2$ molecules present in the gas phase participate to the reaction at 90 K, as indicated by the relative intensities of the bands related to two kinds of carbon dioxide.

In a previous work [13a], by comparing Au/TiO_2 and Au/ZrO_2, we proposed that gold step sites at the borderline with titania, in proximity of the support oxygen-vacancy defects, are involved in both adsorption and reactive activation of CO and O_2.

Being the amount of gold step sites the same on the two samples, a key role must be played by the oxygen vacancies of the support in order to explain the different behaviour of the two samples. Grunwaldt et al. [19] showed very different activities of Au/TiO_2 and Au/ZrO_2 catalysts prepared by the same method (adsorption of Au colloid onto the support). In particular, the authors observed a 100% CO conversion on Au/TiO_2 at 353 K, whereas the Au/ZrO_2 catalysts was inactive, thus strenghtening in some extent the hypothesis that the nature of the support plays a very important role in the catalytic activity of gold catalysts.

The FTIR spectra of CO – $^{18}O_2$ interaction on Au/ZrO_2 at room temperature show that the catalyst is still inactive also at this temperature. As for the carbon dioxide production, only the usual carbonylic bands already observed at 90 K and some additional bands of carbonate-like species on the support are detected (results not reported). On the contrary, after interaction of CO with preadsorbed $^{18}O_2$ on Au/TiO_2 at room temperature (Fig. 4b) the reaction occurs very quickly. In particular, a very weak CO absorption band is observed and a triplet of strong bands at 2351, 2334 and 2315 cm^{-1}, related to different CO_2 isotopomers, grows up. The big effect of the temperature on the CO oxidation reaction rate and on the appearance of carbon dioxide isotopomers in the case of gold supported on titania can be taken as an indication that, in addition to the direct CO –O_2 reaction occurring at 90 K on gold step sites, and strictly related to the amount of these sites, other reactions occur. These reactions involve the exchange between oxygen coming from the gas phase and the oxygen atoms of the support. Moreover, these reactions may occur with a rate related to the oxygen vacancies amount at the surface of the n-type oxidic semiconductors. Wahlstrom et al. [20] suggested that the charge transfer from the conduction band to an anti-bonding level of the oxygen molecules adsorbed on TiO_2 (110) oxygen mono-ionized vacancies may be relevant. In particular, on the basis of STM data, the coordinated surface migration of the adsorbed oxygen and of the oxygen vacancies can be at the origin of an enhanced O_2 supply rate to the catalytic active sites of gold particles. It has been shown that there is a correlation between the O_2 hopping rates and the density of vacancies: the higher the density of surface oxygen vacancies, the larger the O_2 hopping rate observed at a given temperature. Furthermore, the activation energy of the oxygen diffusion, extracted from Arrhenius plots, is close to a half of the ionization energy of the surface oxygen vacancy donor levels. This finding indicates that a model based on a diffusion process of electronic origin and rate-limited by the conduction band electron density population is needed in order to explain the experimental data. The simple model illustrated in the paper [20], using the equations of solid state physics [21], can explain the exponential increase with the temperature of CO oxidation observed on gold supported on oxidic semiconductor, in particular on TiO_2 [2] and on ZnO [22], as a consequence of the increase of the electronic population of the conduction band. An opposite behaviour is observed on some insulating irreducible oxides with the increase of the temperature [23] and it can be related to the decreased concentration of activated CO molecules at the surface of the gold particles and to the lack of contribution from the support in the activation of oxygen.

The role of the support on the catalytic properties of gold nanoparticles in the CO oxidation has been widely discussed in many experimental and theoretical works reviewed in ref [24] and presented in some very interesting theoretical and experimental works in the last year. It has been shown [25] that on perfect surfaces the adhesion to the support of gold particles is negligible, while oxygen vacancies contribute substantially to the interface energy, determining also the size and the shape of the gold particles and, consequently, the number of active sites for CO adsorption. Moreover, the dynamic of oxygen vacancies present at the surface of the n-type semiconductor oxides can play a relevant role in the oxygen activation and therefore in the dependence of the catalytic activity from the nature of the support.

ACKNOWLEDGMENTS
The financial support of the project PRIN 2003 "Nanostructured multifunctional materials with improved photocatalytic activity" and of INCA (Interuniversity Consortium "Chemistry for the Environment") is acknowledged.

REFERENCES

[1] G. C. Bond, D. T. Thompson, Catal. Rev. Sci. Eng., 41 (1999) 319, and references therein.
[2] F. Boccuzzi, A. Chiorino, M. Manzoli, P. Lu, T. Akita, S. Ichikawa, and M. Haruta, J.Catal, 202 (2001) 256.
[3] B. Schumacher, V. Plzak, M. Kinne, R. J. Behm, Catal. Lett., 89 (2003) 109.
[4] T. Tabakova, V. Idakiev, D. Andreeva, I. Mitov, Appl. Catal. A: General, 202 (2000) 91.
[5] M. Haruta, Cattech, 6 (2002) 102, and references therein.
[6] M. Maciejewski, P. Fabrizioli, J.D. Grunwaldt, O.S. Beckert, A. Baiker, PCCP 3 (2001) 3846, and references therein.
[7] M. Haruta, Catal. Today, 36 (1997) 153, and references therein.
[8] M. Sakurai, S. Tsubota and M. Haruta, Appl. Catal. A 102 (1993) 125.
[9] F. Boccuzzi, A. Chiorino, M. Manzoli, D. Andreeva, T. Tabakova, J. Catal. 188 (1999) 176.

[10] T. Tabakova, F. Boccuzzi, M. Manzoli, J.W. Sobczak, V. Idakiev, D. Andreeva, Appl. Catal B, 49 (2004) 73.

[11] T. Tabakova, F. Boccuzzi, M. Manzoli, D. Andreeva, Appl. Catal. A, 252 (2003) 385.

[12] S. Carrettin, P. Conception, A. Corma, J. M. Lopez Nieto, V. F. Puntes, Angew. Chem. Int. Ed., 43 (2004) 2538.

[13] a) M. Manzoli, A. Chiorino, F. Boccuzzi, Surf. Sci., 532-535 (2003) 377; b) F. Boccuzzi, A. Chiorino, M. Manzoli, Surf.Sci., 502-503 (2002) 513.

[14] C. Morterra, V. Bolis, B. Fubini, L. Orio, Surf. Sci., 251/252 (1991) 540.

[15] a) C. Binet, M. Daturi, J. C. Lavalley, Catal. Today, 50 (1999) 207; b) A. Badri, C. Binet J. C. Lavalley, J. Chem. Soc., Faraday Trans. 92 (1996) 1603.

[16] V. A. Bondzie, S. C. Parker and C. T. Campbell, J. Vac. Sci. Technol. A, 17 (1999) 1717.

[17] Chang, S. C., Hamelin, A., and Weaver, M. J., Surf. Sci., 239 (1990) L543.

[18] Q. Fu, A. Waber, M. Flytzani-Stephanopoulos, Catal. Lett., 77 (2001) 87.

[19] J.-D. Grunwaldt, M. Maciejewski, O. S. Becker, P. Fabrizioli, A. Baiker, J. Catal., 186 (1999) 458.

[20] E. Wahlstrom, E.K. Vestergaard, R. Schaub, A. Ronnau, M. Vestergaard, E. Laegsgaard, I. Stensgaard, F. Besenbacher, Science, 303 (2004) 511.

[21] N. W. Ashcroft and N. D. Mermin, Solid state physics, Holt, Rinehart and Winston, New York, 1976.

[22] M. Manzoli, A. Chiorino, F. Boccuzzi, Appl. Catal. B, 52 (2004) 259.

[23] M. Date, M. Okumura, S. Tsubota, M. Haruta, Angew. Chem. Int. Ed.,43 (2004) 2129-2132.

[24] M. C. Kung, C. K. Costello and H. H. Kung in Catalysis vol. 17 (The Royal Society of Chemistry, 2004) p. 152.

[25] N. Lopez, J.K. Norskov, T.V.W. Janssens, A. Carlsson, A. Puig-Molina, B.S. Clausen, J.D. Grunwaldt, J. Catal., 225 (2004) 86.

Studies in Surface Science and Catalysis 155
A. Gamba, C. Colella and S. Coluccia (Editors)

Cu-MCM-22 Zeolite: A Combined X-ray Powder Diffraction and Computational Study of the Local Structure of Extra-Framework Copper Ions

M. Milanesio[a], G. Croce[a], A. Frache[a],

[a] Dipartimento di Scienze e Tecnologie Avanzate, Università del Piemonte Orientale, Via

Bellini 25/G, 15100 Alessandria, Italy; FAX: + 39 0131 287416; Tel.:+ 39 0131 360226.

A.J.S. Mascarenhas[b], E.C. Oliveira[b]

[b] Instituto de Química, Universidade Estadual de Campinas, CP 6154, CEP 13084-971,

Campinas, SP, Brasil; FAX: + 55 19 3788 3023; Tel.:+ 55 19 3788 3095.

Abstract

Local structure and site distribution of extra-framework copper ions in Cu-exchanged MCM-22 zeolite have been determined by synchrotron radiation X-ray powder diffraction analysis combined with *ab-initio* molecular orbital (DFT) calculations. Three Cu sites (S_I, S_{II} and S_{III}) in 6-membered rings and one site (S_{VI}) in a 5-membered ring close to the interlamellar region, were located inside the MCM-22 supercage, whereas no Cu ions were found within the sinusoidal channels. The existence of two families of Cu sites (in 5- and 6-membered rings respectively) with different steric hindrance and adsorptive capacity, as indicated by previous FTIR study of adsorbed NO [A. Frache, *et al.*, Langmuir, 18 (2002) 6875], was fully confirmed in this study.

Keywords: MCM-22 zeolite, MWW, *Ab-initio* Calculations, X-ray Powder Diffraction, Synchrotron Radiation, Copper Location, Extra-Framework Sites.

1. Introduction

MCM-22 zeolite (IZA code MWW [1]) was discovered in 1990 by Mobil researchers. The as-prepared material, or precursor, is a layered solid that transforms into a zeolitic structure upon calcination. The layers display 10-membered ring sinusoidal channels in the precursor sample [2-4]. The surface of the layers shows external pockets, that form large cavities when the layers condense into the final three-dimensional structure [3]. The calcined MCM-22 presents two non-intersecting pore systems, both accessible through 10-membered rings [5,6]. One of them is made of a two-dimensional system of sinusoidal channels. The other channel system consists of large supercages delimited by 12-membered rings in the widest part. This particular combination of large- and medium-pore channels in the air-exposed calcined material was previously proposed on the basis of the results of model reactions [7] and then refined by X-ray powder diffraction (XRPD) synchrotron radiation measurements [5].

Cu-exchanged MCM-22 proved to be a very interesting catalyst in NO_x decomposition, even in the presence of water in the reactant stream [8,9] and is envisaged as the most useful reaction for abatement of these pollutants. MCM-22 has been used in various important industrial processes, and one application is known for Cu-MCM-22 as additive to fluid cracking catalysts [10]. The first information on the adsorption properties and reactivity of Cu-exchanged MCM-22 were obtained by FTIR studies of adsorbed CO and NO [11]. These indicated the presence of at least two Cu^{2+} sites, the precise coordination number and geometry of which could not be determined by infrared spectroscopy alone. Upon reduction of Cu^{2+} in vacuum, the NO adsorption showed two distinct Cu^+ sites, which converted rapidly into Cu^{2+} sites in the presence of NO at room temperature. Infrared spectra of CO adsorbed on these samples showed an absorption assigned to cationic Cu-clusters present only in over-exchanged samples.

Several cation sites have been proposed in the literature for metal-exchanged MCM-22 by the use of different investigation techniques. Prakash *et al.* proposed five different cation sites for Pd(I) in MCM-22 through an ESR and ESEM spectroscopy study [12,13]. All these sites can be grouped into two families: Cu ions in S_I, S_{II} and S_{III} sites are inserted into 6-membered rings while Cu ions in S_{IV} and S_V sites are inserted into 5-membered ring windows of the framework (Figure 1). Finally, it is worth recalling the results of an ^{129}Xe-NMR study suggesting that at low Xe pressure, gas adsorption is observed only in the large cages of MCM-22 [14], where a stronger acidity was located [15].

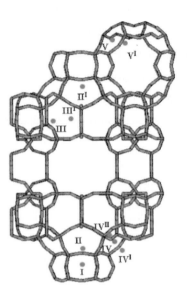

Figure 1: Cation sites as reported in the literature [10].

Cu sites in zeolites different from MCM-22 were widely investigated in the last decade, and the existence of two types of copper (I) sites, with different coordination states, was found in MFI and ferrierite zeolites by using both experimental and theoretical approaches [16]. Combined multi-technique experimental studies were needed, however, to confirm that different Cu species are present in MFI. Among these techniques, X-ray absorption spectroscopy has been very often used [17,18].

To our knowledge, structural investigations of Cu sites in zeolites using X-ray [19] and neutron [20,21] diffraction techniques, have been obtained only for Cu-Y zeolites.

In this work, high-resolution synchrotron radiation X-ray powder diffraction analysis and *ab-initio* quantum chemical cluster calculations are employed for the first time to clarify the structural features of extra-framework Cu sites in MCM-22 zeolite.

2. Experimental

2.1 Sample synthesis

Zeolite MCM-22 was prepared by hydrothermal treatment of a gel with the following composition: $4.44Na_2O:30SiO_2:Al_2O_3:17,76HMI:889H_2O$ (where HMI =

hexamethyleneimine), at 150°C and 60 rpm, for 7 days [22]. The obtained material was thoroughly washed, dried and calcined at 853 K under dry oxygen. The calcined sample was first exchanged with a solution of $NaNO_3$ 0.1 M and subsequently exchanged with a solution 0,01 M of $Cu(NO_3)_2.3H_2O$, for 24 h at room temperature. The concentration of the copper solution was adjusted to obtain the desired Cu^+/Al ratios (see Table 1 first 2 columns). In order to obtain over-exchanged samples, after this period the pH was adjusted to 7.5 with NH_4OH solution [23]. The material was washed, dried and calcined at 773 K under oxygen for 6 h. Cu-exchanged samples were named as Cu(n)-MCM-22 were (n) is the percentage of Cu exchange.

2.2 Data collection and refinement

The samples were ground, loaded into borosilicate capillaries, treated for 5h under vacuum at 423 K, and sealed. High-resolution X-ray powder diffraction data were collected at the Swiss-Norwegian beamline BM1b of ESRF [24] on the Cu-MCM-22 samples with different Cu-exchange levels (see Table 1) to locate the cation sites. The parent H-MCM-22, i.e. the sample before Cu-exchange, was used as reference material. The *P6/mmm* structure of the calcined MCM-22 exposed to air, determined by Leonowicz et al. [5], was used as starting model, and the Rietveld refinement of all the structures was carried out using the GSAS program [25]. The copper ion location was obtained by difference Fourier syntheses. Graphical manipulation and output have been carried out using MOLDRAW [26] and MAESTRO [27] softwares.

2.3 Computational details - *Ab-initio* molecular orbital calculations were performed with the Jaguar software [28] employing the DFT method based on Becke's [29] three parameters hybrid functional and Lee-Yang-Parr's [30] gradient-corrected correlation functional. The Los Alamos Effective Core Potential (ECP), named LACVP [31] in the Jaguar software, was used for Cu, Si and Al atoms, whereas the 6-31G(d) basis set [32] was used for the remaining atoms. The dangling bonds resulting in the cluster model after extraction from the periodic crystal structure of MCM-22 were saturated with H atoms. The positions of the H atoms, added where the cluster was cut from the periodic crystal structure, were fixed in the geometry optimisation to maintain the original overall shape of the MCM-22 cavity.

3. Results and discussions

3.1 Rietveld refinement and copper location - At first, a calcined, vacuum-sealed H-MCM-22 sample was compared to the calcined sample exposed to air to set up the refinement strategy and to check if the sample preparation was successful. The calculated and observed patterns for the calcined, vacuum-sealed sample (Figure 2a) were in good agreement indicating i) that no residual electron density (except for a very small peak located at the center of the supercage, present in all collected data sets and named SP -spurious peak- in Table 1) was present in the MCM-22 channels and ii) that the sample preparation was effective in eliminating organic residues. Conversely, in the calcined sample exposed to air (Figure 2b) the residual curve showed many peaks due to extra-framework matter, mainly constituted by water molecules.

Three situations could be envisaged, concerning the Cu exchanged samples: i) at Cu^+/Al molar ratios smaller than 1, isolated Cu ions would be most probable; ii) at Cu^+/Al molar ratios slightly larger than 1, cationic aggregates were likely to form; iii) at high Cu excess, separate dense Cu phases were formed. Previous studies [33] showed that the sample pretreatments, either under argon or oxygen, leads to different reactivity towards NO. The study of the possible sites for copper exchange in the samples listed in Table 1 should help to clarify this point. At first, particular attention was devoted to define the maximum amount of copper that can be introduced into the MCM-22 channels.

(a)   **(b)**

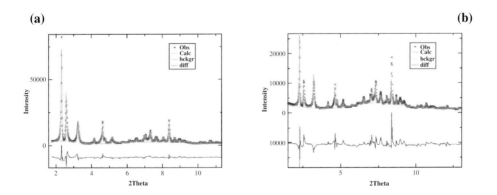

Figure 2: Graphical output after the full profile Rietveld refinements of the XRPD patterns from the vacuum-sealed (a) and exposed to air (b) calcined samples.

Over-exchanged samples with $Cu^+/Al \geq 2$ revealed the presence of CuO particles, indicating that the saturation of cationic sites for copper was achieved for Cu(150)-MCM-22.

After full profile refinement of the Cu-MCM-22 samples, Fourier syntheses helped to locate the extra-framework electron density peaks listed in Table 1, together with their labeling, which were interpreted as copper sites. These peaks were not observed in the calcined, vacuum-sealed H-MCM-22 and are different from the peaks found in the sample exposed to air [5], which were assigned to water molecules.

Table 1

Copper content, and resulting populated Cu sites and lattice parameters (Å) after the Rietveld refinement employing only Si and O atoms of the zeolite framework [5]. Peak labeling as reported in Figure 3.

Sample name	Cu^+/Al	Occupied copper	Lattice parameters	
	mol/mol	sites	a/Å	c/Å
H-MCM-22 (Vacuum)	0	SP*	14.193(1)	25.024(2)
Cu(50)-MCM-22	0.5	SP, S_{II}	14.190(1)	25.014(2)
Cu(100)-MCM-22	1	SP, S_I, S_{II}	14.188(1)	25.027(2)
Cu(150)-MCM-22	1.5	SP, S_I, S_{II}, S_{III}, S_{VI}	14.181(1)	25.017(2)

*SP = spurious peak

The number of extra-framework electron density peaks increased together with the Cu content indicating that the control of Cu exchange at the different levels was successful. The c lattice parameter variations were not correlated to the copper content, whereas the a parameter became smaller with the increase of copper insertion (last two columns of Table 1). This was explained by the fact that the c distance is correlated to the interlamellar distance, which could not be affected by the presence of a small amount of copper in the supercage. The graphical inspection of the structures, obtained after Rietveld refinements, indicates that the most significant effect of the presence of copper ions (in particular S_I and S_{II}) is the contraction of the 6-membered rings, which lay on planes parallel to (0 0 1) and this is consistent with the shortening of the a and b axes.

Cu sites were only found within the supercage, independently of the copper content; sites IV and V proposed by Prakash *et al.* [12] were not observed. This is in agreement with the higher accessibility of the larger cages with respect to the smaller MCM-22 sinusoidal channels, already demonstrated by ^{129}Xe-NMR [14]. Moreover, three over four sites are coordinated to the 6-membered rings, and the Cu site inserted into a 5-membered ring appears only at high Cu loadings.

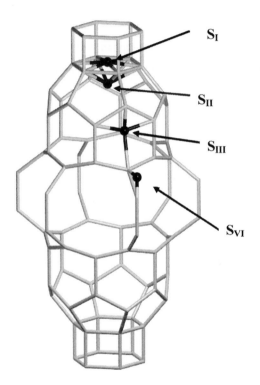

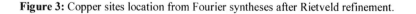

Figure 3: Copper sites location from Fourier syntheses after Rietveld refinement.

The sites determined in this work are found in different parts of the supercage. S_I, S_{II} and S_{III} are within the narrower part of the cavity, whereas S_{VI} is in the wider one. Moreover, Cu ions have different possibility to access these sites: S_I is embedded in the hexagonal prism (sterically this is the most hindered site), S_{II} is on top of the hexagonal prism but already inside the supercage, S_{III} lies in the lateral part of the narrow section of the supercage and finally S_{VI} seems to be the most accessible site because it is located in the widest section of the supercage. These sites can be divided in two families: i) S_I, S_{II} and S_{III} are in 6-membered rings in the narrow part of the supercage in relatively hindered positions and ii) S_{VI} in a 5-membered ring in the large part of the supercage in a sterically accessible position. The different steric hindrance of these two Cu families is consistent with the existence of two families of NO adsorption sites in Cu-exchanged MCM-22 zeolites, as observed by FTIR[†] [11].

As a general observation, the residual extra-framework electron density, obtained by the difference Fourier syntheses, is in accordance with literature data, as three Cu sites over the five suggested by Prakash *et al.* [12] were found in this work. Therefore, the approximate Cu ion locations were successfully determined by the present X-rays diffraction experiment. Nevertheless reliable geometry features of the Cu ion sites could not be obtained by XRPD analysis only, mainly because of the low Cu content of these samples (Cu/Si molar ratios equal 0.017 to 0.05). Moreover, the P6/mmm symmetry, adopted for the MCM-22 framework, is a mandatory choice to obtain a stable Rietveld refinement of XRPD data, but is certainly higher than that of the Cu-MCM-22 structure. Therefore, some theoretical calculations were carried out to study the geometry and energetic features of the Cu exchange into the supercage of the MCM-22 zeolite.

3.2 Theoretical calculations – The geometry and energetic features of Cu-exchange in 5- and 6-membered rings were investigated by *ab-initio* MO calculations, employing the cluster models depicted in Figure 4, where the relevant Cu-O distances are also reported. The starting Cu-site models were obtained after the Rietveld refinements of the XRPD data. A geometry optimization was carried out, as described in the experimental section. The model in Figure 4a resulted to be the most stable (ΔE = -62 kJ/mole) after geometry optimization. Moreover, optimization of the model reported in Figure 4b suggested that the 5-membered ring is not well suited to host the Cu ion, because two over five oxygen atoms have their electronic lone pairs pointing towards the external part of the cavity and are not available for Cu coordination. Conversely, the 6-membered ring shown in Figures 4a and 4c is capable of hosting Cu ions with the shortest Cu-O distances observed for the O atoms bonded to the Al

---

[†] FTIR indicates the existence of the two families also for low Cu loading samples, where the population of the S_{VI} sites is probably too low to be detected by XRPD.

site. Cu ion in site S_1 shows a 3-fold coordination, since the Cu-O contact not shown in Figure 4c are larger than 2.5 Å.

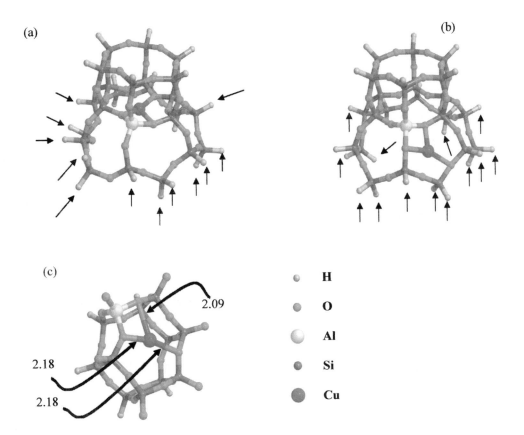

Figure 4: Cluster models employed in the theoretical calculation with the Cu atom inserted in a 6- (a) or 5-membered ring (b). Arrows indicate the hydrogen atoms the positions of which were fixed during the geometry optimization. Relevant Cu-O distances (Å) are also reported in the enlarged view of the Cu insertion in a 6-membered ring (c).

Table 2

Average values of the Cu-O distances (in Å) found in the CCDC database [34], after searching for the Cu-O-Si fragment.

Type of search	Number of Cu-O distances	Mean	Minimum	Maximum
All structures	139	1.95(7)	1.66	2.35
Ordered structures	42	1.92(6)	1.83	2.01

Furthermore, the Cu-O distances obtained after the geometry optimisations are in agreement with those reported in the literature (compare the distance reported in figure 4c with the literature data reported in Table 2), taking into account that the method employed for the theoretical calculations can overestimate these distances.

4. Conclusions

The full profile Rietveld refinement of the XRPD patterns collected on Cu-MCM-22 samples with different Cu exchange levels ($Cu^+/Al < 2$) allowed the location of extra-framework Cu ions. Three Cu sites (S_I, S_{II} and S_{III}), located in 6-membered rings, and one (S_{VI}), located in a 5-membered ring, were detected inside the MCM-22 supercage. No Cu ions were found within the sinusoidal channels. This is in agreement with the more difficult access to the sinusoidal channels, suggested by ^{129}Xe-NMR experiments [14]. Over-exchanged samples with $Cu^+/Al \geq 2$ revealed the presence of CuO particles. Theoretical calculations indicated that Cu ions are always closer to the oxygen atoms bonded to an Al site. Moreover, the different spatial demands of each of the Cu sites supports the existence of two families of copper sites which have a different adsorption capacity toward CO and NO in Cu-exchanged MCM-22 zeolites, as suggested by IR studies [11].

The combined computational and X-ray diffraction structural analysis allowed to define for the first time the local structure of copper ions, and their distribution within the MCM-22 zeolite cages.

Acknowledgements

This work was funded by the Italian MURST in the frame of the "*Progetti di Rilevante Interesse Nazionale*" (PRIN, Cofin2002") and the Brazilian FAPESP ("*Fundação de Amparo à Pesquisa no Estado de São Paulo*", 02/01100-4). The authors are deeply indebted to Dr. W. van Beek of the Swiss-Norwegian beamline at the ESRF for the precious support during the XRPD data collection and refinement. Profs. H. O. Pastore (Un. Campinas, Brasil), L. Marchese and D. Viterbo (Un. Piemonte Orientale, Alessandria, Italy) are acknowledged for useful discussion and assistance during the experiments.

References

[1] W.M. Meier, D.H. Olson, C.H. Baerlocher, Atlas of Zeolite Structure Types, Elsevier, London, **1996**.

[2] M.K. Rubin, P. Chu, US Patent 4 954 325 (1990).

[3] A. Corma, V. Fornés, S. B. Pergher, Th. L. M. Maesen, J. G. Burglass, Nature, 378 (1998) 353.

[4] S.L. Lawton, M.E. Leonowicz, R.D. Partridge, P. Chu, M.K. Rubin, Micropor. Mesopor. Mater., 23 (1998) 109.

[5] M.F. Leonowicz, J.A Lawton, S.L. Lawton, M.K. Rubin, Science, 264 (1994) 1910.

[6] S. Nicolopoulos, J.M. González-Calbet, M. Vallet-Regi, A. Corma, C. Corell, J.M. Guil, J. Pérez Pariente, J. Am. Chem. Soc., 17 (1995) 8947.

[7] A. Corma, C. Corell, A Martinez, J. Pérez-Pariente, Stud. Surf. Sci. Catal., 84A (1994) 1671.

[8] Y. Li, J.N. Armor., Appl. Catal. B, 1 (1992) L21.

[9] B.I. Palella, R. Pirone, G. Russo, A. Albuquerque, H.O. Pastore, M. Cadoni, A. Frache, L. Marchese, Catal. Commun., 5 (2004) 191.

[10] R.P.L. Absil, E. Bowes, G.J. Green, D.O. Marler, D.S. Shihabi, R.F. Socha, U. S. Patent 5 085 762 (1992).

[11] A. Frache, M. Cadoni, C. Bisio, L. Marchese, A.J.S. Mascarenhas, H.O. Pastore, Langmuir, 18 (2002) 6875.

[12] A.M. Prakash,. T. Wasowicz and L. Kevan, J. Phys. Chem. B, 101 (1997) 1985.

[13] T. Wasowicz, A.M. Prakash, L. Kevan, Microporous Materials, 12 (1997) 107.

[14] F. Chen, F. Deng, M. Cheng, Y. Yue, C. Ye, X. Bao, J. Phys. Chem. B, 105 (2001) 9426.

[15] G. Sastre, V. Fornes, A. Corma, J. Phys. Chem. B, 104 (2000) 4349.

[16] P. Nachtigall, M. Davidova, M. Silhan, D. Nachtigallova, Stud. Surf. Sci. Catal., 135 (2001) 177.

[17] C. Lamberti, S. Bordiga, M. Salvalaggio, G. Spoto, A. Zecchina, F. Geobaldo, G. Vlaic, M. Bellatreccia, J. Phys. Chem. B, 101 (1997) 344.

[18] P. Da Costa, B. Moden, G.D. Meitzner, D.K. Lee, E. Iglesia, Phys. Chem. Chem. Phys., 18 (2002) 4590.

426

[19] G. Turnes Palomino, S. Bordiga, A. Zecchina, G.L. Marra, C. Lamberti, J. Phys. Chem. B, 104 (2000) 8641.

[20] A.J. Fowkes, R.M. Ibberson, M.J. Rosseinsky, Chem. Mater, 14 (2002) 590.

[21] R.M. Haniffa, K. Seff, Micropor. Mesopor. Mater., 25 (1998) 137.

[22] A.J.S. Mascarenhas, H.M.C. Andrade and H.O. Pastore, Stud. Surf. Sci. Catal., 135 (2001) 322.

[23] M. Iwamoto, H. Yahiro, Y. Torikai, T. Yoshoka, N. Mizuno, Chem. Lett., (1990) 1967.

[24] Web site: www.esrf.fr.

[25] A.C. Larson, and R.B. Von Dreele, "General Structure Analysis System (GSAS)", Los Alamos National Lab. Report LAUR (2000), 86-748. (http://www.ccp14.ac.uk/solution/gsas/index.html).

[26] P. Ugliengo, D. Viterbo, G. Chiari, Z. Krist., 207 (1993) 9.

[27] MAESTRO. Version 6.0.105, MMshare Version 1.2.013. Schrödinger, Inc. 1500 SW First Avenue, Suite 1180, Portland, OR 97201-5881, Copyright 1999-2003.

[28] Jaguar 5.5, Version 5.5, Schrödinger Inc., Portland, OR, (2003).

[29] A.D. Becke, J. Chem. Phys., 98 (1993) 5648.

[30] C. Lee, W. Yang, R.G. Parr, Phys. Rev. B: Condens. Matter, 37 (1988) 785.

[31] P.J. Hay, W.R. Wadt, J. Chem. Phys., 82 (1985) 299.

[32] W.J. Hehre, L. Radom, P.v.R. Schleyer, J. A. Pople. Ab initio molecular orbital theory. Wiley, New York. 1986.

[33] A.J.S. Mascarenhas, H.O. Pastore, H.M.C. Andrade, A. Frache, M. Cadoni, L. Marchese, Stud. Surf. Sci. Catal., 142A (2002) 343.

[34] F.H. Allen, Acta Cryst., B58 (2002) 380.

Studies in Surface Science and Catalysis 155
A. Gamba, C. Colella and S. Coluccia (Editors)

Vanadium and niobium mixed-oxide catalysts obtained via sol-gel: preparation and catalytic behaviour in oxidative dehydrogenation of propane

P. Moggi[a], S. Morselli[a], C. Lucarelli[a], M. Sarzi-Amadè[a] and M. Devillers[b]

[a] Department of Organic and Industrial Chemistry, University of Parma, Parco Area delle Scienze 17/A, 43100 Parma, Italy

[b] Unité de Chimie des Matériaux Inorganiques et Organiques, Université Catholique de Louvain, Place Louis Pasteur 1/3, B-1348 Louvain-la-Neuve, Belgium

1. INTRODUCTION

The development of efficient catalysts for the selective functionalization of alkanes is a major application area in the research on bulk and supported mixed oxides [1]. Among the various catalysts proposed for the oxidative dehydrogenation (ODH) of propane, mixed oxide catalysts containing vanadium are the most frequently studied. Extensive work is focused on mixed oxides of magnesium and vanadium [2-7], and on vanadium oxides supported on different materials such as silica, alumina, boria, titania [8-9]. The superior performance of vanadium oxide-based systems arises from the specific activity of V-O bonds in the C-H bond activation of alkanes, which is the rate-limiting step of the reaction. After the first C-H bond of alkane is broken, a surface alkyl species is formed. This intermediate can react either by breaking another C-H bond at the β-position, thereby generating the dehydrogenation product, or by forming a C-O bond, resulting in oxygen containing products (including CO_x). The reactivity of lattice oxygen to form C-O bonds with the adsorbed hydrocarbon intermediate must be taken into great consideration, to avoid the complete oxidation of the alkane to carbon oxides. At this regard, it has been suggested that the reactivity of lattice oxygen is related to the strength of the metal-oxygen bond, which can also be viewed in terms of reducibility of the metal cations involved in the M-O-V bonds [10]. Pure V_2O_5 is active in many partial oxidation reactions, because it is characterised by easily reducible V-O-V bonds; nevertheless, the selectivity to the desired product is often low. Mixed oxides of magnesium and vanadium lead to dominant propene selectivity, because they are characterised by less reducible lattice V-O-Mg bonds [11]. Since the strength of the niobium-oxygen bond is greater than that of the vanadium-oxygen bond, niobium oxide was examined as catalyst in propane ODH, to see if it gave any improvement compared to the vanadium compounds. It was found to be a very selective catalyst for propane ODH, but with rather low activity [12]. An increase in the catalytic activity, without compromising the high selectivity to propene, was obtained by promoting niobium oxide with vanadium [13-14]. It was suggested that the activity increase could be related to the formation of surface V-O-V ensembles, which have more reactive bridging oxygen than the V-O-Nb ensembles (the former being easier to reduce than the latter). The high selectivity to propene was probably related to the presence of V-O-Nb bonds. Further experiments on niobia-supported vanadium oxide catalysts clearly confirmed that active sites containing vanadium are definitely necessary, to provide a significant catalytic activity in propane ODH. Moreover, selectivity to propene strongly depended on the catalyst nature, being in relationship with the optimisation of the sorption properties of propene, to prevent total oxidation [15]. The key point to reach high catalytic performances seemed to be the ability to control the nature and interdispersion of the mixed-oxide phases. A preparation method, which causes the vanadium to be distributed homogeneously at the surface and in the bulk, was preferred over a method which deposits vanadium only at the surface, possibly in large clusters [13]. Sol-gel methods are known to be particularly powerful to achieve molecular scale dispersion in mixed oxides. As far as pure niobia is concerned, an extended sol-gel method based on the hydrolysis of a modified Nb-alkoxide precursor obtained from $Nb(OEt)_5$ and acetylacetone was reported to be an interesting method to produce nanocrystalline mesoporous films of the T-phase of Nb_2O_5 [16]. In addition, the implementation of a classical sol-gel route based on metal alkoxides to prepare Nb-rich (Nb/V = 6 or 10) amorphous V-Nb oxides, displaying high surface areas, was reported [17]. In a previous paper, we already reported preliminary results dealing with the development of a non-hydrolytic sol-gel route to synthesise Nb-V and Nb-V-Si catalysts [18]. In a subsequent work, the more commonly used hydrolytic sol-gel method was adopted as a viable alternative to the non-hydrolytic one [19]. Very encouraging results were obtained in the ODH of propane to propene with Nb/V catalysts prepared by both sol-gel ways [20].

The present contribution focuses on the optimisation of the hydrolytic sol-gel preparation method, to achieve the highest possible control of the molecular scale interdispersion of the mixed-oxide phases. More particularly, the effects of various promoters (HCl, HNO_3, oxalic acid and citric acid) and higher Nb/V ratio on the hydrolytic sol-gel preparation are investigated. The aim is to study the possible relationship between the nature and interdispersion of the oxide phases present in the Nb-V systems, and the resulting catalytic performances in the ODH of propane.

2. EXPERIMENTAL

Nb/V gels with an atomic ratio Nb/V of 1:1 were prepared starting from $Nb(OPr^i)_5$ and $VO(OPr^i)_3$ as metal precursors. The syntheses were conducted by adding dropwise a mixture of water, sol-gel promoter and 2-propanol to a stirred mixture of $Nb(OPr^i)_5$ and $VO(OPr^i)_3$ in 2-propanol, leading each time to a stable sol, which turned into a gel or a gelatinous precipitate within few days. HCl, HNO_3, oxalic acid and citric acid as sol-gel promoters were adopted. In one case, no sol-gel promoter was added. Two samples with Nb/V ratio of 1:1 and 9:1 were prepared starting from $NbCl_5$ and $VO(OPr^i)_3$ as precursors. A mixture of water and 2-propanol was added dropwise to a stirred mixture of precursors in 2-propanol, leading each time to a stable sol, which turned into a gel or a gelatinous precipitate within few days. A sample with a Nb/V ratio of 4.5:1 was prepared by adding a mixture of water and 2-propanol to a stirred mixture of $Nb(OPr^i)_5$ and $VO(OPr^i)_3$ in 2-propanol, leading to a stable sol, which turned into a gel in few minutes.

The xerogels were all activated in air at 350°C for 15 h, then at 400°C for 2 h, finally at 550°C for 4 h. They were characterised by surface area determinations, X-ray diffraction (XRD), RAMAN spectroscopy and scanning electron microscopy (SEM). Powder X-ray diffraction patterns were measured on a Philips PW 3710 diffractometer using the Cu K_α radiation (λ =1.54178 Å). The crystalline phases were identified by reference to the powder diffraction data files (JCPDS-ICDD). The BET specific surface area measurements were carried out on a Micromeritics Pulse Chemisorb 2705 analyser using nitrogen at 77 K. Samples were previously outgassed under helium at 473 K. Raman spectroscopy was performed on a DILOR-JOBIN YVON-SPEX spectrometer, model Olympus DX-40, equipped with a He-Ne ($\lambda = 632.8$ nm) laser. SEM micrographs were taken with a Philips XL 30 ESEM instrument equipped with BSE and SE detectors. The ODH catalytic experiments were generally performed with 0.3 g of catalyst, at atmospheric pressure, in the temperature range 400-500°C, at a space velocity of 90 ml min^{-1} g$_{cat}$$^{-1}$. The gas feed composition was 10% C_3H_8, 10% O_2 and 80% He (total flow rate 30 ml min^{-1}). Some experiments were also performed at the higher space velocity of 300 ml min^{-1} g$_{cat}$$^{-1}$.

3. RESULTS AND DISCUSSION

3.1. Structural, spectroscopic and morphological characterisations

The sol-gel preparation of the 1:1 Nb/V systems led to the formation of gelatinous precipitates with the only exception of the sample obtained by adding citric acid as promoter, which turned into a gel. The sol-gel preparation of 4.5:1 and 9:1 Nb/V systems led in both cases to the formation of yellow coloured gels. From these preliminary results, it was concluded that a high Nb amount or the presence of a V complexing agent such as citric acid are necessary conditions for preparing Nb/V mixed gels, otherwise the solution of Nb and V precursors is not stable upon water addition, and a yellow gelatinous precipitate readily forms, independently of the water amount and/or the precursors concentration adopted.

Table 1
Results of XRD analysis and surface area (m^2 g^{-1}) determination

Sample	Crystalline phases	BET Specific Surface Area
1:1 Nb/V	$NbVO_5$; $Nb_{18}V_4O_{55}$; V_2O_5	5.1
1:1 Nb/V (HCl)	$NbVO_5$; $Nb_{18}V_4O_{55}$; V_2O_5	4.1
1:1 Nb/V (HNO$_3$)	$NbVO_5$; $Nb_{18}V_4O_{55}$	4.3
1:1 Nb/V (citric acid)	$NbVO_5$	4.1
1:1 Nb/V (oxalic acid)	$NbVO_5$; $Nb_{18}V_4O_{55}$	4.9
1:1 Nb/V (NbCl$_5$)	$NbVO_5$; $Nb_{18}V_4O_{55}$; V_2O_5	2.2
4.5:1 Nb/V	Nb_2O_5	10.8
9:1 Nb/V (NbCl$_5$)	Nb_2O_5	2.2

The results of XRD measurements and surface area determinations on the prepared samples after the calcination treatments are reported in Table 1. As it can be seen, the systems were characterised by low surface areas whatever the additive used, the 4.5:1 Nb/V sample showing the highest value of 10.8 m^2 g^{-1}. The crystalline phases $NbVO_5$ [46-0046] and $Nb_{18}V_4O_{55}$ [46-0087] were recognised in the XRD patterns of all the precipitated 1:1 Nb/V systems; moreover, in agreement with the overall stoichiometry, crystalline V_2O_5 [09-0387] was detected as a third minor phase in the samples: (i) 1:1 Nb/V prepared without adding sol-gel promoter, (ii) 1:1 Nb/V (HCl) (Fig. 1a) and (iii) 1:1 Nb/V (NbCl$_5$). In contrast, $NbVO_5$ [46-0046] containing only small amount of $Nb_{18}V_4O_{55}$ [46-0087] (shoulder at $2\theta \sim 22°$) was detected in the gel-derived 1:1 Nb/V (citric acid) system (Fig. 1b). The presence of crystalline V_2O_5 in the 1:1 Nb/V (HCl) sample is evidenced in Fig. 2, where the XRD patterns of this sample and of the gel-derived 1:1 Nb/V (citric acid) system are superimposed to the XRD pattern of V_2O_5 [09-0387], in the ranges of 2θ 14-24° and 26-34°.

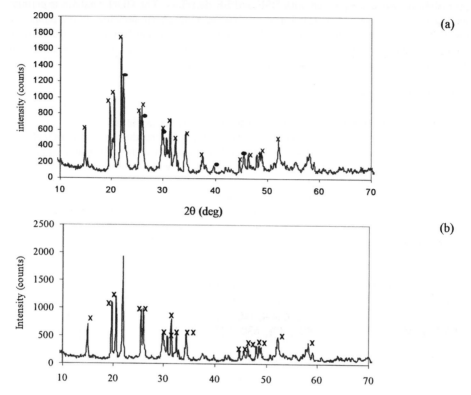

Fig. 1. XRD pattern of: (a) the 1:1 Nb/V system prepared by adding HCl as sol-gel promoter; (b) the 1:1 Nb/V system prepared by adding citric acid as sol-gel promoter. (x) $NbVO_5$ [46-0046]; (•) $Nb_{18}V_4O_{55}$ [46-0087]

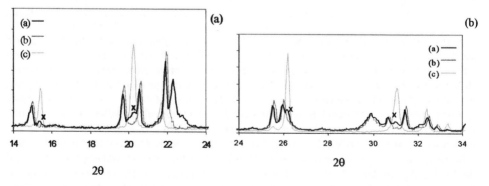

Fig. 2. Comparison of XRD patterns: (a) 1:1 Nb/V (HCl ; (b) 1:1 Nb/V (citric acid); (c) crystalline V_2O_5 [09-0387]

The Raman data are reported in Table 2. The heterogeneity of precipitated 1:1 Nb/V systems was evidenced by recording Raman spectra at different positions on the surface of each sample. Several phase compositions were detected, in agreement with XRD analysis.

Table 2
Raman results

Sample	Raman shift (cm^{-1})	Phases
1:1 Nb/V	995; 701; 528; 481; 404; 302; 283	V_2O_5
	779; 732; 631; 374; 342; 316; 229	$NbVO_5$
	970	$Nb_{18}V_4O_{55}$
1:1 Nb/V (HCl)	994; 700; 529; 479; 404; 304; 284	V_2O_5
	1018; 779; 734; 630; 375; 341; 230	$NbVO_5$
	970	$Nb_{18}V_4O_{55}$
1:1 Nb/V (HNO$_3$)	996; 700; 529; 479; 404; 304; 284	V_2O_5
	1018; 986; 946; 778; 734; 630;	$NbVO_5$
	375; 341; 316; 230	
1:1 Nb/V (citric acid)	1019; 988; 946; 778; 731; 698	$NbVO_5$
	(sh); 628; 374; 341; 317; 230	
1:1 Nb/V (oxalic acid)	994; 701; 528; 481; 403; 303; 284	V_2O_5
	1016; 987; 946; 781; 727; 632;	$NbVO_5$
	374; 340; 316; 228	
	970	$Nb_{18}V_4O_{55}$
1:1 Nb/V (NbCl$_5$)	996; 703; 528; 479; 404; 303; 284	V_2O_5
	1020; 984; 340	$NbVO_5$
4.5:1 Nb/V	987; 898; 630; 533; 470; 313 (sh);	H-Nb_2O_5 [23]
	259; 233	
	1016; 970; 732; 322	$Nb_{18}V_4O_{55}$
9:1 Nb/V (NbCl$_5$)	993; 899; 835; 667; 622;542; 470;	H-Nb_2O_5 [23]
	350 (sh); 310 (sh); 259; 235	

Microcrystalline V_2O_5 was present in all the precipitated samples. The Raman bands at 284, 303, 483, 528, 703 and 996 cm^{-1} coincided with the main features of Raman spectra of crystalline V_2O_5 assigned in the literature [21]. The Raman bands at 1020, 988, 946, 778, 731, 698 (sh), 628, 374, 341, 317, 290 and 230 cm^{-1} of these systems coincided with the main features of Raman spectra of the 1:1 Nb/V (citric acid) sample (Fig. 3), in which the practically pure $NbVO_5$ phase had been detected by XRD. These bands were systematically evidenced in the 1:1 Nb/V (citric acid) sample by recording spectra at different positions, thus confirming the homogeneity of the catalyst. $NbVO_5$ is expected to contain NbO_6 octahedral units sharing corners with VO_4 tetrahedral units [22]. In particular, in the (100) plane the structure can be described as zigzag chains of corner-shared NbO_6 octahedra sharing four oxygens with the VO_4 tetrahedra, which point up and down alternatively. On the basis of the spectroscopic data previously reported for the structurally-related $NbPO_5$ compound [23], Raman bands due to the VO_4 units were assigned as follows: 1020 cm^{-1} (weak), 988 cm^{-1} (strong), 374 cm^{-1} (medium), 230 cm^{-1} (medium). These bands are associated respectively to the stretching (antisymmetrical and symmetrical) and bending (in and out-of-plane) modes of terminal V=O bonds [24]. The Raman bands at 778 and 340 cm^{-1} were associated to the stretching and bending modes of V-O bonds in VO_4 tetrahedra, whereas the Raman bands at 946, 731, 628, 317 and 290 cm^{-1} were attributed to the stretching and bending modes of Nb-O bonds in slightly and highly distorted NbO_6 octahedral units [22-25], the bands at 946, 630 and 290 cm^{-1} being characteristic of distorted NbO$_6$ octahedral structures connected by sharing corners [24,25]. Of a difficult interpretation was the shoulder at about 700 cm^{-1}. According to spectroscopic studies on amorphous niobium and vanadium single oxides [22], it could be assigned to stretching vibrations of a residual polymeric amorphous mixed oxide deriving from the xerogel precursor, not destroyed by calcination.

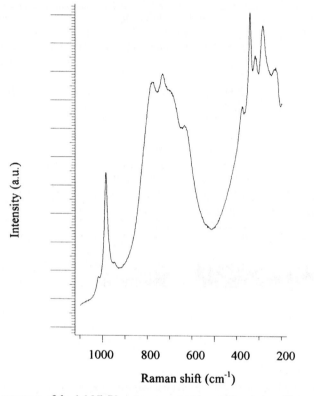

Fig. 3. Raman spectrum of the 1:1 Nb/V system prepared by adding citric acid as sol-gel promoter

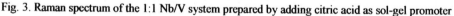

The Raman spectra of the 1:1 Nb/V (citric acid) sample are in quite good agreement with those recently reported in the literature by Briand et al. and Ballarini et al. [26,27]. The main differences are found in the region between 400 and 600 cm^{-1}, in which the authors report the presence of two bands at about 530 and 475 cm^{-1} that are absent in our spectra. Since the broad band at 700 cm^{-1} is also reported in both studies, it is supposed that the bands at 530 and 475 cm^{-1} and the one at 700 cm^{-1} might be related to the polymeric functionality of a less ordered mixed oxide. Raman spectroscopy is particularly sensitive for detection of ordering in structures since ordering gives rise to strong Raman bands, while broad bands are seen in disordered structures. In contrast to Raman spectroscopy, XRD does not reflect such local disorder so much.

The Nb$_{18}$V$_4$O$_{55}$ phase was not easy to identify by Raman analysis owing to the superimposition of its Raman bands to those of NbVO$_5$ and V$_2$O$_5$ phases. Nevertheless, the presence of a Raman band at 970 cm^{-1} in the spectra of samples in which the Nb$_{18}$V$_4$O$_{55}$ phase had been detected by XRD, and not in the pattern of the 1:1 Nb/V (citric acid) sample in which the practically pure NbVO$_5$ phase had been detected, led to consider this band as indicative for this phase.

SEM investigations confirmed the results given by XRD and Raman analyses, by evidencing the presence of different crystalline phases in the 1:1 Nb/V precipitated system prepared without adding a sol-gel promoter. Fig. 4a shows the SEM image of the micrometer-sized crystallites of NbVO$_5$, which was the predominant phase, uniformly distributed. Fig. 4b shows the SEM image of needle-shaped, thin crystallites of V$_2$O$_5$, less abundant and mainly localised on the surface of the catalyst particles. In Fig. 5, the SEM image of the sol-gel derived 1:1 Nb/V (citric acid) sample is reported. The system was homogeneous over the entire area examined. No segregation or surface decorations by extra phases were detected.

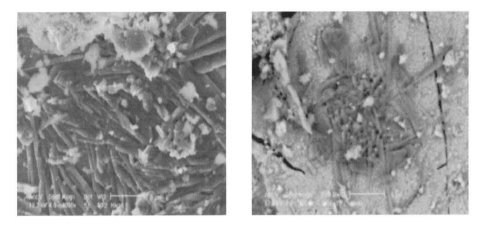

(a)   (b)

Fig. 4. SEM images of the 1:1 Nb/V system prepared without the addition of a sol-gel promoter: (a) NbVO$_5$ crystallites; (b) V$_2$O$_5$ crystallites.

Fig. 5. SEM image of the 1:1 Nb/V (citric acid) system.

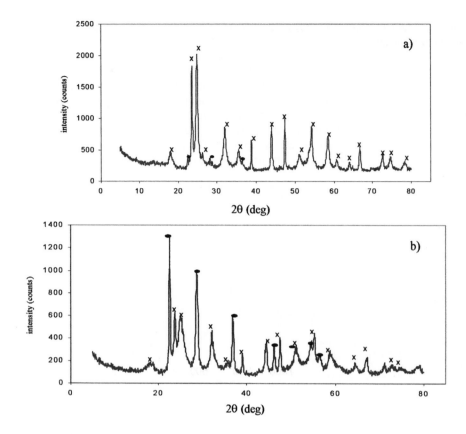

Fig. 6. a) XRD pattern of the 9:1 Nb/V system; b) XRD pattern of the 4.5:1 Nb/V system. (x) Nb_2O_5 [32-0711] (•) Nb_2O_5 [27-1003]

The XRD spectra of the 4.5:1 Nb/V and 9:1 Nb/V systems after the thermal treatment evidenced the presence of crystalline Nb_2O_5 in both samples. In the 9:1 Nb/V system (Fig. 6a), two Nb_2O_5 phases were identified: JCPDS [32-0711], corresponding to the metastable M form of Nb_2O_5 and predominant in the pattern, and [27-1003], less abundant, corresponding to the low-temperature T orthorhombic form of Nb_2O_5. The M-form should be regarded as a less ordered precursor of the high temperature monoclinic H-form [28]. In the 4.5:1 Nb/V system (Fig. 6b), the Nb_2O_5 [27-1003] phase was predominant in the pattern, while the Nb_2O_5 [32-0711] phase was detected in a minor amount. The peaks corresponding to the orthorhombic phase were slightly shifted towards higher values of theta. It was hypothesised that diffusion of some V^{5+} ions had occurred into the structure of Nb_2O_5 [27-1003], causing the shift of the XRD peaks. In both 4.5:1 Nb/V and 9:1 Nb/V systems, there was no evidence of V-containing crystalline phases. The presence of microcrystalline V_2O_5 was excluded in both samples by Raman analysis. The Raman spectra of the 9:1 Nb/V system showed the characteristic bands of the $H-Nb_2O_5$ phase (Fig. 7) [23]: 993, 893, 667, 622, 542, 470, 309, 259, and 235 cm^{-1}. These bands were related to the M-form detected in the sample by XRD. The ν_1 (A_1) Raman active mode of the NbO_4 tetrahedron present in the structure of $H-Nb_2O_5$ was also detected at 835 cm^{-1}. Nb^{5+} ions seldom occur in tetrahedral coordination. In the

YNbO$_4$ compound, containing NbO$_4$ tetrahedra, the ν_1 Raman band appears at 830 cm^{-1}, which is close to that assigned for tetrahedra in our 9:1 Nb/V sample [23].

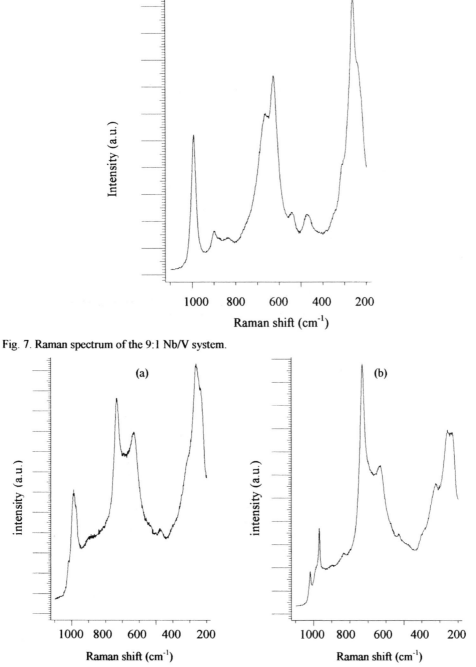

Fig. 7. Raman spectrum of the 9:1 Nb/V system.

Fig. 8. Raman spectrum of the 4.5:1 Nb/V system collected at two different points (a) and (b).

436

In the 4.5:1 Nb/V system, the presence of two different phases was evidenced by recording Raman spectra at different positions. In one pattern (Fig. 8a), the main bands of a Nb_2O_5 phase at about 990, 630, 260 and 230 cm^{-1} were detected. In a second one (Fig. 8b), sharp bands at 1018, 970, 730 and 320 cm^{-1} evidenced the presence of a microcrystalline mixed phase containing vanadium and niobium, most probably $Nb_{18}V_4O_{55}$.

SEM investigations pointed out the homogeneous nature of the 4.5/1 Nb/V system, and evidenced the presence of two similar but different crystalline phases (Fig. 9).

Fig. 9. SEM image of the 4.5:1 Nb/V system.

3.2. Catalytic results

The catalytic results obtained with the 1:1 Nb/V prepared systems at 450°C and space velocity of 90 ml $min^{-1}g_{cat}^{-1}$ are reported in terms of C_3H_8 conversion and propene selectivity in Fig. 10.

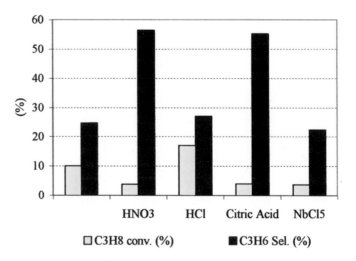

Fig. 10. Catalytic performances of 1:1 Nb/V catalysts in propane ODH. P = 1atm; T = 450°C; gas feed = 30 ml min^{-1} (10% C_3H_8, 10% O_2 and 80% He); catalyst = 0.3g.

The 1:1 Nb/V samples obtained respectively without adding a sol-gel promoter and by adding HCl as sol-gel promoter gave the highest conversions of C_3H_8. This high activity was counterbalanced by a low selectivity to propene. The samples prepared by adding HNO_3 or citric acid as promoter showed rather poor activity, but they were more selective to propene. It was concluded that the presence of segregated crystalline V_2O_5 in the former samples contributed to increase the propane conversion to mainly CO_x products.

In Fig. 11, the catalytic results collected on the gel-derived 1:1 Nb/V (citric acid), 4.5:1 Nb/V and 9:1 Nb/V samples and on the precipitated 1:1 Nb/V sample at 450°C and space velocity of 90 ml $min^{-1}g_{cat}^{-1}$ are compared. The 9:1 Nb/V sample containing the lowest amount of V behaved similarly to pure Nb_2O_5, showing rather poor activity and high selectivity to propene, about 50%. Both activity and selectivity increased only slightly when a higher amount of V was adopted in the 1:1 Nb/V (citric acid) system, containing V mainly as Nb-V mixed oxide. This evidenced that the V-O-Nb functionality in the $NbVO_5$ compound is rather similar to the Nb-O-Nb one. The presence of small amount of V-O-V bonds as crystalline V_2O_5 was sufficient to increase significantly the catalyst activity in the precipitated 1:1 Nb/V sample; the system however was poorly selective to propene. The 4.5:1 Nb/V sample showed both higher activity and selectivity to propene than the precipitated 1:1 Nb/V one. This catalyst was also that displaying the highest specific surface area. The system contained an intermediate amount of V in solid solution with Nb oxide. Since no crystalline or microcrystalline V_2O_5 was detected, it could be hypothesized that a peculiar V-O-Nb functionality existed in this sample, more reducible than the V-O-Nb functionality in the $NbVO_5$ phase, and less reducible than the V-O-V one. The 4.5:1 Nb/V sample was tested also at the higher space velocity of 300 ml min^{-1} g_{cat}^{-1}. The results are reported in table 3. A very good yield of propene, comparable to the best yields reported in the literature for various catalysts [29], was obtained at 550°C, as the right compromise between activity and selectivity performances.

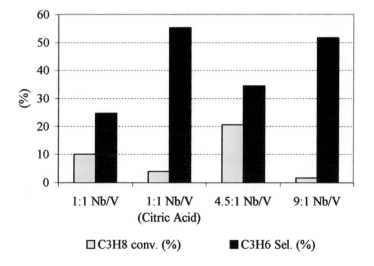

Fig. 11. Catalytic performances of 1:1, 4.5:1 and 9:1 Nb/V catalysts in propane ODH. P = 1atm; T = 450°C; gas feed = 30 ml min^{-1} (10% C_3H_8, 10% O_2 and 80% He); catalyst = 0.3g.

438

Table 3
Catalytic performances of the 4.5:1 Nb/V sample in propane ODH. P = 1atm; gas feed = 30 ml min^{-1} (10% C_3H_8, 10% O_2 and 80% He); space velocity = 300 ml min^{-1} g_{cat}^{-1}. X_{C3H8} = propane conversion; S_{C3H6} = selectivity to propene; Y_{C3H6} = yield of propene

T (°C)	X C_3H_8 (%)	S C_3H_6 (%)	Y C_3H_6 (%)
400	3.9	72.6	2.8
450	8.6	55.7	4.8
500	12.7	55.9	7.1
550	19.4	47.4	9.2

4. CONCLUSIONS

The hydrolytic sol-gel method was adopted in this work to prepare Nb/V mixed oxides systems as potential catalysts for propane ODH.

Several 1:1 Nb/V systems were prepared by varying sol-gel promoter and Nb precursor. It was found that the addition of citric acid as sol-gel promoter stabilises the solution of Nb and V precursors after the addition of water, favouring the direct formation of a gel. As a result, a homogeneous and well-interdispersed 1:1 Nb-V system was obtained by this method. In all the other cases, precipitation of a yellow solid occurred, leading after calcination to heterogeneous systems, in which the presence of segregated V_2O_5 was detected.

Two systems with higher amount of Nb, 9:1 Nb/V and 4.5:1 Nb/V, were also prepared by hydrolytic sol-gel method. In both cases, the addition of citric acid was not necessary. Sols containing higher amounts of Nb precursor were stable and turned rapidly into gels. In both 9:1 Nb/V and 4.5:1 Nb/V samples, there was no evidence of crystalline phases containing V. Only the presence of crystalline Nb_2O_5 was detected by XRD. In the 4.5:1 Nb/V sample, however, XRD pattern suggested that some diffusion of V^{5+} ions had occurred into the structure of orthorhombic Nb_2O_5 phase, and Raman patterns evidenced the presence of microcrystalline $Nb_{18}V_4O_{55}$.

A Nb-V 'synergistic effect' in the ODH of propane seemed to characterise the 4.5:1 Nb/V sample, which showed both higher activity and selectivity to propene than the 1:1 Nb/V samples. The 9:1 Nb/V sample containing the lowest amount of V was found to behave similarly to pure Nb_2O_5, showing rather poor activity and high selectivity to propene, about 50%. Both activity and selectivity were found to increase only slightly with the 1:1 Nb/V (citric acid) system, containing V mainly as Nb-V mixed oxide. The presence of small amount of V-O-V bonds as crystalline V_2O_5 increased significantly the catalyst activity of precipitated 1:1 Nb/V samples.

Acknowledgments
This work was performed within the frame of a Concerted Research Action of the 'Communauté Française de Belgique'. The authors also acknowledge the financial support from the Ministero dell'Università e della Ricerca Scientifica e Tecnologica (Rome) and the Belgian National Fund for Scientific Research (Brussels). M. Sarzi-Amadè was recipient of a Socrates grant for a three-months stay in Louvain-la-Neuve.

REFERENCES
[1] G. Centi and P. Mazzoli, Catal. Today 28 (1996) 351.
[2] M. A. Chaar, D. Patel, M. C. Kung and H.H. Kung, J. Catal. 105 (1987) 483.
[3] M. A. Chaar, D. Patel, M. C. Kung and H.H. Kung, J. Catal. 109 (1988) 463.

[4] M. J. Philips and M. Ternan, Proc. 9th Int. Congr. Catal. 4 (1988) 199B1.

[5] D. Siew Hew Sam, V. Soenen and J. C. Volta, J. Catal. 123 (1990) 417.

[6] A. Corma, J. M. Lòpez Nieto and N. Paredes, J. Catal. 144 (1993) 425.

[7] X. Gao, P. Ruiz, Q. Xin and B. Delmon, J. Catal. 148 (1994) 56.

[8] G. Colorio, J. C. Védrine, A. Auroux and B. Bonnetot, Appl. Catal. A 137 (1996) 55.

[9] T. Blasco and J. M. Lopez Nieto, Appl. Catal. A 157 (1997) 117.

[10] H. H. Kung, Adv. Catal. 40 (1994) 1.

[11] F. Cavani and F. Trifirò, Catal. Today 24 (1995) 307.

[12] R. H. Smiths, K. Seshan and J. R. H. Ross, J. Chem. Soc., Chem. Commun. (1991) 558.

[13] R. H. Smiths, K. Seshan, H. Leemreize and J. R. H. Ross, Catal. Today 16 (1993) 513.

[14] J. R. H. Ross, R. H. H. Smits and K. Seshan, Catal. Today 16 (1993) 503.

[15] T.C. Watling, G. Deo, K. Seshan, I.E. Wachs and J.A. Lercher, Catal. Today 28 (1996) 139.

[16] F. Lenzmann, V. Shklover, K. Brooks and M. Grätzel, J. Sol-Gel Sci. Technol. 19 (2000) 175.

[17] M. Catauro, C. Pagliuca, L. Lisi and G. Ruoppolo, Thermochim. Acta 381 (2002) 65.

[18] F. Barbieri, D. Cauzzi, F. De Smet, M. Devillers, P. Moggi, G. Predieri and P. Ruiz, Catal. Today 61 (2000) 353.

[19] P. Moggi, G. Predieri, D. Cauzzi, M. Devillers, P. Ruiz, S. Morselli and O. Ligabue, Stud. Surf. Sci. Catal. 143 (2002) 149.

[20] P. Moggi, M. Devillers, P. Ruiz, G. Predieri, D. Cauzzi, S. Morselli and O. Ligabue, Catal. Today, 81 (2003) 77.

[21] T. R.Gilson, O. F. Bizri and N. Cheetham, J. Chem. Soc. Dalton Trans. 291 (1973)

[22] J. M. Amarilla, B. Casal and E. Ruiz-Hitzky, Mater. Lett. 8 (1989) 132

[23] A. A. McConnell, J. S. Anderson and C. N. R. Rao, Spectrochim. Acta 32A (1976) 1067.

[24] I. R. Beattie and T. R. Gilson, J. Chem. Soc. A (1969) 2322.

[25] M. A. Vuurman and I. E. Wachs, J. Phys. Chem. 96 (1992) 5008.

[26] N. Ballarini, F. Cavani, C. Cortelli, C. Giunchi, P. Nobili, F. Trifirò, R. Catani and U. Cornaro, Catal. Today 78 (2003) 353.

[27] L. E. Briand, J-M Jehng, L. Cornaglia, A. M. Hirt and I. E. Wachs, Catal. Today 78 (2003) 257.

[28] H. Schäfer,R. Gruehn and F. Schulte, Angew. Chem. Internat. Edit. 5, (1966) 40.

[29] S. Albonetti, F. Cavani and F. Trifirò, Catal. Rev. Sci. Eng. 39 (1997) 413.

Studies in Surface Science and Catalysis 155
A. Gamba, C. Colella and S. Coluccia (Editors)

An EPR study of the surface reactivity of CaO
and a comparison with that of MgO

M.Cristina Paganini, Mario Chiesa, Paola Martino, Stefano Livraghi, Elio Giamello*.

Dipartimento di Chimica IFM, Università di Torino and NIS, Nanostructured Interfaces and Surfaces Centre of Excellence, Via P. Giuria 7, I - 10125 Torino (Italy) Fax ++39 011 6707855 E-mail elio.giamello@unito.it

1. INTRODUCTION

The catalytic properties and the surface reactivity of Calcium Oxide are relatively less known than those of the isostructural Magnesium Oxide, the most studied among the alkaline-earth oxides. It is generally believed that the properties of CaO are extremely close to those of MgO as the two oxides have the same structure (NaCl), similar morphological habit and high ionicity. Magnesium Oxide, in fact, has become in the past two decades a reference model oxide for groups of surface scientists (single crystals, thin layers)[1] and surface chemists (polycrystalline materials)[2] as well as theoretical chemists [3] while CaO is relatively less explored. Actually an interesting debate on the surface chemistry of CaO appeared in the litterature in recent years concerning the discussion about its basicity in comparison to that of MgO. Tanabe and co-workers [4] showed in a series of experimental papers that CaO is more active than MgO in the conversion of benzaldehyde into benzyl benzoate and demonstrated that the order of base strength in the group of alkaline earth oxides is BaO>SrO>CaO>MgO. The reason for the different surface basicity between CaO and MgO has been analysed in detail by Pacchioni et al. [5] by means of theoretical calculations and rationalized on the basis of simple electrostatic arguments. The lower Madelung potential at the surface of CaO (due to the higher lattice constant with respect to that of MgO) implies a reduced stability of the oxide anion. As a result the electron cloud of an O^{2-} ion on CaO is more spatially diffuse compared to that of a similar ion on MgO explaining the increased basicity and a potential higher reactivity.

The nitrogen chemistry during reduction of NO over alkaline earth oxide surfaces has also been studied [6] and CaO has been used as a model substrate to study basic reactions both by experiment [7, 8] and theory [9, 10].

We have investigated the reactivity of CaO by Electron Paramagnetic Resonance (EPR) which revealed particularly suited to describe, by means of various paramagnetic probes, the complex surface chemistry of this oxide.

Though strong analogies between MgO and CaO are easily observed, the higher basicity of CaO shows up in terms of a specific and higher reactivity. This is the case, for instance, of the interaction of CaO surface with nitric oxide. The interest for the interaction of NO with CaO is twofold. NO is a component of the automotive exhaust and thus understanding its reactivity with CaO is important to increase the knowledge on catalytic and non catalytic devices for the exhaust treatment. Moreover, NO is a paramagnetic molecule which reacts or coordinates upon a number of surface ions giving rise to EPR signals whose features can be related to the nature of the adsorption site.

Calcium oxide is also a good candidate for the activation of alkanes at moderate temperatures. The catalytic interaction with alkanes is relevant to many important industrial process, e.g., steam reforming, combustion, partial oxidation, methanation, and isomerization [11]. The activation of methane takes place via homolitic cleavage of the C-H bond at the surface of basic oxides doped with alkaline metals and containing O^- ions. Other possible active sites for homolytic dissociation are thought to be the electrophylic oxygen species such as O_2^- and O_2^{2-} which are able to abstract H atoms from RH [12-15]. The most used systems are lithium promoted magnesium oxide Li/MgO and sodium promoted calcium oxide Na/CaO [16,17,18]. An alternative possibility is that the C-H bond cleavage occurs heterolitically on coordinatively unsaturated metal-oxygen pair sites to produce carbanions, which then subsequently react with oxygen to form oxidized products [19]. This type of cleavage has been proved to act also in the case of surface oxidation of alkenes and of some aromatics like toluene [20,21] on MgO. Because of the weak acidity of the C-H bonds in alkanes ($pK_a > 40$), sites that are able to heterolytically activate methane and other alkanes (if any) must be much more strongly basic than those capable of promoting the same C-H bond breaking on alkenes and toluene.

In the present paper we investigate by EPR spectroscopy the interaction of hydrogen and of methane with the thoroughly dehydrated surface of CaO in order to compare the results with those reported on MgO. Aim of the paper is to verify the effects of both the increase of basicity passing from MgO to CaO and the existence of peculiar features of this latter oxide not observed for MgO and not simply amenable to the different basicity. Various probe molecules have been used to verify this second point. EPR was used as the high sensitivity of the technique allows to monitor even tiny amounts of paramagnetic species.

2. EXPERIMENTAL

The experiments were carried out using CaO obtained via slow thermal decomposition of commercial high-purity $CaCO_3$ (ex-Aldrich). The activation of the sample, to obtain a thoroughly dehydroxylated surface, was performed at 1173 K under a residual pressure of 10^{-5} mbar. The surface area of the resulting oxide is about 80 m^2 g^{-1}. Traces of Mn^{2+} ions in the bulk of CaO are however practically unavoidable and produce an EPR signal with the typical manganese sextet centred nearby the free electron g value. The signal will always appear in the EPR spectra and has been used as internal standard for g value.

To produce trapped electrons centres, the activated samples were exposed respectively to H_2, and CH_4 (100 mbar) and irradiated using a 500 W *Oriel Instruments* UV lamp, incorporating a Hg/Xe arc lamp (250 nm to > 2500 nm), in conjunction with a water filter to avoid IR components.

UV irradiation of the sample was carried out at 77 K. After 15 minutes irradiation, the sample develops a blue coloration, indicating the formation of surface color centers.

High purity gases (H_2, CH_4 - Praxair, ^{14}NO - Matheson) have been used in all experiments.

The EPR spectra were recorded at room temperature and at 77 K on a Bruker EMX spectrometer operating at X-band frequencies.

3. RESULTS AND DISCUSSION

3.1. UV-irradiation of CaO in the presence of hydrogen and methane.

Calcium oxide after activation does not exhibit any EPR signal except for the weak sextet of Mn^{2+} always found in such an oxide and related to a uniform doping by traces of

Mn^{2+}. UV irradiation of the solid in the presence of H_2 produces a pale blue powder and, simultaneously, the EPR spectrum in Fig.1a can be observed. The signal and the coloring are both due to the formation of surface trapped electron centers. The spectrum is analogous to that obtained on MgO using the same procedure and is characterized by the presence of hyperfine coupling of the trapped electron with a H nucleus ($I = \frac{1}{2}$) generating doublets of lines [22]. Like in the case of MgO the spectrum is heterogeneous as different centers are present, characterized by slightly different values of the protonic coupling constant. The perpendicular coupling constants however are higher than in the case of MgO ranging between 2 and 4 Gauss.

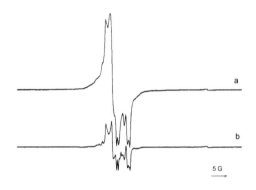

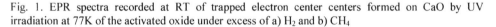

Fig. 1. EPR spectra recorded at RT of trapped electron center centers formed on CaO by UV irradiation at 77K of the activated oxide under excess of a) H_2 and b) CH_4

A similar phenomenon is observed if CaO is irradiated in the presence of methane (100 mbar). The sample undergoes coloring and the EPR spectrum in Fig. 1b is given rise. Though the spectrum appears quite similar to the previous one its features are slightly different in qualitative and quantitative terms. In other words the various trapped electron centers are present in different abundance in the two cases and some centers, not observed by activation in hydrogen atmosphere are given rise when activation is performed in methane atmosphere. A detailed assignment of all the species present at the surface must be based on the simulation of the experimental signals and is beyond the purposes of the present paper. More interesting is to outline that two distinct mechanisms can lead to the formation of trapped electrons (on both CaO and MgO) both involving the activation of two relatively inert molecules like hydrogen and methane in mild conditions. The former one is based on the homolitic activation of the molecule by surface trapped holes localized on O^{2-} ions (in chemical terms a trapped hole can be seen as an O^- ion). These are formed by charge separation consequent to the UV irradiation and reacts with H_2 or CH_4 (in general terms RH) according to the following mechanism:

$$RH_{(gas)} + O^-_{(surf)} \rightarrow OH^-_{(surf)} + R^\bullet \qquad (R = H, CH_3) \qquad (1)$$

$$R^\bullet + O^{2-}_{(surf)} \rightarrow OR^-_{(surf)} + e^-_{(trp)} \qquad (2)$$

Reaction (1) represent the homolytic splitting and reaction (2) the ionisation of the fragment with consequent electron trapping. The occurrence of reaction (2) indicates that electron trapping sites are available at the surface.

The second mechanism is based on heterolytic activation of the RH molecule which can be put into evidence by a particular experiment. If RH is adsorbed on CaO and than the gas phase is carefully evacuated, irradiation of the solid produce again (though to a lesser extent) trapped electrons. Fig. 2 reports the spectrum obtained after contact (adsorption + desorption) of methane with the surface and successive irradiation. The spectrum is easily amenable to the presence of trapped electrons. Since no gas phase is present during the experiment the mechanism of formation of trapped electrons must be as follows:

$$R\text{-}H + O^{2-}_{(surf)} \quad \leftrightarrow \quad R^- + HO^-_{(surf)} \qquad (3)$$

$$R^- + O^{2-}_{(surf)} + h\nu \rightarrow RO^-_{(surf)} + 2\,e^-_{(trp)} \qquad (4)$$

Reaction (3) is a typical acid – base reaction and, as the acidity of RH is very low, requires an extremely efficient basic site for proton abstraction. Alternatively the formation of R^- can be proved by contacting with oxygen the solid with preadsorbed RH. The formation of adsorbed superoxide O_2^- ions by electron transfer from R^- indicates once again the presence of these adsorbed anions at the surface. By this method it has been shown that the basicity of CaO is higher, as expected, than that of MgO as the same reactions described before occur, for CaO, also using less acidic molecules as ethane and propane, while for MgO they do not take place with these two molecules [23].

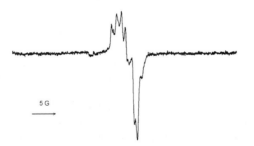

Fig. 2. EPR spectrum of color centers generated by UV irradiation of the sample previously contacted with methane and subsequently evacuated.

Summarizing, the surface of CaO (similarly to that of MgO) is capable of acivating RH in two alternative ways. The former is based on the homolytic splitting of the molecule due to trapped holes (O⁻ ions) photogenerated at the surface by irradiation (reactions 1 and 2) while the latter is based on heterolytic activation of the substrate by an extremely basic surface O^{2-} ion (reaction 3). This mechanism, in particular, does not require irradiation and CaO is able to split not only hydrogen and methane (also activated by MgO) but also ethane, propane, and ciclohexane, all molecules with extremely low acidity not activated by magnesium oxide. A first result of the present investigation agrees with other experimental findings and theoretical forecasts confirming that the basicity of CaO is higher than that of MgO.

3.2 Reactivity of CaO with hydrogen, oxygen and nitric oxide.

The differences between the two alkali-earth oxides are not limited to the superior basicity of Calcium Oxide with respect to Magnesium Oxide. A second issue concerns the presence of surface sites on CaO which are not observed on MgO. This aspect has been investigated comparing the reactivity of the two bare surfaces with simple molecules.

When molecular hydrogen is adsorbed on MgO reaction 3 (with R = H) is known to occur on few particularly active sites. This reaction in the case of hydrogen has been followed by infrared spectroscopy which confirmed the formation of hydride and hydroxyl groups [24, 25]. Reaction (3) does not form paramagnetic species and therefore does not origin EPR signals. In fact, when H_2 is adsorbed on MgO without irradiation no EPR signal is observed and reaction (3) is the only reaction occurring. A different behaviour is observed if molecular hydrogen is adsorbed in the dark at the surface of CaO as the weak spectrum in Fig. 3 is given rise.

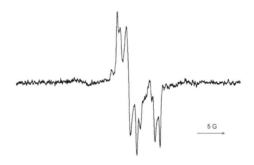

Fig. 3. EPR spectrum of color centers formed on CaO by contact of the sample with hydrogen in the dark.

The spectrum is easily assigned by comparison with the spectrum in Fig. 1 to surface trapped electrons . This result can be understood hypothesizing that both reaction (1) and (2) take place on CaO without any need of irradiation. This is possible only in the case that some O⁻ centres (the centres originating the sequence of reactions) are present at the surface of CaO after activation *in vacuum*.

This idea is confirmed by reaction of the two oxides with O_2 at room temperature. Once again MgO does not react with molecular oxygen whereas by CaO-O_2 interaction the signal in Fig. 4 is originated. Such a signal, already observed by Cordischi et al.[26], is assigned to ozonide O_3^- ions. The ozonide is an ionic radical species whose spectrum is rather insensitive to the local environment so that its spectral shape is almost the same when adsorbed on different solids. The presence of an ozonide on CaO is explained by the following reaction:

$$O_{2(gas)} + O^-_{(surf)} \rightarrow O_3^-{}_{(surf)} \tag{5}$$

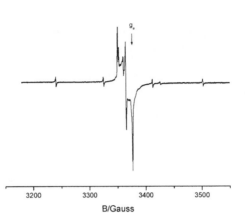

Fig. 4. EPR spectrum at 77K of ozonide species generated by the interaction between molecular oxygen and the bare surface of activated CaO.

Both adsorption of hydrogen and of oxygen indicate the presence of O^- ions at the surface of CaO. A question however arises immediately and concerns the fact that O^- (which is a paramagnetic species) is not observed by EPR on the freshly activated surface. The EPR spectrum of surface O^- ions is well known and can be observed, for instance on MgO, after direct irradiation of the solid and electron-hole separation or when trapped electron centres react with N_2O forming the surface ion and releasing nitrogen [27]. The lack of detectabilty of O^- on CaO is not unknown to surface chemistry. More than ten years ago, for instance, F. Freund et al.[28] identified the presence of O^- on bare and Na^+ doped CaO kept at high temperature. The presence of such an ion was revealed by CDA (Charge Distribution Analysis) technique at T>770K while at room temperature the species seems to disappear (and in fact is not monitored by EPR). The authors explain the fact hypothesising the presence of "dormant" O^- in the form of pairs, in practice true surface (diamagnetic) peroxides or similar peroxo species.

This interesting hypothesis has been verified by reaction of bare CaO with NO. If the pressure of the adsorbed gas is kept under 0.5 mbar a typical axial signal arises (Fig. 5) with narrow lines and the hyperfine structure originated by one N atom interacting with the unpaired electron. The spectrum is actually composed by two slightly different axial signals corresponding to species having the same nature but differing for the location on the surface and for the coordinative environment. The spectrum is destroyed at higher NO pressure and substituted by another signal elsewhere discussed [8]. The signal recorded at low NO pressure, very different from that of molecularly adsorbed NO [29], is compatible only with the formation of a radical species including three oxygen atoms. A similar species was observed in γ-irradiated KNO_3 and assigned to NO_3^{2-} radical ions. No other molecular or ionic radical species containing both N and O atoms has spin-Hamiltonian parameters comparable to those observed in the present case. The species corresponding to the spectrum in Fig.5, however, is not exactly equal to that isolated in the ionic matrix. The difference can be explained invoking the fact that a surface chemisorbed species interacts (and shares electron spin density) with the whole solid. This is not the case of the matrix isolated species.

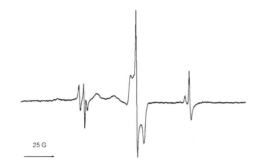

Fig. 5. EPR spectra recorded at RT of 0.4 mbar of NO adsorbed on CaO.

The formation of a surface NO_3^{2-} radical ions can thus be explained in terms of reactivity of NO with a sort of surface peroxide according to:

$$NO_{(gas)} + O_2^{2-}{}_{(surf)} \rightarrow NO_3^{2-}{}_{(surf)} \tag{6}$$

The whole reactivity of CaO with H_2, O_2, NO can now be discussed in the following terms. Few anionic defective sites are present at the surface of CaO whose origin could be due to an alteration of the oxide stoichiometry (sodium traces are practically unavoidable even in high purity materials) or to a particular propensity of the solid itself to non stoichiometry. These oxygen species can be seen as true surface peroxides (O_2^{2-}) or similar, less defined, peroxo-like species and are diamagnetic. This explains the absence of EPR signal on bare CaO and confirms the hypothesis of "dormant" O^- species [28]. At least a fraction of the surface peroxide species reacts with NO at low pressure leading to NO_3^{2-} whose formation cannot be explained unless the presence at the surface of a pair of oxygen ions bearing two negative charges (a peroxide group) is admitted. Some of the peroxide groups are probably constituted by less tightly bound oxygen atoms (for this reason we mentioned the peroxo-like groups) and can be split by an incoming molecule (for example O_2) capable of reaction with the single O^- ion. This is the case of reactivity with hydrogen and oxygen which can thus be understood in terms of

$$O^- \text{-----} O^-{}_{(surf)} \rightarrow 2\,O^- \tag{7}$$

$$2O_{2(gas)} + 2\,O^-{}_{(surf)} \rightarrow 2\,O_3^-{}_{(surf)} \tag{8}$$

The peroxo species, in other words, is split in O^- species, the driving force of the process being the successive reaction with other molecules (O_2, H_2). This explains why the surface reactivity typical of O^- is observed but not the typical spectrum of this monoatomic radical species. The presence of ozonide groups at the surface of CaO has been hypothesized, in similar experimental conditions, by Zecchina et al. [30] on the basis of a broad absorption at 23.500 cm^{-1} in UV-Vis reflectance spectra.

4. CONCLUSIONS

Two main differences characterize the surface of CaO from that of the homologous widely investigated MgO. The former one is due to the higher basicity of Calcium Oxide which, by the way, is expected on the basis of general considerations and of theoretical calculations. The second, not expected, difference consists in the presence of O$^-$ and O$_2^{2-}$ groups at the surface which trigger a peculiar reactivity of CaO (not observed for MgO) with hydrogen, oxygen and nitric oxide, all leading to the paramagnetic entities (trapped electrons, ozonide ions and NO$_3^{2-}$ radical ions respectively) described in the present paper.

REFERENCES

[1] P.A.Cox, A.A. Williams, Surf. Sci., 175 (1986) L782.
[2] E. Giamello, M.C. Paganini, D.M. Murphy, A.M. Ferrari, G. Pacchioni, J. Phys. Chem. B, 101 (1997) 971.
[3] G. Pacchioni, in "The Chemical Physics of Solid Surfaces", Vol. 9, D. P. Woodruff (Ed.), Elsevier, Amsterdam, (2001).
[4] Hattori H.; Maruyama K.; Tanabe K., J. Catal., 44 (1976), 50.
[5] G. Pacchioni, J. M. Ricart, F. Illas, J. Am. Chem. Soc., 116 (1994) 10152.
[6] F. Acke, I. Panas, J Phys. Chem. B, 102 (1998), 5127.
[7] F. Acke , I. Panas, J Phys. Chem. B, 102 (1998) 5158.
[8] M.C. Paganini, M. Chiesa, P. Martino, E. Giamello, J. Phys. Chem. B, 106 (2002) 12531.
[9] Snis A., Panas I., Sur. Sci., 477 (1998) 412.
[10] C. Di Valentin, A. Figini, G. Pacchioni, Surf. Sci., 556 (2004) 145.
[11] Y. Amenomiya, V. Birss, M. Goledzinowski, J. Galuzska , A. Sanger, Catal. Rev.- Sci. Eng., 32(3) (1990) 163.
[12] R.S. Liu, M. Iwamoto, J. Lunsford, J.Chem.Soc.Chem. Commun., (1982) 78.
[13] H.F. Liu, R.S Liu, K.Y. Liew, R.E. Johnson, J. Lunsford, J.Am.Chem.Soc., 106 (1984) 4117.
[14] T. Ito, J.X. Wang, C.H Lin., J. Lunsford, J.Am.Chem. Soc., 107 (1985) 5062.
[15] D.J. Driscoll, W.Martir, J.X. Wang, J. Lunsford, J.Am.Chem.Soc., 107 (1985) 58.
[16] J.X. Wang, J. H. Lunsford., J. Phys. Chem., 90, 1986, 5883.
[17] C. H. Lin, T. Ito, J.X. Wang, J.H. Lunsford, J.Am. Chem. Soc., 109 (1987) 4808.
[18] C. H. Lin, J.X. Wang, J.H. Lunsford, J.Catal., 111 (1988) 302.
[19] V.R.Choudhary, V.H. Rane, J.Catal., 130 (1991) 411.
[20] E. Garrone, E. Giamello, S. Coluccia, G. Spoto, A. Zecchina in Proceedings of the 9th International Conference on Catalysis M.J: Phillips and M. Ternan eds., 4 (1988) 1577.
[21] E. Giamello, Catal. Today, 41 (1998) 239.
[22] M. C. Paganini, M. Chiesa, E. Giamello, S. Coluccia, G. Martra, D. M. Murphy, G. Pacchioni. Surf. Sci., 421 (1999) 246.
[23] M. C. Paganini, M. Chiesa, P. Martino, E. Giamello, E. Garrone J. Phys. Chem. B. 107 (11) (2003) 2575.
[24] S. Coluccia, F. Boccuzzi, G. Ghiotti, C. Morterra, J. Chem. Soc. Faraday Trans. I, 79 (1983) 913
[25] O. Diwald, T. Berger, M. Sterrer, E. Knozinger, Stud. Surf. Chem. Catal., 140 (2001) 237.
[26] D.Cordischi, V.Indovina, M.Occhiuzzi, J. Chem. Soc. Faraday Trans. I, 74 (1978) 883.
[27] M. Chiesa, M.C. Paganini, E. Giamello, D.M. Murphy, Res. Chem. Intermed., 28 (2,3) (2002) 205.

[28] F.Freund, G.C. Maiti, F. Batllo, M. Baerns, J. Chim. Phys., 87 (1990) 1467.
[29] C. Di Valentin, G. Pacchioni, M. Chiesa, E. Giamello, S. Abbet , U. Heiz., J. Phys. Chem. B, 106 (2002) 1637.
[30] A. Zecchina, M.G. Lofthaose and F. Stone, J. Chem. Soc. Faraday Trans. I, 71 (1975) 1476.

Studies in Surface Science and Catalysis 155
A. Gamba, C. Colella and S. Coluccia (Editors)
© 2005 Elsevier B.V. All rights reserved

Kinetics of the Ba^{2+}/Na^+ exchange on a mixed phillipsite–chabazite–rich tuff

F. Pepe[a], D. Caputo[b], B. de Gennaro[b], and C. Colella[b]

[a]Dipartimento di Ingegneria, Università del Sannio,
Piazza Roma 21, 82100 Benevento, Italy

[b]Dipartimento di Ingegneria dei Materiali e della Produzione, Università Federico II,
Piazzale Tecchio 80, 80125 Napoli, Italy

The kinetics of Ba^{2+}/Na^+ ion exchange on a mixed tuff, containing both phillipsite and chabazite, was studied. A fixed bed experimental apparatus was used and the breakthrough curves were determined varying the Ba^{2+} concentration in the inlet solution in the range 0.77–17.5 $eq \cdot m^{-3}$. The experimental results showed that the tuff under consideration is very selective towards Ba^{2+}, with very steep breakthrough curves. A diffusional model, based on the linear driving force (LDF) approximation, was used to interpret the experimental data. The model takes into account both fluid–particle and intra–particle resistances to diffusion, and considers the fact that the tuff contains different zeolites, having different ion exchange properties. Its use allowed to evaluate the internal and the external mass transfer coefficients, and it appeared that, except in one of the experimental runs, their values do not depend on the composition of the liquid phase, confirming the validity of the LDF approximation.

1. INTRODUCTION

Natural zeolitized materials, due to their low cost and wide availability in many areas of the world, have successfully been proposed for a number of cation exchange applications of environmental interest [1, 2] Since wastewaters from nuclear power plants may contain such radioactive species as $^{137}Cs^+$, $^{90}Sr^{2+}$, $^{133}Ba^{2+}$ and $^{60}Co^{2+}$ [3], the use of natural zeolitized materials for the removal of these cations represents a very interesting possibility. Previous investigations [4-6] have shown that phillipsite and chabazite exhibit a good to moderate selectivity towards Ba^{2+}, with equilibrium constants for the Ba^{2+}/Na^+ ion exchange at 25°C equal to 30.6 for phillipsite and to 0.62 for chabazite. However, the design of industrial ion exchange processes using natural zeolitized materials poses a number of problems, mainly due to (a) the complex interactions among the different mass transfer resistances involved, and (b) the fact that natural materials often contain different kinds of zeolites, which exhibit different ion exchange properties.

Despite the fact that a relatively large number of papers have been published concerning the kinetics of ion exchange processes for the removal of fission products, scarce attention has been specifically paid to the use of natural materials. Robinson et al. [7] dealt with the uptake of fission products (mainly Sr^{2+} and Cs^+) by a synthetic chabazite. They presented a thorough analysis of the literature on the subject and modeled their results by taking into account both "micropore" diffusion within the zeolite crystals, and "macropore" diffusion in the inert clay binder. A detailed modeling of the kinetics of the uptake of fission products by synthetic chabazite was also proposed by DePaoli and Perona [8], who successfully applied the pore–diffusion model proposed by Yao and Tien [9]. More recently, Latheef et al. [10] described the uptake of Cs^+ ions by a synthetic crystalline silicotitanate, also evaluating the Cs^+ diffusivity in the solid phase.

Brigatti et al. [11], on the other hand, dealt with the effect of the simultaneous presence of different zeolites in a natural "mixed" material. They studied the selectivity of a phillipsite–chabazite–rich tuff towards Co^{2+}, Cu^{2+}, Zn^{2+}, Cd^{2+} and Pb^{2+}, concentrating their attention on the fact that the material considered was a complex mixture of different solid phases, with different ion exchange properties. Relatively scarcer attention, on the other hand, was given to the kinetics of the ion exchange process. For what concerns this last aspect, it is important to observe that natural ion exchangers (namely, zeolites and clays) are often characterized by very complex structures, which prevents from an easy interpretation of the kinetic data. Therefore, the use of a lumped parameter model, such as the Linear Driving Force (LDF) model [12] could be more suitable for the description of the kinetics of ion exchange processes. According to this model, both liquid-solid and intra-particle mass transfer rates linearly depend on their respective driving forces, and furthermore equilibrium prevails at the solid-liquid interface. Despite the LDF model is rigorously valid only when a linear isotherm exists, Santacesaria et al. [13] and Ruthven [14] indicated that, even in presence of non-linear isotherms, it gives good results when used for the evaluation of breakthrough curves. Following this suggestion, Pepe et al. [15] satisfactorily applied the LDF model to ion exchange on zeolites, describing the kinetics of the Pb^{2+}/Na^+ exchange process on a phillipsite–rich tuff.

The aim of the present paper is to evaluate the efficiency of the Ba^{2+}/Na^+ exchange on a zeolitic tuff, containing both phillipsite and chabazite, in dynamic conditions (elution through zeolitized tuff beds). The experimentally obtained breakthrough curves will be interpreted by means of a diffusional model, derived from the model proposed by Pepe et al. [15], capable of describing the kinetic process taking place. The model takes into account both equilibrium and mass transfer relationships and is capable of handling the special case in which two different zeolites are present in the tuff. Its use will facilitate the evaluation of the main parameters controlling the rate of the exchange process.

2. EXPERIMENTAL APPARATUS AND TECHNIQUE

The kinetics of the ion exchange process was studied by determining the Ba^{2+} breakthrough curves from a fixed bed experimental apparatus, a sketch of which is reported in Fig. 1. The bed was made of tuff grains, tightly accommodated into a cylindrical glass column, having an internal diameter of 0.01 m. The tuff used was a Campanian Ignimbrite from Tufino (Naples, Italy), containing both phillipsite and chabazite. Its chemical and mineralogical composition has been reported elsewhere [16]. As regards in particular the exchanging phases, phillipsite content was 31%, chabazite 27%, analcime plus smectite 5%. A sample of the original tuff was crushed and then sieved and the 30–50 mesh fraction was retained for ion exchange experiments. The apparent density of the tuff was $\rho_{sol} = 1200$ kg m^{-3}, and its overall cation exchange capacity (CEC), determined by means of the cross–exchange method [17], was $Q = 1.91$ eq. Kg^{-1}.

Before each experiment, 15 g of the selected fraction of the tuff were accommodated within the column, which was repeatedly tapped. In this way, a bed with a length $L = 0.27$ m was prepared, characterized by a bulk density $\rho_b = 700$ kg·m^{-3} and a void fraction $\varepsilon = 42\%$.

Afterwards, the bed material was Na–exchanged by eluting it with a 1 M NaCl aqueous solution, with a flow rate of about 6 ml min^{-1}. The elution process lasted about one week, and it was stopped when the concentrations of all cations (except Na$^+$) in the outlet stream were below 1 g·m^{-3}.

The experiments were carried out by feeding a BaCl$_2$ aqueous solution to the fixed bed by means of a peristaltic pump (Pharmacia Fine Chemicals, P3). Different inlet BaCl$_2$ concentrations were used, namely, 0.74, 1.47, 3.64, 7.28 and 17.71 eq·m^{-3}. On the other hand, the flow rate was kept constant at 6.1 ml·min^{-1}, corresponding to a superficial velocity $u = 7.7$ cm·min^{-1}. Samples of the outlet solution were collected by means of a temporized sample collector (Pharmacia Fine Chemicals, PF3) and were analyzed for Ba^{2+} by atomic absorption spectrophotometry (Perkin Elmer, AA100).

3. MODELING OF THE ION EXCHANGE PROCESS

In order to quantitatively describe the fixed bed ion exchange process, a modeling effort was undertaken. With this aim, the presence of the two ancillary exchangers smectite and analcime (overall constituting about 5% of the tuff) was neglected. In fact, it was demonstrated that they do not influence the tuff exchange properties, because of their limited amount and negligible exchange kinetics (analcime) [18]. Therefore, the tuff was assumed to be a homogeneous solid mixture of an exchanging zeolitic phase (58% w/w) and a non exchanging inert (42% w/w). In turn, the zeolite was assumed to be made of phillipsite (53% w/w) and chabazite (47% w/w). Furthermore, phillipsite and chabazite were assumed to compete for Ba^{2+} in solution, each with its own selectivity and all possible zeolite–zeolite interactions were neglected.

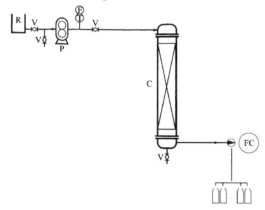

Fig. 1. Sketch of the experimental ion exchange apparatus. C: ion exchange column; R: reservoir; P: peristaltic pump; F: flowmeter; T: thermometer; V: valves; FC: fraction collector

In practice, this was equivalent to considering two "pseudo–tuffs": pseudo–tuff "1" was assumed to be a phillipsitic tuff, constituting a fraction $\alpha_1 = 53\%$ of the total solid, and having a zeolite content of 58%; pseudo–tuff "2", on the other hand, was assumed to be a chabazitic tuff, constituting a fraction $\alpha_2 = 1-\alpha_1 = 47\%$ of the total solid, and again having a zeolite content of 58%.

Since a binary system was considered, only the equations relative to Ba^{2+} transport were taken into account, a stoichiometric relationship being sufficient to give the Na^+ concentration in all the phases involved. Using the LDF model to express the Ba^{2+} exchange rates for pseudo–tuffs 1 and 2 (r_1 and r_2) as a function of liquid and solid compositions, the following equations were written:

$$r_1 = k_{ext}\left(c - c_1^*\right) = k_{int}\left(q_1^* - q_1\right),$$ (1a)

$$r_2 = k_{ext}\left(c - c_2^*\right) = k_{int}\left(q_2^* - q_2\right).$$ (1b)

Here, c is the Ba^{2+} concentration in the liquid bulk, q_i is the average Ba^{2+} concentration in pseudo–tuff i ($i=1, 2$), and c_i^* and q_i^* are the liquid– and solid–side interfacial concentrations relative to pseudo–solid i. Following the hypothesis at the basis of the LDF model, c_i^* was assumed to be in equilibrium with q_i^*. Furthermore, k_{ext} and k_{int} are an external (solid–liquid) and an internal (intra–particle) mass transfer coefficient. However, it has to be observed that k_{ext} essentially depends on the fluid dynamic conditions, and its value can be obtained by means of literature correlations [19]. On the other hand, k_{int} depends on the solid phase diffusivity of the ions involved, and therefore the possibility exists that k_{int} for pseudo–tuff 1 is different from k_{int} for pseudo–tuff 1. However, due to lack of information on this point, this possibility has been disregarded.

In order to evaluate c, q_1 and q_2 as a function of time and position in the fixed bed, Eqs. (1a, b) had to be used in association with the following material balances [20]:

$$\varepsilon D \frac{\partial^2 c}{\partial t^2} - u \frac{\partial c}{\partial z} - \varepsilon \frac{\partial c}{\partial t} - (1 - \varepsilon)a(a_1 r_1 + a_2 r_2) = 0,$$ (2)

$$\rho_b \frac{\partial q_1}{\partial t} = (1 - \varepsilon)a r_1 = 0,$$ (3a)

$$\rho_b \frac{\partial q_2}{\partial t} = (1 - \varepsilon)a r_2 = 0.$$ (3b)

Eq. (2) refers to the liquid phase, and the non-idealities in its flow are taken into account by means of the dispersion model [21]. Eqs. (3a, b), on the other hand, refer to each solid phase. In these equations z is the axial coordinate along the bed, t is the time, D is the axial dispersion coefficient, and a is the (external) solid-liquid contact area.

The initial and boundary conditions for the system of Eqs. (2, 3) depend on the specific problem under consideration. Here, the removal from the liquid phase of a species that initially is not present in the bed is studied. Therefore, the following conditions were used:

$$t = 0, \quad 0 \le z \le L: \quad c = 0, \quad q_1 = q_2 = 0,$$ (4a)

$$t > 0, \quad z = 0: \quad \varepsilon D \frac{\partial c}{\partial z} = u(c_o - c),$$ (4b)

$$t > 0, \quad z = L: \quad \frac{\partial c}{\partial z} = 0.$$ (4c)

These conditions state that at the beginning of each experiment the species being removed from the liquid phase is not present in the column (4a); that its concentration in the solution fed to the fixed bed is constant and equal to c_o (4b); that the fixed bed is a "closed" system, the length of which is L (4c, [21]).

The system made by the partial differential Eqs. (2, 3), together with the constituting Eq. (1) and the boundary conditions (4), was made dimensionless following the approach outlined in [15], and then was integrated at finite differences, using a Fortran program explicitly written for this purpose.

4. RESULTS AND DISCUSSION

Fig. 2 reports the breakthrough curves for the different experiments carried out, as ratio between the outlet and the inlet Ba^{2+} concentrations (c_{out}/c_o) vs. time. As expected, almost no Ba^{2+} could be detected in the effluents at the beginning of each run. Afterwards, c_{out} gradually increased, asymptotically reaching c_o: in these conditions it can be assumed that the zeolite contained in the tuff is in equilibrium with the aqueous solution. The experimental results show that all the curves are rather symmetrical and quite steep, both signs of good selectivity; furthermore, it appears that, at least in the investigated range, the Ba^{2+} concentration in the liquid stream affects only slightly the slope of the breakthrough curves. The data of Fig. 2 also show that, as expected, the higher the Ba^{2+} concentration in the inlet stream to the fixed bed column, the lower the breakthrough time, with an almost inversely proportional dependence on the inlet Ba^{2+} concentration.

Graphical integration of the experimental breakthrough curves allowed to evaluate the CEC of Ba^{2+} for Na^+ in dynamic conditions. The value obtained from the curve concerning the most concentrated solution (17.71 $eq \cdot m^{-3}$) was not considered, because of the uncertainty of the relevant analytical evaluations. The residual four calculated values gave an average of 1.52 $eq \cdot kg^{-1}$, with a standard deviation of 0.10 $eq \cdot kg^{-1}$. The fact that this value is rather far from the experimental CEC value for the tuff (1.91 $eq \cdot kg^{-1}$) indicates that Ba^{2+} can not readily occupy in dynamic conditions all the extraframework cationic sites, in agreement with the equilibrium data, pointing out a good affinity for phillipsite, but an only moderate affinity for chabazite [4, 5].

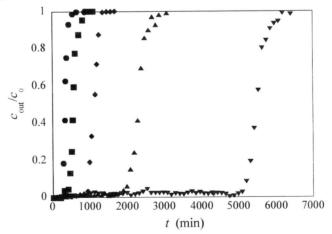

Fig. 2. Barium breakthrough curves obtained eluting Na–exchanged tuff beds with solutions containing 0.74 $eq \cdot m^{-3}$ (▼), 1.47 $eq \cdot m^{-3}$ (▲), 3.64 $eq \cdot m^{-3}$ (♦), 7.28 $eq \cdot m^{-3}$ (■) or 17.71 $eq \cdot m^{-3}$ (●) Ba^{2+}

In order to check the validity of the model described above, integration of Eqs.(2, 3) was carried out for the five available experimental runs (Fig. 3). The equilibrium conditions between the two zeolites and the aqueous solutions were assessed from the available literature data at a total normality $N = 100$ eq·m^{-3} [4, 5], using the procedure outlined in the Appendix. As indicated above, once the boundary conditions and the equilibrium isotherms are assigned, the only parameters of the model are the mass transfer coefficients k_{ext} and k_{int}.

A simple "best fit" procedure led to the conclusion that, in four of the five experimental runs presented, namely, for $c_o = 1.47$, 3.64, 7.28 and 17.71 eq·m^{-3}, a very satisfactory agreement between model and experimental results was reached considering $k_{ext} = 4.6 \cdot 10^{-6}$ m·s^{-1} and $k_{int} = 3.3 \cdot 10^{-5}$ kg·m^{-2}·s^{-1}.

It is interesting to observe that the value of k_{ext} compares very favorably with the estimate of $5.4 \cdot 10^{-6}$ m·s^{-1} given by the correlation of Wilson and Geankoplis [19]. The value of k_{int} indicates a Ba^{2+} solid phase diffusivity in the order of $6 \cdot 10^{-13}$ m^2·s^{-1}. To fit the experimental data relative to $c_o = 0.74$ eq·m^{-3}, it was necessary to use different values of k_{ext} and k_{int}, namely $k_{ext} = 1.3 \cdot 10^{-5}$ m·s^{-1} and $k_{int} = 1.25 \cdot 10^{-5}$ kg·m^{-2}·s^{-1}. The reason for this behavior is not clear: a possible explanation could be a failure of the procedure used for the above mentioned extrapolation of the thermodynamic data from 100 to 0.74 eq·m^{-3}.

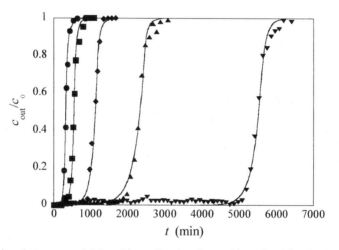

Fig. 3. Comparison between model (continuous lines) and experimental results. (▼) $c_o = 0.74$ eq·m^{-3}; (▲) $c_o = 1.47$ eq·m^{-3}; (◆) $c_o = 3.64$ eq·m^{-3}; (■) $c_o = 7.28$ eq·m^{-3}; (●) $c_o = 17.71$ eq·m^{-3} Ba^{2+}

5. CONCLUSIONS

The possibility of using a zeolitic tuff (Campanian Ignimbrite) for treating wastewaters containing Ba^{2+}, e.g., contaminated by radioactive ^{133}Ba^{2+}, was studied. The tuff resulted very selective for Ba^{2+}, exhibiting very steep breakthrough curves for all concentrations considered. Furthermore, the slope of the breakthrough curves were very little affected by the Ba^{2+} concentration in the liquid stream.

A diffusional model, based on the linear driving force approximation, was used to interpret the experimental data. The model allowed to establish that both the internal and the external mass transfer resistances play a relevant role on the kinetics of the ion exchange process. Eventually, it appeared that, except in one of the experimental runs, the values of the internal and the external mass transfer coefficients do not depend on the composition of the liquid phase, confirming the validity of the LDF approximation.

ACKNOWLEDGEMENTS

The authors wish to thank Dr. P. Aprea for the assistance in carrying out the experimental work.

APPENDIX – PREDICTION OF EQUILIBRIUM ION EXCHANGE ISOTHERMS

In general, the cation exchange between a solution containing A^{z_A+} cations and a zeolite containing B^{z_B+} cations can be described by means of the following reaction:

$$z_B A^{z_A+} + z_A BL_{z_B} = z_A B^{z_B+} + z_B AL_{z_A},$$
(A1)

in which L is a portion of zeolite framework holding unit negative charge [22] and z_A, z_B are the valences of the two cations.

The equilibrium equations corresponding to reaction (A1) is the following:

$$K = \frac{\left[B^{z_B+}\right]^{z_A} x_B^{z_A} \left[AL_{z_A}\right]^{z_B} f_A^{z_B}}{\left[A^{z_A+}\right]^{z_B} x_A^{z_B} \left[BL_{z_B}\right]^{z_A} f_B^{z_A}},$$
(A2)

in which K is the equilibrium constant, x_A and x_B are the activity coefficients for cations A^{z_A+} and B^{z_B+} and f_A and f_B are the activity coefficients in the zeolite phase. Since the total number of equivalents, both in solution and in the solid, is constant during the exchange process, it is useful to express the concentrations which appear in Eq. (A2) as equivalent fractions. Following this approach, and taking cation A^{z_A+} as a reference, one has:

$$K = \left(\frac{N}{Q}\right)^{z_A - z_B} \frac{\left(1 - E_{A(s)}\right)^{z_A} \left(E_{A(z)}\right)^{z_B} x_B^{z_A} f_A^{z_B}}{\left(E_{A(s)}\right)^{z_B} \left(1 - E_{A(z)}\right)^{z_A} x_A^{z_B} f_B^{z_A}} = K_c \frac{f_A^{z_B}}{f_B^{z_A}},$$
(A3)

where $E_{A(s)}$ is the equivalent fraction of A^{z_A+} cations in the solution phase, $E_{A(z)}$ is the fraction of cation sites exchanged with A^{z_A+}, and K_c is the selectivity coefficient, in turn defined as:

$$K_c = \left(\frac{N}{Q}\right)^{z_A - z_B} \frac{\left(1 - E_{A(s)}\right)^{z_A} \left(E_{A(z)}\right)^{z_B} x_B^{z_A}}{\left(E_{A(s)}\right)^{z_B} \left(1 - E_{A(z)}\right)^{z_A} x_A^{z_B}}.$$
(A4)

The prediction procedure proposed by Pabalan [22] allows to evaluate the liquid phase composition (i.e., the value of $E_{A(s)}$) in equilibrium with a zeolite characterized by a given composition (i.e., by a given value of $E_{A(z)}$) at an arbitrary normality N, once a set of equilibrium data is available at a given normality N^*. Such procedure is based on the assumption that the ratio of the activity coefficients in the zeolite phase ($f_A^{z_B}/f_B^{z_A}$) is not sensibly affected by the composition of the aqueous solution (and therefore on the value of N), and only depends on the composition of the zeolite phase.

Using a procedure for the estimation of the activity coefficients in solution (*e.g.* see Caputo et al. [23]), Eq. (A4) allows to calculate the value of K_c corresponding to any couple $[E_{A(s)}, E_{A(z)}]$. These values, in turn, allow to calculate the equilibrium constant K using the following equation [24]:

$$\log(K) = 0.4343(z_B - z_A) + \int \log(K_c) dE_A \ . \tag{A5}$$

Once K and the values of K_c corresponding to each available couple $[E_{A(s)}, E_{A(z)}]$ have been calculated, Eq. (A3) allows to evaluate the ratio (f_A^{zB}/f_B^{zA}) corresponding to each of the available values of $E_{A(z)}$. On the other hand, since such ratio is assumed to be independent on N, the same Eq. (A3) can be used to evaluate $E_{A(s)}$ at a total normality $N \neq N^*$:

$$\frac{\left(1 - E_{A(s)}\right)^{z_A}}{\left(E_{A(s)}\right)^{z_B}} = K\left(\frac{N}{Q}\right)^{z_B - z_A} \frac{\left(1 - E_{A(z)}\right)^{z_A}}{\left(E_{A(z)}\right)^{z_B}} \frac{x_A^{z_B}}{x_B^{z_A}} \frac{f_B^{z_A}}{f_A^{z_B}} \tag{A6}$$

As the activity coefficients x depend on the liquid phase composition, and therefore on $E_{A(s)}$, an iterative procedure is required.

LIST OF SYMBOLS

a	specific surface area of the solid particles, m^{-1}
c	liquid phase concentration, $eq \cdot m^{-3}$
D	dispersion coefficient, $m^2 \cdot s^{-1}$
E	equivalent fraction in the zeolite phase
f	activity coefficient in the zeolite phase
K	equilibrium constant
K_c	selectivity coefficient (see Eq. A3)
k_{ext}	external (solid–liquid) mass transfer coefficient, s^{-1}
k_{int}	internal (intra–particle) mass transfer coefficient, $kg \cdot m^{-3} \cdot s^{-1}$
L	bed length, m
N	total solution normality, $eq \cdot m^{-3}$
Q	ion exchange capacity, $eq \cdot kg^{-1}$
q	solid phase concentration, $eq \cdot kg^{-1}$
r	mass transfer rate, $eq \cdot m^{-2} \cdot s^{-1}$
t	time, s
u	superficial velocity of the liquid, $m \cdot s^{-1}$
x	activity coefficient in solution
z	axial coordinate in the bed, m

Greek symbols

α	fraction of pseudo–tuff "i"
ε	void fraction, dimensionless
ρ_b	bulk density, $kg \cdot m^{-3}$
ρ_{sol}	apparent density, $kg \cdot m^{-3}$

Subscripts and superscripts

*	equilibrium
out	outlet
o	inlet
s	solution
z	zeolite

REFERENCES

[1] C.Colella, In Natural Microporous Materials in the Environmental Technology, (P. Misaelides, F. Macasek, T.J. Pinnavaia and C. Colella (eds.), NATO Sciences Series, Series E: Applied Sciences 362, Kluwer, Dordrecht, The Netherlands, 1999, p. 207.

[2] M. Adabbo, D. Caputo, B. De Gennaro, M. Pansini and C. Colella, Microp. Mesop. Mat., 28 (1999) 315.

[3] M. Sitting, Pollutant Removal Handbook, Noyes Data Corporation, Park Ridge, NJ, 1973, p. 478.

[4] C. Colella, E. Torracca, A. Colella, B. de Gennaro, D. Caputo and M. de'Gennaro, in Zeolites and Mesoporous Materials at the Dawn of the 21st Century (A. Galarneau, F. Di Renzo, F. Fajula and J. Vedrine (eds.), Stud. Surf. Sci. Catal., No. 135, Elsevier, Amsterdam, 2001, p. 148.

[5] B. de Gennaro, A.Colella, P.Aprea and C. Colella, Microp. Mesop. Mat., 61 (2003) 159.

[6] B. de Gennaro and A. Colella, Sep. Sci. Technol., 38 (2003) 2221.

[7] S.M. Robinson, W.D. Arnold and C.H. Byers, AIChE J., 40 (1994) 2045.

[8] S.M. DePaoli and J.J. Perona, AIChE J., 42 (1996) 3434.

[9] C. Yao and C. Tien, Chem. Eng. Sci., 48 (1993) 187.

[10] I.M. Latheef, M.E. Huckman and R.G. Anthony, Ind. Eng. Chem. Res., 39 (2000) 1356.

[11] M.F. Brigatti, G. Franchini, P. Frigieri, C. Gardinali, L. Medici and L. Poppi, Can. J. Chem. Eng., 77 (1999) 163.

[12] E. Glueckauf, Trans Faraday Soc., 51 (1955) 1540.

[13] S. Santacesaria, M. Morbidelli, A. Servida, G. Storti and S. Carrà, Ind. Eng. Chem. Proc. Des. Dev., 21 (1982) 446.

[14] D.M. Ruthven, Principles of Adsorption and Adsorption Processes, Wiley and Sons: NY, 1984.

[15] F. Pepe, D. Caputo and C. Colella, in Oxide–Based Systems at the Crossroads of Chemistry (A. Gamba, C. Colella and S. Coluccia (eds.), Stud. Surf. Sci. Catal., No. 140, Elsevier: Amsterdam, 2001, 369.

[16] B. de Gennaro, P. Aprea and C. Colella, in Atti 6° Congresso Naz. AIMAT (Associazione Italiana Ingegneria dei Materiali), T. Manfredini, (ed.), AIMAT, Bologna, 2002, 4 pp. (CD-Rom).

[17] M. Pansini, C. Colella, D. Caputo, M. de'Gennaro and A. Langella, Microp. Mesop. Mater., 5 (1996) 357.

[18] R.M. Barrer and L. Hinds, J. Chem. Soc., 1879 (1953).

[19] E.J. Wilson and C.J. Geankoplis, Ind. Eng. Chem. Fundam., 5 (1966) 9.

[20] W.L. McCabe, J.C. Smith and P. Harriot, Unit Operations of Chemical Engineering, 4th Edition, McGraw Hill: NY, 1985.

[21] O. Levenspiel, Chemical Reaction Engineering, 3rd Edition, Wiley and Sons: NY, 1999.

[22] R.T. Pabalan, Geochim. Cosmochim. Acta, 58 (1994) 4573.

[23] D. Caputo, B. de Gennaro, P. Aprea, C. Ferone, M. Pansini and C. Colella, This volume.

[24] G.L. Gaines and H.C. Thomas, J. Chem. Phis., 21 (1953) 714.

Studies in Surface Science and Catalysis 155
A. Gamba, C. Colella and S. Coluccia (Editors)
461

Carbonate formation on sol-gel bioactive glass 58S and on Bioglass® 45S5

A. Perardi, M. Cerruti and C. Morterra

Department of Chemistry IFM and Centre of Excellence NIS, University of Turin, Consortium INSTM, Research Unit of Turin University, via P. Giuria 7, 10125 Torino, Italy

1. INTRODUCTION

Bioactive glasses are silica-based bone regenerative materials[1, 2]. They are used for bone replacement in non-load-bearing implants or as surface coating for other materials.

When bioactive glasses are immersed in aqueous solutions such as body fluids, they develop a porous structure through depletion and disruption of the glassy network, cations are released from the bulk, solution pH increases, and a calcium phosphate-rich layer precipitates on their surface. This amorphous layer further crystallizes in the form of hydroxy apatite (HA), and is later transformed into hydroxy carbonate apatite (HCA) by carbonate incorporation[3]. HCA is the same inorganic component of bone tissue, and this favours the formation of a biological bond between the glass and the host tissue[4-6].

The first bioactive glass discovered by Hench in 1969 was Bioglass® 45S5 (composition shown in Table 1)[2]. Bioglass® 45S5 is made by fusion, is non-porous and has a low specific surface area. A second generation of bioactive glasses were synthesized by sol-gel technique[7, 8]. These glasses do not have Na_2O in their composition, and have specific surface area ~2 orders of magnitude higher than Bioglass®. For this reason, they react much faster than traditional glasses[9, 10].

In the present paper, we will study the final step involved in bioactive glass reactivity, i.e., surface carbonation. Carbonation is also relevant during glasses shelf-aging, as discussed by Cerruti et al. in [11].

We will follow the process of carbonation with FT-IR transmission spectroscopy, using CO_2 and H_2O as probe molecules. IR spectroscopy has been used in the past to study carbonation, since it is possible to discriminate between the different types of surface carbonates by analysing the CO_n stretching region[12, 13].

$$
\begin{array}{c}
O \\
\parallel \\
C \\
\parallel \\
O \\
\downarrow \\
M
\end{array}
$$

Scheme 1. Schematic example of linear coordination of CO_2 on a coordinatively unsaturated (CUS) surface cation.

462

When CO_2 interacts with the surface of metal oxides, it can show either acid or basic behavior. In the presence of coordinatively unsaturated (CUS) surface cations, CO_2 can linearly coordinate on them, since it is a weak Lewis base (see Scheme 1). The frequency of the asymmetric ν_{CO} stretching mode falls in the 2400-2300 cm^{-1} range, depending on the specific cation that CO_2 is bound to.

If sufficiently basic surface species are present (like basic O^{2-} ions, OH^- ions, or CUS cation-anion pairs), CO_2 can coordinate on them, and many possible types of carbonate-like species can be formed (see scheme 2)[14].

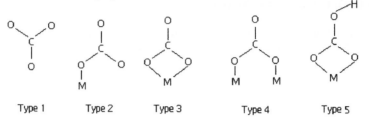

Scheme 2. Surface carbonates formed when CO_2 interacts with a surface possessing sufficiently basic sites.

The specific type of carbonate structure formed can be identified by analyzing the splitting of the two high-ν carbonate vibrations deriving from the degenerate ν_3 mode of the free CO_3^{2-} ion (OCO asymmetric stretching). The degeneration of this mode is removed when the carbonate ion is bound to the surface of a material: two different components are observed, with a spectral splitting proportional to the asymmetry of the surface carbonate species formed and dependent also on the bond angles produced. For the "free" or purely ionic carbonate (type 1 in scheme 2), the vibrations are still degenerate ($\Delta\nu_3=0$, $\nu_3=1420-1470$ cm^{-1}). For monodentate carbonates (type 2), $\Delta\nu_3$ is usually equal to or slightly larger than 100 cm^{-1}. For bidentate carbonates (type 3), $\Delta\nu_3$ is ~300 cm^{-1}, whereas for bridged (or "organic") carbonates (type 4) $\Delta\nu_3$ is usually 400 cm^{-1} or higher. For bicarbonates (hydrogen-carbonates, type 5), $\Delta\nu_3$ is ~200 cm^{-1} and, in addition, there is a sharp δ_{COH} mode at ~1250 cm^{-1}. Of course, these are only indicative values, since they also depend on the polarizing power of the cation and on other factors, such as the presence of water molecules and/or of other cations outside the main coordination sphere[15].

A large number of surface sites available for interactions with probe molecules is usually present only on high surface area materials. For this reason, only these materials are normally used to get reasonably intense adsorption bands. Our first experiments were therefore carried out on a sol-gel synthesized bioactive glass (58S). These results have been reported by Cerruti et al. in [11], and will be used here for comparison. In the present paper, though, we will show that quite intense bands are formed also on the low surface area Bioglass® 45S5 when carbonation is carried out in the proper conditions. In order to identify the types of carbonates formed, we will analyze in parallel the carbonation of Ca-doped silica and of Na-doped silica used as reference systems.

2. MATERIALS AND METHODS

Bioactive glasses: bioactive glass 58S was synthesized via sol gel by NovaMin Technology Inc. (Alachua, Florida, USA). The sol-gel procedure consisted in hydrolysis and polycondensation of tetraethyl orthosilicate (TEOS), triethyl phosphate (TEP) and $Ca(NO_3)_2 \cdot 4H_2O$, using HCl as catalyst. After aging and drying at low temperature, the glasses were stabilized at high temperature (873-973K) for many hours[6]. The sample, amorphous (as revealed by XRD and TEM analysis), was in a powder form. Bioglass® 45S5 was also supplied by NovaMin Technology Inc., and was produced by fusion. Bioglass® particles had an average diameter of ~2 μm.

Pure and Ca- and Na-doped silica systems used as reference samples: Aerosil200 (A200), obtained by flame pyrolysis of $SiCl_4$, was supplied by Degussa (Frankfurt A.M., Germany). In order to obtain the Ca- and Na-doped silica samples (termed A200/Ca and A200/Na, respectively), titrated solutions of either $Ca(NO_3)_2$ or $NaHCO_3$ were used in amount just sufficient to wet, but not overwet, all the powder. Then a thermal treatment in air at 873K for 2 hrs was carried out, in order to decompose and eliminate the nitrates.

The composition, surface area and porosity of both bioactive glass 58S and Bioglass® 45S5 and of all the reference samples are reported in Table 1.

Table 1
Composition, surface area and porosity of bioactive glasses and reference samples

	Composition (mol%)				BET surface area (m^2g^{-1})	Average pore diameter (Å)
	SiO_2	CaO	P_2O_5	Na_2O		
58S	60	36	4	-	142	85
45S5	46.1	26.9	2.6	24.4	~2	-
A200	100	-	-	-	198	-
A200/Ca	96	4	-	-	160	340
A200/Na	80	-	-	20	170	n.d.

IR measurements: the powders were compressed in the form of self-supporting pellets of some 10 mg cm^{-2}. All spectra were obtained with a FTIR Spectrometer (Bruker IFS 113v, equipped with a MCT criodetector). The home-made quartz infrared cell, equipped with KBr windows, was connected to a conventional vacuum line (residual pressure $\approx 10^{-5}$ Torr) and allowed to perform in strictly *in-situ* conditions both thermal treatments on the sample pellets, and probe molecules adsorption/desorption cycles on the activated samples. All IR spectra were recorded at beam-temperature (BT), i.e., the temperature reached by (white) sample pellets in the IR beam. BT is estimated to be some 20–30 K higher than the actual room temperature (RT).

3. RESULTS AND DISCUSSION

Fig. 1a shows the IR spectra of bioactive glasses 58S and 45S5, and of the reference pure and Ca or Na doped silica systems outgassed at RT for 30 minutes.

Pure silica presents a very simple spectrum, with a sharp peak for isolated SiOH hydroxyls at 3745 cm^{-1}, very weak signals in the 2700-3500 cm^{-1} range, corresponding to H-bonded OH-bearing species, and very weak overtone modes of bulk SiOSi vibrations at 1980, 1880 and 1630 cm^{-1}. In A200/Ca and A200/Na, the same hydroxyl peak can be detected, together with a broad tail observed in the 3000-3500 cm^{-1} range for the H-bonded hydroxyls. The latter is broader and centered at lower wavenumber for A200/Na with respect to both A200 and A200/Ca, indicating that a stronger hydrogen bond system is present on A200/Na.

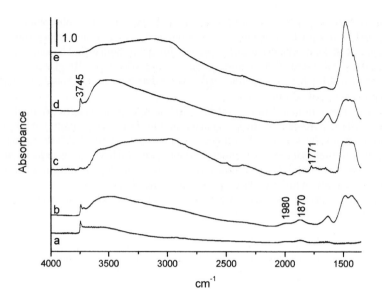

Fig. 1a. Absorbance spectra of Aerosil 200 (a), Ca-doped silica (b), Na-doped silica (c), bioactive glass 58S (d) and Bioglass® 45S5 (e), activated in vacuo for 30 minutes at room temperature.

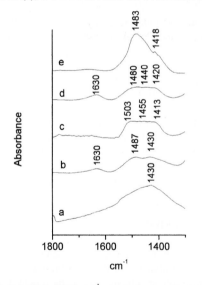

Fig. 1b. Absorbance spectra in the 1300-1800 cm^{-1} spectral range for (a) CaCO$_3$, (b) Ca-doped silica, (c) Na-doped silica, (d) 58S bioactive glass and (e) Bioglass® 45S5, activated for 30 min at RT.

In A200/Ca and A200/Na a band at 1630 cm^{-1} is also observed, relative to cations coordinated indissociated water. Both model systems also show relatively strong bands between 1400 and 1500 cm^{-1}(see Fig. 1b): these bands indicate the presence of carbonates (asymmetric CO$_n$

stretching mode(s); see before) formed by carbonation on active CUS cationic sites. A double carbonate band is observed in the 1430-1495 cm^{-1} range for calcium carbonates on A200/Ca. The small splitting of this band corresponds to the splitting expected for monodentate carbonates. However, it could also be ascribed to ionic carbonates, analogous to those found in CaCO$_3$ (Fig. 1b, spectrum a). In the spectrum of CaCO$_3$, only one peak is visible. The small splitting observed in the spectrum of Ca-doped silica is due to the asymmetry in the carbonate ions brought about by its binding to the surface. Na-doped silica shows a multiple band, also including high frequency components (at 1771 cm^{-1}). This indicates that some bidentate carbonates are also present on this sample.

Spectra for bioactive glasses recall those of doped-silica model systems. In particular, 58S presents the double carbonate band also observed on Ca-doped silica, whereas on Bioglass® 45S5 a single band (with a shoulder at 1418 cm^{-1}) is detected. Also the broad band of H-bonded hydroxyl groups is similar to that of Ca-doped silica for 58S, and to that of Na-doped silica for 45S5. This indicates that the H-bond strength for hydroxyl groups belonging to 45S5 is much higher than that of hydroxyl groups belonging to 58S. On Bioglass® 45S5, the peak at 3745 cm^{-1} relative to free silanols is not observed. Moreover, the band at 1630 cm^{-1} indicative of the presence of coordinated water is very weak, due to the very low surface area of this glass (see Table 1).

The spectrum of bioactive glass 58S activated at 673 K is represented in Fig. 2, as an example of the effect of a high activation temperature on either bioactive glasses or reference doped silica systems. It can be noted that carbonate bands in the 1400-1500 cm^{-1} range have been almost completely deleted by the high temperature treatment, leaving only a small residue. Activation also eliminated all H-bonded hydroxyls, leaving only the sharp peaks at 3745 cm^{-1} and 3570 cm^{-1} for free silanols and CaOH, respectively. No more coordinated water is present on the sample (the peak at ~1630 cm^{-1} is no longer visible).

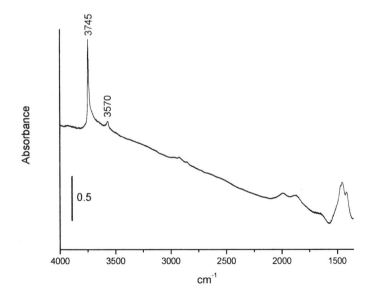

Fig.2. Absorbance spectrum of bioactive glass 58S activated at 673 K.

466

All these modifications indicate that the high temperature treatment cleaned up the surface of the glass. Probe molecules can be allowed on this clean surface, because the CUS surface cations are now available for interactions. Fig. 3 shows the adsorption of CO_2 on bioactive glasses and doped silica reference samples activated at high temperature (a treatment at 573 K was sufficient to remove all the carbonates from the surface of 45S5 and A200/Na, whereas a treatment at 673 K was necessary in the case of 58S and A200/Ca). The spectra are shown as differentials with respect to the background spectrum of the sample activated at high temperature (e.g., see the spectrum shown in Fig.2 for 58S).

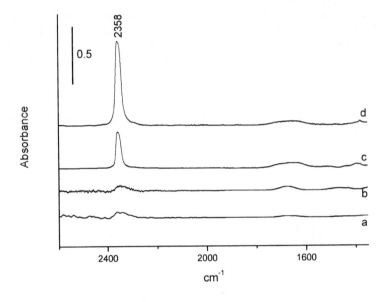

Fig. 3. Admission of ~20 Torr CO_2 on (a) Na-doped silica, (b) Bioglass® 45S5, (c) Ca-doped silica , (d) 58S bioactive glass activated at high temperatures (573 K for Na-doped silica and Bioglass® 45S5 and 673 K for Ca-doped silica and 58S). The spectra presented in the figures are the difference between those recorded after and before CO_2 admission.

When carbon dioxide alone is allowed on the surface of A200/Na (Fig. 3, spectrum a), linear coordination can hardly be observed. Evidently, Na^+ surface sites are much weaker Lewis acid surface sites than Ca^{++} cations (they have similar dimensions but only one positive charge, and this makes them weaker electron-acceptors). No linearly coordinated CO_2 can be observed on 45S5 Bioglass® (spectrum b), even though both Na^+ and Ca^{++} cations are present in its composition. The main reason for this is thought to be the specific surface area, dramatically lower than that of all the other samples (see Table 1): this implies that there are not sufficient surface sites available to interact with CO_2 and to lead to a detectable IR signal of coordinated CO_2 molecules. On A200/Ca (spectrum c) and 58S bioactive glass (spectrum d), a peak at ~ 2358 cm^{-1} is formed, indicative of linear coordination of CO_2 on surface calcium cations.

In all cases, only very limited modifications (if any) are observed in the spectral range characteristic of carbonate-like bands (~1700-1500 cm-1), thus indicating that the basicity of surface O^{2-} ions is in all cases not sufficient for CO_2 to manifest its acidic behaviour.

It was previously observed [11] that, in order to obtain carbonation of bioactive glass 58S, the contemporary presence of CO_2 and an excess of water vapour was necessary. This was related to the creation of a liquid-like water monolayer on the surface of the bioactive glass and, consequentially, to the formation of CO_3^{2-} ions. These ions could react with surface Ca^{++} cations, thus producing ionic surface carbonates, similar to those observed in the bulk of $CaCO_3$ (see Fig.1b).

Fig. 4 shows the results of the admission of CO_2 and an excess of water vapour on bioactive glasses 58S and 45S5 and on the two reference silica samples. In these conditions, linear coordination of CO_2 was not observed on any sample and, as expected, carbonates were formed quite fast on 58S (spectrum a), and increased over time (a'). A double band with peaks centered at 1430 and 1490 cm^{-1} was observed, similar to that formed upon CO_2 and water contact with Ca-doped silica (spectrum b). The similarity of behaviour between this two samples indicate that carbonate formation on bioactive glass 58S is ruled by the reactivity of surface Ca^{++} cations only, and is not appreciably influenced by the contemporary presence of phosphorous in the composition of the glass.

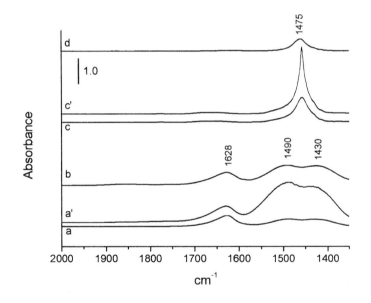

Fig. 4. Admission of both 5 torr CO2 and 15 Torr H2O on samples treated at 573 and 673 K as reported in Fig.3. IR spectra are presented as differences with respect to the spectra of the samples activated at 573 and 673 K before the contact with probe molecules. Spectra were recorded after 10 seconds and 1 week for 58S (spectra a and a', respectively), 1 hr for Ca-doped silica (b), 10 seconds and 30 minutes for 45S5 (c, c'), and 20 minutes on Na-doped silica (d).

When both CO_2 and an excess of water were allowed on 45S5, the sudden formation of a single band centered at 1475 cm^{-1} was observed (spectrum c), which increased in intensity over time (c'). Such a relevant carbonate formation on 45S5 was unexpected, because of the extremely low surface area exposed by this material (see Table 1 and the discussion above for linear CO_2 uptake). This observation implies that the process does not occur only on the surface of the material, but involves also inner cations present in some of the layers beneath the surface. This may also explain why the carbonation process goes on as time passes: more and more inner layers are involved, and more carbonates can be formed, until all CO_3^{2-} ions available are consumed.

The shape and position of the carbonate peak formed on 45S5 is quite similar to that observed upon CO_2 and H_2O contact with Na-doped silica (spectrum d). This is quite interesting, since sodium and calcium should be present in the same amount in Bioglass® 45S5 (see Table 1). The formation of sodium carbonates alone indicates that surface layers of Bioglass® 45S5 are enriched in sodium. Such inhomogeneity in the structure of the material is probably related to Na^+ ions migration during samples aging [16].

4. CONCLUSIONS

Carbonate formation on a sol-gel synthesized bioactive glass (58S) and on a traditional fused glass (Bioglass® 45S5) has been studied by means of infrared spectroscopy, using water and carbon dioxide as probe molecules to simulate the carbonation process. Silica samples doped with Ca and Na have been also analyzed for comparison.

It was observed that both CO_2 and an excess of water vapour are necessary to form carbonates on both bioactive glasses and reference silica systems: this indicates that CO_3^{2-} ions are formed through dissolution of CO_2 in a liquid-like water monolayer on the sample surface, and that they react with Ca^{++} or Na^+ cations, giving rise to ionic carbonates. Carbonates formation at spectroscopically detectable levels also on a very low surface area material such as Bioglass® 45S5 indicates that not only surface cations are involved in the carbonation process, but also some cations belonging to the inner glass layers. This is confirmed by the fact that carbonation increases during time and the proces keeps going for a very long time.

Mostly calcium-bound carbonates are formed on 58S bioactive glass, whereas only sodium-bound carbonates are formed on Bioglass® 45S5, even though both sodium and calcium are present in the latter sample. This indicates that the surface layers of Bioglass® are enriched in sodium.

REFERENCES

[1] L.L. Hench, J. Am. Ceram. Soc., 74 (1991) 1487.
[2] L.L. Hench, R.J. Splinter, W.C. Allen, T.K. Greenlee, J. Biomed. Mater. Res. Symp. 2-part I(1971) 117.
[3] A.E. Clark and L.L. Hench, J. Biomed. Mater. Res., 10 (1976) 161.
[4] M.M. Pereira, A.E. Clark, L.L. Hench, J. Biomed. Mater. Res. 28 (1994) 693.
[5] T. Nakamura, T. Yamamuro, S. Higashi, T. Kokubo and S. Ito, J. Biomed. Mater. Res. 19 (1985) 685.
[6] O.H. Andersson and I. Kangasniemi, J. Biomed. Mater. Res. 25 (1991) 1019.
[7] R. Li, Sol-gel processing of bioactive glass powders, Ph.D. dissertation, University of Florida, 1991.
[8] R. Li, A.E. Clark, L.L. Hench, J. Appl. Biomater. 2 (1991) 231.
[9] H. Oonishi, S. Kutrshitani, E. Yasukawa, H. Iwaki, L.L. Hench, J. Wilson, E. Tsuji and T. Sugihara, J. Clin. Orthop. Related Res. 334 (1997) 316.
[10] D.L. Wheeler and K.E. Stokes, In Vivo Evaluation of Sol-Gel Bioglass®, part I. Trans 23rd Annual Meeting of the Soc. Biomater. New Orleans, LA, 1997.

[11] M. Cerruti and C. Morterra, Langmuir, 20 (2004) 6382.

[12] R.A. Brooker, S.C. Kohn, J.R. Holloway and P.F. McMillan, Chem. Geol. 174 (2001) 241.

[13] A. Stoch, W. Jastrzebski, A. Brozek, B. Trybalska, M. Cichocinska and E. Szarawar, J. Mol. Struct. 511-512 (1999) 287.

[14] G. Busca and V. Lorenzelli, Mater. Chem. 7 (1982) 89.

[15] H.J. Freund and M.W. Roberts, Surf. Science Reports 25 (1996) 225.

[16] R.G. Newton and S. Davison, Conservation of Glass, Butterworths, London, 1989.

Studies in Surface Science and Catalysis 155
A. Gamba, C. Colella and S. Coluccia (Editors)
© 2005 Elsevier B.V. All rights reserved

Acidity properties of CHA-zeolites (SAPO-34 and SSZ-13): an FTIR spectroscopic study

Laura Regli[a], Silvia Bordiga[a], Adriano Zecchina[a], Morten Bjørgen[b], Karl Petter Lillerud[b]

[a] Centre of excellence NIS, Dipartimento di Chimica IFM Università di Torino, Via P. Giuria 7, 10125 Torino

[b] Department of Chemistry, University of Oslo, P.O. Box 1033, N-0315 Oslo, Norway

ABSTRACT

Zeolite structures based on the chabazite topology, such as H-SAPO-34, possess unique shape-selectivity properties for converting methanol into light olefins. In addition to the topology, zeolite acidity is inherently linked to catalyst activity and selectivity. The acidic properties of high silica Chabazite (H-SSZ-13) have attracted much attention the last decade and conclusions drawn so far have essentially been founded on quantum chemical methods. An experimental based benchmark of the acidity of H-SSZ-13 has hitherto not been available. In this work, transmission FTIR spectroscopy provides a description of the different acidic sites of H-SAPO-34 by using water and methanol as probe molecules and compare the results with those obtained with the zolitic homologue: H-SSZ-13. The results demonstrate that both materials are strongly acidic, essentially having two distinct sets of Brönsted sites that show a different accessibility. However all sites interact with bases forming strong H-bonded adducts; moreover, in presence of high water and or methanol loadings, clear evidences of protonation have been found only in case of H-SSZ-13.

1. INTRODUCTION

Since 1976 many studies in the field of heterogeneous catalysis were concentrated on investigation of MTO (methanol to olefins) and MTG (and methanol to gasoline) processes [1]. Among acidic catalysts, zeolites have attracted main attention due to their properties of high activity and selectivity [2, 3, 4]. In particular, the most commonly used catalyst for this kind of processes is H-SAPO-34, a microporous aluminium phosphate with CHA structure, where some phosphorus is replaced into the framework by silicon [5]. Chabazite framework is determined by hexagonal prisms connected through bridges originating a structure characterized by cages delimited by 8-members ring windows. A peculiarity of this framework is to have all sites exposed into the big cages.

In the literature a lot of information about H-SAPO-34 collected by NMR and FTIR spectrosopies is available, [6, 7, 8, 9, 10, 11, 12], while experimental investigations on H-SSZ-13 are nearly absent due to the difficulties to obtain this material [13]. Data obtained for

CHA zeolites refer only to samples characterized by a very high aluminium content, as these materials can be easily obtained from FAU zeolites [14].

In order to better explain reactivity towards small organic molecules, a comparative study between H-SAPO-34 and the zeolitic homologue H-SSZ-13 (Si/Al = 11.7) has been performed. In particular, in this work we have investigated by FTIR spectroscopy the reactivity of H-SAPO-34 and H-SSZ-13 towards H_2O and CH_3OH.

2. EXPERIMENTAL

H-SAPO-34 was synthesized using standard procedure while a special procedure has been followed to obtain H-SSZ-13 [15]. In this way it is possible to obtain a material with a low aluminium content. The key-point is the use of a special template and to perform a calcination of the sample under control of the temperature. To remove template the sample was heated in an oxygen flow at 573 K for 12 h and then, after a slow increasing of temperature, at 773 K for 12 h again. FTIR transmission spectra have been collected using compressed self supported zeolite wafers. In order to remove all impurities adsorbed on sample surfaces, wafers have been pre-treated at 773 K under high vacuum before to start IR measurement. Controlled dosages of H_2O and CH_3OH have been put in contact with the pellets using a vacuum manifold connected with IR cell. The spectra have been recorded on Bruker IFS 28 FTIR spectrometers, equipped with cryogenic MCT detectors at 2 cm^{-1} resolution.

3. RESULTS AND DISCUSSION

3.1. Structure and spectroscopic properties of H-SSZ-13 and H-SAPO-34

Both H-SSZ-13 and H-SAPO-34 belong to the CHA family [16]. CHA structure can be described by imagining two six-member rings interconnected by four-member rings to form prisms. Connecting these prisms by a four-member ring, a cage delimitate by eight T centres is generated. The final structure is a framework characterized by the presence of big cages delimited by 8 T sites windows and little cages (prisms). The structure contains only one unique T site and, consequently, there are only four different oxygens: three of them are part of an 8 member ring window, while the last one is part of a 6 member ring unit (see O3 in Fig. 1). An interesting point would be to know if there is a difference in stability among all the sites and if the protons distribution observed in the two materials is similar or not.

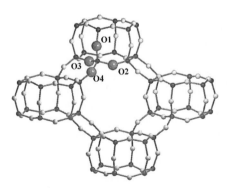

Fig. 1: Schematic representation of CHA structure. The four different oxygen atoms in the asymmetric unit, are evidenced by labels: O1-O4

In Fig.2 FTIR spectra of H-SAPO-34 (curve a) and H-SSZ-13 (curve b) samples are reported. The spectra have been obtained after heating samples at 773 K in vacuum for 1 h.

The IR spectrum of H-SAPO-34 is dominated by a strong absorption centred at 3627 cm⁻¹ and by a second component, less intense at 3606 cm⁻¹. For sake of simplicity we will refer to the first band as to HF component and to the second absorption to LF band.

As far as H-SSZ-13 it is concerned, in the strong Brönsted sites region, a complex absorption due to two contributions (HF and LF signals) is visible. In particular we observe a maximum at 3616 and a shoulder at 3603 cm⁻¹. These signals have been assigned to stretching modes of OH groups of two different Brönsted acidic sites existing in both catalysts [5, 6, 13]. We observe that the spectra line shapes are very similar for both samples, being the HF component more intense. Major differences between the two materials are evident in the high frequency region where H-SAPO-34 spectrum shows only very weak peaks at 3748, 3742 and 3676 cm⁻¹ assigned to ν(OH) of Si-OH, Al-OH and P-OH groups respectively. Conversely, H-SSZ-13 spectrum shows a complex strong absorption with a maximum at 3740 cm⁻¹ and a shoulder at 3712 cm⁻¹ that can be assigned to ν(OH) mode of free and terminal silanols. Finally, the shoulder centred at 3500 cm⁻¹, clearly visible in H-SSZ-13, has been explained in terms of hydroxylated nests.

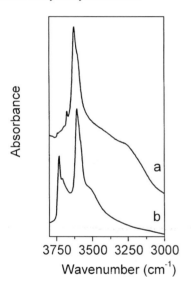

Fig. 2 IR spectra of H-SAPO-34 (curve a) and H-SSZ-13 (curve b) outgassed in vacuum at 773 K.

Previous studies, based on powder neutron diffraction data have suggested that LF and HF bands are due to protons mainly located on two different sites, namely O1 and O2 in H-SSZ-13 and O1 and O4 in H-SAPO-34 [5, 13] with a very similar population for both sites. A recent IR study performed on these materials has shown that the abundance of the two kind of sites is not equivalent, being HF site more abundant [17] and more accessible. These properties have suggested that HF component is associated to protons located on T centre of 8-members ring and oriented to centre of big cage, while LF band has been associated with protons located on T centre of 6-members ring and not directly exposed to the big cage window. Note that these two materials, even if characterized by a compositional difference, show very similar spectroscopic features.

3.2. Interaction with H₂O

3.2.2 H-SAPO-34 and H₂O

In Fig. 3a) background subtracted spectra obtained after increasing dosages of H₂O on H-SAPO-34 are reported, while in Fig. 3b) the effect of subsequent pumping is illustrated.

474

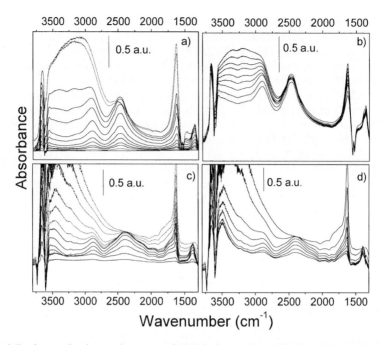

Fig. 3 Background subtracted spectra of H₂O in interaction with H-SAPO-34 (parts a) and b)) and H-SSZ-13 (parts c) and d)). In parts a) and c) increasing loading of water are represented: low coverage (solid curves) and high coverage (dashed curves) are reported. In parts b) and d) spectra of outgassing system are reported.

Interaction with H₂O molecule causes a complete change of H-SAPO-34 spectrum profile. Each curve corresponds to an increasing H₂O loading, till to reach the equilibrium with vapor pressure. The spectra have been divided into two series: solid curves refer to low H₂O loading while dashed curves indicate high loading of water. In case of solid curves, ratio between probe molecules and Brönsted sites is about 1:1, while dashed lines refer to > 1 ratio.

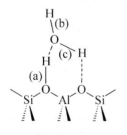

H-bonded
Scheme I

Starting from very low coverages we observe the appearance of a sharp band at 3700 cm⁻¹ due to ν(OH) of a nearly free hydroxyl (group (b) in scheme I), the parallel erosion of HF and LF bands due to strong Brönsted sites and the contemporary growth of a complex absorption in the 3500-1300 cm⁻¹ region. These phenomena are due the formation of strong H-bond between slightly basic probe molecule and acidic site of the zeolite. Stretching frequencies of Brönsted sites are perturbed and red-shifted generating an absorption in which it is possible to recognize a maximum around 2400 cm⁻¹. In the mean time, δ(OH) and γ(OH) modes, blue-shifted in consequence of the formation of H-bonded adducts, originate first overtone components (2δ (OH) and 2γ(OH)) which overlap perturbed ν(OH). The result is the formation of A, B and C bands. This phenomenon is called Fermi Resonance and the holes are called Evans windows [18, 19, 20,

21]. In this case we observe two clear maxima at 2880 and 2400 cm^{-1}: A and B components respectively. C band is less visible because it grows below δ(OH) of H$_2$O (1630 cm^{-1}). The band associated to the δ(H$_2$O) mode is distorted in the low frequency side because of the appearance of a couple of negative components at 1552 and at 1524 cm^{-1} in correspondence of a doublet assigned to combinations and overtone modes of bulk vibrations. This suggests that these bands selectively respond to interactions with strong ligands.

At high H$_2$O loading, Evans window is only partially filled: this happens because not all Brönsted sites are deprived of their own proton. This indicates that water clusters are not basic enough to extract H$^+$ from the acidic sites. Previous studies have already found that in SAPO-34 there is an equilibrium between protonated and H-bonded water in this material [12].

Part b) of Fig. 2 reports the effect of outgassing at room temperature. We observe that the interaction with water is quite strong as the amount of reversible water is low. This observation is consistent with the high ionicity of the matrix which makes the framework hydrophilic.

3.2.1 H-SSZ-13 and H$_2$O

In Fig. 3c) background subtracted spectra obtained after increasing dosages of H$_2$O on H-SSZ-13 are reported, while in Fig. 3d) the effect of subsequent pumping is illustrated. Also in this case the spectra associated to low H$_2$O loadings are reported as solid lines, while spectra associated to high H$_2$O loadings are reported as dashed curves.

Interaction with H$_2$O molecule causes similar effects to those previously described in case of H-SAPO-34. Major features will be briefly illustrated in the following:

i)   Growth of a band at 3700 cm^{-1} assigned to ν(O-H$_{b,water}$) modes of nearly unperturbed O-H bonds in H$_2$O, (group (b) in Scheme 1).

ii)   Appearance of a maximum around 3540 cm^{-1} due to ν(O-H$_{c,water}$) vibration (i.e. water OH bonds involved in weak H-bonds to zeolite oxygens)

iii)   Erosion of all Brönsted sites (HF and LF components)

iv)   Formation of A, B and C components centered at 2875, 2380 and 1670 cm^{-1}. Note that in this case B component is stronger then A one and that C band is clearly visible on the left side of δ(OH) mode of H$_2$O.

v)   Formation of a band at 1380 cm^{-1} due to perturbed δ(OH) mode of hydroxyls involved in H-bonds.

At high water loading, spectra profile change showing the filling of Evans windows. This is the consequence of the transformation of neutral H-bonded species in favour of ionic ones. The reaction is illustrated in Scheme II. In presence of a water cluster the extraction of the proton from the framework is observed. In this

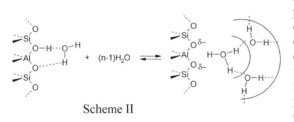

Scheme II

condition H-bonds are weaker and for this reason we don't have any Fermi Resonance effect.

Desorption of water at 300 K is performed connecting IR cell to vacuum line and reducing equilibrium pressure of H$_2$O. The results, reported in part d), show that the interaction with water is largely reversible. However water is not completely eliminated, but it is possible to affirm that H-SSZ-13 behaves as an hydrophobic zeolite.

3.3 Interaction with CH₃OH

3.3.1 H-SAPO-34 and CH₃OH

Effects of interaction between H-SAPO-34 and methanol are shown in Fig. 4a) and Fig. 4b). Spectra are background subtracted. Curves in Fig. 4a) can be divided in two families: first one is represented by solid drawn curve and the second one is presented using dashed curves.

Solid drawn curves illustrate situation in which ratio between CH_3OH molecules and Brönsted acidic sites is 1:1.

Methanol is characterized by a proton affinity only slightly higher than to that of water (PA_{H2O} = 166.5 kcal/mol; PA_{CH3OH} = 181.9 kcal/mol) and for this reason we expect to see a comparable reactivity. Another similarity with water is represented by the presence of an OH group in methanol which implies the formation of more than one H-bonds. In particular we expect a medium-strong H-bond between the oxygen of the methanol and the hydrogen of the Brönsted sites and a very weak H-bond due to the interaction between CH_3OH hydroxyls and the weakly negative oxygen of the zeolitic framework (see scheme III) [20, 21, 22]. Upon interaction of CH_3OH, IR spectrum of H-SAPO-34, presents all features corresponding to the formation of strong H-bonds. In particular the A B C triplet, caused by Fermi Resonance is visible (maxima located at 2930, 2600 and 1660 cm⁻¹ and minima at 2670 and 2080 cm⁻¹). In respect to water we observe that component C is more intense, as the barycentre of the broad absorption due by the stretching of bond (1) is shifted to lower frequency. In this condition we are in presence of a potential curve with a very flat minimum [20, 22]. Simultaneously, evidence of the perturbation of ν(OH) mode of bond (2) appears in the range 3680-3640 cm⁻¹, while an absorption at about 3540 cm⁻¹, is associated to a very weak H-bond between OH group of methanol and oxygen of framework. Bands at 3000, 2955, and 2850 cm⁻¹ are due to ν(CH₃), while a component at 2919 cm⁻¹ can be assigned to an overtone of δ(CH₃) mode at 1450 cm⁻¹ enhanced by Fermi Resonance effect [22]. In 2000-1300 cm⁻¹ region the spectra are quite complex due to the formation of a negative component at 1570 cm⁻¹. This behaviour can be explained by considering that CH_3OH perturbs framework vibrations. The negative band at 1450 cm⁻¹, corresponds to an Evans window due to the superposition of the C band with a δ(CH₃) mode.

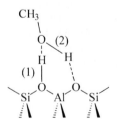

Scheme III

When we increase CH_3OH loadings, the ratio between adsorbed molecules and Brönsted site becomes >1. In this condition an intensification of all components is observed and a signal at 3360 cm⁻¹ becomes recognizable. This component is probably due to ν(OH) of methanol liquified into the zeolite channels. These features indicate that in working conditions we don't observe any significant deprotonation of methanol.

In part b) of Fig. 4 effects of outgassing at room temperature (full lines) and subsequent heating at 573 K for 30' are shown (bold curve). It is clear that interaction of CH_3OH with Brönsted sites is only partially reversible as even after a prolonged sample outgassing all bands assigned to formed adducts are still present. In this condition only physisorbed methanol have been removed.

Upon heating at 573 K, we observe some changes in the IR spectrum which can be associated to some transformation of the adsorbed species. As indicated in literature [22, 23], $(CH_3)_2O$ could be a reaction product, following the reaction scheme reported below:

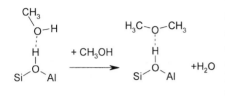

Scheme IV

The formation of $(CH_3)_2O$ seems confirmed by the appearance of peaks at 3000, 2955 and 2850 cm^{-1} assigned to $\nu(CH_3)$ groups of $(CH_3)_2O$ while bands at 2896 and 2860 cm^{-1} could be assigned to the overtone bands of $\delta(CH_3)$ enhanced by Fermi Resonance effects. However it is clear that the conversion of methanol in methylether has been only partial.

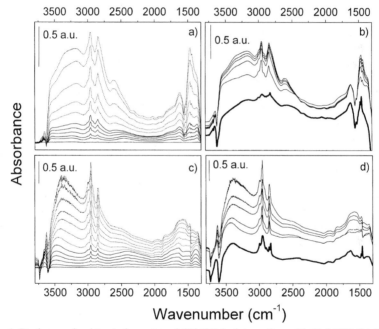

Fig. 4. Background subtracted spectra of CH_3OH in interaction with H-SAPO-34 (parts a) and b)) and H-SSZ-13 (parts c) and d)). Increasing loading of CH_3OH are represented in parts a) and c): in solid curves low loadings and in dashed curves high loadings are reported. In parts b) and d) spectra of outgassing samples (solid curves) and heated samples (bold curves) are reported.

3.3.2 H-SSZ-13 and CH₃OH

The behavior of H-SSZ-13 upon CH$_3$OH interaction is illustrated in Fig. 4c) and Fig.4d). Spectra are reported after background subtraction. In Fig. 4c) solid curves refer to low coverage of methanol, while dashed curves represent high coverage situation.

Main features related to the interaction between H-SSZ-13 and methanol are illustrated in the following:

i)   As in the case of H-SAPO-34 formation of ABC triad is observed (maxima at 2940, 2400 and 1600 cm^{-1}) and minima at 2720 and 1975 cm^{-1}. These signals are due to perturbation of ν(OH) of Brönsted sites interacting with oxygen of methanol.

ii)   A band at around 3500 cm^{-1} is ascribable to the stretching of a second H-bond formed by the hydrogen of the hydroxyl group of methanol and the oxygen of H-SSZ-13 framework (OH$\cdots$O).

iii)   Signals at 3000, 2956, 2920 and 2854 cm^{-1} are assigned to ν(CH$_3$): peaks at 3000, 2958 and 2854 cm^{-1} are due to asymmetric stretching modes of CH$_3$ while 2916 cm^{-1} components probably can be ascribable to an overtone of δ(CH$_3$) mode at 1450 cm^{-1} enhanced by Fermi resonance effect.

iv)   Negative component appears at 2050, 1890 and 1580 cm^{-1} in consequence of subtraction of background because of modifications of framework vibrations due to presence of methanol.

v)   Negative band at 1450 cm^{-1} corresponds to an Evans window due to the superposition of the C band with a δ(CH$_3$) mode.

When loading of CH$_3$OH is increased (see Fig. 4c) dashed curves), and the ratio between methanol molecules and Brönsted site becomes >1, many spectroscopic features change. In particular we observe the formation of a broad absorption ranging covering all the spectral range. Evans windows at 2700 cm^{-1} and 1975 cm^{-1} appear completely filled and this is due to disappearance of H-bonded species in favour of the formation of ionic species.

The interaction methanol/H-SSZ-13 is very strong and the effect of outgassing at room temperature is very modest as reported in Fig. 4d). In the same picture the effect of sample heating at 537 K is shown (bold curve in Fig. 4d). In this case we observe clear formation of (CH$_3$)$_2$O as testified by appearance of bands at 2896 and 2860 cm^{-1} assigned to ν_{sym}(CH$_3$) of hydrogen-bonded (CH$_3$)$_2$O [22].

4.   CONCLUSIONS

FTIR spectroscopy has allowed to follow H-SAPO-34 and H-SSZ13 activity toward H$_2$O and CH$_3$OH. Both the materials are characterized by two active sites associated with two distinct IR bands (HF) (LF) components. Both hydroxyls are acidic and interact with bases even if the (HF) components seems to be more reactive because it is consumed before the LF one.

By comparing the activity of the two materials in respect of the same probe, H-SAPO-34 appears less acidic that H-SSZ-13: in fact, interaction of water and methanol on H-SAPO-34 doesn't show extensive evidence of proton transfer while it has been observed that H-SSZ-13 can be deprotonated at high loadings.

[1] S. L. Meisel, J.P. McCullogh, C. H. Lecthaler, P. B. Weisz, ChemTech 6 (1976) 86

[2] J. A. Rabo, R. J. Pellet, P. K. Coughlin, E. S. Shamson, H. G. Karge, J. Weitkamp, (Eds.), Zeolites as catalysts and detergent builders, Elsevier: Amsterdam 1989.

[3] Stöcker, M. Microporous Mesoporous Mater. 29 (1999) 3-48.

[4] J. Chen, P. A.Wright, S. Natarajan, J. M. Thomas, Stud. Surf. Sci. Catal. 84 (1994) 1731.

[5] L. Smith, A. K. Cheetam, L. Marchese, J. M. Thomas, P. A. Wright, J. Chen, E. Gianotti, Catal. Lett. 41 (1996) 13-16

[6] B. A. Aufdembrink, D. P .Dee, P. L. McDaniel, T. Mebrahtu, T. L. Slager, J. Phys. Chem. B 107 (2003) 10025.

[7] A. Frache, E. Gianotti, L. Marchese, Catal. Today 77 (2003) 371.

[8] W. Song, D. M. Marcus, H. Fu, J. O. Ehresmann, and J. F. Haw, J. Am. Chem. Soc. 124 (2002) 3844

[9] B. Arstad and S. Kolboe, J. Am. Chem. Soc. 123 (2001) 8137

[10] L. Smith, A. K. Cheetam, R. E. Morris, L. Marchese, J. M. Thomas, P. A. Wright, J. Chen, Science 271 (1996) 799

[11]L. Marchese, J. M. Thomas, P. A. Wright, J. Chen, Catal. Lett. 41 (1996) 13

[12] L. Marchese, J. Chen, P. A. Wright, J. M. Thomas, J. Phys. Chem. B 97 (1993) 8109

[13] L. J. Smith, A. Davidson, and A. K. Cheetam, Catal. Lett. 49 (1997) 143

[14] B. A. Aufdembrink, D. P. Dee, P. L. McDaniel, T. Mebrahtu and T. L. Slager J. Phys. Chem B 107 (2003) 10025

[15]S. I. Zones, US Patent 4 544 538 1985 and private communication Zones, S. I.

[16] W. M. Meier, D. H. Olson, C. H. Baerlocher, Atlas of Zeolite Structure Types (1996)

[17 S. Bordiga, L. Regli, D. Cocina, C. Lamberti, M. Bjorgen, K.P. Lillerud, J. Am. Chem. Soc. submitted

[18] A. Jentys, G. Wareka, M. Derewinski and J. Lercher, J. Phys. Chem. 93 (1989) 4837.

[19] L. M. Parker, D. M. Bibby, G. R. Burns, Zeolites 11(3) (1991) 293.

[20] C. Pazè, S. Bordiga, C. Lamberti, M. Salvataggio and A. Zecchina, J. Phys. Chem. B 101 (1997) 4740

[21] A. Zecchina, F. Geobaldo, G. Spoto, S. Bordiga, G. Ricchiardi, R. Buzzoni and G. Petrini, J. Phys. Chem. B 100 (1996) 16584.

[22] A. Zecchina, S. Bordiga, G. Spoto, D. Scarano, G. Spanò, F. Geobaldo, J. Chem. Soc. Faraday Trans. 92 (1996) 4863

[23] A. G. Pelmenschikov, G. Morosi, A. Gamba, A. Zecchina, S. Bordiga, E. A. Paukshtis, J. Phys. Chem 97 (1993) 11979

Studies in Surface Science and Catalysis 155
A. Gamba, C. Colella and S. Coluccia (Editors)

The role of surfaces in hydrogen storage

Giuseppe Spoto[a,b], Silvia Bordiga[a,b], Jenny G. Vitillo[a], Gabriele Ricchiardi[a] and Adriano Zecchina[a,b*]

[a]Dipartimento di Chimica IFM and NIS Centre of Excellence, Università di Torino, Via Pietro Giuria 7, 10125 Torino, Italia.

[b]INSTM UdR Torino Università, Torino, Italia.

ABSTRACT

This review deals with the main materials employed so far for hydrogen storage and it is specifically focused on the role of surface phenomena in their storage performance. Surface properties are relevant in all classes of materials: when dihydrogen is stored in the form of hydrides, the structure, texture and reactivity of the surfaces have large influence on the kinetics of charge/discharge cycles. In the storage of molecular hydrogen, surface-molecule interactions are responsible for the storage properties of the materials.
A better understanding of the fundamental aspects of the chemistry of hydrogen at surfaces is needed in order to design improved storage materials.

1. INTRODUCTION

The demand of clean fuels alternative to hydrocarbons is increasing and hydrogen is emerging as a viable choice. If due consideration is made on the fact that the energy problem is certainly one of the main issues of the modern times [1], the importance of solving all the problems associated with hydrogen-based energy cycle is clearly emerging [2].

However one of the main concerns about a hydrogen-based energy economy is the efficient storage and transport of this highly flammable gas [3].

The solution of this problem is very difficult because it must satisfy a certain number of requirements: a) high hydrogen content per unit mass of the material used for the storage; b) high hydrogen content per unit volume; c) moderate pressure P and temperature T for the storage (P preferably lower than 400 MPa, i.e. the pressure that can be reached by a simple compressor and T not too far from that of the ambient); d) easy hydrogen release and e) environmentally friendly by-products.

Many strategies have been followed or suggested in recent years to solve this problem. The most important ones are: 1) storage in metals and alloys; 2) storage in complex hydrides (alanates, borides); 3) storage by trapping in clathrates (ice and others); 4) storage in microporous materials (carbons, zeolitic materials, metal-organic frameworks, polymers) and 5) others. By others we intend less common strategies which have been only proposed or that are at very preliminary research stages.

As all systems (usually solids) involve adsorption of hydrogen from the gas phase in the storage processes, common problems encountered in these strategies concern:

a) the dissociative interaction of dihydrogen with the surface of metals, alloys and complex hydrides (governing the enthalpy and speed of charging and discharging processes);

b) the interaction strength of hydrogen in the molecular form with the internal surfaces of micropores and/ or of cages of entrapping materials (governing the stability of the molecular adducts and the optimal temperature of storage).

Charging and discharging problems have been prevalently tackled in an empirical way, by transferring the experiences accumulated in other areas (surface science, catalysis, etc.) to the hydrogen storage materials.

However, due to the peculiar properties of dihydrogen in term of thermodynamic properties and chemical reactivity, a more basic approach to the problem of hydrogen activation and bonding seems preferable.

On the basis of the introductory remarks, a special attention will be devoted to the indirect and direct role of the surfaces on the hydrogen adsorption- desorption processes. In our opinion, this key aspect of the hydrogen storage topic has never been extensively treated and merits a specific contribution. It will be concluded that the role of the surface properties is higher than usually recognized even for storage systems like metals, metal alloys and precursors of complex hydrides where the role of surfaces in only occasionally mentioned.

The key role of surface properties is particularly emerging when the use of high surface area materials (micro and mesoporous) as hydrogen adsorbers is specifically considered. For the reasons outlined before, the part devoted to the high surface area materials (either as such or suitably functionalised) will represent the most extended part of this contribution. It will be documented that the role of high surface area materials in solving the problems of an efficient hydrogen storage could be higher than usually recognized.

The final chapter will be devoted to the conclusions and to the perspectives.

2. STORAGE IN METAL HYDRIDES AND THE ROLE OF SURFACES

This strategy is based on the formation of bulk metal hydrides and involves (in the case of adsorption/absorption from the gas phase) dissociation of the hydrogen molecule on the surface of metals or alloys particles or (in the case of electrolytic deposition) the neutralization of H_3O^+ (Fig. 1).

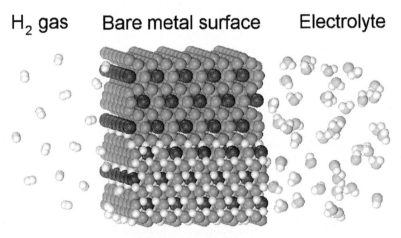

H$_2$ gas   Bare metal surface   Electrolyte

Metal hydride

Fig. 1 Schematic representation of two possible charging processes for hydrogen storage in metal hydrides. The example refers to the LaNi$_5$ alloy. The H atoms in the interstices between the metal atoms could be derived from the dissociation of initial surface hydrogen molecules (left) or from the dissociation of water molecules (right). The O, H and metal atoms are represented as black, dark grey and light grey spheres respectively.

If the case of storage from the gas phase, the first step to be taken into account is the dissociative hydrogen adsorption on the external surfaces with formation of surface hydrides. The surface atomic hydrogen then penetrates into the lattice and forms multiple chemical bonds with the metal bulk atoms [4,5]. In this process (adsorption and successive formation of bulk hydrides), chemical forces are always involved. This strategy is facing several difficulties including the cost of metals and alloys, the low uptake by weight and, sometimes, unfavourable adsorption-desorption kinetics. Notice that the kinetics of the whole process is surface-mediated: hence the role of surfaces is of paramount importance in this strategy. That this is the case is shown by the observation that dispersion of the metals and alloys by milling or by supporting the metal particles on high surface area solids is beneficial [6].

To discuss this point more deeply, let us underline that the dissociation of the hydrogen molecule on metal surface occurs easily only when the d-metals are involved. In fact, only surface frontier orbitals with d-character can bind hydrogen and lead to its surface dissociation without the overcome of a high surface barrier. This is verified for the prototype Pd/H_2 system and for all the active alloys (which invariably contain transition metal atoms). As dissociation and recombination on the surface are governing the speed of charging and discharging processes, the role of d-orbitals in surface phenomena is understood. Also the poisoning effect of surface impurities is foreseen. The considerations outlined before explain why metals like Mg and Li, whose hydrides are stable and are characterized by high hydrogen content, cannot be easily charged and discharged. In fact, in absence of d-type frontier orbitals, hydrogen dissociation occurs only at high T. Active surface sites for hydrogen dissociation and recombination on Mg and Li can be generated by milling because of the formation of highly reactive surface defects. This however introduces remarkable difficulties in the reproducible realization of hydrogen charging-discharging cycles.

From this it is emerging that the problems of reducing the dimension of the metal particles to increase the rate of hydrogen adsorption and desorption and to control surface poisoning by impurities occupy a central position in the future developments. On this basis it is evident that more studies of the effect of high surface area supports on the hydrogen uptake by dispersed metal particles are necessary.

3. STORAGE IN COMPLEX HYDRIDES: SURFACE EFFECTS

This strategy consists in the storage of hydrogen under the form of complex hydrides like sodium-aluminium hydrides $NaAlH_4$ (alanates) and sodium- boron hydrides [2] $NaBH_4$ whose theoretical storage capacity is in the 7-13 mass% range. Also $LiAlH_4$, $LiBH_4$ and other complex hydrides have been considered. The formation of complex hydrides involves formation of Al-H and B-H chemical bonds which are disrupted during desorption. This means that the involved enthalpies are considerable. One of the difficulties encountered with these systems is often represented by their partial irreversible character which makes them good candidate only for one time applications. It must be mentioned however that also in this case the reduction of the grain size by the use of suitable supports or by milling favours multiple use. Also the addition of suitable catalysts (like TiN) is beneficial because it favours the hydrogen dissociation [7]. From these consideration the role of high surface area materials is evidenced. If the problems of the partial irreversibility and stability are solved and the role of the catalyst understood at the molecular level, these systems may have applications of vital importance since they are characterized by the highest hydrogen content per unit mass.

4. STORAGE UNDER MOLECULAR FORM ON HIGH SURFACE AREA MATERIALS

This strategy is based on the adsorption of dihydrogen on high surface area solids and can involve either weak dispersion forces and/or stronger overlap forces.

Inspection of the relevant literature allows to safely conclude that the binding energy of dihydrogen with covalent surfaces is in the 3-5 kJ mol^{-1} interval. This binding energy is similar to that of dihydrogen with neutral atoms and molecules and suggests that molecular hydrogen complexes are highly unstable and can be formed only at $T < 77$ K. An exception to this conclusion could be represented by corrugated surfaces, where the experimental results seem to suggest that hydrogen adsorbed in nanopockets and nanochannels of microporous materials could be more stabilized. In this chapter we shall mainly discuss high surface area carbons (either amorphous or crystalline), metal-organic frameworks (MOF) and zeolites.

4.1. Storage in carbon nanotubes and microporous carbons: role of dispersive forces

Among the microporous solids proposed so far, microporous carbons [4,8] (Fig. 2a) and nanotubes [9-11] (Fig. 2b) merit special attention, because they have been thoroughly studied and because their properties are in many cases paradigmatic for the illustration of the role of surfaces in molecular adsorption for hydrogen storage.

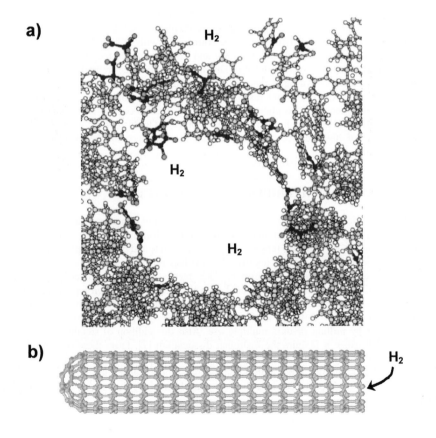

a)

H_2

H_2

H_2

b)

H_2

Fig. 2 a) Carbon film generated by classical molecular dynamics simulation of a cluster beam deposition process (adapted from ref. 12). b) Model of nanotube. The possible positions occupied by hydrogen molecules are also qualitatively showed.

Because of their structural simplicity we start with carbon nanotubes which have recently attracted a great deal of attention and generated large hopes in the scientific community. However, it must be mentioned that the initial results obtained on the adsorption of hydrogen on carbon nanotubes have been severely revised [2,10,11].

For the discussion to come we shall assume that hydrogen adsorption on these systems can be either chemical or physical (molecular).

Molecular adsorption usually involves very low interaction energies (about 10 kJ mol^{-1}) and occurs abundantly only at low T ($\leq$ 200 K) for moderate hydrogen pressures. This type of interaction should be energetically similar to that found or calculated for the hydrogen/graphite system.

In the case of nanotubes both the external and internal surface areas must be considered. Some researchers have calculated that external surfaces have lower potential to physically adsorb H$_2$.

The ideal nanotube for this strategy should possess two properties: surface area $\cong$ 1000 m^2 g^{-1} and adsorption enthalpy definitely larger than 5-7 kJ mol^{-1} (which is the estimated figure for H$_2$ interaction with the graphite layers) [11].

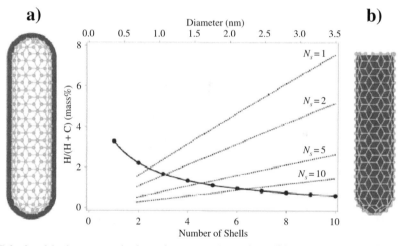

Fig. 3 Calculated hydrogen uptake in carbon nanotubes. The solid curve represents the hydrogen physisorbed as a monolayer at the surface of a single wall nanotube as a function of the tube diameter. The dotted curves represent the hydrogen condensed in the nanotube (density as liquid hydrogen at –253 °C) as a function of the number of shells. Adapted from ref. 8, page 707.

The various possibilities concerning hydrogen adsorption in the carbon nanotubes are represented in Fig. 3.

As for internal surfaces is concerned, it has been claimed that the adsorption of dihydrogen on the internal curved walls close to the tip involve up to 30 kJ mol^{-1} adsorption enthalpy [13]. This data have not been experimentally confirmed. A chemical interaction involving hydrogen dissociation and formation of C-H bonds has also been hypothesised. However this hypothesis has never been confirmed for high purity nanotubes.

486

The amount of adsorbed hydrogen is roughly proportional to the surface area [14]. However a specific role of microporosity on the adsorptive capacity has been also put into evidence. The authors of ref. 14 have concluded that porous carbons with surface area in the 2000-3000 $m^2 g^{-1}$ and suitably tuned microporosity could be considered with favour for hydrogen storage. In this respect, it must be mentioned that the actual synthetic capacity in the field of nanostructured carbon is primitive and that crystalline microporous carbons, although certainly stable [15,16], have never been synthesized.

4.2. Storage in metal-organic frameworks: role of dispersive interaction with the internal surfaces

Along this research direction new perspectives have been disclosed by the initial observation that MOF-5 (MOF = Metal-Organic Framework) [17] (Fig. 4) can store large amounts of hydrogen at 78 K (about 4.5% by weight). This figure has been recently reduced [18] to about 2%. The high storage capacity of MOF-5 is likely the result of the very high surface area (2500 to 3000 $m^2 g^{-1}$).

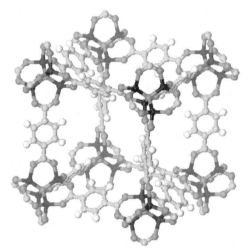

Fig. 4 Structure of MOF-5. It consists of OZn_4 units connected by organic linkers (1,4-benzenedicarboxylate). The elements are coded on a grey scale according to their atomic number: H (white) < O < Si < Zn (black). The presence of nanovoids is well evident.

O. Hübner et al. [19] have calculated the interaction energy of dihydrogen with aromatic systems and in particular with the aromatic ring of the dilithium terephthalate: the obtained value (4.30 kJ mol^{-1}) is higher than that calculated for benzene ring and lower than that estimated for graphene layers (7.2 kJ mol^{-1}) [11]. The higher value calculated for graphene has been attributed to the long range part of the dispersion forces. These long range contributions should play an important role in determining the interaction of hydrogen with the sites located on the micropores walls of MOF-5: hence an adsorption energy slightly larger than 4.30 kJ mol^{-1} is, at least in principle, expected [20]. From all these considerations the need of further information on the interaction of dihydrogen with MOF-5 sites is clearly emerging. For the reasons which will be discussed in the next paragraph, we expect that IR spectroscopy investigations should be able to give information on the structure of adsorbed hydrogen.

4.3. Storage in zeolitic materials: role of dispersive, polarization and overlap interactions

Zeolitic materials (Fig. 5) have also been tested for hydrogen storage [21,22].

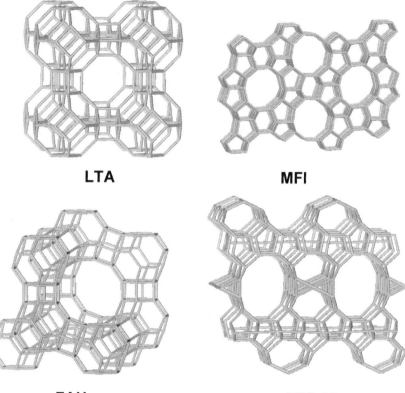

LTA   **MFI**

FAU   **ETS-10**

Fig. 5 Different zeolitic frameworks. The cavities diameter is ranging from 4.0 Å to 8.0 Å.

The interaction between molecular hydrogen and the zeolitic structure can be separated into two main components: the interaction with the surface atoms of the channels and cavities and the interaction with charge balancing cations hosted into the internal volume. The interaction with the cations can be approximated in first approximation with the interaction between an hydrogen molecule and a bare cation. The binding energies obtained for the complexes with alkaline cations at MP2/aug-cc-pVQZ level [23] are in the range of 4-30 kJ mol^{-1} going from Rb$^+$ to Li$^+$. From this, it is inferred that the lower is the radius of the cation the higher is the interaction energy: this results clearly indicates that electrostatic (polarization) forces are dominating. An interaction energy of about 30 kJ mol^{-1} (for the Li$^+$) is very interesting from hydrogen storage point of view because Li$^+$H$_2$ complexes are expected to be stable at RT under moderate pressures. Unfortunately the introduction of the Li$^+$ as counterion in zeolitic frameworks strongly affects the energy of interaction. In fact, the effective field sensed by dihydrogen is not only that due to the cation contribution as the shielding effect of the negative zeolitic surface must be also taken into account. It has for instance reported that for H$_2$ in Li-ZSM-5 $\Delta H^0_{ads} = 6.5$ kJ mol^{-1}, only [24].

IR spectroscopy of adsorbed hydrogen has proved to be a powerful mean to investigate the perturbation induced by the adsorption process on the vibrational spectrum of dihydrogen molecule [25-27]. This perturbation is usually accompanied by the appearance of IR activity and by a shift $\Delta \tilde{\nu}$ of the H-H stretching mode towards lower frequency. As the shift $\Delta \tilde{\nu}$ increases with the interaction energy, it is evident that the IR spectroscopy of adsorbed hydrogen can give information about the forces sensed by the molecule in the cavities. From this the great utility of IR spectroscopy in giving information on the adsorbing sites and clustering is inferred. In particular a detailed analysis of the spectra of hydrogen adsorbed on a large variety of zeolitic materials characterized by: i) channel diameters ranging in a wide interval, ii) different concentration, radii, charges and chemical properties of the balancing cations and iii) absence of aluminium in the structure (purely siliceous materials) can give information of the role of dispersive, polarization and chemical (overlap) forces in stabilizing the hydrogen adducts so helping the effort toward a rational design of optimised microporous structures.

Examples of IR spectra of adsorbed hydrogen on several types of sodium exchanged zeolitic materials (NaA, NaY, Na-ETS-10) and in a purely siliceous material (silicalite) are shown in Fig. 6a.

By comparing the results obtained on silicalite and on sodium-exchanged solids it can be easily appreciated that the Na^+ cations polarize the hydrogen molecule and that the extent of polarization is maximum in NaA, which is the materials with the smallest pores. The detailed description of the spectra is outside the limited scope of this review.

In Fig. 6b, the IR spectra of hydrogen adsorbed in a molecular form on Cu-ZSM-5 are also represented. We want to attract the attention of the reader on the ν(HH) frequency of the $Cu^I(H_2)$ adducts which is shifted downward of about 1000 cm^{-1} with respect to the gas phase [28].

This is indicative of the presence of chemical overlap forces which make the complex stable at RT. As the hydrogen molecule is readily desorbed simply by pumping at RT, it is inferred that we are in presence of an ideal material for hydrogen storage. Unfortunately the concentration of active Cu^I is not high enough to make this system sufficiently attractive for practical purposes.

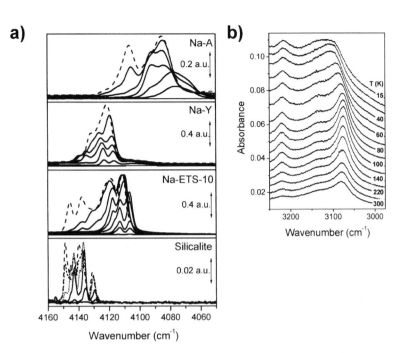

Fig. 6 a) From top to the bottom: IR spectra of H$_2$ adsorbed at 20 K on NaA, NaY, Na-ETS-10 and silicalite (the different curves correspond to increasing filling conditions). Unpublished results obtained in our laboratory. b) IR spectra of H$_2$ adsorbed on Cu^I counterions of Cu-ZSM-5 [28].

5. STORAGE OF H$_2$ UNDER MOLECULAR FORM BY TRAPPING IN CLATHRATES

5.1. Trapping in water clathrates

Hydrogen-bonded H$_2$O frameworks are constituted by polyhedron cages which can guest hydrogen molecules at temperatures and pressures at which they would otherwise exist as free molecules [29]. The synthesis of these compounds occurs under high hydrogen pressure and cooling. The formation of the clathrate can be represented as trapping of hydrogen molecules into the cages of the nascent framework of H$_2$O.

490

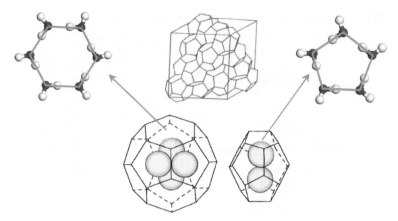

Fig. 7 The clathrate hydrate crystal structure. The hydrogen molecules are represented as spheres. Adapted from ref. 29, page 2248.

Once trapped in the cages, the hydrogen molecules cannot escape because the windows are too small to allow hydrogen diffusion. The interaction with the surface atoms of the cavities is essentially of the dispersive type: consequently the interaction energy is very weak. The release of dihydrogen occurs as the H_2O framework collapses because of the temperature increase. The clathrate strategy is at the infancy and more hydrogen bonded clathrates wait to be investigated. Recently, synthesis conditions of hydrate clathrates has been performed under milder conditions [30] and this justifies the optimism manifested by many authors about their future employment for hydrogen storage purpose.

5.2. Trapping in inorganic cages (silica, cordierite, etc.)

Zeolitic materials are also known to trap molecular hydrogen following a mechanism similar to that illustrated before for classical clathrates. The first observation of this property has been made for cordierite, a zeolitic material characterized by very small cavities (Fig. 8).

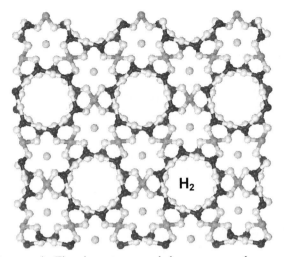

Fig. 8 Cordierite framework. The elements are coded on a grey scale according to their atomic number: O (white) < Mg < Al < Si (black). The possible position of trapped H_2 molecules is indicated.

The windows connecting the cavities are too small to allow diffusion of dihydrogen at RT. Also the size of the counterions balancing the negative charge of the framework plays a role in this trapping process. However, if the zeolitic material is heated at high temperature in presence of high hydrogen pressure, the windows undergo sufficient dilatation to allow hydrogen diffusion into the small cages. Successive fast cooling is than sufficient to allow trapping of a small fraction of hydrogen molecules inside the framework. Trapped hydrogen is then released by moderate heating. What is known about cordierite suggests that this system is certainly a not promising material for hydrogen storage since high temperatures are needed to allow diffusion. However the principle of trapping hydrogen in zeolitic cavities has stimulated several investigations on A and Y zeolites [22,31]. From these studies a moderate storage capacity have been demonstrated (particularly for A zeolite).

6. CONCLUSIONS AND PERSPECTIVES

On the basis of the research made so far on the various storage systems we can conclude that the problem of the storage of hydrogen is far from being solved. Many solids can store large amounts of hydrogen. However the charging and discharging processes are still very problematic. We think that the role of surfaces is essential and that new studies must be devoted to this aspect of the hydrogen storage saga.

REFERENCES

1.   S. Pacala, R. Socolow, Science 305 (2004) 968-972.
2.   J.A. Turner, Science 305 (2004) 972-974.
3.   A. Züttel, Naturwissenschaften 91 (2004) 157-172.
4.   L. Schlapbach, A. Züttel, Nature 414 (2001) 353-358.
5.   A.M. Seayad, D.M. Antonelli, Adv. Mater. 16 (2004) 765-777.

492

6.   R.C. Bowman, B. Fultz, MRS Bull. 27 (2002) 688-693.
7.   B. Bogdanović, G. Sandrock, MRS Bull. 27 (2002) 712-716.
8.   A. Züttel, S. Orimo, MRS Bull. 27 (2002) 705-711.
9.   M. Hirscher, et al., Appl. Phys. A 72 (2001) 129-132.
10.   J.S. Arellano, L.M. Molina, A. Rubio, J.A. Alonso, J. Chem. Phys. 112 (2000) 8114-8119.
11.   T. Heine, L. Zhechkov, G. Seifert, Phys. Chem. Chem. Phys. 6 (2004) 980-984.
12.   Y. Hu, S. B. Sinnott, J. Mater. Chem., 14 (2004), 14, 719-729.
13.   Y. Okamoto, Y. Miyamoto, J. Phys. Chem. B 105 (2001) 3470-3474.
14.   M.G. Nijkamp, J.E.M.J. Raaymakers, A.J. van Dillen, K.P. de Jong, Appl. Phys. A 72 (2001) 619-623.
15.   A.L. Mackay, H. Terrones, Nature 352 (1991) 762-762.
16.   T. Lenosky, X. Gonze, M. Teter, V. Elser, Nature 355 (1992) 333-335.
17.   H. Li, M. Eddaoudi, M. O'Keeffe, O.M. Yaghi, Nature 402 (1999) 276-279.
18.   J.L.C. Rowsell, A.R. Millward, K.S. Park, O.M. Yaghi, J. Am. Chem. Soc. 126 (2004) 5666-5667.
19.   O. Hübner, A. Glöss, M. Fichtner, W. Klopper, J. Phys. Chem. A 108 (2004) 3019-3023.
20.   S. Bordiga, J.G. Vitillo, G. Ricchiardi, L. Regli, D. Cocina, A. Zecchina, B. Arstad, M. Bjørgen, K.P. Lillerud, J. Am. Chem. Soc. (2005) submitted.
21.   H.W. Langmi, A. Walton, M.M. Al-Mamouri, S.R. Johnson, D. Book, J.D. Speight, P.P. Edwards, I. Gameson, P.A. Anderson, I.R. Harris, J. Alloys Compd. 356 (2003) 710-715.
22.   J. Weitkamp, M. Fritz, S. Ernst, Int. J. Hydrogen Energy 20 (1995) 967-970.
23.   J.G. Vitillo, A. Damin, A. Zecchina, G. Ricchiardi, J. Chem. Phys. (2005) accepted.
24.   C. Otero Areán, O.V. Manoilova, B. Bonelli, M. Rodríguez Delgado, G. Tunes Palomino, E. Garrone, Chem. Phys. Lett. 370 (2003) 631-635.
25.   V. Kazansky, A. Serykh, Microporous Mesoporous Mat. 70 (2004) 151-154.
26.   M. Sigl, S. Ernst, J. Weitkamp, H. Knözinger, Catal. Lett. 45 (1997) 27-33.
27.   A. Zecchina, C. Otero Areán, G. Tunes Palomino, F. Geobaldo, C. Lamberti, G. Spoto, S. Bordiga, Phys. Chem. Chem. Phys. 1 (1999) 1649-1657.
28.   G. Spoto, E. Gribov, S. Bordiga, C. Lamberti, G. Ricchiardi, D. Scarano, A. Zecchina, Chem. Commun. (2004) 2768-2769.
29.   W.L. Mao, H.K. Mao, A.F. Goncharov, V.V. Struzhkin, Q.Z. Guo, J.Z. Hu, J.F. Shu, R.J. Hemley, M. Somayazulu, Y.S. Zhao, Science 297 (2002) 2247-2249.
30.   L.J. Florusse, C.J. Peters, J. Schoonman, K.C. Hester, C.A. Koh, S.F. Dec, K.N. Marsh, E.D. Sloan, Science 306 (2004) 469-471.
31.   D. Fraenkel, J. Shabtai, J. Am. Chem. Soc. 99 (1977) 7074-7076.

Studies in Surface Science and Catalysis 155
A. Gamba, C. Colella and S. Coluccia (Editors)

493

Characterization of nanosized gold, silver and copper catalysts supported on ceria

T. Tabakova[a], F. Boccuzzi[b], M. Manzoli[b], A. Chiorino[b] and D. Andreeva[a]

[a]Institute of Catalysis, Bulgarian Academy of Sciences, 1113 Sofia, Bulgaria

[b]Department of Chemistry I. F. M., University of Torino, Via P. Giuria 7, Torino, 10125, Italy and INCA, Interuniversity Consortium Chemistry for the Environment, Viale della Libertà 5/12 Marghera (Venezia) Italy

A comparative study of CO adsorption on nanosized Au, Ag and Cu supported on ceria has been performed. HRTEM images combined with EDS analysis have shown presence of highly dispersed metallic particles on the surface of the catalysts. FTIR results indicate that IB metals cause a strong modification of the surface properties of ceria, which leads to the appearance of more co-ordinatively unsaturated sites on the surface. Reduced Au/CeO_2 activates CO molecules on negatively charged nano-gold particles near to the oxygen vacancies on the ceria surface. FTIR spectra of CO adsorbed on oxidized Cu/ceria show an easy generation of metallic copper even at 90 K. Ag/ceria catalyst is almost completely unable to adsorb and activate CO molecules due to the low adsorption energy for CO on silver.

1. INTRODUCTION

Nanoparticles can be useful tools for both fundamental studies and applications in a wide range of disciplines, due to their structure and properties. Gold in its bulk form is chemically inert and it has been considered as one of the least catalytically useful metals because of its chemical inertness and because of the difficulty to obtain a high dispersion on common support materials. Nanotechnology has transformed gold from a marginal catalyst into a very effective one with unique properties. In the last 15 years, it has been widely proved that it is possible to prepare gold nanoparticles (with size below 10 nm) deposited on metal-oxide supports and, in such conditions, gold exhibits catalytic activity that is radically different from bulk gold; in fact nanosized gold particles supported on metal oxides drastically enhance catalytic activity in a number of reactions [1]. Haruta and co-authors firstly established an exceptionally high CO oxidation activity of supported nano-gold catalysts even at a temperature as low as 200 K [2]. The astounding growth in gold nanoparticles research, as reflected by a great number of publications and patents, promises actually new applications for gold nanotechnology in a wide variety of reactions with respect to pollution control, chemical processing and fuel cells [3]. Recent studies concerning Au/ceria catalysts have shown a high and stable performance in the low-temperature water-gas shift reaction [4]. The renewed interest in the water-gas shift (WGS) reaction ($CO + H_2O \rightarrow CO_2 + H_2$) is due to its importance in the production of pure hydrogen. Hydrogen is one of the most important fuels for the use in fuel cell systems. Moreover, the WGS is a critical step in fuel processors for the preliminary CO clean-up and the additional hydrogen generation prior to the CO preferential oxidation.

Copper-based catalysts are also well-known as low-temperature CO shift catalysts and are widely used in the industrial field. The nature of these catalysts and the role which copper plays in the catalytic processes are subjects of great interest. FTIR investigations of CO adsorption on Cu/ZnO and on Cu/TiO_2 have revealed a close similarity in the behaviour of copper and gold catalysts [5].

The aim of this work is to study the electronic state of nanosized Au, Ag and Cu supported on ceria and to examine the nature of CO species adsorbed on the metals. FTIR spectroscopy is the most appropriate technique to get informations on the nature of the surface active sites in order to clarify the WGS reaction mechanism. The knowledge of the reaction mechanism can give important information for the selection of highly active catalysts.

2. EXPERIMENTAL

2.1. Catalysts preparation

The samples were prepared by deposition-precipitation of $HAuCl_4.3H_2O$ (pH 7.0) or $AgNO_3$ (pH 9.0) or $Cu(NO_3)_2.3H_2O$ (pH 7.0) with K_2CO_3 at 333 K on ceria previously prepared and suspended in water by ultrasound. The resulting precipitates were aged 1 h at 333 K, filtered and washed until no Cl^- or NO_3^- could be detected. Furtherly, the precipitates were dried in vacuum at 353 K and calcined in air at 673 K for 2 hours. All the samples were synthesized in a "Contalab" laboratory reactor (Contraves-AG, Switzerland) enabling the complete control of the reaction parameters: pH, temperature, stirrer speed, reactant feed flow, etc. The IB metals loading for each of catalyst was 3 wt.%. "Analytical grade" chemicals were used in catalysts preparation.

2.2. Catalysts characterization

HRTEM analysis has been performed using a Jeol JEM 2010 (200kV) microscope equipped with an EDS analytical system Oxford Link. The powdered samples were ultrasonically dispersed in isopropyl alcohol and the obtained suspensions were deposited on a copper grid, coated with a porous carbon film. As for the Cu/CeO_2 sample, a gold grid was employed in order to perform the EDS analysis.

The FTIR spectra have been taken on a Perkin-Elmer 1760 spectrometer (with a MCT detector) with the samples in self-supporting pellets introduced in a cell allowing thermal treatments in controlled atmospheres and spectrum scanning at controlled temperatures (from 90 to 300 K). The experiments were performed on samples preliminarily heated up to 673 K in dry oxygen and cooled down in the same atmosphere (oxidized sample) or reduced in hydrogen at different temperatures (373 or 523 K) and outgassed at RT (reduced sample). Band integration and curve fitting have been carried out by "Curvefit", in Spectra Calc (Galactic Industries Co.) by means of Lorentzian curves.

3. RESULTS AND DISCUSSION

HRTEM measurements showed that ceria is highly crystalline and it consists of particles with an average size of 4.5 nm. The deposition of IB metals does not influence the size of ceria crystallites. The analysis of the fringes observed in the micrographs revealed that the support has a cubic structure and that it mainly exposes the (111) face. The presence of very highly dispersed gold clusters (d about 1 nm) (inset of Fig. 1a) has been evidenced by EDS analysis (not shown for the sake of brevity).

Moreover, gold particles with size of about 10 nm and, in some few cases, also around 25 nm were detected on Au/CeO_2, as shown in Fig. 1a. HRTEM combined with EDS allowed us to assume that in this catalyst there is a bimodal gold particle size distribution.

On the contrary, copper is very highly and uniformly dispersed on Cu/CeO_2 (Fig. 1b). It is very difficult to distinguish the copper particles because of the low contrast difference between the metal and the support. Anyway, the presence of copper is revealed by EDS (inset of Fig. 1b), confirming a high dispersion of the metal. No copper agglomerates have been observed.

The surface of Ag/CeO_2 is covered by a AgO_x layer, in accordance with the FTIR data that will be illustrated later on. Since silver is unstable under the electronic beam of the microscope, an agglomeration into metallic silver particles occurs during the first few minutes of exposure, thus producing a lot of small silver particles that coalesce to form big particles by increasing the time under the electronic beam (not shown for the sake of brevity).

The adsorption of CO at 90 K on the oxidized Au/ceria catalyst (Fig. 2, fine curve) produces a very weak band at 2100 cm^{-1} which is assigned, on the basis of previous works of gold supported on other oxides [6], to CO chemisorbed on Au^0 step sites of metallic particles. Two bands at 2151 and 2170 cm^{-1}, due to CO on Ce^{4+} cations with different co-ordinative unsaturation [7] have been observed in the carbonylic range, too.

The interaction with the CO molecules even at 90 K gives rise to the bands in the 1800-800 cm^{-1} region (not shown). Particularly, the bands observed at 1510 and 1320 cm^{-1} and at 1480, 1385, 1356, 1061 and 853 cm^{-1} are ascribed to carboxylate and carbonate species, respectively [8]. The formation of carbonate and carboxylate species is an indication of CO oxidation by the surface oxygen species and also of a concomitant ceria reduction even at 90 K.

Quite unexpectedly, the band at 2100 cm^{-1}, related to CO adsorbed on metallic gold, is very weak in comparison with the intensity observed on gold samples supported on other oxides [9]. Gold particles with size of 3 nm are no more covered by oxygen after calcination at 573 K, as revealed by a previous study [10]. Only the particles with size smaller than 2 nm are oxidized. The very weak interaction between the metal and the CO molecule can be explained by the presence of the very small gold clusters (d ≤ 1-2 nm), that are covered by oxygen species, and of the big Au particles which expose terraces unable to adsorb CO.

The intensity of the bands at 2151 and 2170 cm^{-1}, related to CO adsorbed on the Ce^{4+} sites, is stronger than the intensity of the band at 2148 cm^{-1} observed on pure ceria (Fig. 2, bold curve). This difference is an indication of the appearance of larger amounts of co-ordinatively unsaturated sites (c.u.s.) on the ceria surface when gold is deposited.

The ratio between the integrated areas of the band at 2151 cm^{-1} on Au/CeO_2 and of the band at 2148 cm^{-1} on pure ceria is 6.5, while the ratio between the BET surface areas of Au/CeO_2 and of CeO_2 is only 1.2. Therefore, the increase of intensity of the absorption bands of CO on Ce^{4+} observed on Au/CeO_2 could not be only related to the increase of the surface area but also to a gold-induced modification of ceria.

Nanosized gold particles cause modification of the surface properties of ceria and, as a result of this modification, cerium is reduced (Ce^{3+}) and uncoordinated (c.u.s.) near the very small gold clusters.

The spectra of Au/CeO_2 reduced at 373 K and at 523 K in contact with 2.5 mbar CO at 90 K are shown in Fig. 2b (fine and bold curves, respectively).

496

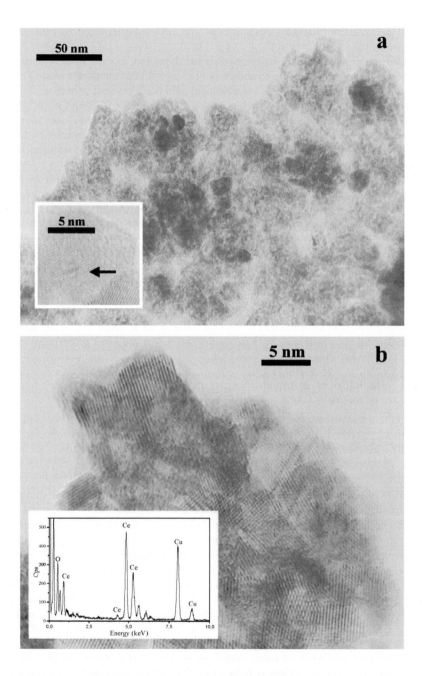

Figure 1. (a): HRTEM image of Au/CeO$_2$. Inset: detail showing a gold particle with size of 1 nm. Images taken at an original magnification of 120,000 and 800,000, respectively. (b): HRTEM image of Cu/CeO$_2$. Image taken at an original magnification of 800,000. Inset: EDS analysis of the image reported in section (b). The presence of the Au signal in the EDS spectrum is due to the employed grid.

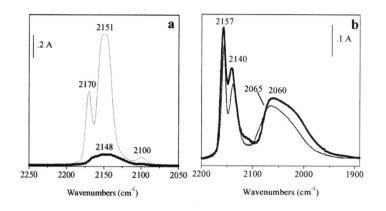

Figure 2. FTIR absorption spectra of 2.5 mbar CO adsorbed at 90 K on oxidized Au/CeO$_2$ (section a, fine curve), on oxidized CeO$_2$ (section a, bold curve), on Au/CeO$_2$ reduced at 373 K (section b, fine curve) and on Au/CeO$_2$ reduced at 523 K (section b, bold curve) in the carbonylic region. All the spectra have been normalised on the weight of the pellets.

The band observed at 2157 cm^{-1} could be assigned to CO adsorbed on Ce^{3+} sites [11]. The band at 2140 cm^{-1} showed low stability to the outgassing at 90 K and it was not detected at RT. This band corresponds to liquid-like CO. The broad band at 2060 cm^{-1} can be assigned to CO adsorbed on Au$^{\delta-}$ sites exposed at the surface of very small gold clusters interacting with Ce^{3+}. The formation of negatively charged gold sites after reduction has already been observed by us for Au/α-Fe$_2$O$_3$ and Au/TiO$_2$ [6]. In the case of Au/CeO$_2$, HRTEM images and EDS analysis reveal that a fraction of gold is highly dispersed. As previously reported, these very small clusters are covered by adsorbed oxygen at the end of the calcination treatment in air. By reduction in H$_2$, the oxygen adsorbed on the gold clusters and some of the surface oxygen atoms of ceria react, giving rise to the formation of water and oxygen vacancies and/or Ce^{3+} defects on ceria. The presence of these defects allows an electron transfer from the support to the gold particles and leads to the localisation of electron density on small gold clusters. This phenomenon is more evident for Au/CeO$_2$ reduced at 523 K, as shown by the enhanced intensity and by the red-shift of the band at 2060 cm^{-1} (Fig. 2b, bold curve).

The integrated area of the absorption at 2060 cm^{-1} (Fig. 2b, bold line) is 18.31. This value is significantly larger (more than 20 times) than the value obtained for the band at 2100 cm^{-1}, 0.84, (Fig. 2a, fine line). This difference appears as a confirmation that the small gold clusters result covered by adsorbed oxygen after the oxidation treatment and that the terraces exposed by the big particles are not able to adsorb CO.

The spectra of CO adsorbed at 90 K on oxidized Cu/ceria are reported in fig. 3 in the 2200-2000 cm^{-1} region (section a) and in the 1800-800 cm^{-1} range (section b).

Bands at 2172, 2152 and 2089 cm^{-1} are present after CO adsorption (Fig. 3a). The band at 2089 cm^{-1} can be attributed to carbonylic species adsorbed on metallic copper particles.

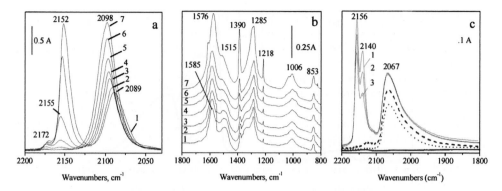

Figure 3. FTIR absorption spectra of oxidized Cu/CeO$_2$ in contact with 4 mbar CO during a gradual increase of the temperature from 90 up to 300 K (curve 1-7) in the carbonilyc region (section a) and in the carbonate region (section b) and of reduced Cu/CeO$_2$ in contact with 4 mbar at 90 K (curve 1); 2.5 mbar at 90 K (curve 2); 1 mbar at 90 K (curve 3); 4 mbar at RT (dashed curve); after outgassing at RT (bold curve) and after readmission of 4 mbar CO at RT (dotted curve).

According to literature data, concerning IR investigation of ceria-based samples, the bands at 2172 and 2152 cm^{-1} are ascribed to CO adsorbed on-top on Ce^{4+} sites of various coordinative unsaturation [12]. During the gradual heating of the sample up to RT in the presence of 4 mbar CO the bands at 2152 and 2172 cm^{-1} simultaneously disappear (Fig. 3a, curve 7). The band at 2089 cm^{-1} blue-shifts to 2098 cm^{-1} and its intensity strongly increases. The rise in intensity of this band can be connected with the formation of more Cu^0 sites as a result of the reduction of the very small CuO$_x$ particles that are highly dispersed on the surface.

The intensity of the band at 2098 cm^{-1} decreases nearly twice when the gas phase is evacuated at RT for 15 min (not shown). The partial irreversibility to the outgassing indicates that a fraction of the sites produced by the increase of the temperature is represented by Cu$^+$ whose carbonyls are known to show significantly higher stability than Cu^0 carbonyls. Obviously, the reversible component of the band at 2098 cm^{-1} is related to CO on Cu^0 sites. Finally, bands at 1585, 1500, 1390, 1340, 1285, 1218, 997, 853 cm^{-1} grow up simultaneously in the low frequency region after CO adsorption at 90 K (Fig.3b, curve 1). The formation of different carbonate, bicarbonate and inorganic carboxylate structures is an indication of fast Cu^{2+} reduction even at low temperature.

The spectra of Cu/ceria reduced at 473 K in contact with different CO pressures at 90 and 300 K in the carbonylic stretching region are shown in Fig. 3, section c. The admission of CO at 90 K produces bands at 2067, 2140 and 2156 cm^{-1}. The band at 2067 cm^{-1} can be assigned, analogously to reduced Au/CeO$_2$, to the appearance of similar negatively charged copper species after reduction in H$_2$. Bands with similar low frequencies have been observed on reduced Cu/ZnO [13]. The authors have suggested that under very reducing atmospheres oxygen vacancies are formed at the surface of ZnO. These sites have a relatively higher electron density. A study of Li et al. has showed that the surface oxygen of ceria is substantially weakened by the presence of copper and its reduction temperature is lowered by several hundred degrees [14]. Our experiments confirm also the easy reducibility of ceria in the presence of copper. In fact, after reduction, we observed a well-defined weak band at 2127 cm^{-1} in the spectrum of Cu/ceria reduced at 473 K. According to Binet et al. [7] the band arises from the forbidden $^2F_{5/2} \rightarrow {}^2F_{7/2}$ electronic transition of Ce^{3+} located at subsurface defective lattice sites. An electron transfer from these defective sites can explain the frequency shift of the Cu-CO band.

The decrease of the CO pressure from 4 to 1 mbar at 90 K (Fig. 3c, curves 1, 2, and 3) leads to the simultaneous decrease of the intensity of the bands at 2156 and 2140 cm^{-1}. They completely disappear at 300 K (dashed curve). On the contrary, the band at 2067 cm^{-1} does not show any shift; it only slightly decreases at RT. The outgassing at the same temperature causes the disappearance of this band (bold curve) and a band at the same frequency and with almost the same intensity is produced after the readsorption of CO at this temperature. This is a very important feature and indicates that copper, highly dispersed on ceria, is very stable, even more than gold.

In fig. 4 the absorption spectra of CO adsorbed at 90 K on oxidized (fine curve) and reduced (bold curve) Ag/ceria catalyst and on pure ceria (dotted curve) are reported. An intense band at 2149 cm^{-1} and a weaker one at 2168 cm^{-1} are observed in the spectrum of the oxidized catalyst. A very weak band at 2100 cm^{-1} could be seen, too. On the basis of our previous results concerning Au/ceria and Cu/ceria, the bands at 2149 and at 2168 cm^{-1} can be assigned to CO adsorbed on Ce^{4+} sites. The increased intensities of the bands in respect to the ones on ceria alone could be an indication, as on the previously discussed samples, of a modification of the support by the metal deposition procedure. However, a contribution to the band at 2168 cm^{-1} of CO adsorbed on Ag$^{\delta+}$ sites cannot be excluded.

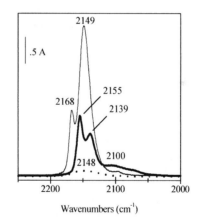

Figure 4. FTIR absorbance spectra of 5.0 mbar CO adsorbed at 90 K on oxidized Ag/CeO$_2$ (fine curve), reduced Ag/CeO$_2$ (bold curve) and on ceria alone (dotted curve) in the carbonylic region.

The inlet of CO at 90 K on Ag/CeO$_2$ catalyst reduced at 523 K (bold curve) produces in the carbonylic region a strong band at 2155 cm^{-1} that can be assigned to CO adsorbed on Ce^{3+} sites; another one at 2139 cm^{-1} due to liquid-like CO and a broadening from the low frequency side. The reduction of the CO pressure causes a simultaneous decrease of the intensity of all the components and at RT only a very weak band at 2121 cm^{-1} is present (not shown). The broad and weak absorption at lower frequencies can be tentatively related to Ag^0 sites, possibly stabilised by ceria and able to adsorb small amounts of CO.

4. CONCLUSIONS

The following conclusions may be drawn:
- The deposition-precipitation of IB metals on ceria causes strong modification of the support which leads to the appearance of more co-ordinatevely unsaturated sites on the surface.
- The reduction in hydrogen of Au/ceria and Cu/ceria catalysts results in the formation of negatively charged gold or copper sites and ceria defects on the surface. An electron transfer from the oxygen vacancies of the support to the very small metallic clusters occurs.
- FTIR spectra of CO adsorbed on oxidized Cu/ceria show an easy generation of metallic copper sites even at 90 K. CO molecules are adsorbed on highly dispersed Cu^0 sites exposed on the surface of Cu/ceria catalyst.
- The Ag/ceria catalyst is almost completely unable to adsorb and activate CO molecules due to the very low strength of the Ag-CO bond.

ACKNOWLEDGEMENTS

T. Tabakova gratefully acknowledges a NATO Outreach Fellowship granted by CNR, – Italy. M. Manzoli, A. Chiorino and F. Boccuzzi gratefully aknowledge the Ministero dell'Istruzione, dell'Universita' e della Ricerca Scientifica for the financial support in the project "Materiali multifunzionali nanostrutturati con migliorata attività fotocatalitica" (PRIN 2003)."

REFERENCES

References
[1] M. Haruta, M. Date, Appl. Catal. A, 222 (2001) 427.
[2] M. Haruta, N. Yamada, T. Kobayashi, S. Iijima, J. Catal., 115 (1989) 301.
[3] Ch. Corti, R. Holliday, D. Thompson, Gold Bull., 35 (2002) 111.
[4] D. Andreeva, V. Idakiev, T. Tabakova, L. Ilieva, P. Falaras, A. Bourlinos, A. Travlos, Catal. Today, 72 (2002) 51.
[5] F. Boccuzzi, A. Chiorino, J. Phys. Chem., 100 (1996) 3617.
[6] F. Boccuzzi, A. Chiorino, M. Manzoli, D. Andreeva, T. Tabakova, J. Catal., 188 (1999) 176.
[7] C. Binet, M. Daturi, J. C. Lavalley, Catal. Today, 50 (1999) 207.
[8] F. Bozon-Verduraz, A. Bensalem, J. Chem. Soc., Faraday Trans., 90 (1994) 653.
[9] F. Boccuzzi, A. Chiorino, Stud. Surf. Sci. Catal., vol. 140 (2001) 77.
[10] F. Boccuzzi, A. Chiorino, M. Manzoli, P. Lu, T. Akita, S. Ichikawa, M. Haruta, J. Catal., 202 (2001) 256.
[11] A. Badri, C. Binet. J. C. Lavalley, J. Chem. Soc., Faraday Trans., 92 (1996) 1603.
[12] A. Bensalem, J.-C. Muller, D. Tessier, F. Bozon-Verduraz, J. Chem. Soc., Faraday Trans., 92 (1996) 3233.
[13] N-Y. Topsoe, H. Topsoe, J. Mol. Cat. A, 141 (1999) 95.
[14] Y. Li, Q. Fu, M. Flytzani- Stephanopoulos, Appl. Catal. B, 27 (2000) 179.

Studies in Surface Science and Catalysis 155
A. Gamba, C. Colella and S. Coluccia (Editors)

Conjugated molecules in nanochannels: nanoengineering for optoelectronics

R. Tubino[a], E. Fois[b], A. Gamba[b], G. Macchi[a], F. Meinardi[a] and A. Minoia[a,b]

[a]INFM and Dipartimento di Scienza dei Materiali, Università di Milano Bicocca, via R.Cozzi 53, I-20125 Milano (Italy)

[b]DSCA, Università dell'Insubria and INSTM, Via Lucini 3, I-22100 Como, (Italy)

1. INTRODUCTION

Close-packed confinement of conjugated oligomers within parallel-aligned nanochannels of oxide hosts permits the possibility of addressing a number of novel solid-state photophysical and electronic properties. They arise from the peculiar supramolecular architecture, imposed by the host system, in which the guest molecules are organized in arrays of weakly interacting molecular wires.

The properties of organic semiconductors consisting of films of polyconjugated molecules depend on both the chemical species of the constituting units and their arrangement in the crystal structure. The methods of crystal engineering allow to design and control, in the solid state, the supramolecular architecture thus allowing for a fine tuning of the macroscopic properties. Besides substrate evaporation, another way of producing molecular assemblies with controlled spacial organisation is the incorporation of active molecules (light emitters, non-linear optics (NLO) molecules...) into a proper host matrix [1]. In particular, the incorporation of polyconjugated molecules possessing a 'long' molecular axis in the nanochannels of suitable host compounds yield a supramolecular structure in which the arrangement of the guest molecules is dictated by the surrounding host. The nanochannel diameter is the parameter that plays a key role in determining the properties of the host-guest compounds since it dictates the number and size of molecules that can be accommodated within each channel. The aim of this paper is to briefly review the fabrication process, the structure and the optical properties of inclusion compounds, in which the guest is constituted by rod-like polyconjugated molecules inserted into a variety of nanochannel-forming inorganic oxides.

2. MOLECULAR EXCITONS

The optical properties of conjugated molecules generally depend on their state of aggregation since intermolecular interactions modify their energy level position. The intermolecular excited state coupling is given by:

$$\Delta = \iint \rho_A(r_1)\rho_B(r_2)\left(|r_1 - r_2|\right)^{-1} dr_1 dr_2 \qquad (1)$$

502

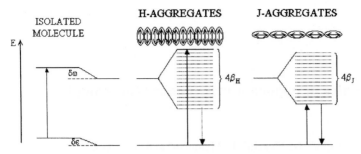

Fig.1. Schematic view exciton formation in molecular aggregates

where ρ_A and ρ_B are the transition densities of neighbouring molecules at r_1 and r_2. This expression can be exactly evaluated if the molecular wavefunctions are known. In the case of intermolecular distances much larger than the length of the conjugated molecule, this equation is well approximated by a simple dipole-dipole interaction term [2, 3].

The theory of molecular excitons, initially introduced by Frenkel [4], has been mainly developed by Peierls [5], Wannier [6] and Davydov [7]. For a good discussion on the optical properties of molecular crystals the reader should consult the book of Pope [8] and references therein. In the following, we will focus only on the implications of the molecular excitons theory that are useful in the comprehension of the optical properties of the hybrid nanostructures discussed here.

In the simple case of an aggregate of N molecules in a one-dimensional lattice with M molecules per unit cell, by considering only nearest-neighbour interactions and neglecting surface effects by applying periodic boundary conditions, the localised electronic levels of the individual molecules (which are degenerate at large intermolecular distances) are modified as shown schematically in Fig. 1, according to the relation [8]:

$$E(k) = E_0 + \delta\omega_0 + 2\beta\cos(kd) \qquad (2)$$

where k is the exciton wavenumber, E_0 the molecular electronic transition energy, d the nearest-neighbour separation, β the interaction energy between translationally equivalent neighbouring molecules and $\delta\omega_0 = \delta\varepsilon\text{-}\delta\omega$, as shown in Fig. 1, represents the difference in interaction energy between the molecule and the surrounding upon molecular excitation. The ground state shifts by an amount $\delta\varepsilon$, whereas the excited state shifts by $\delta\omega$ and splits into a band of N states of 4β width.

In Fig. 1 the dipolar interaction approximation has been used for pictorial clarity. For a parallel dipole arrangement (the so-called H-aggregate) only the electronic transition between the ground state and the upper level of the exciton band is dipole-allowed, whereas for a head-to-tail dipole arrangement (the so-called J-aggregate) only the transition between the ground state and the lowest level of exciton band is allowed. As a consequence in H-aggregate, when the intermolecular distances are close enough to give strong intermolecular interactions, the optical absorption is blue-shifted with respect to the isolated molecule and the transition from the bottom of the exciton band, from which according to the Kasha's rule [9] emission takes place, is dipole-forbidden. On the contrary, the molecular exciton model predicts that in a J-aggregate both absorption and emission are red-shifted with respect to the isolated molecule with allowed transitions to (and from) the bottom of exciton band. In general the crystal phase of rod-like conjugated chromophores, such as for example

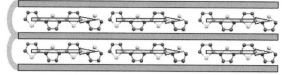

Fig. 2. Pictorial view of a dye-loaded nanochannel structure. The horizontal arrows on molecules show the molecular transition dipole moments.

oligophenylenes, oligothiophenes, oligophenylene-vinylenes…, which spontaneously forms by solvent evaporation or vapour deposition, is the so-called herring bone structure, which consists of weakly interacting planes in which the molecules are organized in a two-dimensional H-aggregate with the transition dipoles perpendicular to the herringbone plane. This type of structure, in which the close packing among the molecules allows large charge carrier mobility - usefully exploited in the fabrication of organic transistors [10] - has the drawback of being weakly emissive even when the constituting molecules have large luminescence quantum yield when isolated. For this reason, in order to obtain emissive solid-state aggregates, one has to fabricate supramolecular structure with a J-type aggregation.

Fig. 2 shows a pictorial view of a hybrid nanostructure obtained by inserting an organic chromophore into the parallel nanochannels of a suitable host. It is expected (and indeed it has been found) that the luminescence quantum yield of this host-guest system will be greatly enhanced by this peculiar supramolecular organization consisting of linear arrays of rod-like active molecules.

3. FABRICATION AND SPECTROSCOPIC PROPERTIES OF INCLUSION COMPOUNDS

3.1 Host matrices

In order to constrain molecules in a one-dimensional fashion, suitable host matrices are needed. In Tab. 1, the porous oxides we used for inclusion compounds preparation are listed. We focussed our attention to hosts with small-sized (< 10 Å) nanochannel system as zeolite LTL, AlPO$_4$ and ETS-10: they offer a different chemical and geometrical environment for guest dyes (presence/lack of free cations, shape of channels, etc...). Otherwise, mesoporous silicate MCM-41 is useful to include larger molecules (e.g. porphyrins, see Section 3.2), having a tailorable pores diameter, up to hundreds of Angströms.

Table 1
Host matrices chemical and geometrical properties

	Chemical formula	Channel dimension (Å)
LTL	$K_9Si_{27}O_{72}Al_9$	7.1 x 7.1[1]
ETS-10	$(Na_{12}K_4)Si_{40}Ti_8O_{104}$	4.9 x 7.6
MCM-41	SiO_2 amorph.	36 x 36
AlPO$_4$-5	$Al_{12}P_{12}O_{48}$	7.3 x 7.3

[1] diameter can reach 12.1 Å inside the channel

3.2 Guest chromophores

Tab. 2 summarizes the properties of guest molecules employed in our work. Among rod-like dyes, oligothiophenes (OTs) [11, 12] appear particularly interesting because of their

Table 2
Guest chromophores properties

	Molecular structure	λ_{abs} (eV)	λ_{em} (eV)	Dimensions (Å)
OTs		-[1]	-[1]	(> 7) x 4
DMe4T		3.26 (380 nm)	2.58 (481 nm)	14 x 6
Anthracene		3.47 (357 nm)	3.10 (400 nm)	9 x 4
TPPS		2.87 (432 nm)	1.93 (642 nm)	14 x 14

[1]These values are strictly related to the conjugation length, i.e. the number of thiophene units.

applications in photonic and non-linear optics [13]. 3,3'''-dimethyl-quaterthiophene (DMe4T) is an interesting case study concerning torsionable linear molecules: the two methyl groups at the external thiophene rings give to the molecule more degrees of freedom, letting it to assume different conformational structures. Otherwise, anthracene, due to the aromatic π-bond extending over its whole length, is a fully planar dye. Tetrakis(p-sulfonatophenyl) porphyrin (TPPS) is a larger discoid conjugated system, both of scientific and applicative interest (electron transfer and light harvesting phenomena in biological functions involve monomeric and aggregated porphyrins). It is therefore possible to exploit these different kinds of chromophores as optical probes in order to investigate host-guest interactions.

3.3 Inclusion strategies

The choice of inclusion technique is related to the thermal properties of the guest molecule and to its diffusion rate within the host channels: inorganic oxides can easily tolerate higher temperature than organic compounds which can undergo thermal degradation.

In order to achieve the best control of chromophore loading, we performed the fabrication of inclusion compounds by a sort of physical vapour deposition exploiting the high surface area of inorganic host matrices (up to ~1000 m^2/g for MCM-41 and ~600 m^2/g for LTL). The host powder and the guest molecules are loaded in two separate chambers and dried in mild vacuum; then they are put into contact and heated until the chromophore starts to sublimate and the porous oxide reaches a homogenous colour [14]. The sample is then repeatedly washed to remove the molecules adsorbed on the outer surfaces of the host.

3.4 Spectroscopic properties

Photoluminescence (PL) and optical absorption spectroscopy are useful tools for investigating the intermolecular (e.g. aggregates formation) and host-guest interactions. Fig. 3 shows PL

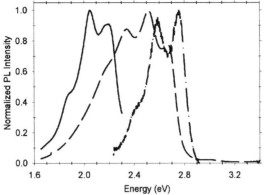

Fig. 3. Comparison between normalized PL spectra of DMe4T: polycrystalline powder (solid line), $\sim 10^{-5}$ M solution in chloroform (dash-dotted line) and LTL inclusion compound (dashed line) collected at 77 K.

spectra of DMe4T loaded zeolite LTL compared to isolated molecule in dilute solution and polycrystalline powder. DMe4T is known to crystallize in an herringbone fashion, an example of two-dimensional H-type aggregation, where interactions between transition moments lead to the formation of a wide excitonic band (see Section 1): the purely electronic transition between the bottom of excitonic band and the ground state is dipole-forbidden; however, transitions toward the vibronic levels of the former are generally allowed. Accordingly to the theory described above, polycrystalline powder PL spectrum is shifted to lower energy in respect to that of isolated molecule and shows a well-resolved vibronic progression. In the case of non-interacting molecules dispersed in an inert matrix, emission of guest chromophores is expected to be almost identical to that of isolated molecules in dilute solution; on the contrary, DMe4T-loaded LTL PL spectrum is red-shifted of ~ 0.1 eV, being the mark of some sort of host-guest interactions, e.g. reduced conformational freedom, dipolar or atomic-specific interactions, aggregation effect, etc... A deeper understanding of these phenomena is needed for a successful tuning of host-guest systems optical properties.

4. CONFORMATIONAL ANALYSIS OF OLIGOTHIOPHENES INCLUDED IN ZEOLITES

Apart from the extent of the intermolecular coupling previously discussed, also the conformation of the active conjugated molecule inside the nanochannel controls the wavelength of absorption and emission by the chromophore. As a matter of fact deviation from planarity reduces the degree of π electron delocalization, thus producing an increase of the HOMO–LUMO separation. The possibility of predicting the changes in conformation of the chromophores and their mutual interactions upon their insertion in the oxide framework is therefore essential to control the properties of the resulting hybrid nanostructure.

In the following we will provide a brief outline on how a combination of *ab initio* and classical molecular dynamics approaches can provide great support in understanding the electronic properties and dynamical processes of the active molecules, notably oligothiophenes, inside the matrix. We will focus our attention on the calculation of the

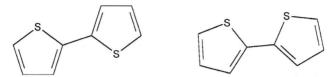

Fig. 4. Trans-planar (right) and cis-planar (left) conformations for 2,2'-bithiophene.

conformation of the 2,2'-bithiophene (2T), that is depicted in Fig. 4 in its *trans-planar* and *cis-planar* conformation.

The complexity of the host-guest system and the necessity to study its equilibrium properties required the use of classical molecular dynamics technique (MD) instead of *ab initio* calculation.

We have employed the CVFF augmented central valence force field (FF)[15] to simulate the isolated molecule, the zeolite and the host-guest system. All atoms were free to move in the simulations. 2T was considered a rigid body, with the torsion angle the only the degree of freedom. Coulomb interaction[15] were calculated via the Ewald method. The torsional part of the 2T FF was modified, as described below. Molecular dynamics simulations were performed by using the DL_POLY package[16].

4.1 Isolated molecule
The 2T torsion energy profile in gas phase is well known, being under study at least by a decade [17]: the absolute minimum corresponds to a trans-distorted (also known as anti-gauche) conformation, with a torsional angle (S-C-C-S) about of 150° with a statistical weight about of 56%, at the temperature of 371 K. There is also a relative minimum, cis-distorted (syn-gauche) for a torsion of about of 40°, with statistical weight about of 44% and with energy of 0.75 kcalmol^{-1} above the energy of absolute minimum.

Several force field (FF) for MD, such as MM2 [18], Dreiding [19], UFF [20] and AMBER [21] have been developed during the years, and they all can predict quite well the most stable conformations for isolated molecules, however they badly reproduce the torsion barrier. To obtain a reliable torsion potential, which is central to predict the conformational properties, we have carried out *ab initio* calculations for the energy profile at B3LYP 6-31G** [22] level of theory, as a function of the torsion angle θ. The calculated barrier is 2.05 kcalmol^{-1} for the cis/trans conversion, while it amounts to 2.70 kcalmol^{-1} for the trans/cis. From the *ab initio* energies we have subtracted the contribution to the energy due to (non torsion) atom-atom interactions as calculated via the CVFF force field. The calculated differences were fitted, as function of θ, to the Fourier series of Eq. (3), truncated at the 5th term :

$$V(\vartheta) = \sum_{n=0}^{4} V_n (1 + \cos(n\vartheta)) \qquad (3)$$

We have produced two different force fields for 2T, one completely apolar, and one polar FF, where point charges, obtained from the *ab initio* calculations, were included.

4.2 LTL Zeolite framework

We have chosen as host system the zeolite LTL that is actually used in the fabrication of inclusion compounds, since it contains parallel nanochannel running through the length of the zeolite microcrystals. Zeolite LTL, whose structure is depicted in Fig. 5, is a framework aluminosilicate that has general composition $Na_3K_6[Al_9Si_{27}O_{72}](H_2O)_{21}$.

The crystallographic cell is hexagonal (a=b=18.40Å, c=7.52Å). Extra framework cations are present to balance negative charges introduced on the framework by aluminium atoms. In all our calculations, with periodic boundary conditions, we have simulated only the dehydrated zeolite. Indeed, dehydration is an important step in the preparation of host/guest composites.

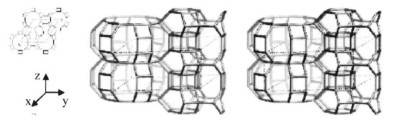

Fig. 5. Lateral view of the schematic structure for the zeolite LTL. Elliptical cells, repeated along the Z direction, form the channels. The elliptical cells have the smaller diameter about of 7 Å and the larger one of about 12 Å.

Moreover we have simulated systems where extra framework cations are all potassium ones. It is important to notice that these charges produce a strong local electric field within the zeolite nanochannels. As the channels develop along the c axis, we have adopted simulation cell that are multiple of the crystallographic cell along such direction, therefore we have simulated only one independent channel.

4.3 Host-guest system

While the gas phase (at 371 K) behaviour of 2T is known, its behaviour when confined in zeolitic channels is yet unknown. We have performed a series of simulations, at different temperatures, both for the isolated 2T molecule and for 2T encapsulated in LTL channel. Moreover the effects of the zeolitic electric field have been checked by comparing the polar and the non polar 2T force fields. We have monitored the torsion behaviour of 2T by inspecting the S-S distance along the trajectories. A value of such a distance close to 4.5Å corresponds to a trans–planar configuration while a value of 3.2 Å indicates a cis-planar geometry.

For the isolated molecule both FF predict the existence of two conformers characterized by θ close to the experimental ones with the correct distribution of the population of the two stable conformers. As a first step in studying properties of confined 2T we have studied the low loading regime, where only host/guest interaction are relevant. When simulating the 2T/zeolite system, periodic boundary conditions have been adopted. In order to minimize the interaction of 2T with its image (guest/guest interactions), we have inserted one 2T molecule inside a hexagonal simulation box of (a=b=18.40Å, c=6×7.52Å).

Upon insertion of 2T into the charged zeolite framework, the effects of the atomic charges are quite noticeable. Both the force fields predict a most stable s-gauche-trans conformation with a torsional angle of about 160°, corresponding to a S-S distance about of 4.5Å, for 2T when it is included in LTL zeolite, but they predict a different molecular behavior: polar force field, in which electrostatic interactions between molecule and framework are considered, predicts

508

that at room temperature and at 371 K (that was the experimental gas temperature for electron diffraction measurements) when 2T falls in the absolute minimum it remains in s-gauche-trans conformation, as shown in Fig. 6.

Upon raising the temperature up to 500K, 2T still shows an absolute minimum in s-gauche-trans, but the conformation switch is now allowed, because the molecule has enough energy to overcome the cis/trans inter-conversion energy barrier. This behavior is in contrast with results obtained with the neutral FF, which indicate that 2T can overcome the rotational barrier also at room temperature, as displayed in Fig. 7.

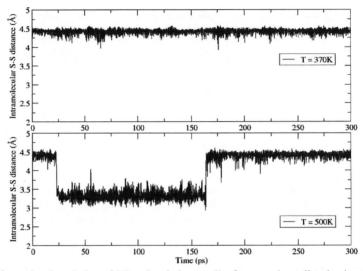

Fig. 6. Conformational evolution of 2T molecule in a zeolite framework predicted using the polar FF. Charges inhibit the intramolecular torsion at room temperature and up to 500 K.

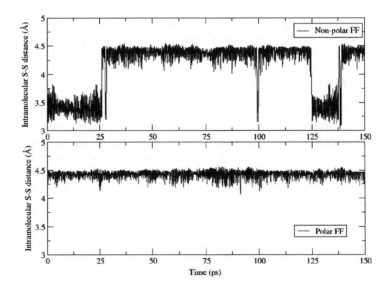

Fig. 7. Different conformational behaviors for 2T predicts form the two FF.

Starting by the syn-gauche conformation, simulations for the included molecule, performed with polar FF, shows a fast cis/trans inter-conversion (Fig. 8): this suggests that the interactions between the internal electrical field of zeolite and the molecule destabilize the relative minimum and stabilize the absolute one respect to the isolated molecule case. Only at temperature of 100K there is not change in the molecule conformation that rests in the relative minimum.

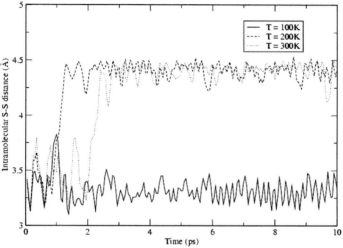

Fig. 8. Starting by syn-gauche conformation, a quick cis/trans interconversion occurs and the molecule rest held in gauche-trans geometry. This suggests that the relative minimum is destabilized while the absolute one is stabilized by the confinement.

In summary the effects of the zeolite charged framework on the bithiophene is to stabilize the trans conformation, hampering the conformation mobility, observed in the gas phase and in solution.

5. CONCLUSIONS

The fabrication of hybrid organic-inorganic nanostructures in which active molecules are organized in linear arrays inserted into the nanochannel of a suitable host is very promising to overcome the problem of spontaneous H aggregation, responsible for the suppression of the luminescence in solid state aggregates. Since a concentration of active molecules up to 10 % can be obtained by loading an inorganic host, this result is achieved by keeping high the molecular density (comparable to a "normal" crystal).

Novel phenomena arising from this one-dimensional ordering of the chromophores (molecular nanowires) are also expected. They include:
- superradiant emission [23] originating from a coherent coupling of the molecular transition moments of the collective state at the bottom of the exciton band. The resulting ultrafast radiative decay rate can be exploited in the fabrication of micro-lasers [24]

- one dimensional photonic energy transfer to funnel the electronic excitation along the nanochannel of the inclusion compound [14,25]
- macroscopic NLO response exploiting the head-to-tail arrangements of suitable push-pull molecules within the channel.

Acknowledgements
The precious help of Arantzazu Zabala Ruiz for the synthesis of zeolite-L microcrystals is gratefully acknowledged.

REFERENCES
[1] D.F.Eaton, A.G.Anderson, W.Tam, Y.Wang, J.Am.Chem.Soc., 109 (1987) 931.
[2] E.G. McRae, M. Kasha, Physical Processes in Radiation Biology, Academic Press, New York, 1964.
[3] M. Kasha, Radiat. Res., 20 (1971) 55.
[4] J. Frenkel, Phys. Rev., 37 (1931) 1276.
[5] R. Peierls, Ann. Phys., 13 (1932) 905.
[6] G.H. Wannier, Phys. Rev., 52 (1937) 191.
[7] A.S. Davydov, Zh. Eksp. Teor. Fiz., 18 (1962) 210.
[8] M. Pope, C.E. Sweneberg, Electronic Processes in Organic Crystals, Clarendon Press, Oxford, 1982.
[9] M. Kasha, Disc. Faraday Soc., 9 (1950) 14.
[10] Z. Bao, Adv. Mater., 12 (2001) 227.
[11] D. Fichou, J.M. Nunzi, F. Charra, N. Pfeffer, Adv. Mater., 6 (1994) 64.
[12] D. Fichou (Ed.), Handbook of Oligo- and Polythiophenes, Wiley-VCH, 1999.
[13] T.A. Skotheim, Handbook of Conducting Polymers, Marcel Dekker, New York, 1986.
[14] M. Pauchard, A. Devaux, G. Calzaferri, Chem. Eur. J., 6 (2000) 3456.
[15] J-R. Hill, C.M. Freeman, L. Subramanian, Reviews in Computational Chemistry, Wiley-WHC, New York 2000, 141.
[16] W. Smith, T.R. Forrester, DL_POLY 2.13 CCLRC, Daresbury Laboratory, Warrington UK, 2001
[17] S. Samdal, E.J. Samuelsen and H.V. Volden, Synth. Met., 59 (1993) 229.
[18] J.C. Tai, J. Lii, N.L. Allinger, J. Comp. Chem., 10 (1989) 635.
[19] S.L. Mayo, B.D. Olafson, and W. A. Goddard, J. Phys. Chem., 94 (1990) 8897.
[20] A.K. Rappé, C.J. Casewit, K.S. Colwell, W.A. Goddard III, and W.M. Skiff, J. Am. Chem. Soc., 114 (1992) 10024.
[21] W.D. Cornell, P. Cieplak, C.I. Bayly, I.R. Gould, K.M. Merz Jr., D.M. Ferguson, D.C. Spellmeyer, T. Fox, J.W. Caldwell, and P.A. Kollman, J. Am. Chem. Soc., 117 (1995) 5179.
[22] A.D. Becke, J. Chem. Phys., 98 (1993) 5648.
[23] F.C. Spano, J. Chem. Phys., 12 (2001) 5376.
[24] Ö. Weiss, J. Loerke, U. Wüstefeld, F. Marlow, F. Schüth, J. Solid State Chem. 167 (2002) 302.
[25] G. Calzaferri, S. Huber, H. Maas, C. Minkowski, Angew. Chem. Int. Ed. 42 (2003) 3732.

Studies in Surface Science and Catalysis 155
A. Gamba, C. Colella and S. Coluccia (Editors)

Effect of anilines as a synthesis component on the hydrophobicity of silica

A. C. Turallas[a], G. Romanelli[a,b], P. Vázquez[a]

[a]Centro de Investigación y Desarrollo en Ciencias Aplicadas "Dr. Jorge J. Ronco" (CINDECA), 47 N° 257 (B1900AJK) La Plata (Argentina).
e-mail: vazquez@quimica.unlp.edu.ar

[b]Laboratorio de Estudio de Compuestos Orgánicos (LADECOR), 1 y 115 (B1900AJK) La Plata (Argentina).

1. INTRODUCTION

In the field of polymeric systems an important goal is to reinforce the polymer matrix and to improve mechanical and thermal properties. Inorganic additives are introduced into polymer systems as fillers or as reinforcing agents. Among the hybrid organic-inorganic powdered materials, the silica receives considerable interest because different degrees of hydrophobic/hydrophilic are determining its dispersion in the polymeric matrix. The synthesis and characterization of hybrid organic–inorganic powdered materials have received great attention in past years. These materials present properties related to the matrix as well as the organic phase, which can be combined in order to obtain new materials. The sol–gel method is an important route to prepare such materials using alkoxysilanes R-Si(OR)$_3$ and tetraethylorthosilicate (TEOS) as precursors. The polycondensation of alkoxysilanes can be described in three steps: (i) hydrolysis, (ii) silanol condensation and (iii) silanol–alcohol condensation.

The relative simplicity and versatility of the sol–gel process, when compared to covalent-bonding methods to obtain hybrid materials, associated with low cost, are responsible for its extensive utilization. The possibility of obtaining different physicochemical characteristics like surface area, particle shape and size, porosity and organic functionalization grade, are some advantages of this process [1-3]. However, a little change in the parameter conditions of the synthesis, like solvent, catalyst or silane amount, can produce great modifications in the final properties of the materials. In this context, the study of the synthesis conditions and the related resulting properties are very important to understand these systems [4].

The silica can have hydrophobic properties when its surface is functionalized. This process requires the presence of silanol groups at the surface. There are many types of silica differing in their surface properties, in particular, the density of silanol groups and the types of silanols (isolated, hydrogen bonded, geminal) vary by techniques of preparation of silica. The purpose of this investigation is to know the grafting process on silica prepared by sol-gel. The principle of the investigation work consists in comparing the grafting of anilines, with different active groups, on silica to obtain different degree of the surface hydrophobicity. For a better understanding of these solids, the characterization of the nature of the species was determined by FT-IR, XRD, DTA-TGA and surface area (S$_{BET}$).

512

2. EXPERIMENTAL
2.1. Silica preparation by sol-gel technique
All the experiences were made with final molar ratio of TEOS/EtOH/H$_2$O equal to 1:1:4. The TEOS-EtOH-AcH sols were stirred at atmospheric pressure at room temperature (r.t.), during 30 min. Then, the hydrolysis process began with slow addition of distilled water. After the water addition, gelation of the sols was carried out at r. t. and the wet gels were then aged in the same medium until dry silica particles were obtained. These solids were washed with ethanol and distilled water. After that, they were dried at r.t. and calcined to 673, 873 and 1073 K.

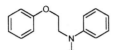

- *N-(2-phenoxyethyl)aniline*
  is named **A42**

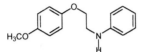

- *N-(2-(4-methoxy)phenoxy)ethyl)aniline*
  is named **A45**

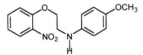

- *N-(2-(2-nitrophenoxyethyl)aniline*
  is named **A57**

2.2. Functionalization of silicas
The (grafting agents) were anilines, prepared in our laboratory [5] (Scheme 1). The functionalized silica was prepared by addition of aniline to a suspension of silica in refluxing toluene, stirred for 5 h. The solid was filtered, washed in a soxhlet apparatus with diethylether and dichloromethane and then dried, at 393 K [6].

2.3. Characterization
Textural properties. N$_2$ adsorption and desorption isotherms at 77 K were carried out using a Micromeritics Accusorb 2100 equipment. *FTIR.* A Bruker IFS 66 equipment, pellets in BrK and a measuring range of 400-4000 cm^{-1} were used to obtain the FT-IR spectra of solids. *X-ray diffraction.* The XRD patterns were obtained by using a Phillips PW-1714 diffractometer with built-in recorder.

3. RESULTS AND DISCUSSION

Sol-gel technology is a widely researched field and the hybrid organic-inorganic materials can be fabricated by carrying out the inorganic sol-gel process in presence of organic species, which are capable of interacting chemically with the metal alkoxides. "Hybrid material" means a disordered physico-chemical system that forms intricate organic and inorganic networks [7]. The degree of linking and cross-linking of the organic and inorganic networks influences the mechanical and chemical properties and can be controlled by varying, for example, the catalyst used [8]. In this work, the acetic acid was used and the influence of stirring was reported in previous paper [9].

Table 1 shows the specific surface areas (S$_{BET}$) of the samples as a function of heat treatment procedures employed. These results show a decrease of S$_{BET}$ for an increase of the temperature.

FTIR spectroscopy has been used to study the structure of the silica and hybrid material.

Figure 1 shows the spectra of the silica prepared by sol-gel at r.t. and calcined to 673, 873 and 1073 K, respectively. For silica, the broad adsorption positioned in the range 3000–3800 cm^{-1} is due to H-bonded silanols. In addition, this band is attributed to the presence of hydroxyl groups of water present on the surface, at r.t. Additionally, bands corresponding to Si-O-Si symmetric and asymmetric vibrations are located at 812 and 1104 cm^{-1}, respectively. The small shoulder at 959 cm^{-1} can be assigned to the Si-OH group. During the heat treatment procedures employed, the silica surface is progressively dehydroxylated. Numerous investigations have correlated the surface dehydroxylation with two bands at 490 cm^{-1} (called D$_1$) and 604 cm^{-1} (D$_2$) that are superimposed on the broad band at about 440 cm^{-1}, which is the most intense signal in the spectrum of v-SiO$_2$ [10].

Table 1. S$_{BET}$ of silica, obtained by sol-gel, at different temperatures

	r.t.	673	873	1073
S$_{BET}$ (m^2/g)	431	312	131	0.6

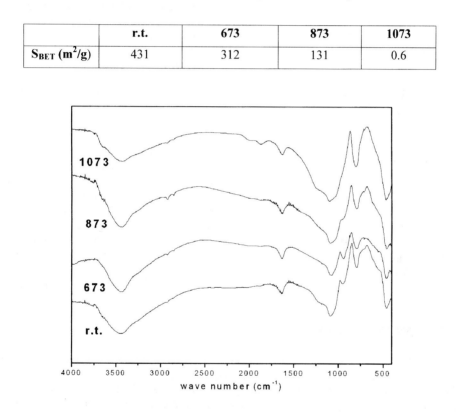

Fig.1. FTIR spectra of silica, obtained by sol-gel, at different temperatures.

In this research, representative substituted N-phenoxyethylanilines were used as functionalizating: N-(2-phenoxyethyl)aniline (**A42**), N-(2-(4-metoxy)phenoxy)ethyl)aniline (**A45**) and N-(2-(2-nitrophenoxyethyl)aniline (**A57**), respectively. The pKa for all anilines are showed in Table 2. The acid dissociation constants of the N-phenoxyethyl-ammonium ions were attained by fluorescence spectrometry [11]. The results showed that, when there is a p-methoxy group at the N-phenyl ring (**A45**), the acidity of the aryl alkyl ammonium ion is

514

lowered compared to the unsubstituted compound (**A42**). The OMe group may produce a decrease electron density at the N atom through proton chelation. The lowering acidity effect increases when the methoxy group changes its position and there is a nitro group in its place (**A57**).

Table 2. pKa of anilines

Aniline	R	R′	pKa
A42	H	H	3.83
A45	H	OMe	3.70
A57	OMe	NO$_2$	3.40

The immobilization of the aniline groups on the matrix surface silica was studied by FTIR. Spectra of silica and functionalized silica with **A42, A45 and A57** anilines, at r.t. and 393 K are presented in Figure 2.

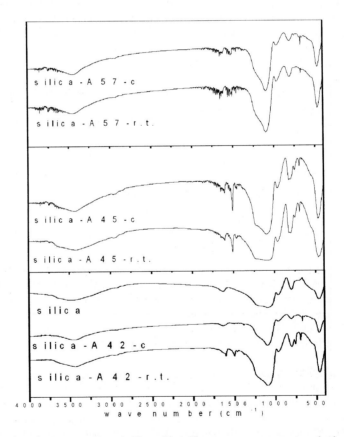

Fig. 2. FTIR spectra of functionalized silica with anilines, at room temperature (r.t.) and calcined at 393 K (c).

The main differences between functionalized silica spectra and that of the pure silica are due to: the N-H stretching band (3300-2600 cm^{-1}, broad peak, overlapped with the stretching bands of silanol of silica and -CH_2- groups), overtone band (1975 cm^{-1}). Two possible vibrational bands: asymmetrical -NH_3^+ bending band (1600 cm^{-1}) and the symmetrical -NH_3^+ bending band (1500 cm^{-1}) [12].

As the occurrence of reactions involving amino groups, with the consequent formation of new species, is unlikely, these additional components could be assigned to amino-ethyl species experiencing some peculiar interaction with the boundary. The complexity of bands due to such species indicate the presence of some heterogeneity, this behaviour could be due to constraints of amino-ethyl species in the silica gel framework [13].

The mode of anchorage of the organic compounds is of prime importance for the characteristics of the final functionalized silica. After functionalization, it was determined the degree of hydrophobicity of silica surface, with water (H_{water}) and wettability, with ethanol ($M_{Ethanol}$), respectively. Table 3 shows these properties for silica synthesized by sol-gel and functionalized with A42, A45 and A57 anilines, at r.t. and calcined (c). In according to functional groups of anilines, the hydrophobicity degree is very good for silica-A42-c but is it inexistent for silica–A57-r.t., respectively. In relation to ethanol wettability, the functionalized silica presents a different behavior that could be due to particular interaction of the surface sites with water than with ethanol molecules.

Table 3. Hydrofobicity degree (H_{water}) and Ethanol wettability ($M_{Ethanol}$)

	H_{water}	$M_{Ethanol}$
Silica	0	1
Silica A42-r.t	0.64	0.42
Silica A42-c	0.88	0.21
Silica A45-r.t	0.16	1.21
Silica A45-c	0.43	1.42
Silica A57-r.t	-0.54	2.00
Silica A57-c	0.31	1.21

In addition these results evidence the preservation of hydrophilic patches on the silica surface. This hypothesis has been taken into account during the functionalization procedure in solvent as toluene, perhaps in polar solvents and different treatments of silica surface this behavior is made by other alternative routes.

The final mechanism for our systems is in study and these results will be analyzed according to different possible mechanisms in order to discriminate between them.

ACKNOWLEDGMENTS
The authors thank to Fundación Antorchas (Project N° 4248-43) for the financial support and Tec. Graciela Valle for FT-IR analysis.

REFERENCES
[1] F. Pavan, W. Magalhaes, M. de Luca, C. Moro, T. Costa, E. Benvenutti, Journal of Non-Crystalline Solids 311 (2002) 54.
[2] R. Makote, M.M. Collinson, Anal. Chim. Acta 394 (1999) 195.

516

[3] M. Collinson, Critical Rev. Anal. Chem. 29 (1999) 28.
[4] F.Pavan, H. Hoffmann, Y. Gushikem, T. Costa, E. Benvenutti, Materials Letters 55 (2002) 378.
[5] J. Jios, G. Romanelli, J.C. Autino, D. Magiera, H. Duddeck, Z. Naturforsch. 57b (2002) 226.
[6] M. Lasperas, T. Lloret, L. Chaves, L. Rodriguez, A. Cauvel, D. Brunel, Studies in Surface Science and Catalysis (Elsevier Science B.V.) 108 (1997) 75.
[7] Z. Sassi, J.C. Bureau, A. Bakkali, Vibrational Spectroscopy 28 (2002) 251.
[8] D. Bersani, P. Lottici, M. Casalboni, P. Prosposito, Materials Letters 51 (2001) 208.
[9] N. Quaranta, P. Vázquez, M. Caligaris, E. Benavidez, Proceedings SILICA 2001 (2001).
[10] A. Burneau, B. Humbert, O. Barres, J.P. Gallas and J. C. Lavalley, "The Colloid Chemistry of Silica", Advances in Chemistry Series (Ed. H. Bergna), American Chemical Society (1994).
[11] J. Autino, L. Bruzzone, G. Romanelli, J. Jios, H. Ancinas, Anal. Quim. Int. Ed. 94 (1998) 292.
[12] J. Brinker, American Chemical Society – Advances in Chemistry Series 234 (1994) 361.
[13] G. Sartori, F. Bigi, R. magi, R. Sartorio, D. Macquarrie, M. Lenarda, L. Storaro, S. Coluccia, G. Martra, J. of Catalysis 222 (2004) 410.

Studies in Surface Science and Catalysis 155
A. Gamba, C. Colella and S. Coluccia (Editors)

Removal of cadmium, zinc, copper and lead by red mud, an iron oxides containing hydrometallurgical waste

M. Vaclavikova[a], P. Misaelides[b], G. Gallios[b], S. Jakabsky[a] and S. Hredzak[a]

[a] Institute of Geotechnics, Slovak Academy of Sciences, Watsonova 45, 043 53 Kosice, Slovakia

[b] School of Chemistry, Aristotle University of Thessaloniki, GR–54124 Thessaloniki, Greece

Red mud, a residue of the alumina production industry which is actually a solid waste, has been studied as a potential sorbent for the removal of toxic bivalent cations (i.e. Cd, Zn, Cu and Pb) from aqueous solutions in the presence of 0.01 M $NaNO_3$. The experimental data were modeled with Langmuir and Freundlich isotherms and fitted quite well. The relatively high uptake indicated that red mud can adsorb considerable amounts of cadmium and zinc from near–neutral aqueous solutions (maximum uptake capacity for cadmium: 68 $mg \cdot g^{-1}$ at pH 6 and ca. 133 $mg \cdot g^{-1}$ for zinc at pH 7). A significant uptake was also observed for copper and lead at pH 6 and 7 respectively which was attributed to precipitation of the respective insoluble hydroxides. TCLP leaching tests before and after the metal removal have shown that read mud is an environmentally compatible material that could be used for the wastewater treatment. Regeneration possibilities have also been observed.

1. INTRODUCTION

The presence of heavy metal ions in the environment is of major concern due to their toxicity to many life forms. Unlike the organic pollutants, metal ions do not degrade into harmless end–products. The increasingly stringent European legislation on the purity of drinking waters and on the concentration of contaminants in wastewaters has created a growing interest in the development of treatment processes for the removal of hazardous metals from aqueous solutions. Among the various methods proposed for this purpose adsorption proved to be of the most promising ones [1, 2]. Several natural (e.g. natural zeolites, bentonites, metal oxides) and synthetic (e.g. synthetic zeolites, resins, metal phosphates and silicates, synthetic oxides/hydroxides/hydroxyoxides) materials have been investigated as sorbents for heavy metal removal from solutions achieving different levels of success [3, 4–9]. Moreover, considerable research work has been done on various industrial waste materials in order to develop suitable sorbents for water treatment; so fly ash [10, 11], blast furnace slug [12], biomass [13, 14] and bagasse fly ash [15], among others have been tested as sorbents for heavy metal removal with various levels of success.

The aim of this work was to investigate the possibility to use red mud for the water and wastewater treatment. Red mud is an insoluble fine–grained residue of alumina production industry (Bayer process: extraction of alumina from bauxite by treatment of the ground ore with hot concentrated base) [16, 17]. The major constituents of red mud are Fe_2O_3, TiO_2, Al_2O_3, CaO, SiO_2 and Na_2O [e.g. 18–20]. Since ca. 1.0–1.5 tones of red mud are produced for each tone of alumina, millions of tones of this caustic material must annually be disposed worldwide [e.g. 21–23]. Red mud, due to its high aluminum, iron and calcium content, has been suggested as an inexpensive adsorbent for the removal of toxic metals (e.g., As, Cr, Pb, Cd, Zn), dyes and other hazardous substances from water and wastewater streams [e.g., 1, 24–33]. It has also been proposed for the immobilization of heavy metals in soils [e.g., 11, 30, 31]. The other potential applications of this interesting industrial waste include the production of building materials and ceramics, the utilization as filler in asphalt roads, the use as a source of various minerals and as iron ore [e.g., 34].

Cadmium, lead, copper and zinc were selected as test cations for the investigation of the sorption properties of red mud in this study. Cadmium and lead are elements occupying high toxicity positions in the EPA tables. On the other hand, copper and zinc are elements taking part in many biological processes but considered as toxic and hazardous to human health at elevated concentrations. The suggested safe level of copper in drinking water tends to be pegged at 1.5 to 2 $mg \cdot l^{-1}$.

2. EXPERIMENTAL

The material used for the experimental work was obtained from the alumina production plant in Ziar nad Hronom, Slovakia and characterized by chemical analysis, neutron activation analysis and powder X–ray diffraction (Powder-XRD, Philips X'Pert Pro X–ray diffractometer, CuK_α radiation). The specific surface area of the material was determined by low temperature nitrogen adsorption using a Gemini 2360 equipment, whereas a Helos and Rodos apparatus (Sympatec GmbH, Germany) with the wet dispersion base Rodos 11 SROV was utilized for the grain size analysis. The magnetic properties of the material were also investigated.

The adsorption of Cd(II)-, Zn(II)-, Cu(II)- and Pb(II)- cations from model aqueous solutions of their nitrate salts (metal concentration: 20–400 $mg \cdot l^{-1}$) was studied using batch–type equilibrium experiments (sorbent concentration: 2 $g \cdot l^{-1}$, contact time: 24 hours, temperature: 25.0±0.1°C, constant ionic strength established by 0.01 M $NaNO_3$). Analytical grade chemicals and distilled water were used for the preparation of the solutions. The pH of the solutions was adjusted with NaOH and HNO_3. The equilibrium concentrations were determined by Atomic Absorption Spectroscopy (AAS, using a Varian Spectr AA–30). The metal uptake by the sorbent was calculated using the initial and equilibrium concentrations according to the formula :

$$Q_{eq} = \frac{C_{init} - C_{eq}}{C_S}, \tag{1}$$

where Q_{eq} ($mg \cdot g^{-1}$) is the amount of heavy metals loaded per weight unit of solid red mud, C_{init} ($mg \cdot l^{-1}$) is initial concentration in solution before the treatment, C_s ($g \cdot l^{-1}$) is sorbent concentration and C_{eq} ($mg \cdot l^{-1}$) is the equilibrium concentration in solution after the treatment. The experimental data were modeled according to the well known Langmuir and Freundlich adsorption equations [35, 36].

The Langmuir adsoption isotherm, which is expressed as

$$Q_{eq} = Q_{max} \frac{KC_{eq}}{1 + KC_{eq}}, \tag{2}$$

where Q_{eq} (mg·g⁻¹) is the amount of metal sorbed per mass unit of adsorbent under equilibrium conditions, C_{eq} (mg·l⁻¹) is the equilibrium metal concentration in solution, Q_{max} (mg·g⁻¹) is the maximum adsorption at monolayer coverage and K is a constant related to energy of adsorption. This isotherm, the simplest physically plausible isotherm, is based on three assumptions: the adsorption cannot proceed beyond monolayer coverage, all sites are equivalent and the surface is uniform and the ability of a molecule to adsorb at a given site is independent of the occupation of neighboring sites. On the other hand, the Freundlich isotherm takes into account a logarithmic fall in the enthalpy of adsorption with the surface coverage. It can describe the equilibrium on the surface on heterogeneous surfaces of sorbents and does not assume the formation of a monolayer. The Freundlich isotherm is expressed as

$$Q_{eq} = K_F C_{eq}^{1/n}, \tag{3}$$

where Q_{eq} and C_{eq} have the same definition in Eq. (2), K_F is a Freundlich constant representing the adsorption capacity and n is a constant depicting the adsorption intensity. The computer code MINEQL+ [37] was used for the calculation of the thermodynamic speciation equilibrium, as a function of pH for the respective metal species. All calculations were performed for metal concentration 100 mg·l⁻¹ and constant ionic strength 0.01 M NaNO₃.

The environmental compatibility of red mud and the leaching ability of its metal loaded forms were tested by means of the U.S. EPA – Toxicity Characteristic Leaching Procedure (TCLP) [38]. For this purpose red mud was agitated for 18 hours in acetic acid (pH 2.88 ± 0.05, leachant/waste ratio: 20, ambient temperature). After the separation of the solid phase, the metal concentration in the leachant was determined by AAS and compared with the EPA classification standards.

3. RESULTS AND DISCUSSION

The chemical composition of the red mud used is given in Table 1 and its XRD pattern in Fig. 1. Its chemical stability is due to the predominance of mineral phases, which are slightly soluble in water (e.g., hematite, calcite, quartz and maghemite). The specific surface area of the sorbent, determined by the BET method, was found to be 18.38 m²·g⁻¹, whereas 90% of its particles had a size smaller than 22 μm. The grain size distribution of sorbent is given in Fig. 2.

Table 2 gives the TCLP test results for the red mud used and the comparison with the EPA classification standards. The TCLP results showed that only negligible amount of trace elements could be leached by the acetic acid solution (pH 2.88 ± 0.05) which did not exceed in any case the U. S. EPA allowable limits. These findings indicate that the red mud we used is an environmentally friendly material and it could be potentially used for wastewater treatment. Brunori et al. [24] have found similar results, as far as the environmental compatibility is concerned, using the Italian law leaching tests as well as the Microtox, ASTM microalgae toxicity and the sea urchin embryo toxicity tests.

The uptake of cadmium by red mud at pH 6 and ionic strength 0.01 M NaNO₃ is shown in Fig. 3. The experimental data are plotted with the solid rectangles while the continuous curve represents the Langmuir model fit. The dotted lines above and below the curve represent the 95% confidence interval. The parameters of the model are also shown in the figure. It is observed that the Langmuir model fits the data surprisingly well ($R^2 = 0.99$). It is noted that one of the basic assumptions of the Langmuir model is the homogeneity of the surface. Red mud, which contains various mineral phases, is inhomogeneous at microscopic level.

Table 1
Main chemical constituents of the red mud

Component	Content [%]
SiO_2	13.35
Fe_{Total}	25.34
FeO	0.72
Fe_2O_3	35.37
TiO_2	8.52
CaO	1.49
MgO	0.58
Cr_2O_3	0.54
MnO	0.54
Al_2O_3	8.37
Na_2O	9.27
K_2O	1.07
PbO	0.01
ZnO	0.01
SO_4^{2-}	1.65
Loss of ignition	9.27

Table 2
TCLP test results of the red mud

Element	Results [mg·l⁻¹]	US EPA Standards [mg·l⁻¹]
Ag	0.03	5.00
As	<2.00	5.00
Ba	5.10	100.00
Cd	0.09	1.00
Cr	0.15	5.00
Hg	<0.01	0.20
Pb	<0.20	5.00
Se	0.002	1.00

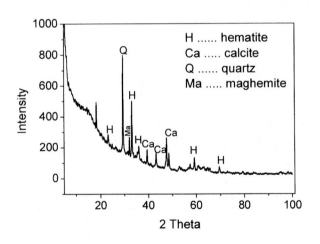

Fig.1. Powder XRD pattern of used red mud

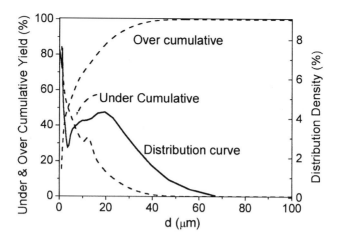

Fig. 2. Grain size of used red mud

However, the good fit of the model leads to the assumption that the Cd cations in solution interact with a local surface, which could be expressed as an average of the different constituents. This local surfaces seems to be, at macroscopic level, homogeneous. The maximum sorption capacity, Q_{max}, is 68 mg Cd/g of sorbent; a value which is above the average capacity of the natural sorbents [39]. However, the obtained results are lower than those presented in ref. [25] for thermally activated red mud (130 mg/g at pH 4 and 30° C) and in ref. [29] (108 and 106 mg·g^{-1} for pH 5 and 5.5, respectively). It should be noted that untreated red mud (as it comes from the factory) was used for this work, while the other investigators treated the red mud chemically and thermally, which increases its cost.

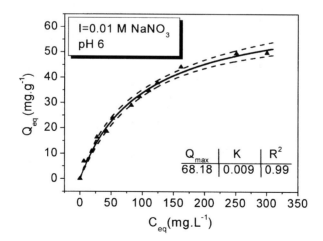

Fig. 3. Cadmium uptake by red mud: Langmuir adsorption model (the dotted lines represent the 95% confidence interval).

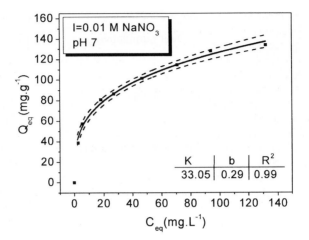

Fig 4. Zinc uptake by red mud: Freundlich adsorption model. (the dotted lines represent the 95% confidence interval).

Thermodynamic calculations with MINEQL+ (for a total cadmium concentration of 100 mg·l⁻¹ and 0.01 M NaNO₃) have shown that cadmium remains in solution mainly as Cd^{+2} up to a pH value of about 8.4, where it starts precipitating as insoluble $Cd(OH)_2$. At pH 9.1 more then 90% of cadmium precipitates. A small amount of $CdNO_3^+$ (ca.1.8%) is also present in the pH region between 2.5 and 8.4. At pH 8.3 to 8.6 $CdOH^+$ appears in about 1.5%. So, the cadmium uptake by the sorbent takes place mainly from its cationic form; it is very unlikely that any surface precipitation would occur under the experimental conditions studied here (i.e. pH 6).

The corresponding data as well as model parameters for the zinc uptake by red mud at pH 7 are shown in Fig. 4. The experimental data for the removal of zinc are better fitted by a Freundlich isotherm. The sorption capacity of red mud for zinc was much higher (ca. 133 mg Zn/g of sorbent) than for cadmium. It is noted that in the case of zinc removal, the final pH of the solution was quite close to the precipitation pH. The Zn precipitation as $Zn(OH)_2$ starts at pH ca. 7.3. However, the accumulation of zinc cations close to the surface of the sorbent can slightly lower the precipitation pH. So, for zinc it is probable that both sorption and surface precipitation take place. This could also explain the agreement of the obtained data with a Freundlich model rather than with a Langmuir, which assumes monolayer coverage. Initially zinc is abstracted from solution by sorption, which should also follow a Langmuir type isotherm. However, as zinc accumulates near the surface the solubility product of zinc oxide is reached and surface precipitation takes place; zinc is further removed from the solution as an insoluble precipitate. Combination of the two mechanisms results in a higher zinc removal from the solution than the Langmuir type sorption can justify and so the experimental data are better fitted with a Freundlich model. Our findings agree quite well with the results presented in ref. [25] for zinc removal by thermally modified red mud (145, 130 and 118 mg·g⁻¹ for 30, 40 and 50°C, respectively).

In the case of lead and copper removal, at pH 7 and 6 respectively, precipitation effects were obvious. Thermodynamic calculations have shown that lead remains in solution up to a pH value of ca. 5.7, where it starts to precipitate as insoluble $Pb(OH)_2$. At pH 6.3 more then 90% of lead precipitates. Copper remains soluble up to pH value of ca. 5.4, where the precipitation of insoluble $Cu(OH)_2$ starts. As precipitation is probably the main mechanism for the removal of these two cations from solution, the results in Fig. 5 are presented as the effect of initial metal concentration on removal efficiency. It is observed that lead is almost completely removed at all initial concentrations studied (20–400 $mg \cdot l^{-1}$). Copper has a different behavior. For initial concentrations 20–180 $mg \cdot l^{-1}$ is completely removed. However, as the concentration increases over 200 $mg \cdot l^{-1}$ an almost linear decrease in removal efficiency is observed. For 400 $mg \cdot l^{-1}$ initial concentration only 50% of copper is removed (Fig. 5). It is noted that under our experimental conditions sorption and precipitation could not be distinguished. Even though, surface precipitation seems to be the main mechanism for cation removal, the presence of the sorbent is essential. While without solid a voluminous and loose dispersion is formed in the presence of the sorbent, a compact precipitate is observed which can be easily separated from the liquid phase.

The exact mechanism of metal removal cannot be identified from this work. Red mud has a surface charge value that is between the corresponding ones of the constituent mineral phases (e.g., the PZC values of SiO_2 and Fe_2O_3 are respectively ca. 2.3 and 8.6). At pH higher than 5 the surface is negatively charged and the metal ions are attracted by electrostatic forces. In the case of cadmium, which mainly exists as cation in the solution, both sorption and ion exchange can take place. In the case of zinc surface precipitation additionally occurs. Copper and lead are mainly removed due to surface precipitation.

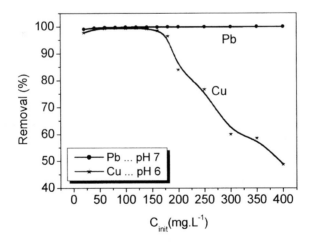

Fig. 5. Pb and Cu removal by used red mud as a function of initial concentration

Leaching experiments of the metal loaded red mud using acetic acid (pH 2.88 ± 0.05) showed that 63% of the sorbed Cd, 75% of the Zn, 57% of the Cu and 46% of the Pb can be removed. These data indicate that the metal loaded material could relatively easily be regenerated, but, on the other hand, special measures should be taken for its final disposal (e.g. stabilization in a more environmental friendly matrix).

4. CONCLUSIONS

Red mud, a solid byproduct of the alumina production industry, was found to be an environmentally compatible sorbent able to remove cadmium, zinc, lead and copper from model aqueous solutions. The best removal efficiency was observed for zinc (ca. 133 $mg \cdot g^{-1}$ or 2.0 $mmol \cdot g^{-1}$ for pH 7 in the presence of 0.01 M $NaNO_3$). Cadmium was also satisfactory removed at pH 6 and 0.01 M $NaNO_3$ with maximum removal efficiency of 68 $mg \cdot g^{-1}$ or 0.6 mmol Cd/g of solid. Copper and lead were almost completely removed at pH values 6 and 7 respectively from initial metal concentrations from 20–200 $mg \cdot l^{-1}$. Modeling of the experimental data has shown that cadmium follows a Langmuir type sorption isotherm while zinc fits better a Freundlich isotherm. Their difference was attributed to the surface precipitation that also takes place in the zinc case. Copper and lead were removed mainly by precipitation.

The TCLP test results indicate that the red mud is generally an environment friendly material that could efficiently be applied to the wastewater treatment technology. The treatment of its heavy metal–loaded forms with acetic acid solutions pointed out its regeneration possibilities but also the necessity of stabilization before its eventual final disposal.

ACKNOWLEDGEMENTS

The work presented by this paper was funded by the Greek–Slovak bilateral cooperation under the Project "Development, testing and application of magnetic sorbents for removal of heavy metals from wastewater and soil".

REFERENCES
[1] V.K. Gupta, M. Gupta and S. Sharma, Wat. Res., 25 (2001) 1125.
[2] S.J.T. Pollard, G.D. Fowler, C.J. Sollars and R. Perry, Sci Total Environ., 116 (1992) 31.
[3] E. Erdem, N. Karapinar and R Donat, J. Colloid Interface Sci., 280 (2004) 309.
[4] O. Abollino, M. Aceto, M. Malandrino, C. Sarzanini and E. Mentatsi, Wat. Res., 37 (2003) 1619.
[5] K.A. Matis, A.I. Zouboulis, G.P. Gallios, T.Erwe, and Ch. Blöcher, Chemosphere, 55 (2003) 65.
[6] L.C.A. Oliveira, R.V.R.A. Rios, J.D. Fabris, K.Sapag, V.K. Garg and R.M. Lago, Applied Clay Sci., 22 (2003) 169.
[7] D. Zamboulis, S.I. Pataroudi, A.I. Zouboulis and K.A. Matis, Desalination, 162 (2004) 159.
[8] F. Pagnanelli, F. Veglio and L. Toro, Chemospere, 54 (2004) 905.
[9] Y. Xu and L. Axe, J. Colloid Interface Sci., 282 (2005) 11.
[10] S.E. Bailey, T.J. Olin, R.M. Bricka and D.D. Adrian, Wat. Res., 33 (1999) 2469.
[11] R. Ciccu, M. Ghiani, A. Serci, S. Fadda, R. Peretti and A. Zucca, Min. Eng., 16 (2003) 187.
[12] S.V. Dimitrova and D.R. Mehandriev, Wat. Res., 32 (1998) 3289.
[13] K.C. Sekhar, C.T. Kamala, N S. Chary, A.R.K. Sastry, T. Nageswara Rao and M. Vairamani, J. Hazard. Mat., 108 (2004) 111.
[14] N. Chubar, J.R. Carvalho and M.J.N. Correia, Colloids and Surfaces A: Physicochem. Eng. Aspects, 230 (2003) 57.
[15] K.T. Park, V.K. Gupta, D. Mohan and S. Sharma, The Environmentalist, 19 (1999) 129.
[16] A. Hind, S. Bhargava and S. Grocott, Colloids and Surfaces, 146 (1999) 349.
[17] E. Lopez, B. Soto, M. Asias, A. Nunes, D. Rubinos and M.T. Barral, Wat. Res., 32 (1998) 1314.
[18] M. Vaclavikova, Heavy Metals Removal from Wastewaters, PhD Thesis, Institute of Geotechnics of SAS Kosice (2003).

[19] M. Kusnierova, P. Fecko, H. Vaskova, A. Luptakova, D. Kristofova, V. Sepelak, E. Boldizarova, in Proceedings of the V International Conference Metallurgy, Refractories and Environment, P. Palfy and E. Vircikova (eds.), Stara Lesna, High Tatras, 2002, pp. 165–169.

[20] J. Pradham, J. Das, S. Das and R.S. Thakur, J. Colloid Interface Sci., 204 (1998) 169.

[21] H. Genc, J.C. Tjell, D. McConchie and O. Schuiling, J. Colloid Interface Sci., 264 (2003) 327.

[22] S.J. Park, D. Seo and C. Nah, J. Colloid and Interface Sci., 251 (2002) 225.

[23] A. Agrawal, K.K. Sahu and B.D. Pandey, J. Resources, Conservation & Recycling, 42 (2004) 99.

[24] C. Brunori, C. Cremisini, P. Massanisso, V. Pinto, and L. Torricelli, J. Hazard. Mat., B117 (2005) 55.

[25] V.K. Gupta and S. Sharma, Environ. Sci. Technol., 36 (2002) 3612.

[26] J. Pradham, S.N. Das and R.S. Thakur, J. Colloid Interface Sci., 217 (1999) 137.

[27] R. M. Enick, E.J. Beckman, C. Shi and J. Xu, Energy & Fuels, 15 (2001) 256.

[28] K. Komnitsas, G. Bartzas and I. Paspaliaris, Min. Eng., 17 (2004) 183.

[29] R. Apak, K. Guclu and M.H. Turgut, J. Colloid Interface Sci., 203 (1998) 122.

[30] S. Wang, Y. Boyjoo, A. Choueib and Z.H. Zhou, Wat. Res., 39 (2005) 129.

[31] C. Namasivayam, R.T. Yamura and D.J.S.E. Arasi, Environ. Geol., 41 (2001) 269.

[32] E. Lombi, F.-J. Zhao, G. Wieshammer, G. Zhang and S.P. McGrath, Environ. Pollut., 118 (2002) 445.

[33] E. Lombi, F.-J. Zhao, G. Zang, B. Sun, W. Fitz, H. Zhang and S.P. McGrath, Environ. Pollut., 118 (2002) 435.

[34] L. Y. Li, J. Environ. Eng. Div. ASCE, 124 (1998) 254.

[35] O. Altin, H.O. Özbelge and T. Dogu, J. Colloid Interface Sci., 198 (1998) 130.

[36] F. Helfferich, Ion Exchange, Dover Publ. Inc., New York, 1995.

[37] MINEQL+ Version 4.01, A Chemical Equilibrium Modeling System, Environmental Research Software (1998).

[38] US EPA Method 1311 – Toxicity Characteristic Leaching Procedure, US Environmental Protection Agency (1994) (www.epa.gov).

[39] G. McKay, Use of Adsorbents for the Removal of Pollutants from Wastewaters, CRC Press, USA (1996) pp. 91–92.

[18] M. Kuschitrona, P. Lucku, H. Vasicava, M. Lopunova, T. Kraselinkot, N. Septela, H. Ishihara-gan, in Proceedings of the Vth International Conference Metallurgy, Refractories and Environment, P. Palfy, eds, V. Vorkova, eds, J. Shint Lesna, Hugi, Vajna, 2002, pp. 165-169.

[19] D. Peacklin, J. Caucer, Das and K.S. Thakur, J. Colloid Interface Sci. 204 (1998) 169.

[20] O. Gonsky, J.J. Pad, D. McConnor and O. Schulling, J. Chem. B Interface 800, 864 (2001) 125.

[21] S. Fitzak, G. Sirvaid, E. Pieto J. Colloid and Interface Sci, 213 (2004) 255.

[22] N. Angenolt, R.K. Sala, eds J.A. Parvez, J. Phys. Chem. Chem. Chem A. Exceub 46, 12 (2002) 46.

[23] E. Seglafi, C. Oksans, Pl. Sivoulanin, N. Mima, and L. Perrelle, J. Dansof Mau, 24 (12) (2001) 65.

[24] V.K. Gupta and S. Sharma, Environ. Sci. Technol. 36 (2002) 39.

[25] J. Prchatka, K.M. Ero and R.S. Thakur, J. Colloid Interface Sci, 31 (1998) 67.

[26] K. Sen, Z. Helerstorn, M. Sel and J. Ne, J. Sep. Sci. 8 Purit. 18 (2002) 66.

[27] A. Kumanotu, G. Remeskawel, Proceedings, Adsorbtion, 12 (2004) 531.

[28] C. Sutter, G. Sinaluk and M. J. Gupt, J. Colloid Interface Sci, 281 (2004) 52.

[29] S. Boyd, M. Thotan, A.Y. Vanchtoum, J. J. Phys. Chem. 1024, 3670 (2001)

Authors

528

STUDIES IN SURFACE SCIENCE AND CATALYSIS

Advisory Editors:
B. Delmon, Université Catholique de Louvain, Louvain-la-Neuve, Belgium
J.T. Yates, University of Pittsburgh, Pittsburgh, PA, U.S.A.

536